T5-BAM-478

REVIEWS in MINERALOGY

Volume 22

THE Al_2SiO_5 POLYMORPHS

DERRILL M. KERRICK

CONTENTS

CHAPTER		PAGE
1	INTRODUCTION	1
2	CRYSTAL STRUCTURES, OPTICAL AND PHYSICAL PROPERTIES	13
3	PHASE EQUILIBRIA	37
4	NON-STOICHIOMETRY	111
5	LATTICE DEFECTS	169
6	Al-Si DISORDER IN SILLIMANITE	187
7	THE FIBROLITE PROBLEM	207
8	METAMORPHIC REACTIONS	223
9	REACTION KINETICS AND CRYSTAL GROWTH MECHANISMS	257
10	ALUMINUM METASOMATISM	311
11	ANATECTIC MIGMATITES, MAGMATIC PEGMATITES AND PERALUMINOUS GRANITOIDS	353

Series Editor: Paul H. Ribbe

MINERALOGICAL SOCIETY OF AMERICA

REVIEWS IN MINERALOGY Volume 22

THE Al_2SiO_5 POLYMORPHS

DERRILL M. KERRICK

The Pennsylvania State University
Department of Geosciences
University Park, Pennsylvania 16802

Series Editor:

PAUL H. RIBBE
Department of Geological Sciences
Virginia Polytechnic Institute & State University
Blacksburg, Virginia 24061

Published by:

THE MINERALOGICAL SOCIETY OF AMERICA
1130 Seventeenth Street, N.W., Suite 330
Washington, D.C., 20036 U.S.A.

Printed by BookCrafters, Inc., Chelsea, Michigan 48118

REVIEWS in MINERALOGY

(Formerly: *SHORT COURSE NOTES*)

ISSN 0275-0279

VOLUME 22, THE Al_2SiO_5 POLYMORPHS

ISBN 0-939950-27-8

ADDITIONAL COPIES of this volume as well as those listed below may be obtained from the MINERALOGICAL SOCIETY OF AMERICA, 1130 Seventeenth Street, N.W., Suite 330, Washington, D.C. 20036 U.S.A.

Volume 1: Sulfide Mineralogy, 1974; P. H. Ribbe, Ed. 284 pp. Six chapters on the structures of sulfides and sulfosalts; the crystal chemistry and chemical bonding of sulfides, synthesis, phase equilibria, and petrology. ISBN# 0-939950-01-4.

Volume 2: Feldspar Mineralogy, 2nd Edition, 1983; P. H. Ribbe, Ed. 362 pp. Thirteen chapters on feldspar chemistry, structure and nomenclature; Al,Si order/disorder in relation to domain textures, diffraction patterns, lattice parameters and optical properties; determinative methods; subsolidus phase relations, microstructures, kinetics and mechanisms of exsolution, and diffusion; color and interference colors; chemical properties; deformation. ISBN# 0-939950-14-6.

Volume 4: Mineralogy and Geology of Natural Zeolites, 1977; F. A. Mumpton, Ed. 232 pp. Ten chapters on the crystal chemistry and structure of natural zeolites, their occurrence in sedimentary and low-grade metamorphic rocks and closed hydrologic systems, their commercial properties and utilization. ISBN# 0-939950-04-9.

Volume 5: Orthosilicates, 2nd Edition, 1982; P. H. Ribbe, Ed. 450 pp. Liebau's "Classification of Silicates" plus 12 chapters on silicate garnets, olivines, spinels and humites; zircon and the actinide orthosilicates; titanite (sphene), chloritoid, staurolite, the aluminum silicates, topaz, and scores of miscellaneous orthosilicates. Indexed. ISBN# 0-939950-13-8.

Volume 6: Marine Minerals, 1979; R. G. Burns, Ed. 380 pp. Ten chapters on manganese and iron oxides, the silica polymorphs, zeolites, clay minerals, marine phosphorites, barites and placer minerals; evaporite mineralogy and chemistry. ISBN# 0-939950-06-5.

Volume 7: Pyroxenes, 1980; C. T. Prewitt, Ed. 525 pp. Nine chapters on pyroxene crystal chemistry, spectroscopy, phase equilibria, subsolidus phenomena and thermodynamics; composition and mineralogy of terrestrial, lunar, and meteoritic pyroxenes. ISBN# 0-939950-07-3.

Volume 8: Kinetics of Geochemical Processes, 1981; A. C. Lasaga and R. J. Kirkpatrick, Eds. 398 pp. Eight chapters on transition state theory and the rate laws of chemical reactions; kinetics of weathering, diagenesis, igneous crystallization and geochemical cycles; diffusion in electrolytes; irreversible thermodynamics. ISBN# 0-939950-08-1.

Volume 9A: Amphiboles and Other Hydrous Pyriboles—Mineralogy, 1981; D. R. Veblen, Ed. 372 pp. Seven chapters on biopyribole mineralogy and polysomatism; the crystal chemistry, structures and spectroscopy of amphiboles; subsolidus relations; amphibole and serpentine asbestos—mineralogy, occurrences, and health hazards. ISBN# 0-939950-10-3.

Volume 9B: Amphiboles: Petrology and Experimental Phase Relations, 1982; D. R. Veblen and P. H. Ribbe, Eds. 390 pp. Three chapters on phase relations of metamorphic amphiboles (occurrences and theory); igneous amphiboles; experimental studies. ISBN# 0-939950-11-1.

Volume 10: Characterization of Metamorphism through Mineral Equilibria, 1982; J. M. Ferry, Ed. 397 pp. Nine chapters on an algebraic approach to composition and reaction spaces and their manipulation; the Gibbs' formulation of phase equilibria; geologic thermobarometry; buffering, infiltration, isotope fractionation, compositional zoning and inclusions; characterization of metamorphic fluids. ISBN# 0-939950-12-X.

Volume 11: Carbonates: Mineralogy and Chemistry, 1983; R. J. Reeder, Ed. 394 pp. Nine chapters on crystal chemistry, polymorphism, microstructures and phase relations of the rhombohedral and orthorhombic carbonates; the kinetics of CaCO, dissolution and precipitation; trace elements and isotopes in sedimentary carbonates; the occurrence, solubility and solid solution behavior of Mg-calcites; geologic thermobarometry using metamorphic carbonates. ISBN# 0-939950-15-4.

Volume 12: Fluid Inclusions, 1984; by E. Roedder. 644 pp. Nineteen chapters providing an introduction to studies of all types of fluid inclusions, gas, liquid or melt, trapped in materials from the earth and space, and their application to the understanding of geological processes. ISBN# 0-939950-16-2.

Volume 13: Micas, 1984; S. W. Bailey, Ed. 584 pp. Thirteen chapters on structures, crystal chemistry, spectroscopic and optical properties, occurrences, paragenesis, geochemistry and petrology of micas. ISBN# 0-939950-17-0.

Volume 14: Microscopic to Macroscopic: Atomic Environments to Mineral Thermodynamics, 1985; S. W. Kieffer and A. Navrotsky, Eds. 428 pp. Eleven chapters attempt to answer the question, "What minerals exist under given constraints of pressure, temperature, and composition, and why?" Includes worked examples at the end of some chapters. ISBN# 0-939950-18-9.

Volume 15: Mathematical Crystallography, 1985; by M. B. Boisen, Jr. and G. V. Gibbs. 406 pp. A matrix and group theoretic treatment of the point groups, Bravais lattices, and space groups presented with numerous examples and problem sets, including solutions to common crystallographic problems involving the geometry and symmetry of crystal structures. ISBN# 0-939950-19-7.

Volume 16: Stable Isotopes in High Temperature Geological Processes, 1986; J. W. Valley, H. P. Taylor, Jr., and J. R. O'Neil, Eds. 570 pp. Starting with the theoretical, kinetic and experimental aspects of isotopic fractionation, 14 chapters deal with stable isotopes in the early solar system, in the mantle, and in the igneous and metamorphic rocks and ore deposits, as well as in magmatic volatiles, natural water, seawater, and in meteoric-hydrothermal systems. ISBN #0-939950-20-0.

Volume 17: Thermodynamic Modelling of Geological Materials: Minerals, Fluids, Melts, 1987; H. P. Eugster and I. S. E. Carmichael, Eds. 500 pp. Thermodynamic analysis of phase equilibria in simple and multi-component mineral systems, and thermodynamic models of crystalline solutions, igneous gases and fluid, ore fluid, metamorphic fluids, and silicate melts, are the subjects of this 14-chapter volume. ISBN # 0-939950-21-9.

Volume 18: Spectroscopic Methods in Mineralogy and Geology, 1988; F. C. Hawthorne, Ed. 698 pp. Detailed explanations and encyclopedic discussion of applications of spectroscopies of major importance to earth sciences. Included are IR, optical, Raman, Mossbauer, MAS NMR, EXAFS, XANES, EPR, Auger, XPS, luminescence, XRF, PIXE, RBS and EELS. ISBN # 0-939950-22-7.

Volume 19: Hydrous Phyllosilicates (exclusive of micas), 1988; S. W. Bailey, Ed. 698 pp. Seventeen chapters covering the crystal structures, crystal chemistry, serpentine, kaolin, talc, pyrophyllite, chlorite, vermiculite, smectite, mixed-layer, sepiolite, palygorskite, and modulated type hydrous phyllosilicate minerals.

Vol. 20: Modern Powder Diffraction, 1989; D.L. Bish & J.E. Post, Eds. 369 pp.

Vol. 21: Geochemistry and Mineralogy of Rare Earth Elements, 1989; B.R. Lipin & G.A. McKay, Eds. 348 pp.

Vol. 22: The Al2SiO5 Polymorphs, 1990; D.M. Kerrick (monograph) 406 pp.

Vol. 23: Mineral-Water Interface Geochemistry, 1990; M.F. Hochella, Jr. & A.F. White, Eds.

THE Al_2SiO_5 POLYMORPHS

FOREWORD

For at least twelve years the Short Course Committee of the Mineralogical Society of America has received suggestions that MSA sponsor a course on the aluminum silicate minerals. But no one came forward to organize it. Thus, when in early 1989, Derrill Kerrick of The Pennsylvania State University offered to write a monograph for **Reviews in Mineralogy** on the Al_2SiO_5 polymorphs *sans* short course, the MSA Council responded positively, but cautiously. This would be only the second monograph in the 21-volume series (see opposite page).

A careful review policy was established and sixteen referees, acknowledged on p. *v*, were recruited to scrutinize the eleven chapters of the manuscript. This book is the product of more than two years' effort by Kerrick. Mrs. Marianne Stern has worked diligently with the author and me this past six months to prepare the final camera-ready copy. Her efforts are greatly appreciated!

Paul H. Ribbe
Blacksburg, VA
September 12, 1990

DEDICATION

MARSHALL E. MADDOCK

and

ROBERT L. ROSE

(San Jose State University)

who epitomize *quality* in college education

PREFACE AND ACKNOWLEDGMENTS

Recently I have been involved in a spectrum of studies involving the aluminum silicates. In addition to their primary importance in metamorphic petrology, the aluminum silicates illustrate a wide variety of experimental, theoretical, and experimental problems. Because the Al_2SiO_5 polymorphs alone offer a pedagogic illustration of many important principles of modern metamorphic petrology, I offered a seminar on the aluminum silicate polymorphs during my sabbatical leave at the Mineralogisch-Petrographisches Institut der Universität, Basel, Switzerland. An updated version was given as a graduate course at The Pennsylvania State University in 1988. This book was primarily inspired by the generally enthusiastic response of the graduate students who attended this course. However, it would not have been written without Professor Martin Frey's arrangement of generous financial support from the University of Basel. I am very grateful to Martin and Ann Frey for their gracious hospitality (including numerous outstanding home-cooked meals) extended to me and my family.

Compared to other volumes in the *Reviews in Mineralogy* series, a book exclusively devoted to three minerals may seem to be narrowly focused and esoteric. However, as discussed in Chapter 1, the aluminum silicate polymorphs are perhaps *the* most important mineral group to metamorphic petrologists. Because these minerals occur in anatectic migmatites and peraluminous granitoids, they are also important in igneous petrology. In spite of their geologic significance, there are a variety of experimental, theoretical, and field problems involving the aluminum silicates. Theoretical problems include the nature and energetics of lattice defects, order/disorder, crystalline (solid) solution, and interfacial energy. The aluminum silicates epitomize the importance of understanding the mechanisms and kinetics of heterogeneous metamorphic reactions. The difficulties in calibration of the pressure-temperature (P-T) phase equilibrium diagram illustrate the pitfalls of hydrothermal experimentation and the need to understand the methodology and uncertainties of calorimetric measurements of thermodynamic data of minerals. Thus, this book covers a wide variety of topics that must be considered in the analysis of metamorphic systems. In so doing, this volume illustrates the fact that modern metamorphic petrology demands an awareness of a wide spectrum of geologic variables and processes.

In concert with the tenor of the Mineralogical Society of America *Reviews in Mineralogy* series, this volume is intended to provide a comprehensive review, summarizing the methods, theories and pitfalls of the various contributions on the aluminum silicates. Hopefully, this book will provide readers with a reasonably in-depth overview, and thus avoid the need for extensive, independent literature reviews. Although a concerted effort was made to give a *balanced* coverage of divergent theories regarding various problems involving the aluminum silicates, this critique nevertheless includes some of the author's biases.

Several sections of this book present the *chronological* development of research on various topics, giving readers historical perspectives on the development of theories, models and biases on various problems regarding the aluminum silicates. As in all fields, several landmark studies have set the tone for the strategy of approach to problems. Although such studies have provided important steps forward in our understanding of natural phenomena, they have had the undesirable effect of entrenching biases and methodology. In this volume I have attempted to point out the deleterious effects of certain parochial approaches, an example being the aluminum immobility concept discussed in Chapter 10.

I am deeply indebted to numerous individuals for their kind efforts in making this volume possible. S.J. Mackwell's assistance in word processing was invaluable. I am very grateful to H.W. Day, C.T. Foster, J.A. Grambling, E.S. Grew, D.A. Hewitt, M.J. Holdaway, D. London, R.C. Newton, R.C. Peterson, S. Raeburn, P.H. Ribbe, G.R. Rossman, J.A. Speer, R.J. Tracy, H.R. Wenk, and R.P. Wintsch for their critical reviews of earlier versions of the manuscript. I thank those who kindly responded to my request for information, preprints, theses, figures and photographs: A.C. Barnicoat, B. Beddoe-Stephens, F.W. Breaks, W.A. Dollase, P. Cěrný, J.-C. Doukhan, M.D. Dyar, R. Elan, T. Flöttmann, C.T. Foster, E.D. Ghent, J.A. Grambling, R.H. Grapes, E.S. Grew, M.J. Holdaway, D. London, V.J. Morand, R.C. Newton, A. Nitkiewicz, G.L. Nord, Jr., J.J. Papike, D.R.M. Pattison, J.M. Rice, J.L. Rosenfeld, M.J. Rubenach, B.L. Sherriff, J.V. Walther, H.R. Wenk, D.F. Weill, R.P. Wintsch, B.W.D. Yardley, and E-an Zen. M.K. Kerrick, S.E. Kerrick, L.M. Miller and S. Raeburn provided considerable assistance in preparation of this book. J.A.D. Connolly computed several diagrams. I am *extremely* indebted to P.H. Ribbe for his countless hours of review and editing of this book. Paul's efforts in editing all of the volumes of the *Reviews in Mineralogy* series are indeed monumental.

Derrill M. Kerrick
University Park, PA
September 12, 1990

THE Al_2SiO_5 POLYMORPHS

TABLE OF CONTENTS

Page

ii Copyright; Additional Copies
iii Dedication; Foreword
iv Preface and Acknowledgments
xi List of Symbols

1. INTRODUCTION

1 FACIES SERIES AND BARIC REGIMES
6 TECTONIC-METAMORPHIC ANALYSIS OF METAMORPHIC BELTS
6 CALIBRATION OF OTHER GEOTHERMOMETERS AND GEO-BAROMETERS
11 RATIONALE FOR TOPICAL ORGANIZATION OF SUCCEEDING CHAPTERS

2. CRYSTAL STRUCTURES, OPTICAL AND PHYSICAL PROPERTIES

13 INTRODUCTION
14 ANDALUSITE
14 Crystal structure
14 Optical properties
21 Thermal expansion, compressibility and elasticity
26 SILLIMANITE
26 Crystal structure
28 Optical properties
38 Thermal expansion, compressibility and elasticity
30 KYANITE
30 Crystal structure
34 Optical properties
34 Thermal expansion

3. PHASE EQUILIBRIA

37 INTRODUCTION
40 EXPERIMENTAL HYDROTHERMAL STUDIES
40 Experiments of Evans (1965)
43 Experiments of Newton (1966a,b)
43 *Kyanite-andalusite equilibrium*
43 *Kyanite-sillimanite equilibrium*
45 *Al_2SiO_5 phase equilibrium diagram of Newton (1966a)*
45 *Critique of Newton's (1966a,b) experiments*
47 Experiments of Richardson et al. (1968) and Richardson et al. (1969)
47 *Kyanite-sillimanite equilibrium*
47 *Kyanite-andalusite equilibrium*
47 *Andalusite-sillimanite equilibrium*
49 *Al_2SiO_5 phase equilibrium diagram of Richardson et al. (1969)*
49 *Critique of the experiments of Richardson et al. (1968) and Richardson et al. (1969)*
55 Experiments of Holdaway (1971)
55 *Kyanite-andalusite equilibrium*
56 *Andalusite-sillimanite equilibrium*
57 *Al_2SiO_5 phase equilibrium diagram of Holdaway (1971)*
57 *Critique of Holdaway's (1971) experiments*
61 Experiments of Brown and Fyfe (1971)
62 *Al_2SiO_5 phase equilibrium diagram of Brown and Fyfe (1971)*
64 *Critique of Brown and Fyfe's (1971) experiments*
65 Experiments of Bowman (1975)
65 *Experiments at 0.5 kbar*
67 *Experiments at 2 kbar*
67 *Al_2SiO_5 phase equilibrium diagram of Bowman (1975)*
70 *Critique of Bowman's (1975) experiments*
71 Experiments of Heninger (1984)
72 *Critique of Heninger's (1984) experiments*
73 Experiments of Bohlen et al. (ms.)
74 SOLUBILITY STUDIES AT ATMOSPHERIC PRESSURE
74 Experimental study of Weill (1966)
77 *Al_2SiO_5 phase equilibrium diagram of Weill (1966)*
77 Experiments of Bowman (1975)
79 Critique of the experimental results of Weill (1966) and Bowman (1975)

80 CALORIMETRIC STUDIES
81 Heat capacities and vibrational entropies
85 Solution calorimetry
88 ANALYSIS OF Al_2SiO_5 PHASE EQUILIBRIA UTILIZING THERMODYNAMICALLY CONSISTENT DATA SETS
97 CALIBRATION OF THE Al_2SiO_5 PHASE EQUILIBRIA WITH MINERAL PARAGENETIC DATA
105 SUMMARY AND CONCLUSIONS
107 OTHER EQUILIBRIA IN THE Al_2SiO_5 SYSTEM
109 EPILOGUE

4. NON-STOICHIOMETRY

111 MAJOR ELEMENT NON-STOICHIOMETRY
111 Crystal chemistry of sillimanite-mullite solid solution
113 Evidence for sillimanite-mullite solid solution
120 Thermodynamic analysis of sillimanite-mullite solid solution
121 MINOR ELEMENT NON-STOICHIOMETRY
121 Transition elements
121 *Kyanite*
124 *Andalusite*
134 *Sillimanite*
137 *Thermodynamic analysis of transition element solid solution*
143 *Partitioning of transition elements between coexisting polymorphs*
156 Zoning of transition elements
162 Boron
165 Hydroxyl
168 Other elements

5. LATTICE DEFECTS

169 POINT DEFECTS
170 Intrinsic point defects
171 Extrinsic point defects
172 LINE DEFECTS
178 PLANAR DEFECTS
178 Stacking faults
179 Antiphase boundaries
181 Twinning and kink bands in kyanite
184 Grain boundaries

6. Al/Si DISORDER IN SILLIMANITE

187 THERMODYNAMIC MODELING
189 PHASE EQUILIBRIUM EXPERIMENTS
196 EXPERIMENTAL HEAT TREATMENT
199 NATURAL SILLIMANITE AND FIBROLITE
199 X-ray diffraction
200 Neutron diffraction
201 Spectroscopic studies

7. THE FIBROLITE PROBLEM

207 INTRODUCTION
207 EXPERIMENTAL HYDROTHERMAL STUDIES
209 CALORIMETRIC STUDIES
211 NON-STOICHIOMETRY
214 LATTICE DEFECTS
214 Al-Si DISORDER
216 GRAIN BOUNDARY ENERGY
220 CONCLUSIONS

8. METAMORPHIC REACTIONS

223 INTRODUCTION
223 THE KYANITE → SILLIMANITE REACTION
230 THE ANDALUSITE → SILLIMANITE REACTION
236 THE KYANITE → ANDALUSITE REACTION
241 THE ANDALUSITE → KYANITE REACTION
243 REACTIONS INVOLVING FIBROLITE
246 Some fibrolite-forming reaction mechanisms
246 Base cation leaching
249 Deformation-induced fibrolitization
252 Aluminum metasomatism
253 The fibrolite ⇄ *sillimanite reaction*
253 RETROGRADE ALTERATION (REPLACEMENT) REACTIONS

9. REACTION KINETICS AND CRYSTAL GROWTH MECHANISMS

257 INTRODUCTION
257 EXPERIMENTAL REACTION KINETICS

261 FIELD EVIDENCE FOR REACTION KINETICS
271 Strain-assisted reactions
273 FIBROLITE METASTABILITY
273 Contact metamorphism
276 Regional metamorphism
296 Implications for an equilibrium model
296 KINETIC MODELING OF POLYMORPHIC REACTIONS
302 CHIASTOLITE: CRYSTAL GROWTH MECHANISMS

10. ALUMINUM METASOMATISM

311 INTRODUCTION
312 Al_2SiO_5-BEARING VEINS AND SEGREGATIONS FORMED BY REPLACEMENT
325 Al_2SiO_5-BEARING VEINS AND SEGREGATIONS FORMED BY CRYSTALLIZATION WITHIN FRACTURES AND CAVITIES
330 Al_2SiO_5-bearing segregations in the Lepontine Alps, Switzerland: a case study
344 Kyanite-bearing veins in eclogites
346 FIBROLITE AND ALUMINUM METASOMATISM
352 EPILOGUE

11. ALUMINUM SILICATES IN ANATECTIC MIGMATITES AND PERALUMINOUS GRANITOIDS

353 INTRODUCTION
353 ANATECTIC MIGMATITES
354 PERALUMINOUS GRANITOIDS
359 MAGMATIC PEGMATITES

363 **REFERENCES**

LIST OF SYMBOLS

A, And = andalusite
A = affinity of reaction
a = unit cell dimension
a_i^α = activity of component i in phase α
B = isotropic equivalent temperature factor
b = Burgers vector
b = unit cell dimension
C_p = heat capacity at constant pressure
C_v = heat capacity at constant volume
c = unit cell dimension
E_A = activation energy barrier of reaction
E_d = energy of disorder
E_{TW} = dislocation tilt wall energy
f_i = fugacity of component i
G_d = Gibbs free energy of disorder
G_T = Gibbs free energy at temperature T
ΔG_f = Gibbs free energy of formation
ΔG_m = Gibbs free energy change of mixing
ΔG_r = Gibbs free energy change of a reaction
$\delta(\omega)$ = phonon density of states
H = enthalpy
ΔH_f = enthalpy change of formation
$\Delta \bar{H}_i$ = change (upon mixing) of partial molar enthalpy of component i
ΔH_r = enthalpy change of reaction
ΔH_v = enthalpy of formation of a vacancy defect
h = Planck constant
K, Ky = kyanite
K_{eq} = equilibrium constant
K_D = distribution coefficient
n = hydration number
n = term in Arrhenius equation
o = superscript referring to 1 bar
P = pressure
p = probability
R = gas constant
R = reaction rate
r_o = inner radius of hollow cylinder coaxial with dislocation
S, Sil = sillimanite
S_c = configurational entropy
S_d = entropy of disorder

S_T = entropy at T
ΔS_r = entropy change of reaction
s_{ij} = elastic compliance
ΔS_{vib} = vibrational entropy
$\Delta \bar{S}_i^{ex}$ = change (upon mixing) of partial molar excess entropy of component i
s = Bragg-Williams order parameter
T = temperature
T_{eq} = equilibrium temperature
T_m = melting temperature
V = molar volume
V_d = volume of disorder
$V_O^{''}$ = oxygen vacancy defect
$V_{Si}^{''''}$ = Si vacancy defect
ΔV_r = volume change of a reaction
$\overline{W}$ = molar dislocation strain energy
W_G^{α} = symmetric regular solution interaction parameter for phase α
X_i^{α} = mole fraction of component i in phase α
x = least refractive index
y = intermediate refractive index
z = greatest refractive index
α = coefficient of thermal expansion
α_i = number of atomic sites for mixing of species i
β = compressibility
μ = shear modulus
μ_i = chemical potential of component i
ν = Poisson's ratio
ρ = dislocation density
σ = principal stress
ω = vibrational frequency
Θ = angular mismatch between lattices of grains separated by a symmetrical dislocation tilt wall
$\square$ = oxygen vacancy

CHAPTER 1

INTRODUCTION

"The importance of the aluminum silicate polymorphs to the metamorphic petrologist cannot be overstated..."

P.H. Ribbe, 1980

The Al_2SiO_5 polymorphs continue to be of *paramount* importance in metamorphic petrology. Their usefulness stems from their abundance in metapelites, the fact that some isograd reactions in metapelites correspond to polymorphic transformations involving the aluminum silicates, and the simple pressure-temperature (P-T) phase equilibrium relations of the Al_2SiO_5 system. Although numerous other thermobarometers have been developed and utilized (especially within the last decade), petrologists still consider the Al_2SiO_5 polymorphs for primary thermobarometry of metamorphic rocks. In fact, the Al_2SiO_5 phase equilibria are used to *calibrate* other geothermometers and geobarometers. The following review highlights some landmark papers exemplifying the petrologic significance of the Al_2SiO_5 polymorphs.

FACIES SERIES AND BARIC REGIMES

Prior to about 1960, the utility of the aluminum silicates for metamorphic thermobarometry was hampered by uncertainty in the calibration of the P-T phase equilibrium diagram involving the polymorphs. This state of affairs is epitomized by Figure 1.1 from the classic Geological Society of America Memoir of Fyfe, Turner, and Verhoogen (1958).

On the basis of his thermodynamic analysis of the Al_2SiO_5 polymorphs, Miyashiro (1961) published a milestone paper in which he accepted the phase equilibrium topology shown in the right of Figure 1.1, and tentatively placed the triple point at 8 kbar and 300°C. Coupling the Al_2SiO_5 phase diagram with the albite = jadeite + quartz equilibrium (Fig. 1.2), he established three major "facies series" characterized by different prograde P-T trajectories. The P-T trajectories of the "kyanite-sillimanite" and "andalusite-sillimanite" facies series (Fig. 1.2) are based on the Al_2SiO_5 phase equilibria. From this analysis, Miyashiro (1961) established his now-famous concept of "paired metamorphic belts", whereby belts containing metamorphic rocks indicative of low pressure metamorphism (trajectory #2 in Fig. 1.2) are juxtaposed with those

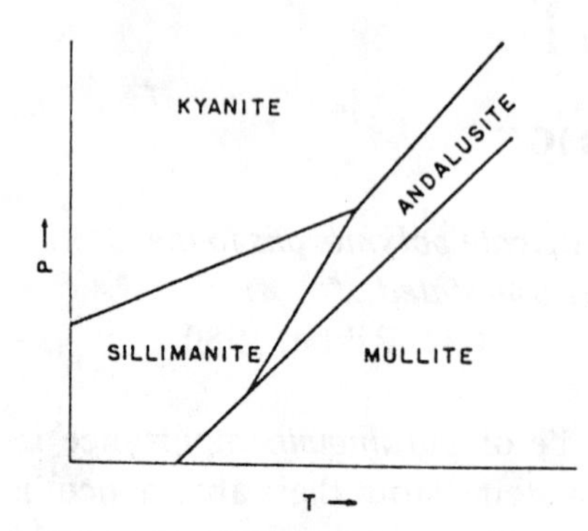

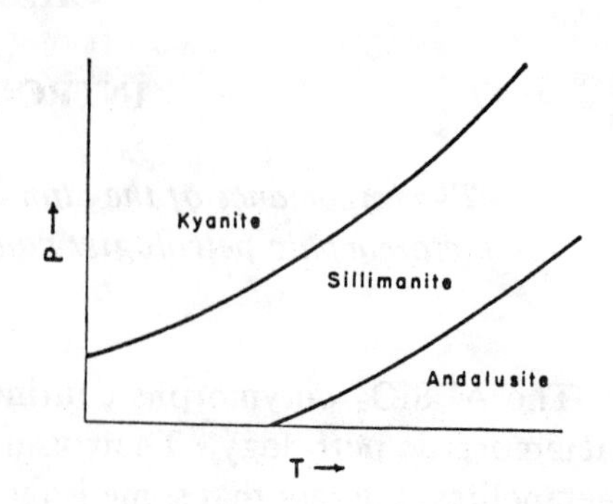

Figure 1.1. Possible P-T phase equilibrium relations for the Al_2SiO_5 polymorphs. (From Fyfe et al., 1958, Figs. 63-65).

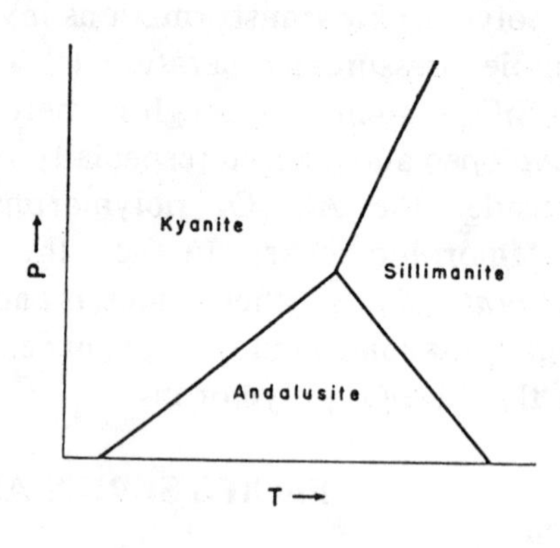

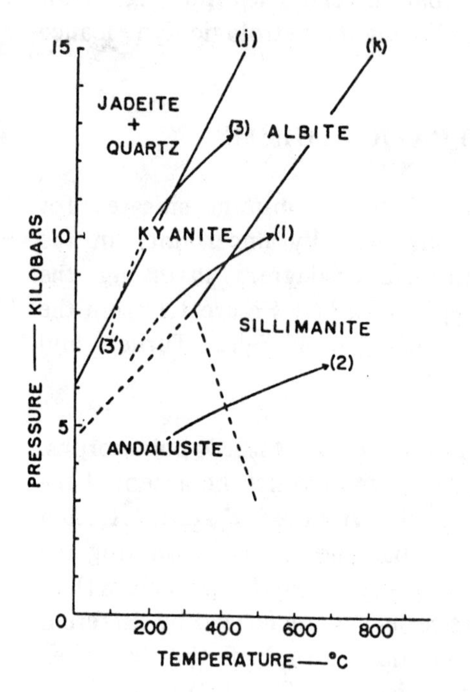

Figure 1.2. Miyashiro's (1961) depiction of P-T gradients for (2) low-pressure ("andalusite-sillimanite"), (1) medium pressure ("kyanite-sillimanite"), and (3) high-pressure ("jadeite-glaucophane") facies series. The albite = jadeite + quartz equilibrium (j) is shown in addition to the three univariant Al_2SiO_5 equilibria (k = kyanite-sillimanite). (From Miyashiro, 1961, Fig. 4).

containing high-pressure mineral assemblages (trajectory #3 in Fig. 1.2). For the last three decades, this concept has played a major role in the analysis of plate tectonics and the interpretation of orogenic belts.

Using a petrogenetic grid involving numerous equilibria in addition to the Al_2SiO_5 polymorphs, Hietanen (1967) established eight facies series, each characterized by a unique prograde P-T path (Fig. 1.3). Based primarily on her studies of metapelites in Idaho, Hietanen (1967) chose a triple point at 5 kbar and 500°C. The Al_2SiO_5 phase equilibria were critical in Hietanen's calibration of the P-T gradients for her various facies series. For example, the Buchan-type and Pyreneean-type facies series (#3 and #4, respectively, in Fig. 1.3) display the prograde index mineral progression: andalusite → sillimanite → K-feldspar; thus, the P-T trajectory of these facies series was constrained to pressures below the triple point and at higher pressures than the invariant point formed by the intersection of the andalusite-sillimanite equilibrium with the muscovite + quartz = Al_2SiO_5 + K-feldspar + H_2O equilibrium.

In an effort to improve baric analysis utilizing mineral assemblages in metapelites, Carmichael (1978) chose a simplified model system which includes the Al_2SiO_5 polymorphic equilibria and two dehydration equilibria involving the aluminum silicates. As shown in Figure 1.4, Carmichael (1978) established six "bathozones". The isobaric boundaries between the bathozones, which he referred to as "bathograds", are coincident with invariant points in the phase diagram. It is important to emphasize that the pressures of *all* of Carmichael's bathozones and bathograds are dependent upon the P-T location of univariant equilibria involving the Al_2SiO_5 polymorphs. His bathograd separating bathozones 3 and 4 (Fig. 1.4) corresponds to the Al_2SiO_5 triple point, and is equivalent to the "triple point isobar" which Thompson and Norton (1968) mapped in New England (Fig. 1.5). Carmichael discussed the geologic utility of the bathozone/bathograd concept. For example, the southwestward progression of bathozones from #1 in central Maine to #6 in western Connecticut (Fig. 1.6) suggests that the Paleozoic metamorphic rocks exposed in southern New England are deeper-seated than the equivalent rocks exposed in northern New England, thereby implying post-Pennsylvanian northward tilting (Carmichael, 1978, p. 784).

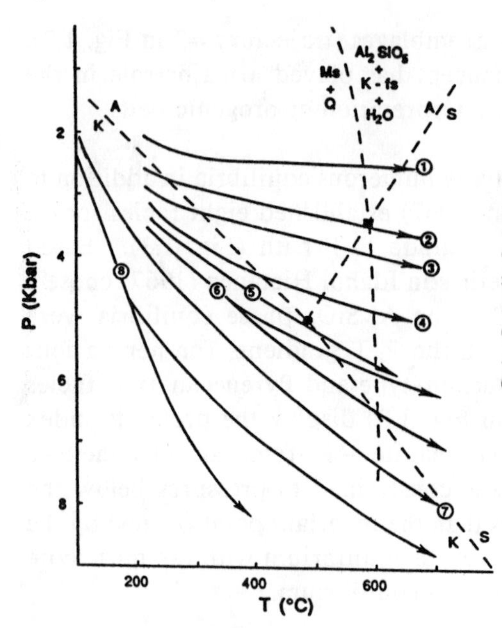

Figure 1.3. P-T gradients (solid lines) for Hietanen's (1967) facies series. The solid circle is the Al_2SiO_5 triple point, and the solid square is the invariant point formed by the intersection of the andalusite-sillimanite equilibrium with the muscovite + quartz = Al_2SiO_5 + K-feldspar + water equilibrium. (Redrawn from Hietanen, 1967, Fig. 1).

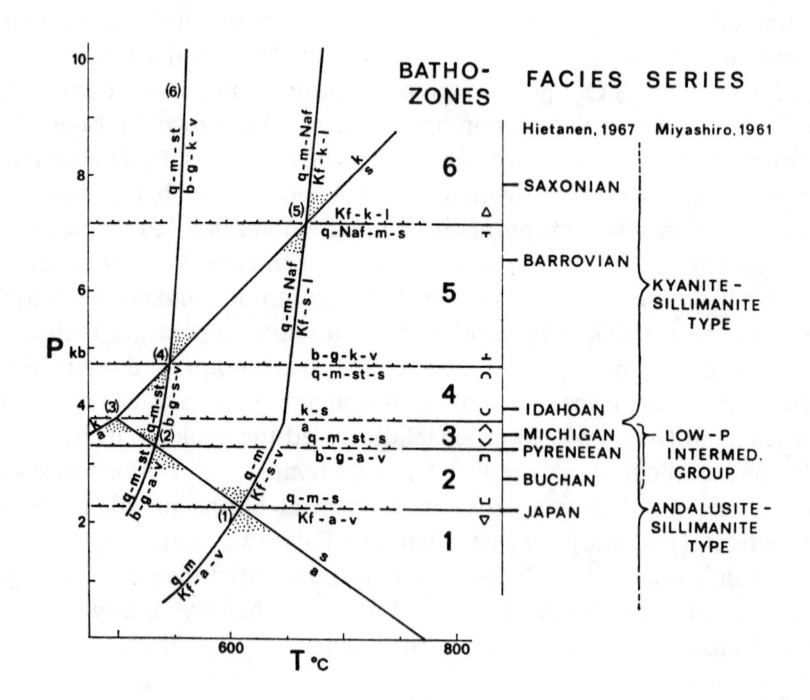

Figure 1.4. Phase equilibrium diagram showing Carmichael's (1978) bathozones and bathograds (horizontal lines separating the bathozones). Abbreviations for the aluminum silicate polymorphs are: a = andalusite, k = kyanite, s = sillimanite. Note that all five invariant points used for determining the pressures of the bathograds involve Al_2SiO_5 polymorphic equilibria. (From Carmichael, 1978, Fig. 2).

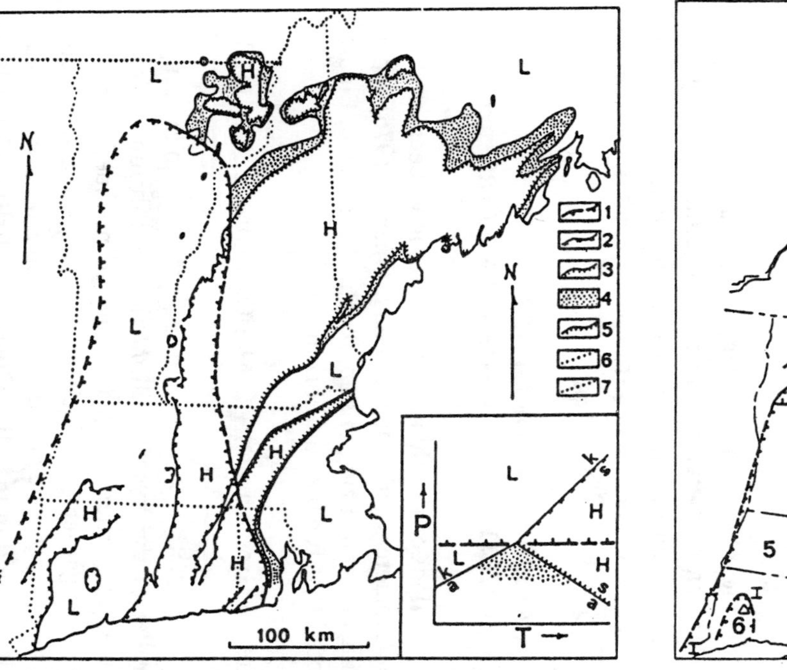

Figure 1.5. Map of New England showing Thompson and Norton's (1968) triple point isobar (heavy dashed line with barbs). The inset shows the correspondence between this isobar, Carmichael's (1978) kyanite-sillimanite bathograd, and the Al_2SiO_5 triple point (H = higher-grade zones, L = lower-grade zones). (1) kyanite-sillimanite bathograd, (2) kyanite-sillimanite isograd, (3) staurolite or andalusite isograd, (4) staurolite zone (with andalusite), (5) andalusite-sillimanite isograd, (6) interstate boundary, (7) international boundary. (From Carmichael, 1978, Fig. 1).

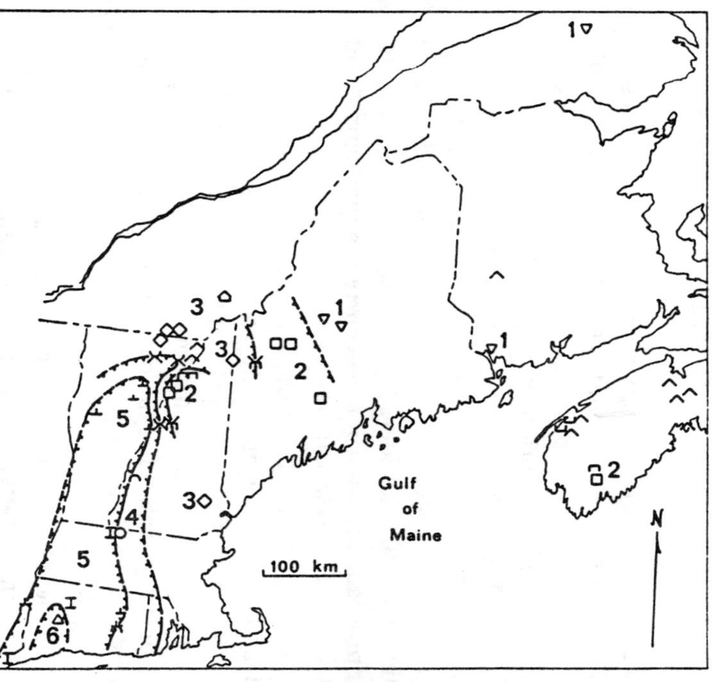

Figure 1.6. Location of Carmichael's (1978) six bathozones in New England. The open symbols refer to the location of key mineral assemblages defining the bathozones (see Carmichael, 1978, Table 1). (From Carmichael, 1978, Fig. 4).

TECTONIC-METAMORPHIC ANALYSIS OF METAMORPHIC BELTS

The P-T stability relations of the aluminum silicates have been useful for analyses of the tectonic-metamorphic evolution of regionally metamorphosed rocks. An example is provided by Tracy and Robinson's (1979) analysis of the tectonic evolution of regionally metamorphosed rocks in central Massachusetts. Structurally, this area consists of the Bronson Hill anticlinorium to the west (containing several mantled gneiss domes) and the Merrimack synclinorium to the east. As shown in Figure 1.7, Tracy and Robinson (1980) considered three major tectonic events. They concluded that prograde metamorphism occurred in the "nappe" and "backfold" stages, whereas retrograde metamorphism was contemporaneous with the "dome" stage (Fig. 1.7). Petrographic evidence reveals that sillimanite replaced andalusite in metapelitic rocks of the Merrimack synclinorium, whereas those of the Bronson Hill anticlinorium display the prograde kyanite → sillimanite transformation. These mineralogical contrasts provide the primary reason for the contrasting P-T trajectories shown in Figure 1.7.

The importance of the aluminum silicates in the analysis of the thermobarometric evolution of metamorphic rocks is epitomized by the fact that the Al_2SiO_5 phase equilibrium diagram is central to the derived P-T-t paths in most chapters of the recently-published book: Evolution of Metamorphic Belts (Daly et al., 1989).

CALIBRATION OF OTHER GEOTHERMOMETERS AND GEOBAROMETERS

The P-T stability relations of the aluminum silicates have been used to calibrate other geothermometers and geobarometers. Ghent's (1976) paper represents a pioneering effort utilizing this approach. As shown in Figure 1.8, Ghent (1976) noted that for the equilibrium:

$$3\ \text{Anorthite}_{ss} = \text{Grossularite}_{ss} + 2\ Al_2SiO_5 + \text{Quartz}\ , \qquad [1.1]$$

isopleths of log K_D (= 3 log $X^{Gar}_{Ca_3Al_2Si_3O_{12}}$ - 3 log $X^{Plag}_{CaAl_2Si_2O_8}$) in the range -1 to -3 are subparallel to the kyanite-sillimanite equilibrium. From electron probe analyses of garnet and plagioclase in metapelitic rocks containing the assemblage: garnet + plagioclase + Al_2SiO_5 + quartz, Ghent et al. (1979) derived an average value of log K_D = -2.0 ± 0.2 for rocks at the kyanite-sillimanite isograd in the Mica Creek area, British Columbia.

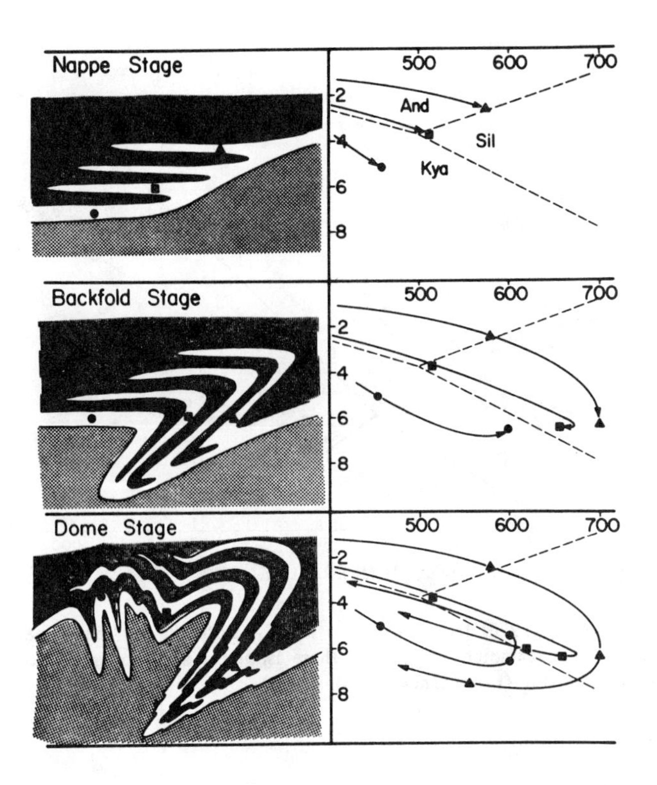

Figure 1.7. Tracy and Robinson's (1980) tectonic-metamorphic synthesis for regional metamorphism in central Massachusetts. The P-T diagrams track three selected points (triangle, circle, and square) initially corresponding with three different structural levels of the nappe stage. (From Tracy and Robinson, 1980, Fig. 6).

To incorporate non-ideal mixing, the equilibrium constant can be formulated as the product: $K_D \times K_\gamma$, where K_D is the distribution coefficient expressed in terms of mole fractions of components in solid solutions, and K_γ is the "non-ideal" term containing activity coefficients of these components. Figure 1.9 shows that with the assumption of ideal mixing ($\log K_\gamma = 0$), equilibrium [1.1] is not coincident with the kyanite-sillimanite equilibrium. Assuming these rocks were metamorphosed at P-T conditions corresponding to the kyanite-sillimanite equilibrium, Ghent et al. (1979) adjusted the value of K_γ so that the P-T equilibrium with log K_D = -2.0 became coincident with that of Holdaway's (1971) kyanite-sillimanite equilibrium (Fig. 1.9). The resultant log K_γ value of -0.4 thus represents an *empirical* calibration of the garnet-plagioclase-Al_2SiO_5-quartz equilibrium. It is important to emphasize that this procedure is dependent upon the accuracy of Holdaway's (1971) location of the kyanite-sillimanite equilibrium. The garnet-plagioclase-Al_2SiO_5-quartz

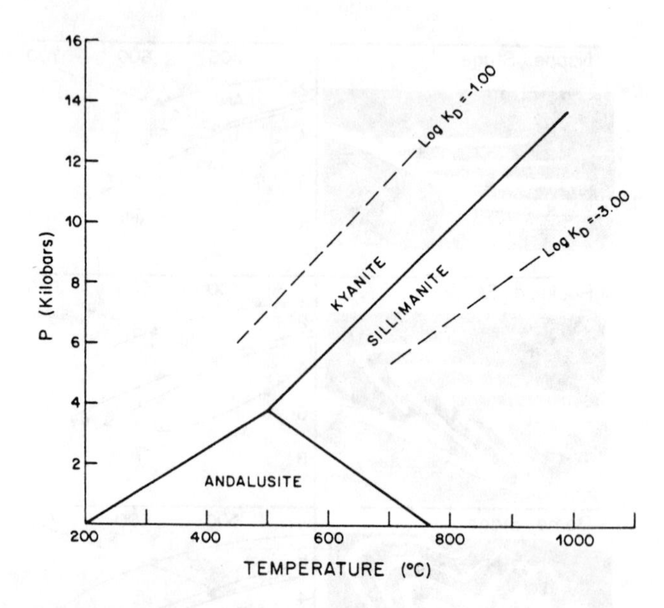

Figure 1.8. Ghent's (1976) location of two log K_D isopleths for the garnet-plagioclase-Al_2SiO_5-quartz geobarometer. Note that these isopleths are subparallel to the kyanite-sillimanite equilibrium. (From Ghent, 1976, Fig. 2).

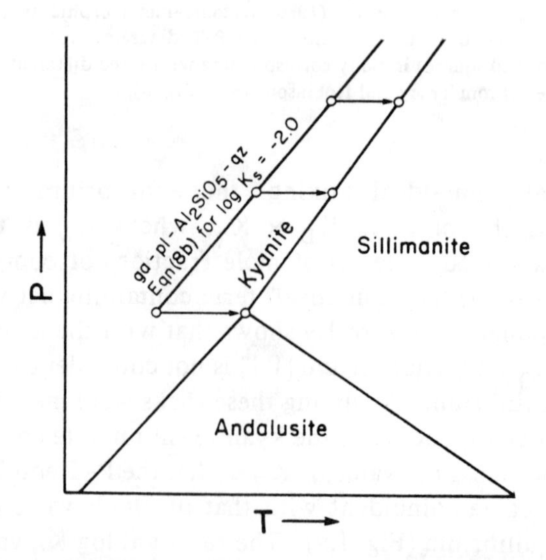

Figure 1.9. Schematic illustration of the isopleth for log K_S (= log K_D) = - 2.0 for the garnet-plagioclase-Al_2SiO_5-quartz barometer. Arrows show the shift to coincidence with the kyanite-sillimanite equilibrium by an "empirical" correction of the garnet-plagioclase-Al_2SiO_5-quartz equilibrium for non-ideal mixing (i.e., adding log K_γ = -0.4). (From Ghent et al., 1979, Fig. 2).

barometer has subsequently been modified. For example, Newton and Haselton (1981) and Ganguly and Saxena (1984) assumed non-ideal mixing in garnet and plagioclase and derived γ_{Gr} and γ_{An} through thermodynamic modeling. Consequently, their equations of state are not based upon a "field" calibration of the non-ideal mixing term (K_γ). However, there are uncertainties in the thermodynamic modeling of silicate solid solutions involved in this equilibrium (e.g., Newton and Haselton, 1981, p. 145). Although the accuracy of this geobarometer has been improved by the experimental study of Koziol and Newton (1988), McKenna and Hodges' (1988) analysis suggests that significant uncertainties (± 2.5 kbar) remain in determining paleopressures. Thus, the "field" calibration of Ghent (1976) and Ghent et al. (1979) of the garnet-plagioclase-Al_2SiO_5-quartz equilibrium remains an important example of a potentially powerful strategy for the analysis of non-ideal mixing.

Using the aluminum silicate phase equilibrium diagram of Holdaway (1971), Hodges and Spear (1982) carried out a field-based calibration of mineral thermobarometers in the Mt. Moosilauke area, New Hampshire. Because this area coincides with the "triple point isobar" of Thompson and Norton (Fig. 1.5), Hodges and Spear (1982) assumed that metamorphism occurred at P-T conditions near the Al_2SiO_5 triple point. They provided an empirical analysis of non-ideal thermodynamic mixing parameters for the garnet-biotite geothermometer and the garnet-plagioclase-Al_2SiO_5-quartz geobarometer. They derived empirical activity coefficients for components in garnet, biotite and plagioclase, which yielded the best agreement between the garnet-biotite geothermometer, the garnet-plagioclase-Al_2SiO_5-quartz geobarometer, and Holdaway's (1971) triple point (Fig. 1.10). It is important to emphasize that the accuracy of Hodges and Spear's (1982) calibration of the garnet-biotite geothermometer and garnet-plagioclase-Al_2SiO_5-quartz geobarometer depend on the accuracy of Holdaway's (1971) triple point. Furthermore, their analysis assumes that metamorphism of rocks in the Mt. Moosilauke area occurred close to the triple point. However, Kerrick and Speer (1988) concluded that the Mt. Moosilauke area was metamorphosed at pressures below the triple point. Hodges and Spear's (1982) paper nevertheless represents an outstanding attempt to provide a consistent, field-based calibration of geothermometers and geobarometers. Furthermore, their contribution illustrates the importance of the Al_2SiO_5 phase equilibrium for *primary* thermobarometry of metamorphic rocks.

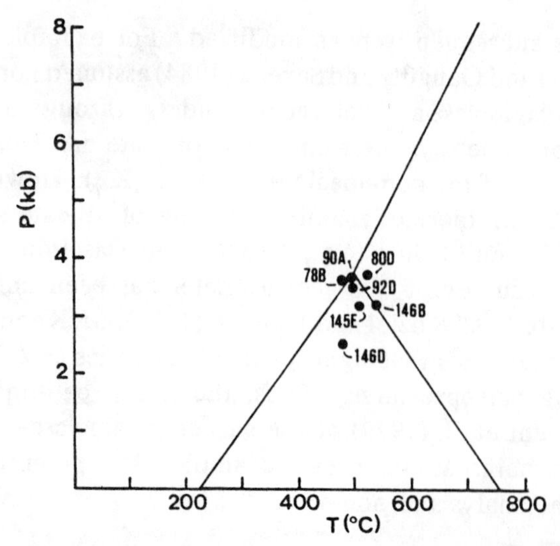

Figure 1.10. Hodges and Spear's (1982) thermobarometric estimates for samples from the Mt. Moosilauke area, New Hampshire. The Al_2SiO_5 stability relations are from Holdaway (1971). (From Hodges and Spear, 1982, Fig. 7).

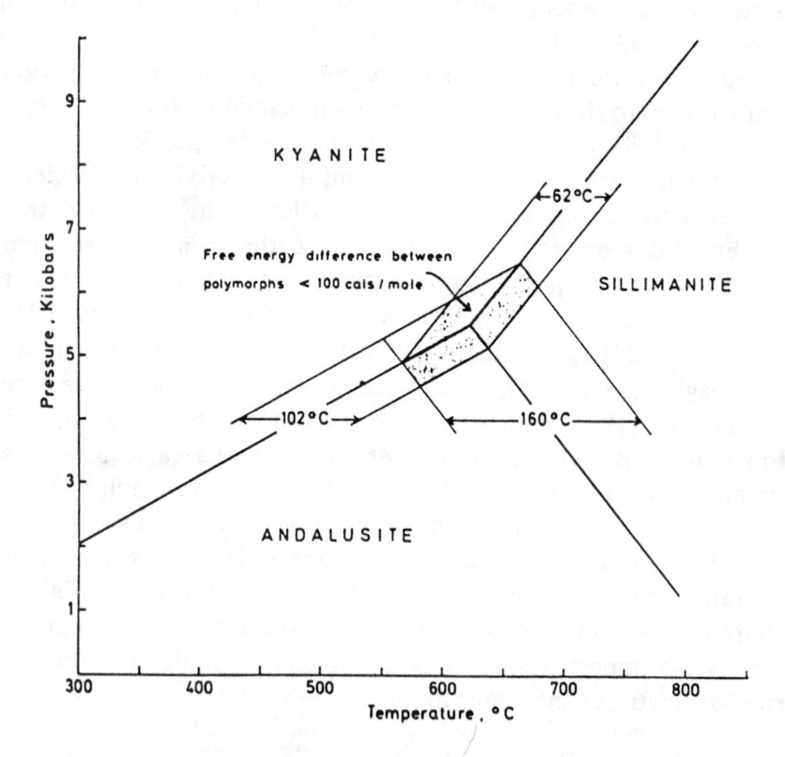

Figure 1.11. Effect of ± 100 cal (= 418 J) variation in ΔG_r of the Al_2SiO_5 equilibria. (From Richardson et al., 1969, Fig. 4).

RATIONALE FOR TOPICAL ORGANIZATION OF SUCCEEDING CHAPTERS

Most petrologic studies involving the aluminum silicates directly utilize *univariant* phase equilibria in the Al_2SiO_5 system. Within the last decade, most petrologists have used Holdaway's (1971) phase diagram for the thermobarometric analysis of rocks containing the Al_2SiO_5 polymorphs. There is, however, considerable uncertainty regarding the calibration of the phase equilibrium diagram and the direct application of this phase diagram to Al_2SiO_5-bearing rocks. The uncertainties in experimental calibration stem from the sluggishness of the three univariant solid-solid reactions. Compared to most other metamorphic equilibria, ΔG_r for transformations involving the Al_2SiO_5 polymorphs is small. Because of the direct correlation between ΔG_r and heterogeneous reaction kinetics (Fyfe et al., 1958; Fisher and Lasaga, 1981), experimental attempts to bracket the Al_2SiO_5 univariant equilibria suffer from the "...plague of the small ΔG's" (Fyfe et al., 1958, p. 22). Of the three polymorphic transformations involved, the andalusite $\rightleftarrows$ sillimanite reaction has the smallest ΔG_r. The experimental difficulties in detecting this kinetically-sluggish reaction are notorious. Accurate derivation of the Al_2SiO_5 phase equilibria through measurements of thermodynamic properties, which is an alternative to hydrothermal experimental studies, is also plagued by the small values of ΔG_r. From measurements of the enthalpies, entropies, and molar volumes of phases, an equilibrium can be computed from

$$\Delta G_r = 0 = \Delta H_r - T\Delta S_r + P\Delta V_r \ .$$

Because of the small values of ΔG_r, relatively small errors in the enthalpies, entropies, and volumes of the phases yield considerable error in the P-T location of the Al_2SiO_5 univariant equilibria. Accordingly, Chapter 3 contains an extensive review and critique of experimental and thermodynamic data bearing on the calibration of the Al_2SiO_5 phase equilibrium diagram.

As illustrated in Figure 1.11, the small values of ΔG_r for the Al_2SiO_5 polymorphic transformations are such that small perturbations in the chemical potentials of the polymorphs can significantly alter the phase equilibrium diagram. Important variables that could significantly perturb the chemical potentials of the aluminum silicates include non-stoichiometry (solid solution), lattice defects, and disorder. The importance of these variables in the Al_2SiO_5 system are respectively considered in Chapters 4, 5 and 6.

The properties, stability relations, and petrologic nature of fibrolite, which has been traditionally considered to be fine-grained sillimanite, have long been an enigma in metamorphic petrology. Because fibrolite is considerably more common than coarse-grained sillimanite, and because many petrologic analyses have been made *assuming* that fibrolite and sillimanite are synonymous, Chapter 7 reviews the "fibrolite problem" in detail.

The remaining chapters explore various petrologic problems, including the nature of reactions involving the aluminum silicates in metapelites (Chapter 8), reaction kinetics and crystal growth mechanisms (Chapter 9), aluminum metasomatism and the controversial problem regarding the mobility of aluminum in metapelites (Chapter 10), and the distribution and significance of the Al_2SiO_5 polymorphs in anatectic migmatites and peraluminous granitoids (Chapter 11).

To maximize the utility of this volume, the references are *comprehensive*, listing many uncited papers as well as referenced ones. In this way readers are provided with a *single*, extensive bibliographic source.

CHAPTER 2

CRYSTAL STRUCTURES, OPTICAL AND PHYSICAL PROPERTIES

INTRODUCTION

The single crystal X-ray refinements of Taylor (1928, 1929), Náray-Szabo et al. (1929), Burnham and Buerger (1961) and Burnham (1963a,b) are fundamental to our understanding of the crystal structures of the Al_2SiO_5 polymorphs. The crystal structures have been summarized by Papike and Cameron (1976), Winter and Ghose (1979), Ribbe (1980), Papike (1987) and Smyth and Bish (1988).

All of the polymorphs share a common crystal structure feature; i.e., chains of AlO_6 edge-shared octahedra parallel to the *c* crystallographic axis (Fig. 2.1). The chains are cross-linked by Si in tetrahedral coordination and by Al with different coordination in each polymorph: Al^{IV} in sillimanite, Al^{V} in andalusite and Al^{VI} in kyanite. These chains account for the prismatic crystal habit parallel to [001] and the well developed {*hk*0} cleavage(s).

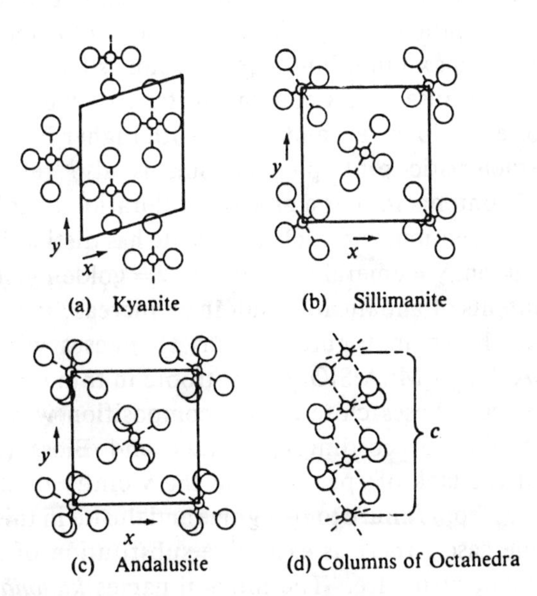

Figure 2.1. Octahedra in the Al_2SiO_5 polymorphs. (a), (b) and (c) show the arrangement of adjacent octahedra, whereas (d) illustrates the octahedral chains parallel to *c*. (From Deer et al., 1982, Fig. 304; copyright Longman Group Ltd.).

ANDALUSITE

Crystal structure

Andalusite consists of edge-shared chains of AlO_6 octahedra that are cross-linked through corner-sharing with double chains consisting of SiO_4 tetrahedra and Al in five coordinated trigonal bipyramids (Figs. 2.2 and 2.3). Penta-coordinated aluminum is unusual amongst silicate minerals. The double chains of alternating tetrahedra and trigonal bipyramids are linked by edge sharing of adjacent bipyramids. As measured by appreciable differences between various Al-O bond distances (Fig. 2.4 and Al_1 in Table 2.1), the AlO_6 octahedra are notably distorted from ideal O_h symmetry. Burnham and Buerger (1961, p. 289) concluded that the distortions of these octahedra result from "...the configurational requirements of the other polyhedra rather than the conditions within the octahedral chains themselves". The Si-O bond distances (Table 2.2) are quite similar.

Optical properties

As summarized in Table 2.3, andalusite is characterized by fairly high relief (refractive indices range from 1.63 to 1.65) and relatively low birefringence (the maximum interference color is first order yellow for a 30 μm section). Because of low contents of transition elements, andalusite is usually colorless. However, andalusite with higher Fe^{3+} content, which produces characteristic pink pleochroism, is also relatively common. Elevated Mn^{3+} contents impart a greenish coloration to andalusite in hand specimen. In thin section, Mn-rich andalusite has distinctive pleochroism: x = yellowish green; y = emerald green, and z = golden yellow. Increasing Fe and Mn contents of andalusite result in an increase in refractive indices (Fig. 2.5). As shown in Figure 2.5 (inset), phases with about 6 mole percent of $FeAlSiO_5$ + $MnAlSiO_5$ are isotropic in thin section. Because of low birefringence, phases close to this composition would appear to be virtually isotropic. Accordingly, Gunter and Bloss (1982, p. 1224) concluded that the lack of specimens in the vicinity of this composition "...possibly arises from failure to recognize andalusite in this near-isotropic form". In some cases, there is extensive substitution of Fe^{3+} and Mn^{3+} into the andalusite structure. The mineral names *kanonaite* and *viridine* have been used for Mn-rich andalusites. However, the most recent decision of the IMA Commission on New Minerals and Mineral Names recommends: (1) kanonaite is reserved for the (fictive) $Mn^{VI}Al^{V}SiO_5$ end member, (2) the name viridine is deleted, (3) *manganian andalusite* is used

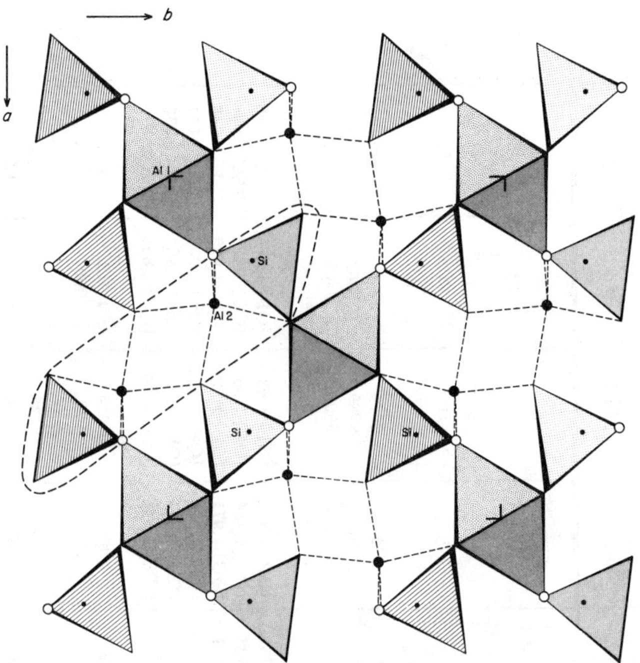

Figure 2.2. Projection of the andalusite structure down the *c* axis. The area enclosed by the oval-shaped dashed line is shown in Figure 2.3. (From Papike, 1987, Fig. 7*c*).

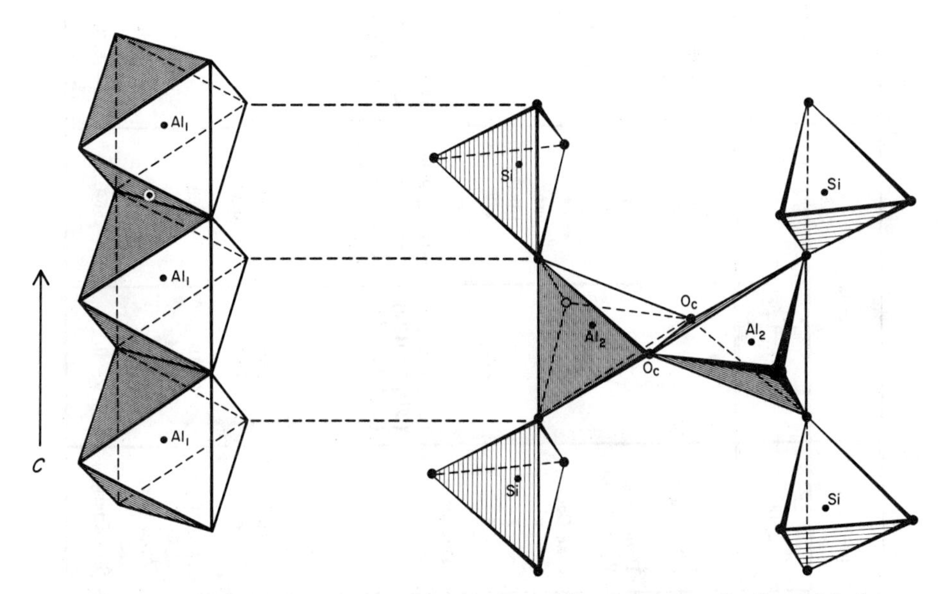

Figure 2.3. Projection down the *a* axis of andalusite showing the key coordination polyhedra. (From Papike, 1987, Fig. 7*d*).

End-Member	Andalusite		Sillimanite		Kyanite			
Site	Al1	Al2	Al1	Al2	Al1	Al2	Al3	Al4
C.N.	6	5	6	4	6	6	6	6
Occupant	Al	Al	Al	Al	Al	Al	Al	Al
Point Sym.	2	m	$\bar{1}$	m	1	1	1	1
Wyckoff Not.	4e	4g	4a	4c	2i	2i	2i	2i
Frac.Coord.								
x	0	.3705	0	.1417	.3254	.2974	.0998	.1120
y	0	.1391	0	.3449	.7040	.6989	.3862	.9175
z	.2419	1/2	0	1/4	.4582	.9505	.6403	.1649
Distances								
	(OA) 1.827 (2)	(OA) 1.816	(OA) 1.914 (2)	(OB) 1.751	(OB) 1.874	(OB) 1.934	(OB) 1.986	(OA) 1.816
	(OB) 1.891 (2)	(OC) 1.840	(OB) 1.868 (2)	(OC) 1.711	(OF) 1.884	(OC) 1.881	(OC) 1.924	(OA) 1.998
	(OD) 2.086 (2)	(OC) 1.899	(OD) 1.954 (2)	(OD) 1.796 (2)	(OG) 1.971	(OD) 1.889	(OE) 1.862	(OB) 1.846
		(OD) 1.814 (2)			(OH) 1.987	(OF) 1.914	(OF) 1.968	(OD) 1.911
					(OK) 1.847	(OK) 1.930	(OF) 1.883	(OE) 1.933
					(OM) 1.848	(OM) 1.925	(OG) 1.885	(OH) 1.875
Mean	1.935	1.836	1.912	1.764	1.902	1.913	1.918	1.896
σ	.121	.036	.039	.041	.062	.023	.050	.065
Poly.Vol.	9.539	5.153	9.175	2.791	8.977	9.136	9.164	8.921
Q.E.	1.0114	--	1.0109	1.0062	1.0155	1.0141	1.0180	1.0139
Ang. Var.	18.0	--	36.4	20.5	47.7	50.2	57.0	42.5
Site Energy	-2490.	-2569.	-2573.	-2526.	-2532.	-2563.	-2543.	-2531.

Table 2.1 (*opposite page*). Aluminum sites in the Al_2SiO_5 polymorphs. Bond distances are in angstroms. C.N. = coordination number, Mean = mean cation-oxygen bond distance, σ = standard deviation of mean cation-oxygen bond distance, Poly. Vol. = polyhedral volume, Q.E. = quadratic elongation of polyhedra, An. Var. = angular variation of polyhedra. *Site Energy* is the electrostatic energy (in kcal) for one mole of sites. The designations for various oxygen atoms may be transformed into more conventional notation by subscripting: OA = O_A, OB = O_B, etc. (From Smyth and Bish, 1988, Table 4.6.2; copyright Allen & Unwin, Inc.).

Table 2.2 (*right*). Silicon sites in the Al_2SiO_5 polymorphs. Abbreviations as in Table 2.1. (From Smyth and Bish, 1988, Table. 4.6.3; copyright Allen & Unwin, Inc.).

End-Member	Andalusite	Sillimanite	Kyanite	
Site	Si	Si	Si1	Si2
C.N.	4	4	4	4
Occupant	Si	Si	Si	Si
Point Sym.	m	m	1	1
Wyckoff Not.	4g	4c	2i	2i
Frac.Coord.				
x	.2460	.1533	.2692	.2910
y	.2520	.3402	.0649	.3317
z	0	3/4	.7066	.1892
Distances				
1	(OB) 1.646	(OA) 1.640	(OD) 1.631	(OA) 1.640
2	(OC) 1.618	(OC) 1.573	(OE) 1.643	(OC) 1.629
	(OD) 1.630 (2)	(OD) 1.645 (2)	(OH) 1.621	(OG) 1.627
			(OM) 1.647	(OK) 1.649
Mean	1.631	1.626	1.636	1.636
σ	.011	.035	.011	.010
Poly.Vol.	2.211	2.203	2.241	2.243
Q.E.	1.0043	1.0013	1.0012	1.0018
Ang. Var.	16.4	3.4	4.8	7.1
Site Energy	-4404.	-4426.	-4443.	-4458.

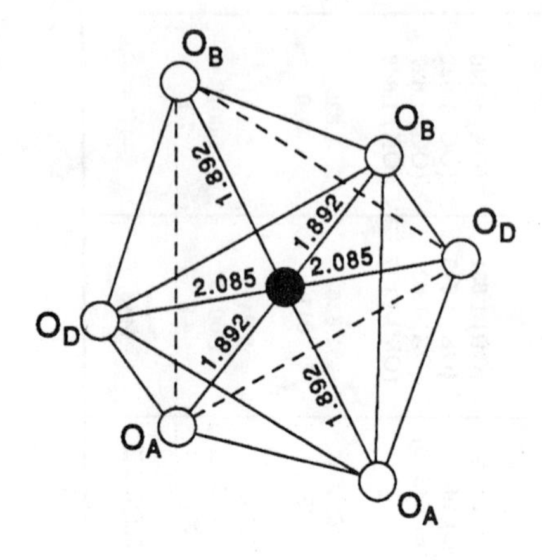

Figure 2.4. Geometry of the distorted octahedron in andalusite. The bond distances are in angstroms. (Redrawn from Hålenius, 1978, Fig. 2).

Table 2.3. Summary of optical and physical properties of andalusite. The principal refractive indices are designated: α, β and γ, whereas the crystallographic axes are x, y and z. Other abbreviations are: D = density, H = hardness, O.A.P. = optic axial plane, δ = birefringence, Z = number of formula units per unit cell. (From Deer et al., 1982, p. 759; copyright Longman Group Ltd.).

Orthorhombic (−)

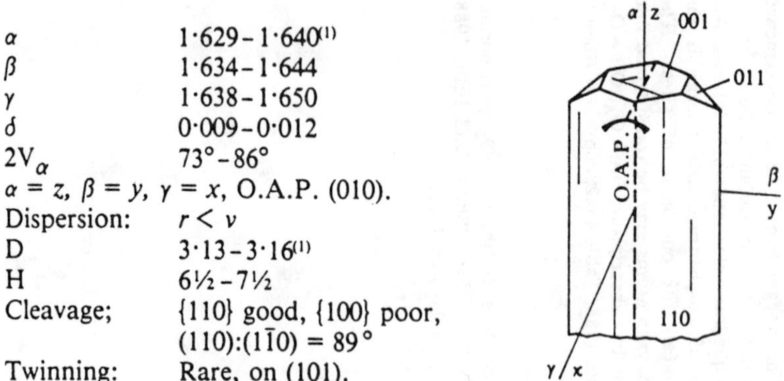

α	1·629 – 1·640[1]
β	1·634 – 1·644
γ	1·638 – 1·650
δ	0·009 – 0·012
$2V_\alpha$	73° – 86°
$\alpha = z$, $\beta = y$, $\gamma = x$, O.A.P. (010).	
Dispersion:	$r < v$
D	3·13 – 3·16[1]
H	6½ – 7½
Cleavage;	{110} good, {100} poor, (110):(1$\bar{1}$0) = 89°
Twinning:	Rare, on (101).
Colour:	Usually pink, but may be white or rose-red; also grey, violet, yellow, green or clouded with inclusions; in thin section normally colourless, but may be pink or green.
Pleochroism:	In coloured varieties weak, with α rose-pink, β and γ greenish yellow.
Unit cell:	*a* 7·79, *b* 7·90, *c* 5·56[1] Å; *V* 324 $Å^3$. $Z = 4$. Space group *Pnnm*.

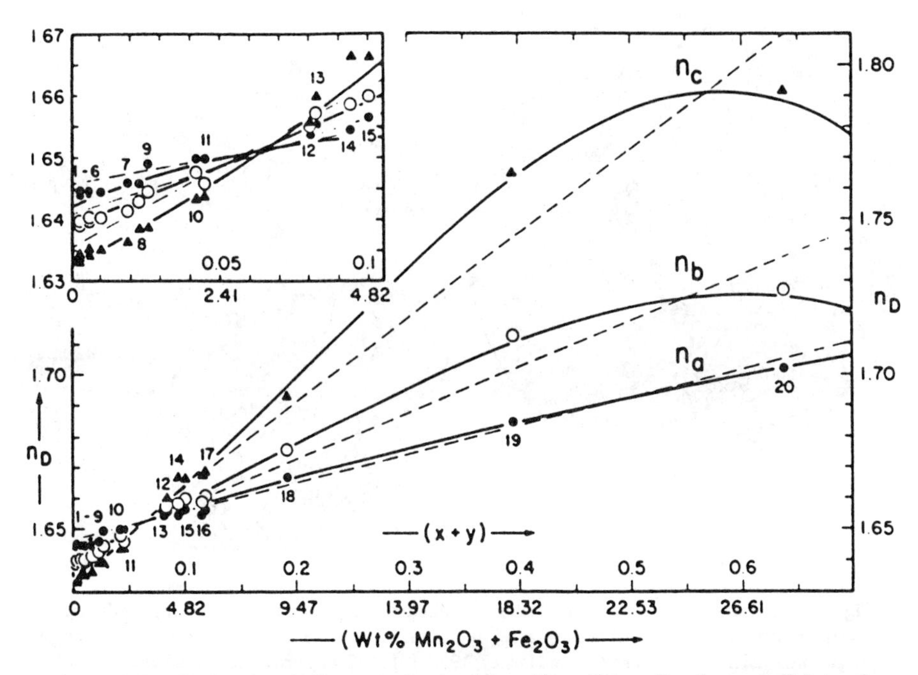

Figure 2.5. Refractive indices obtained with sodium light vibrating parallel to the crystallographic axes (*a*, *b*, and *c*) for a variety of compositions in the andalusite-kanonaite solid solution series. The inset in the upper left is an expanded view of the lower left area of the main plot. The compositional variable (x + y) on the abscissa is the fractional occupancy of Fe^{3+} + Mn^{3+} in the octahedral site. Note that the curves for n_a, n_b, and n_c cross at (x + y) ≈ 0.06; thus, crystals of this composition are isotropic. (From Gunter and Bloss, 1982, Fig. 3).

for compositions with Mn^{VI} < Al^{VI} and *aluminous kanonaite* for Mn^{VI} > Al^{VI}, and (4) *ferrian manganian andalusite* refers to phases with the andalusite structure that contain appreciable Mn^{3+} *and* Fe^{3+} (Gunter and Bloss, 1982).

Andalusite most commonly occurs in metamorphosed pelitic rocks. Two crystal habits of andalusite are common in this lithology: (1) anhedral, poikiloblastic andalusite, and (2) chiastolite.

Poikiloblastic andalusite typically displays irregular, anhedral crystal shapes. "Spotted" hornfelses characterized by poikiloblasts of andalusite (Fig. 2.6) and/or cordierite are common in pelites that have been subjected to contact metamorphism and low-pressure regional metamorphism. Although andalusite and untwinned cordierite typically have similar anhedral, poikiloblastic habits, they are readily distinguished by marked differences in refractive indices. That is, the low refractive indices of cordierite result in low relief, whereas andalusite has higher relief due to

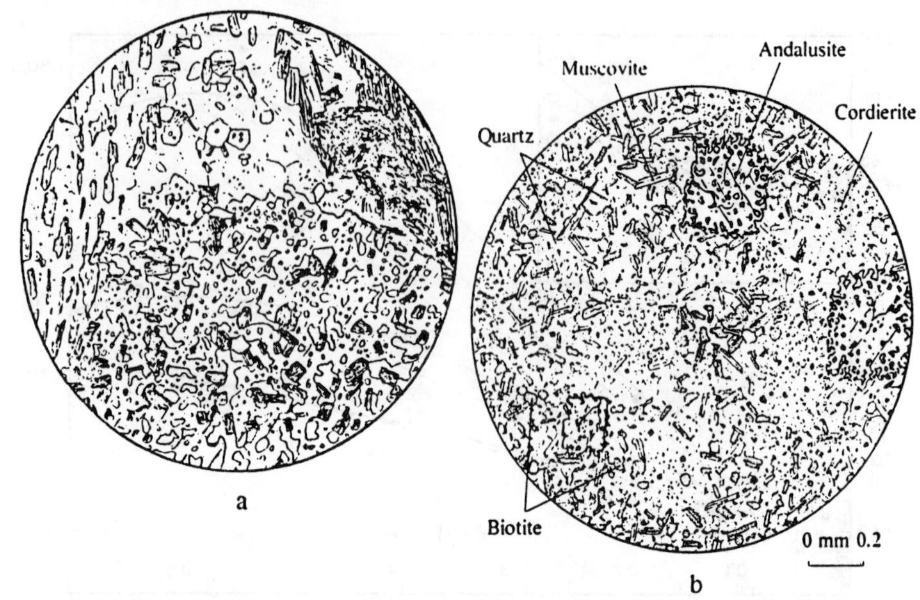

Figure 2.6. Examples of the poikiloblastic, anhedral habit of andalusite. (a) Andalusite-biotite schist near Fyvie, Aberdeenshire, Scotland. The lower half of the sketch shows a single andalusite poikiloblast. (From Harker, 1950, Fig. 108; copyright Methuen and Co. Ltd.). (b) Spotted hornfels from the Onawa aureole, Maine. The spots consist of anhedral andalusite poikiloblasts. (From Best, 1982, Fig. 12-6; copyright W.H. Freeman and Co.).

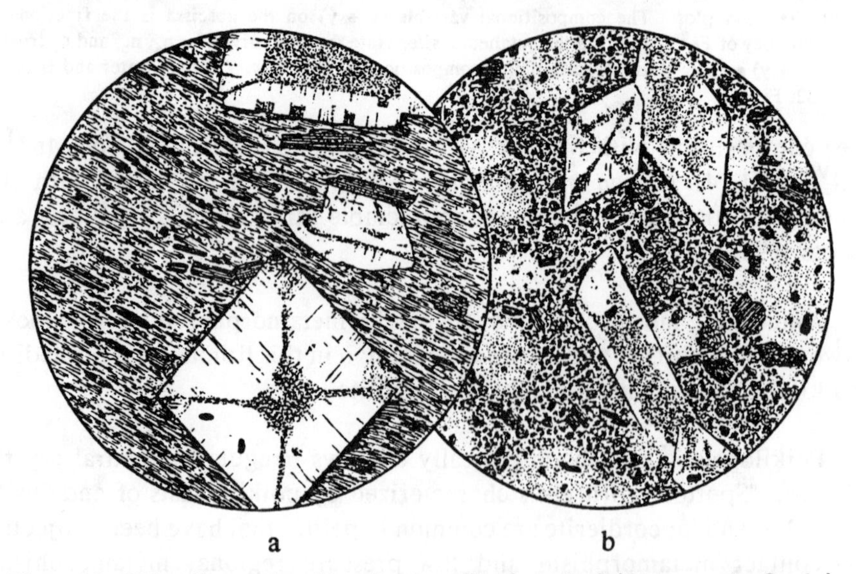

Figure 2.7. Chiastolite porphyroblasts in metapelites from the Skiddaw aureole, Cumbria, England. (a) Note the characteristic thin, orthogonally-disposed, inclusion-rich bands and the square-shaped, inclusion-rich central portion of the square crystal (lower half of sketch). The elongated crystal in (b) is typical of chiastolite with the *c* axis at a low angle to the plane of the section. [(a): from Harker, 1964, Fig. 7; copyright Methuen & Co. Ltd. (b): from Nockolds et al., 1978, Fig. 33.10; copyright Cambridge University Press].

higher refractive indices.

Chiastolite is characterized by elongate crystals bounded by {110} prism faces. In (001) sections, chiastolite is distinguished by characteristic "X"-shaped, inclusion-rich bands that intersect the exterior of the crystal at the edges of crystal faces (Figs. 2.7a). In some cases, inclusion-rich areas are also arranged in the form of a square that is in the center of the crystal (Figs. 2.7a and 2.8a). In (001) sections, chiastolite shows either square cross sections (Fig. 2.7a) or a distinctive cruciform pattern (Figs. 2.8a and 2.9). In sections cut perpendicular to the *c* axis, chiastolite porphyroblasts have dimensions up to 2-3 cm across, whereas in sections cut parallel to *c*, chiastolite attains lengths (parallel to *c*) of 10-15 cm.

Thermal expansion, compressibility and elasticity

Winter and Ghose (1979) carried out the most recent and detailed study of the thermal expansion of andalusite. As shown in Figure 2.10a, the temperature variation in unit cell dimensions and cell volume is nearly linear up to 1000°C. The relative increase in cell edges with increasing temperature is $a > b > c$ (Fig. 2.10a). Winter and Ghose (1979, p. 581) attributed the relatively small increase in *c* to the octahedral chains and to the "...fully extended chains of alternating Si tetrahedra and Al_2 trigonal bipyramids". They noted that the increase in individual bond lengths with temperature is a function of bond length. Accordingly, for the Al_1 octahedral site (Fig. 2.4), the thermal expansion of the long Al_1-O_D bonds exceeds that of Al_1-O_A and Al_1-O_B. In the (001) plane, the Al_1-O_D bonds are symmetrically disposed about a 30° angle to the *a* axis, thereby explaining the relatively large increase in the *a* cell dimension with increasing temperature.

From single-crystal X-ray measurements utilizing a diamond anvil apparatus, Ralph et al. (1984) determined the unit-cell dimensions and crystal structure of andalusite at 12, 25, and 37 kbar. As shown in Figure 2.11, the relative decrease of unit cell dimensions with increasing pressure is: $a > b > c$. This relationship is the inverse of the relative changes in unit cell dimensions upon increasing temperature (Fig. 2.10). The crystal chemical arguments given by Winter and Ghose (1979) for the changes in cell dimensions with increasing temperature can be applied to the changes with increasing pressure. Accordingly, within the Al octahedron, the long Al_1-O_D bond shows a relatively large decrease in length with increasing pressure. Because of the symmetrical disposition of the Al_1-O_D bonds about the *a* axis, the relatively large compressibility of these bonds accounts for the marked decrease in the *a* cell dimension with increasing

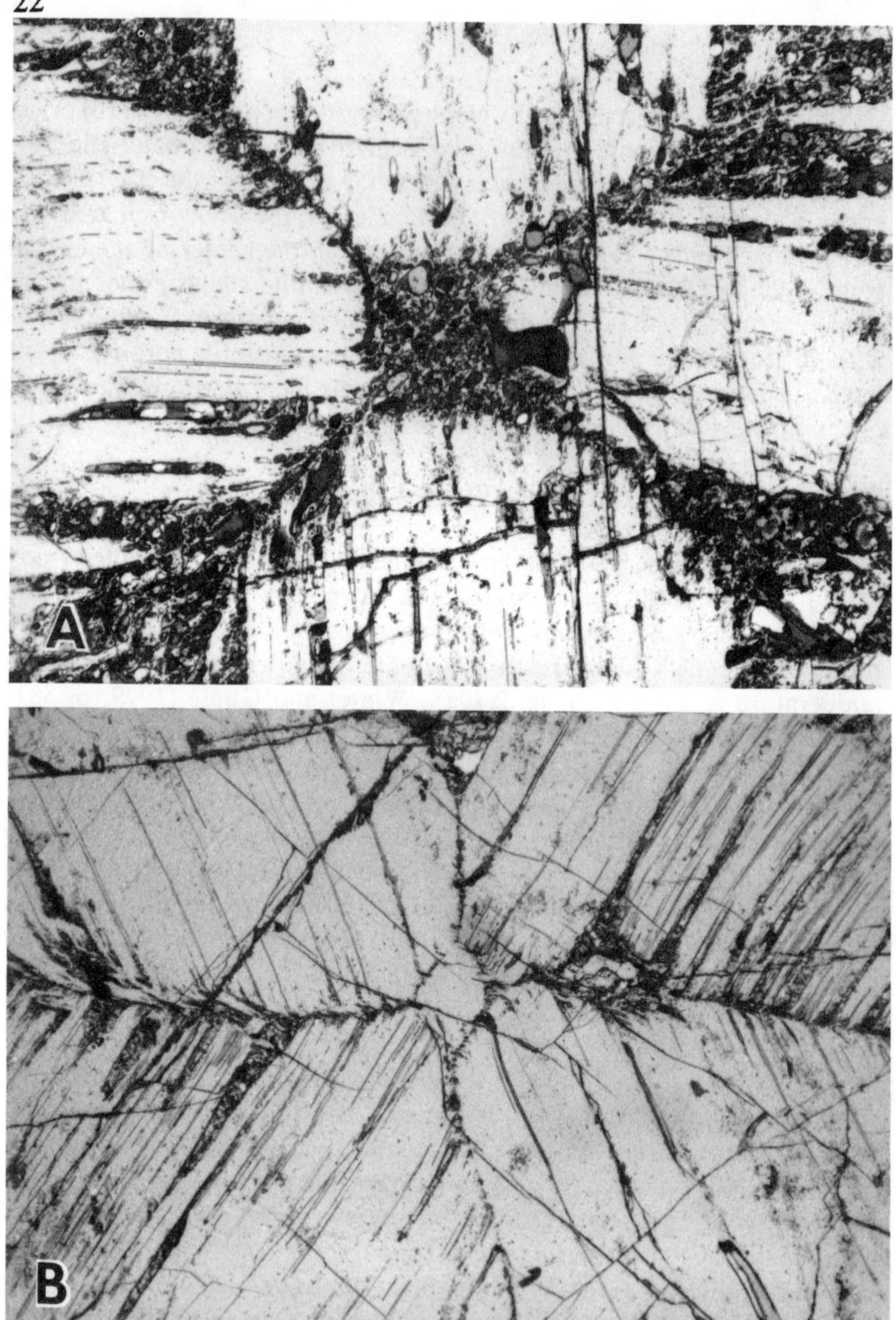

Figure 2.8. Chiastolite porphyroblasts from the Mt. Raleigh pendant, British Columbia. (a) Cruciform chiastolite with square, poikiloblastic core, and with the characteristic "X"-shaped, inclusion-rich cross. Note the elongated quartz inclusions in each of the four sectors. (b) Chiastolite porphyroblast with inclusion-poor core (center) and well-developed, elongated quartz inclusions in each of the four sectors. (From Kerrick and Woodsworth, 1989, Figs. 2a and 2b).

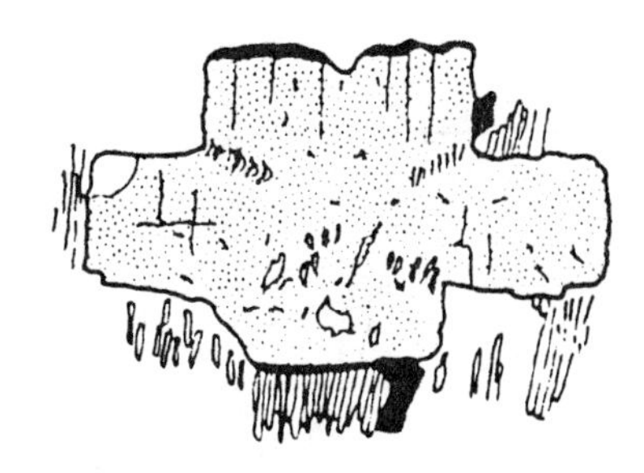

Figure 2.9. Cruciform chiastolite from Fintown, Donegal, Ireland. (From Naggar and Atherton, 1970, Fig. 3f).

pressure. Ralph et al. (1984) attributed the intermediate compressibility along *b* to the relatively long Al_2-O_C bonds parallel to this direction. In contrast to the AlO_5 and AlO_6 polyhedra, the SiO_4 tetrahedron is relatively incompressible.

Interpretations of the compressibility measurements of Ralph et al. (1984) are comparable with the study of Iishi et al. (1979), who carried out rigid-ion, lattice dynamical modeling of polarized Raman and infrared spectra of andalusite. The calculations of Iishi et al. (1979) revealed that the Si-O bonds are nearly four times stronger than the Al-O bonds. Such differences reflect the fact that the Si-O bonds of Si tetrahedra are about 60% covalent in character whereas the Al-O bonds of AlO_5 and AlO_6 polyhedra average 30% covalency. The interpretations of Iishi et al. (1979) are compatible with those derived from the compressibility measurements of Ralph et al. (1984); i.e., the Al polyhedra are considerably more compressible than the Si tetrahedra.

Vaughan and Weidner (1978) measured the elasticity of andalusite by Brillouin scattering (i.e., measurement of the Doppler shift of light that is scattered from thermally-generated acoustic waves). As shown in Table 2.4, the elastic compliance parallel to the *c* axis (s_{11}) is larger than that parallel to the *a* or *b* axes. These results are compatible with Winter and Ghose's (1979) crystal chemical interpretation of the anisotropies of thermal expansion; i.e., the chains of octahedra and the double chains of SiO_4 tetrahedra and AlO_5 trigonal bipyramids are relatively incompliant parallel to *c*. As shown in Figure 2.12, the curve of compressional velocity in the *a*-*b* plane shows maxima that are 30° from the *b* axis. These maxima are consistent with the orientation of the octahedra in the *a*-*b* plane. In this plane, the rigidity of andalusite is larger than that of sillimanite (Fig. 2.12). Vaughan and Weidner (1978) attributed the larger rigidity of andalusite to the presence of voids between the octahedral chains (Fig. 2.2).

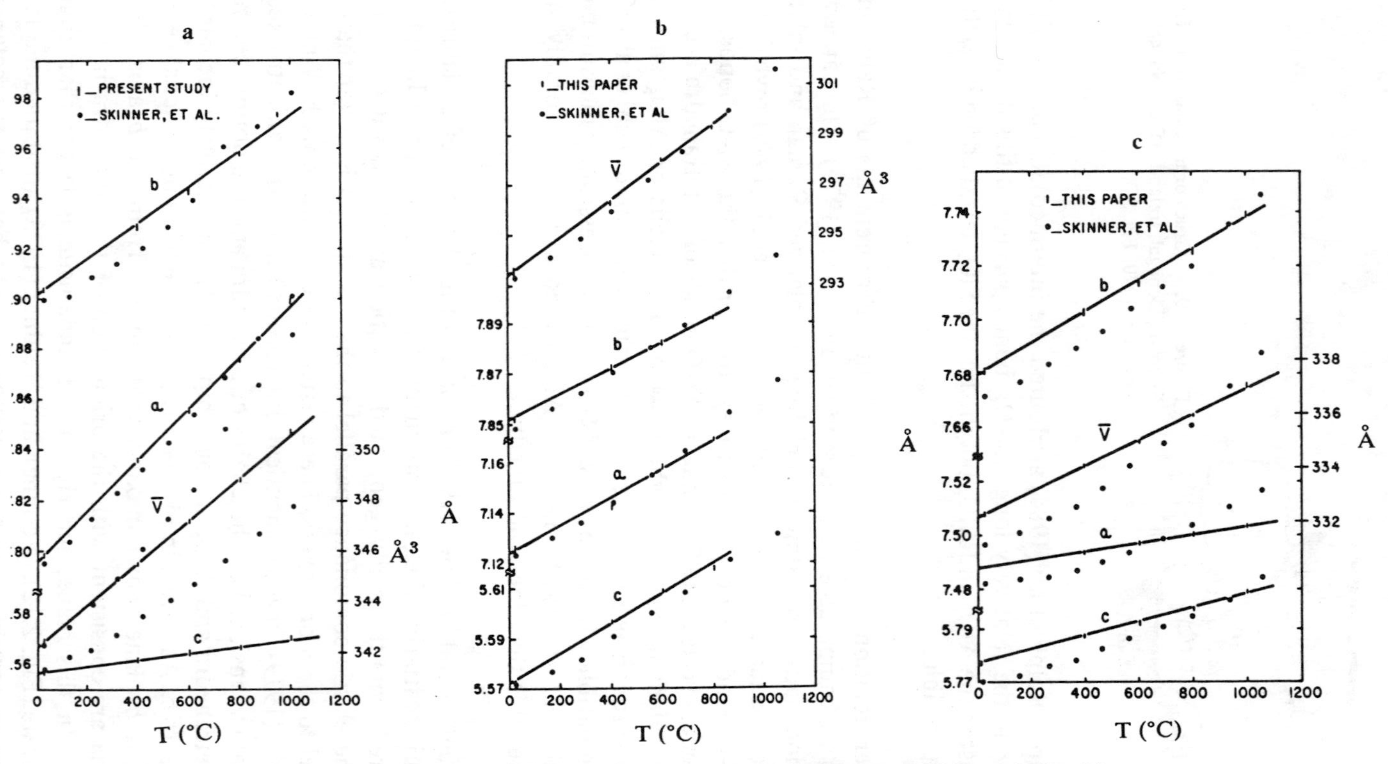

Figure 2.10. Variation of the unit cell volume and the lengths of the unit cell edges of (a) andalusite, (b) kyanite and (c) fibrolitic sillimanite, as a function of temperature. The data of Skinner et al. (1961) are shown for comparison. (From Winter and Ghose, 1979. Fig. 1).

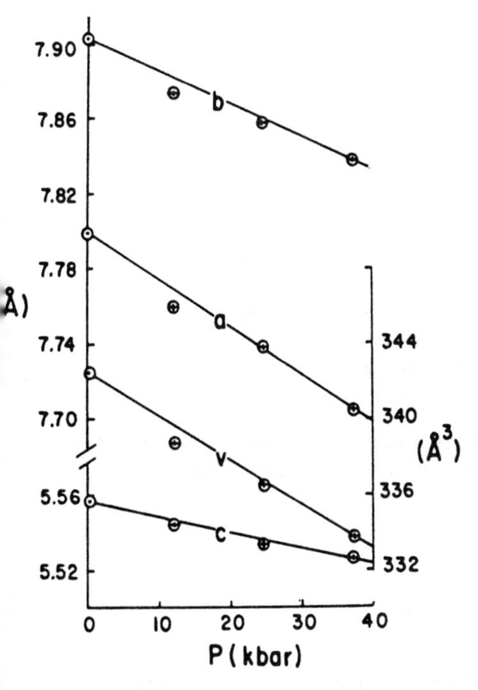

Figure 2.11. Variation in the unit cell dimensions and cell volume of andalusite with pressure. (From Ralph et al., 1984, Fig. 2).

Table 2.4. Elastic constants (c_{ij}) and elastic compliances (s_{ij}) of andalusite and sillimanite determined from Brillouin scattering data. (From Vaughan and Weidner, 1978, Table 3).

Index (ij)	*Andalusite*			*Sillimanite*		
	c_{ij} [a]	error[d]	s_{ij} [b]	c_{ij} [a]	error[d]	s_{ij} [b]
11	2.334	0.021	0.532	2.873	0.033	0.413
22	2.890	0.027	0.399	2.319	0.040	0.653
33	3.801	0.025	0.323	3.884	0.039	0.359
44	0.995	0.010	1.005	1.224	0.019	0.817
55	0.878	0.012	1.139	0.807	0.013	1.240
66	1.123	0.027	0.890	0.893	0.020	1.192
23	0.977	0.033	−0.0708	1.586	0.076	−0.2344
13	1.162	0.030	−0.1360	0.834	0.051	−0.0274
12	0.814	0.051	−0.1039	0.947	0.033	−0.1498

	Voigt	*Reuss*	*Voigt*	*Reuss*
Bulk modulus[a]	1.659	1.580	1.751	1.664
Shear modulus[a]	1.004	0.977	0.951	0.878
Comp. velocity[c]	9.763	9.573	9.652	9.350
Shear velocity[c]	5.651	5.572	5.417	5.204

[a] c_{ij} and bulk and shear moduli are in megabars
[b] s_{ij} are in inverse megabars
[c] Velocities are in km/s
[d] Errors refer to the rms derivation in the particular c_{ij} when the velocity errors reported in Table 3 are propagated through to the elastic constants

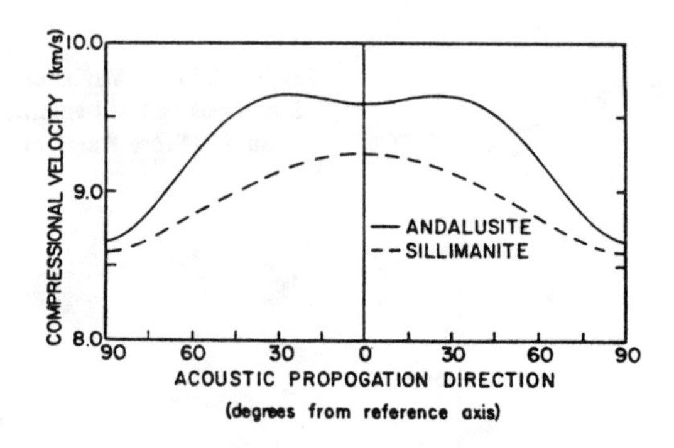

Figure 2.12. Compressional velocity in the *a-b* plane of andalusite and sillimanite. The reference axis is *b* in andalusite and *a* in sillimanite. (From Vaughan and Weidner, 1978, Fig. 6).

SILLIMANITE

Sillimanite is distinguished by two contrasting textural habits: coarse, prismatic *sillimanite* and acicular *fibrolite*. Numerous workers have questioned the general assumption that, with the exception of grain size, sillimanite and fibrolite are identical. Because the "fibrolite problem" is reviewed in Chapter 7, the following discussion omits some details of the comparison of the properties of fibrolite and sillimanite (e.g., unit cell parameters).

Crystal structure

The structure of sillimanite is similar to that of andalusite. That is, sillimanite contains chains of edge-shared octahedra parallel to *c* that are cross-linked by double chains of tetrahedra containing Si and Al (Figs. 2.13 and 2.14). The tetrahedral chains are linked to adjacent polyhedra through corner-sharing. The octahedra are elongated parallel to the Al_1-O_D direction (Figs. 2.13), which is 30° from the *b* axis. Within the double tetrahedral chains, Al_2 and Si atoms are displaced toward the bridging O_C oxygen atoms. In fact, Burnham (1963a, p. 147) noted that the Al_2-O_C and Si-O_C distances are "abnormally short". The [001] chain structure of sillimanite accounts for the crystal habit elongate parallel to *c*. The good {010} cleavage of sillimanite *may* reflect the ease of breaking bonds involving the (O_D) oxygen atoms linking the octahedral and tetrahedral chains (Fig. 2.14).

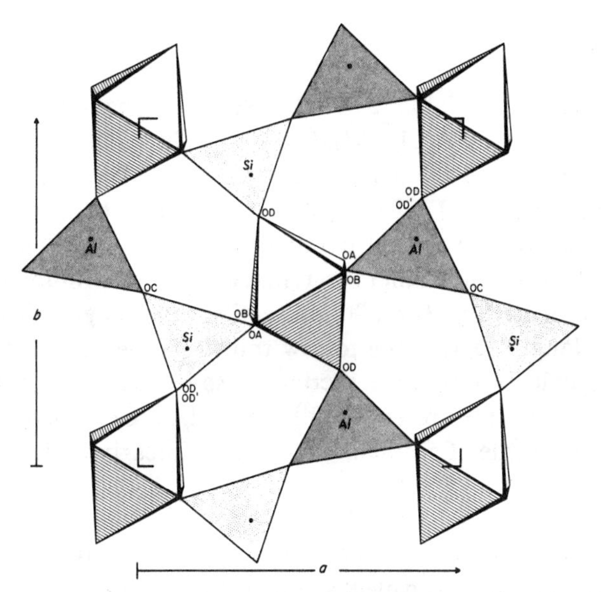

Figure 2.13. Projection of the sillimanite crystal structure down the *c* axis. (From Papike, 1987, Fig. 7*a*).

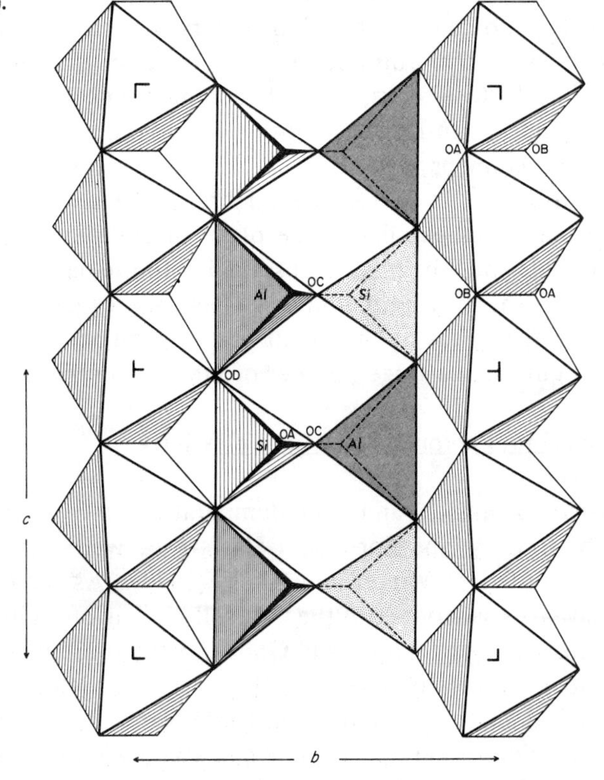

Figure 2.14. Projection of the sillimanite crystal structure down the *a* axis. (From Papike, 1987, Fig. 7*b*).

Optical properties

As summarized in Table 2.5, sillimanite is characterized by high refractive indices, low 2V (21-30°), and strong dispersion (r > v). Sections cut normal to the *c* axis are particularly diagnostic because of the nearly square shape [(110)^(1$\bar{1}$0) = 88°] and the prominent {010} cleavage (Fig. 2.15). For elongate crystals cut parallel to *c*, sillimanite is distinguished from andalusite in having higher birefringence (the maximum interference color is second order blue for a 30 μm section) and length-slow character (andalusite is length fast). Due to low transition element contents, most sillimanite is colorless in thin section. However, some sillimanite is colored because of the presence of Fe^{3+} and/or Cr^{3+} (Rossman et al., 1982). The optic angle of sillimanite decreases linearly with increasing Fe content (Fig. 2.16).

As the name implies, *fibrolite* refers to the fibrous variety of sillimanite. Petrographic and electron microscopy examination of fibrolite (Figs. 2.17, 2.18 and 2.19) reveal acicular crystals with large aspect ratios. Some fibrolite has sub-micron crystal diameters. Kerrick and Speer (1988) subjectively considered fibrolite to be restricted to crystals less than 10 μ in diameter. In schistose rocks, fibrolite commonly occurs as myriads of interlocking crystals forming anastamosing folia paralleling foliation (Fig. 2.17b). In hand specimens, these folia impart a characteristic "silky" sheen to foliation surfaces. In some cases, fibrolite aggregates form small nodules with long axes parallel to the plane of foliation ("faserkiesel"). Fibrolite aggregates may be readily observed with the naked eye in a thin section held in front of a light source. These aggregates form brightly illuminated, whitish areas as the result of the composite reflection and refraction of light from a large number of fibrolite crystals.

Thermal expansion, compressibility and elasticity

The thermal expansion and high-temperature crystal chemistry of fibrolite from Brandywine Springs, Delaware (a well-known fibrolite locality) were studied by Winter and Ghose (1979). As shown in Figure 2.10c, the relative expansivities of the unit cell dimensions with increasing temperature are: *b* > *c* > *a*. Winter and Ghose (1979) attributed these to the presence of open tunnels parallel to *c* (Fig. 2.13), which allow adjacent double tetrahedral chains to rotate in the (001) plane. As shown in Figure 2.20, the double chains marked **A** rotate clockwise whereas those marked **B** rotate counterclockwise. This enhances expansion of *b* and inhibits expansion of *a*. Another factor contributing to the larger expansion along

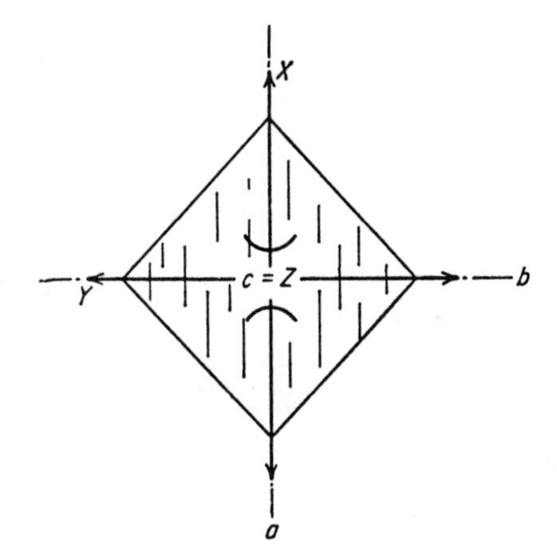

Figure 2.15. Schematic illustration of the optical properties of a crystal of prismatic sillimanite cut parallel to (001). Note the characteristic square crystal shape and the well developed {010} cleavage. (From Kerr, 1977, Fig. 15-21; copyright McGraw-Hill Inc.).

Table 2.5. Summary of the optical and physical properties of sillimanite. Symbols as in Table 2.3. (From Deer et al., 1982, p. 719; copyright Longman Group Ltd.).

Orthorhombic (+)

α	1·653 – 1·661
β	1·657 – 1·662
γ	1·672 – 1·683
δ	0·018 – 0·022
$2V_\gamma$	21° – 30°

$\alpha = x$, $\beta = y$, $\gamma = z$;
O.A.P. (010), $Bx_a \perp (001)$

Dispersion:	$r > v$ strong.
D	3·23 – 3·27.
H	6½ – 7½.

Cleavage:	{010} good, uneven transverse fractures.
Colour:	Normally colourless or white, also yellow, brown, greyish green, bluish green; colourless in thin section.
Pleochroism:	In thick sections coloured varieties may be pleochroic with α pale brown or pale yellow, β brown or greenish, γ dark brown or blue.
Unit cell:	a 7·48, b 7·67, c 5·77 Å; V 331 Å³. Z = 4. Space group *Pbnm*.

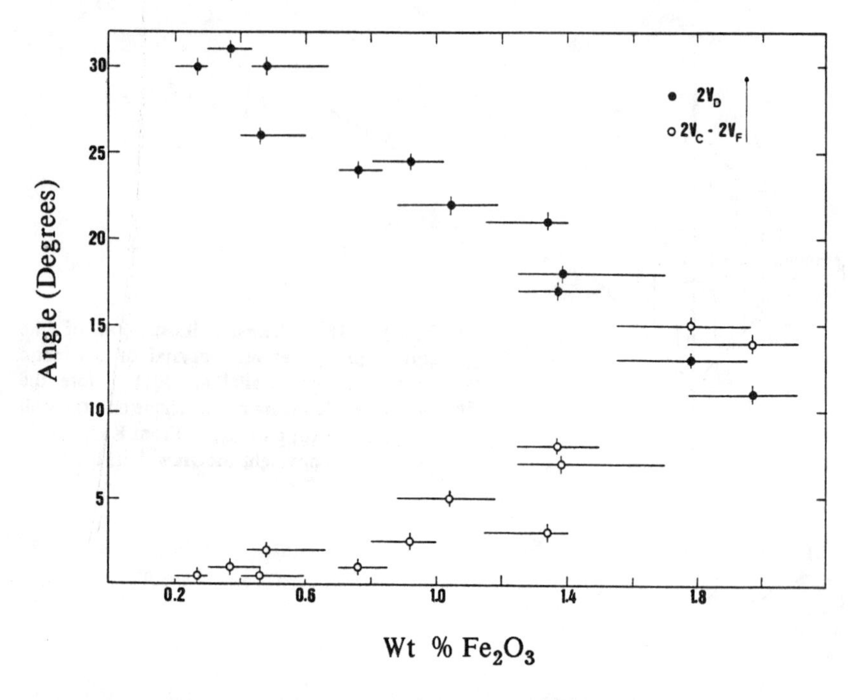

Figure 2.16. Variation of the optical angle (filled circles) and dispersion (open circles) of sillimanite with Fe_2O_3 content. (From Evers and Wevers, 1984, Fig. 4.).

b relative to *a* is the Al_1-O_D bonds oriented 30° on either side of the *b* direction.

Vaughan and Weidner (1978) determined the elasticity of a sillimanite single crystal using Brillouin scattering. As with andalusite, sillimanite is stiffest parallel to the *c* axis because of the [001] octahedral and tetrahedral chains. In contrast to andalusite, c_{11} in andalusite is greater than c_{22}. "This interchange is in keeping with the interchange of the orientation of the octahedra relative to the *a* and *b* axes... This would suggest that the octahedra are controlling the elasticity in these directions, and thus are more compliant than the lower coordinated polyhedra" (Vaughan and Weidner, 1978, p. 142).

KYANITE

Crystal structure

Kyanite can be considered as having a distorted close-packed arrangement of oxygen atoms. The much greater density of kyanite

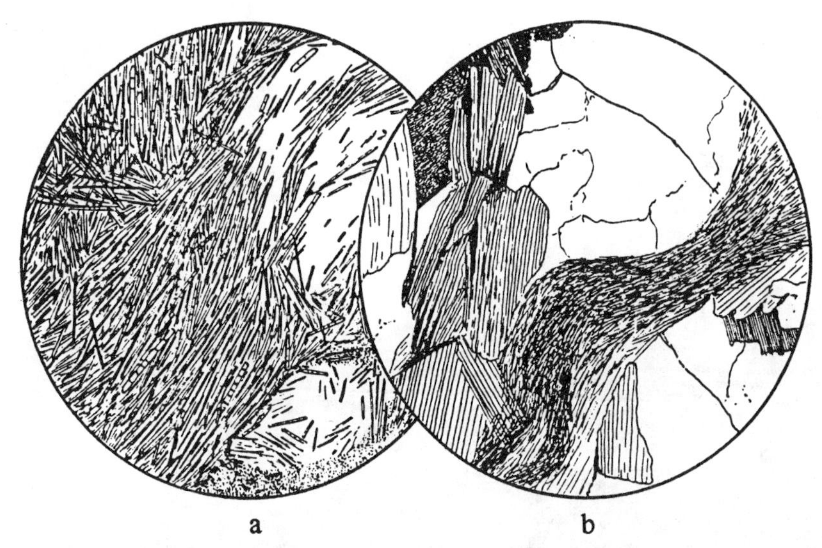

Figure 2.17. Fibrolite in gneisses from Glen Clova, Scotland. (a) The left half of the sketch shows a typical intergrowth of myriads of fibrolite crystals. Inclusions of fibrolite in quartz occur in the right half. (b) Sinusoidal-shaped bundle of fibrolite intergrown with mica. (From Harker, 1964, Fig. 106; copyright Methuen & Co. Ltd.).

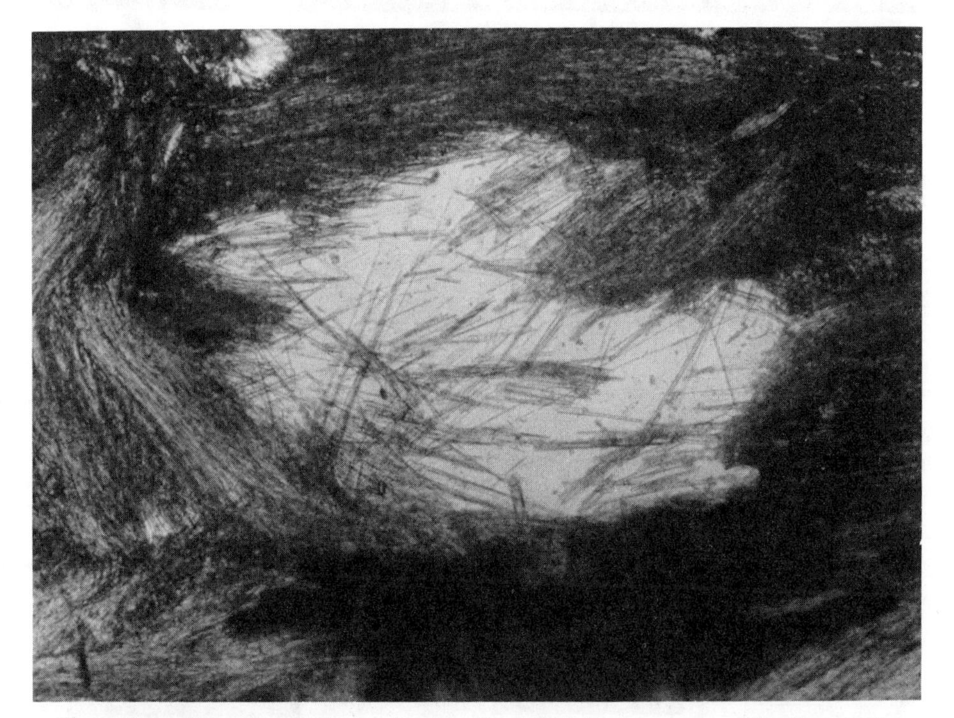

Figure 2.18. Photomicrograph showing fibrolite. Myriads of intergrown fibrolite crystals occur in the peripheral areas of the photo. Individual fibrolite needles are surrounded by quartz in the central portion of the photo. (Photograph by A. Nitkiewicz).

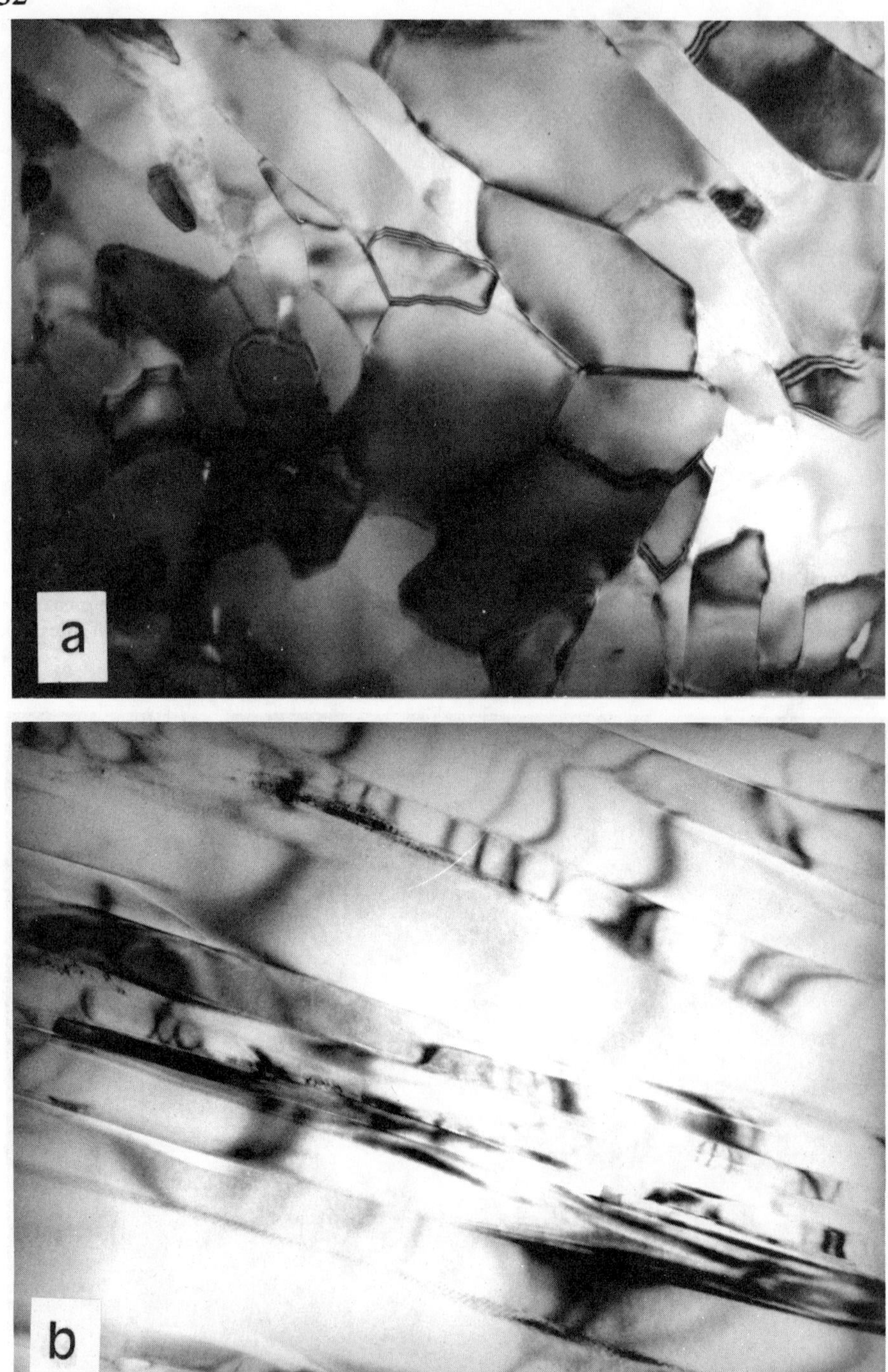

Figure 2.19. Interlocking fibrolite crystals as viewed with transmission electron microscopy. (a) section cut at high angle to the *c* axis, (b) section cut at a low angle to the *c* axis. (Photograph by G. L. Nord, Jr., U. S. Geological Survey, Reston, Virginia).

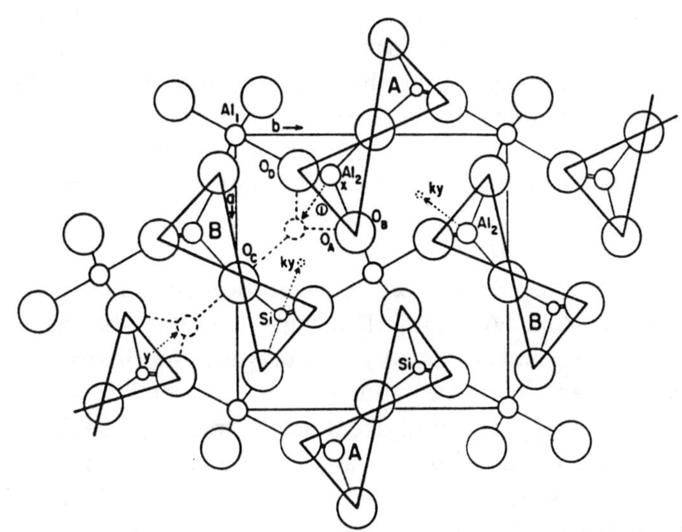

Figure 2.20. Projection of the sillimanite crystal structure down the *c* axis. The heavy lines mark sets of linked tetrahedra (A and B). According to Winter and Ghose (1979), rotation of these tetrahedra with increasing temperature is responsible for the relative thermal expansion of the *a* and *b* unit cell dimensions (see text). The dashed circles and arrows represent Winter and Ghose's (1979) hypothesized shifts of Al_2 and Si atoms in the transformation of sillimanite to andalusite and to kyanite. The shifts of the Al_2 atoms labeled x and y depict the sillimanite → andalusite transformation, whereas atomic shifts for the sillimanite → kyanite transformation are denoted by "ky" at the ends of dashed arrows. (From Winter and Ghose, 1979, Fig. 8).

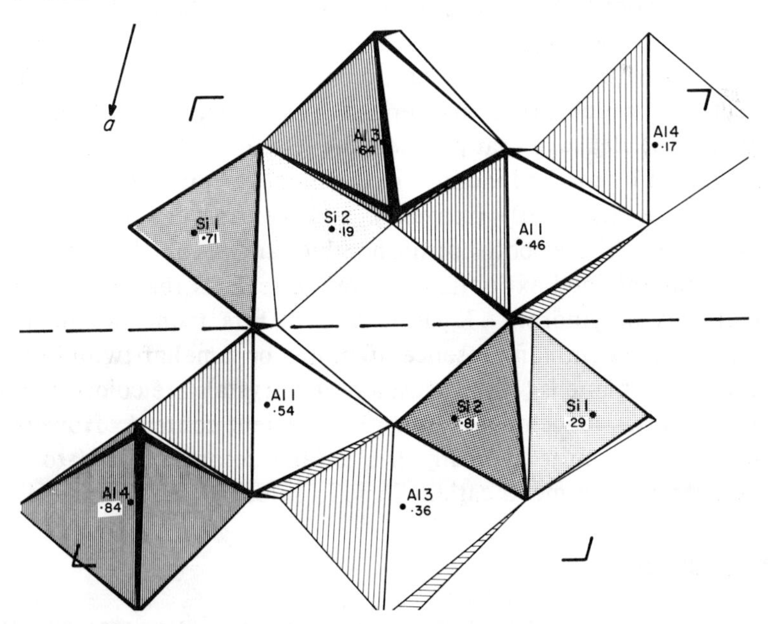

Figure 2.21. Projection of the kyanite structure down the *c* axis. The dashed line represents Burnham's (1963b) hypothesized location of a (100) cleavage plane. The fractional numbers give height in relative coordinates along *c*. (From Papike, 1987, Fig. 7e).

compared to the other two polymorphs is in part attributable to this feature. Chains of octahedra, forming a "zig-zag" pattern parallel to the *c* axis, are cross-linked by Si tetrahedra and additional Al octahedra (Fig. 2.21). Vacant octahedral sites form open channels parallel to [001]. Burnham (1963b) hypothesized that the good {100} cleavage of kyanite occurs along planes containing the least number of oxygen atoms (Fig. 2.21). There are four distinct Al sites, each characterized by different average Al-O bond distances (Table 2.1). There are two distinct tetrahedral silica sites: Si_1 and Si_2 both with identical average Si-O bond distances (Table 2.2). Al_1 and Al_2 occupy the edge-sharing octahedral chains, whereas Al_3 and Al_4 occupy sites between the chains. The Al_1 and Al_3 octahedra share five edges with neighboring octahedra, whereas Al_2 and Al_4 share four edges with adjacent octahedra. Because the polyhedral volumes and site energies of the Al_2 and Al_3 octahedra are nearly identical (Table 2.1), these sites may be considered as virtually identical from a crystal chemistry standpoint.

Optical properties

As summarized in Table 2.6, kyanite is characterized by high relief (it has the highest refractive indices of the Al_2SiO_5 polymorphs) and moderate birefringence (the maximum interference color is first-order red for a 30 μm section). Sections cut parallel to (001) show well-developed (100) and (010) cleavages - the non-orthogonal intersection of these cleavages [(100)^(010) = 79°] is particularly diagnostic (Table 2.6). In (100) sections, which are characterized by the presence of one well-developed (010) cleavage that parallels the length of the crystals, kyanite is distinguished by marked inclined extinction (the γ indicatrix axis is 30° to the trace of the (010) cleavage). In contrast to andalusite and sillimanite, which are both orthorhombic and thus display parallel extinction, the inclined extinction of kyanite reflects the triclinic crystal system of this polymorph. Kyanite also differs from andalusite and sillimanite by the common presence of simple or lamellar twinning. The commonest twin plane is (100). Most kyanite crystals are colorless in thin section as a consequence of low transition element content. However, in hand specimen, kyanite typically has a blue coloration due to minor amounts of transition metal cations in solid solution (see Chapter 4).

Thermal expansion

The thermal expansivities in the *a*, *b*, and *c* directions of kyanite are considerably more alike than those of andalusite and sillimanite (Fig. 2.10). Because of the lack of continuous chains of tetrahedra, the thermal

Table 2.6. Summary of the optical and physical properties of kyanite. Symbols as in Table 2.3. (From Deer et al., 1982, p. 780; copyright Longman Group Ltd.).

Triclinic (−)

α	1·710–1·718
β	1·719–1·724
γ	1·724–1·734
δ	0·012–0·016
$2V_\alpha$	78° - 83°

γ':z on (100) = 27° - 32°, on (010) = 5° - 8°;
α':x on (001) = 0°-3°: Bx_a nearly ⊥ (100).

Dispersion:	$r > v$ weak.
D	3·53 - 3·65
H	5½ - 7, variable.
Cleavage:	{100} perfect, {010} good, {001} parting. (001):z = 85°.

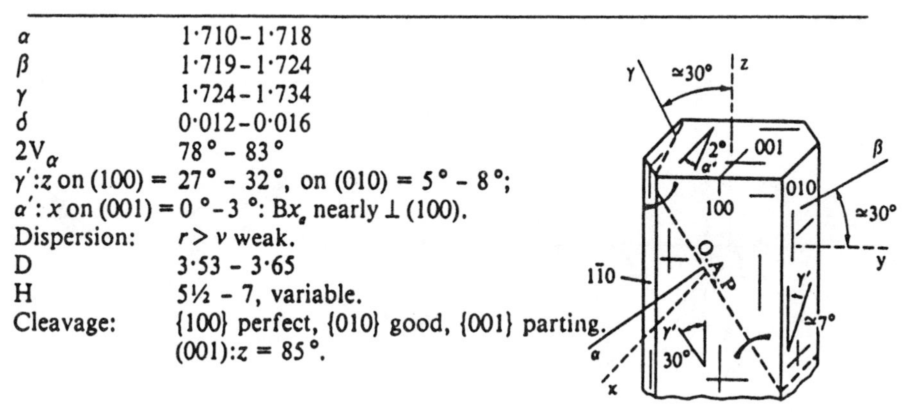

Twinning:	Lamellar on (100), twin axis ⊥ (100) or ‖y or z; multiple on (001), by pressure.
Colour:	Blue to white, also grey, green, yellow, pink or black; colourless to pale blue in thin section.
Pleochroism:	Weak; in thick sections α colourless, β violet-blue, γ cobalt-blue.
Unit cell:	a 7·12, b 7·85, c 5·57 Å. α 89·98°, β 101·12°, γ 106·01°. Z = 4 Space group $P\bar{1}$.

expansion in the c direction is larger in kyanite than andalusite or sillimanite (Fig. 2.10). Because the octahedral chains in kyanite share several different edges, there is no clearly dominant thermal expansion direction in the a-b plane.

CHAPTER 3

PHASE EQUILIBRIA

INTRODUCTION

"I sometimes think our close geological relatives must think that experimentalists are a little peculiar in their obsession over the Al_2O_3-SiO_2 system. There are other systems that can perhaps yield more exact information about the same problems involving rocks, and our progress in this system does not exactly promote confidence in our abilities."

W.S. Fyfe (1969, p. 291)

In spite of the pre-eminent importance of the Al_2SiO_5 polymorphs for metamorphic thermobarometry, considerable uncertainty shrouds the calibration of P-T phase equilibria involving these phases. As with the above quote, this problem is epitomized by the opening statement of Holdaway's (1971) landmark paper: "The Al_2SiO_5 phase diagram is perhaps the most studied and least well defined silicate phase diagram". Newton (1987) provided the most recent review of the Al_2SiO_5 stability relations. He focused on the experimental calibrations of Richardson et al. (1969) and Holdaway (1971), and concluded that no cogent arguments can be made that would permit a clear choice between the phase diagram of Richardson et al. (1969) versus that of Holdaway (1971). The consequences of this dilemma for metamorphic thermobarometry are significant. For example, in their thermobarometric analysis of the contact aureole of the Lilesville Pluton in North Carolina, Evans and Speer (1984, p. 297) concluded: "Contact metamorphism was isobaric at 4.0-5.1 or 2.0-3.5 kbar depending on the choice of aluminosilicate triple point". Within the last decade, most petrologic studies have accepted Holdaway's (1971) phase diagram; however, some studies (e.g., Droop and Treloar, 1981) diplomatically presented the diagrams of Richardson et al. (1969) and Holdaway (1971), or a compromise diagram with a triple point between these two calibrations (Froese and Gasparini, 1975; Greenwood, 1976; Pattison, 1989; Morand, ms.).

The purpose of this chapter is to review comprehensively and critique existing data bearing on the P-T calibration of equilibria involving the aluminum silicate polymorphs. This analysis benefits from experiments carried out in the author's hydrothermal laboratory, a cooperative experimental study with A.L. Montana of the University of California,

Los Angeles (UCLA), and S.R. Bohlen of the U.S. Geological Survey (USGS), Menlo Park, California, and calorimetric measurements obtained in cooperative research with B.S. Hemingway, R.A. Robie, and H.T. Evans, Jr., of the U.S. Geological Survey (USGS), Reston, Virginia, and L. Topor, O.J. Kleppa, and R.C. Newton of the University of Chicago.

Zen (1969) comprehensively reviewed studies bearing on the Al_2SiO_5 phase relations. Zen's (1969) P-T diagram (Fig. 3.1) epitomized the disparate calibrations of the Al_2SiO_5 phase equilibria. He concluded that several experimental studies were unreliable because of failure to demonstrate equilibrium (e.g., synthesis experiments) and incorrect calibration of pressure. Many experimental investigations discredited by Zen were also questioned by Richardson et al. (1969), Newton (1969), and Holdaway (1971). Thus, this review focuses on those studies which appear to have survived critical analysis and therefore provide apparently credible data for calibration of the Al_2SiO_5 phase equilibria.

The erroneous early experimental determinations of the Al_2SiO_5 P-T phase "equilibria" had deleterious consequences for the geological utility of these phase diagrams. For example, Bell's (1963) phase diagram (Fig. 3.2), which exemplifies the pitfalls of attempting to determine equilibria by *synthesis* from oxides and/or gels, resulted in erroneous geologic interpretations. Most important was the recognition that Bell's P-T diagram yielded depths of burial for kyanite-bearing metapelites that were significantly larger than those indicated by geologic estimates of the depth of burial (Rutland, 1965, p. 123). Clark (1961) suggested that *tectonic overpressure* was a possible reason for these discrepant depths of burial. Tectonic overpressure arises where the mean pressure $\bar{P}$ [= $(\sigma_1 + \sigma_2 + \sigma_3)/3$] is larger than the lithostatic pressure P_L (= ρgh) resulting from the weight of the overlying rocks. Defined as $\bar{P} - P_L$, tectonic overpressure is dependent upon rock strength. The resultant dilemma is epitomized by Rutland's (1965, p. 136) conclusion "...the required [tectonic] overpressures seem to be very unlikely for most kyanite-grade metamorphism". Subsequent studies have shown that rock strength is insufficient to support tectonic overpressures exceeding a few hundred bars.

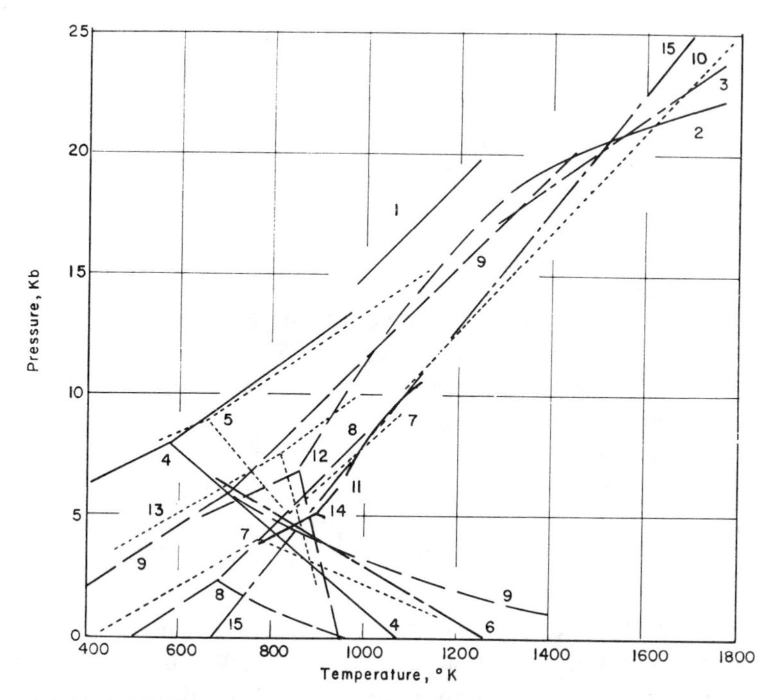

Figure 3.1. Zen's (1969) summary of the determinations of the Al_2SiO_5 phase equilibria. The numbers correspond to the following studies: (1), Griggs and Kennedy (1956), (2) Clark et al. (1957), (3) Clark (1961), (4) Bell (1963), (5) Khitarov et al. (1963), (6) Evans (1965), (7) Newton (1966a,b), (8) Weill (1966), (9) Holm and Kleppa (1966), (10) Matsushima et al. (1967), (11) Richardson et al. (1967) and Richardson et al. (1969), (12) Althaus (1967), (13) Pugin and Khitarov (1968), (15) Richardson et al. (1968b). (From Zen, 1969, Fig. 1).

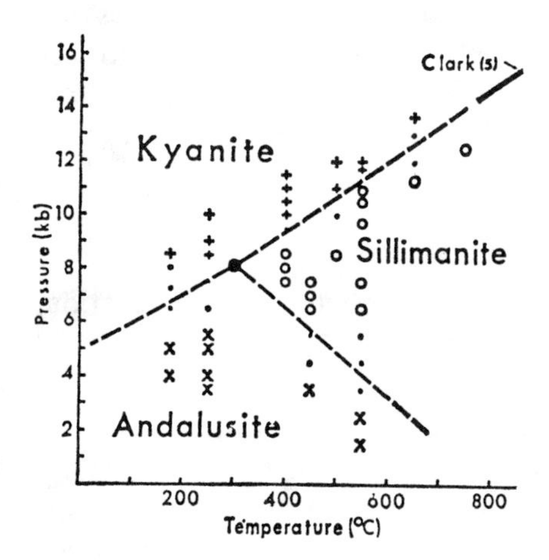

Figure 3.2. Bell's (1963) hydrothermal synthesis experiments of the Al_2SiO_5 polymorphs. The "+" symbols are runs where kyanite formed, the open circles are those where sillimanite formed, the "x" symbol refers to runs where andalusite formed, and the small dots are runs with inconclusive results. The solid portion of the kyanite-sillimanite equilibrium is from Clark (1961). (From Bell, 1963, Fig. 1).

EXPERIMENTAL HYDROTHERMAL STUDIES

"Few metamorphic systems have presented as much difficulty to the experimental worker as Al_2O_3-SiO_2."

G.C. Brown and W.S. Fyfe (1971, p. 227)

In the following detailed review and critique of experimental studies on the Al_2SiO_5 phase equilibria, readers should be aware of the errors in experimental "brackets" arising from precision and/or accuracy errors in measurement of temperature and pressure. Experimentalists commonly give estimates of temperature uncertainties based on fluctuations in run temperatures arising from imprecise temperature control, coupled with estimates of accuracy errors in thermocouple calibration. They less commonly publish estimates of pressure uncertainties. Sources of precision and accuracy errors in run pressures and temperatures are discussed in several studies (e.g., Bell and Williams, 1971; Johannes, 1978; Kerrick, 1987a). These uncertainties must be added to the pressure and temperature limits of experimental brackets. For example, let as assume that an experimentalist reports an *uncorrected* equilibrium bracket of 700-710°C using the two mean temperature values recorded from a series of temperature measurements during the runs defining the bracket. With a reasonable thermocouple accuracy of ±5°C, the bracket should be reported as 695-715°C. This correction has not been universally made in published reports of experimental studies. To avoid confusion, I have not made such corrections in the following discussion of the limiting temperatures and pressures defining experimental brackets. Readers planning to use experimental bracketing data reported in this book should consult the original references and, if necessary, make such corrections.

Experiments of Evans (1965)

Using weight changes of single Al_2SiO_5 crystals to monitor reaction, Evans (1965) studied the equilibrium:

$$\text{Muscovite} + \text{Quartz} = Al_2SiO_5 + \text{K-feldspar} + \text{Water} \quad .$$

From his experimental data (Table 3.1), he concluded that the P-T slope of this equilibrium with sillimanite (curve 3 in Fig. 3.3) is steeper than the equilibrium with andalusite (curve 2 in Fig. 3.3). This relationship is mandated by a Schreinemakers analysis and the Clapeyron equation (see Yardley, 1989, Fig. A.3). The invariant point defined by these two equilibria coincides with a point on the andalusite-sillimanite equilibrium.

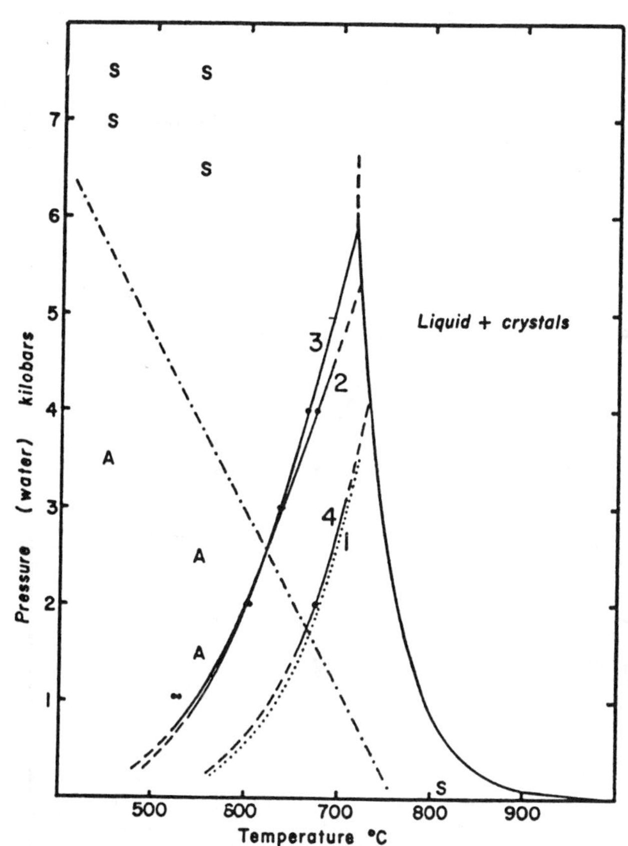

Figure 3.3. Summary of Evans' (1965) experimental determination of muscovite dehydration equilibria. The numbered equilibria are as follows: (1) muscovite = K-feldspar + corundum + water, (2) muscovite + quartz = K-feldspar + andalusite + water, (3) muscovite + quartz = K-feldspar + sillimanite + water, (4) muscovite + quartz = K-feldspar + kyanite + water. The filled circles are Evans' (1965) experimentally-determined equilibrium points. Note the coincidence of the invariant point formed by the intersection of equilibria (2), (3), and the andalusite-sillimanite equilibrium (dot-dash line). (From Evans, 1965, Fig. 10).

Table 3.1. Summary of Evans' (1965) experimental determination of muscovite equilibria. Equilibria (2), (3), and (4) represent the muscovite + quartz dehydration equilibrium with andalusite, sillimanite, and kyanite, respectively. (From Evans, 1965, Table 3).

Water pressure	Temperature °C			
	System: muscovite + quartz			System: muscovite
	(2)	(3)	(4)	
1050	$525 {+20 \atop -10}$	$530 {+20 \atop -10}$		
2000	600	605	675	680
3000	640 ± 15	635		710
4000	675 ± 15.	665		
	Reaction leading to mullite formation at 2000bars: 680 ± 20			

Reactions (2), (3), and (4) leading to andalusite, sillimanite, and kyanite, respectively.

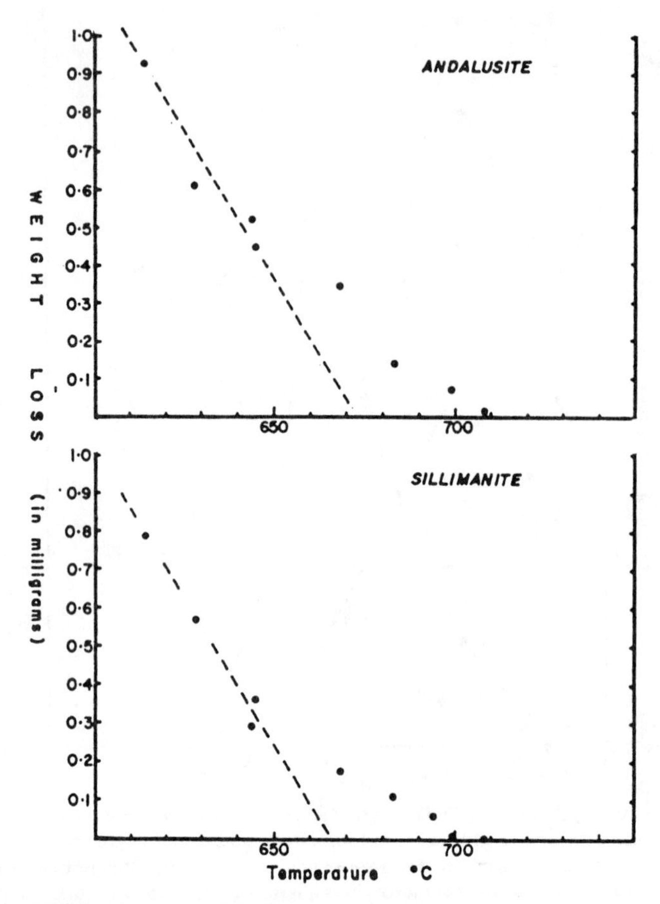

Figure 3.4. Weight changes andalusite (*top*) and sillimanite (*bottom*) single crystals in experiments at 4 kbar with run charges containing water and powdered muscovite + quartz + adularia. (From Evans, 1965, Fig. 9).

In considering errors in the experimental brackets, Evans concluded that this invariant point occurs at 600-640°C and 2-3 kbar. His determination of the andalusite-sillimanite equilibrium has been used in subsequent analyses of Al_2SiO_5 phase equilibria (Holdaway, 1971; Robie and Hemingway, 1984). However, in reviewing Evans' (1965) data, Fyfe (1967, p. 73) argued that because of the small entropy differences between andalusite and sillimanite, it would be "almost impossible" to locate precisely the andalusite-sillimanite equilibrium from experimental determination of equilibria (2) and (3) of Evans (1965). Of the four isobars investigated by Evans (1965), the largest difference in the equilibrium temperatures of reactions (2) and (3) is 10°C at 4 kbar (Table

3.1). As shown in Figure 3.4, his equilibrium temperatures were determined by *linear* extrapolation (to zero weight change) of straight lines fitted to his data at T < 650°C. The data shown at T > 650°C (Fig. 3.4) were excluded because of the complication introduced by melting in the run charges. The limited number of data points at T < 650°C, coupled with the *possibility* that the weight change data could instead be fitted with a curved line, bring into question the accuracy of the equilibrium temperatures of reactions (2) and (3) derived from the data in Figure 3.4. In recognizing such uncertainties, Evans (1965) placed a reasonable temperature error of ±15°C on his derived equilibrium temperatures at 3 and 4 kbar (Table 3.1). Applying this temperature uncertainty to equilibria (2) and (3), there is complete overlap of the error bands of these equilibria over the P-T range shown for these equilibria in Figure 3.3. Accordingly, uncertainties in the experimental location of equilibria (2) and (3) indeed support Fyfe's (1967) exclusion of Evans' (1965) bracket on the andalusite-sillimanite equilibrium. The interpretation of Evans' (1965) bracket is also clouded by the fact that he did not demonstrate reaction reversibility. Nevertheless, the validity of his equilibrium temperature at 2 kbar is supported by subsequent experimental reversals of this equilibrium (see Schramke et al., 1987).

Experiments of Newton (1966a,b)

Kyanite-andalusite equilibrium. Newton (1966a) used a piston cylinder apparatus in an experimental investigation of the kyanite-andalusite equilibrium at 700-800°C. Pressures were calibrated using well-characterized phase transitions (the melting curve of LiCl, and the Bi(I)-Bi(II) and calcite-aragonite phase transitions). Starting materials consisted of a powdered mixture of kyanite and andalusite, and reaction direction was determined by comparing X-ray peak heights of andalusite versus kyanite in the run products with those of the starting material. Conclusions regarding reaction direction from X-ray data were supported by optical examination of run products. As shown in Figure 3.5, runs in the kyanite stability field yielded well-developed overgrowths on the kyanite "seed" crystals.

Kyanite-sillimanite equilibrium. Newton (1966b) investigated the kyanite-sillimanite equilibrium at 750°C using the same experimental apparatus and experimental techniques as in the kyanite-andalusite study. Reaction direction was monitored by comparing the ratios of the intensities of the $02\bar{1}$ reflection of kyanite to the 120 reflection of sillimanite. As shown in Figure 3.6, he found a very consistent variation

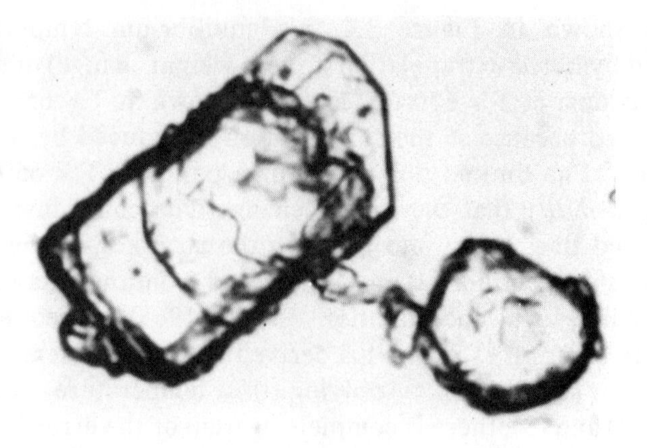

Figure 3.5. Overgrowth of hydrothermally-grown kyanite on a seed crystal of kyanite. This is the run product of Newton's (1966a) experimental determination of the andalusite → kyanite reaction at 800°C and 8.1 kbar. (From Newton, 1966a, Fig. 1).

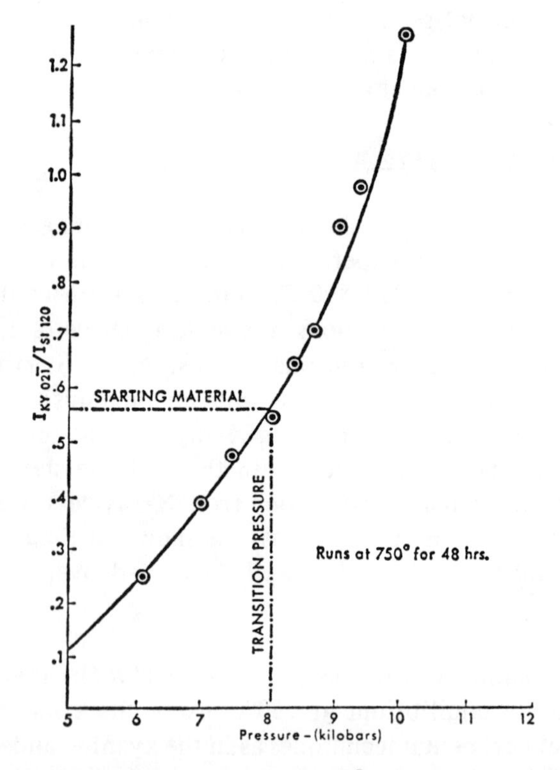

Figure 3.6. Ratio of the intensities of the kyanite (02$\bar{1}$) and sillimanite (120) X-ray peaks as a function of pressure. This summarizes Newton's (1966b) experimental results on the kyanite ⇄ sillimanite reaction at 750°C. (From Newton, 1966b, Fig. 1).

in this ratio as a function of pressure, thereby lending credence to his experimental results.

Al_2SiO_5 phase equilibrium diagram of Newton (1966a). As shown in **Figure 3.7**, Newton (1966a) located the Al_2SiO_5 triple point by intersection of the kyanite-sillimanite and kyanite-andalusite equilibria. The kyanite-sillimanite equilibrium was constrained by Newton's (1966b) 750°C and 8.1±0.4 kbar bracket and by the Clapeyron slope calculated with available entropy and molar volume data. The kyanite-andalusite equilibrium was constrained by Newton's (1966a) brackets on the metastable extension of the kyanite-andalusite equilibrium (Fig. 3.7) and the calculated Clapeyron slope. Compressibility and thermal expansion were not incorporated in the computed Clapeyron slopes. The error bands shown in Figure 3.7 for the kyanite-sillimanite and kyanite-andalusite equilibria were derived from the limits of the experimental brackets. Newton's (1966a) estimate of the error in location of the triple point (stippled polygon in Fig. 3.7) was derived by the intersection of these error bands.

Critique of Newton's (1966a,b) experiments. Newton's (1966a, 1966b) piston cylinder experiments were carried out with a talc pressure medium. Because of frictional drag of the piston and non-hydrostatic stresses within the solid pressure medium, pressure calibration is a notorious problem with this apparatus (Bell and Williams, 1971). Nevertheless, the detailed interlaboratory comparison of piston cylinder experiments on the albite = jadeite + quartz equilibrium by Johannes et al. (1971) lends credence to the accuracy of pressures reported by Newton (1966a,b). In particular, Newton's bracket on this equilibrium at 600°C using talc as a pressure medium is in good agreement with that of Johannes (Johannes et al., 1971, Johannes, 1978) who used an NaCl pressure medium that produces virtually hydrostatic pressure. Furthermore, Newton's experimental bracket on the albite = jadeite + quartz equilibrium at 600°C is in excellent agreement with the 600°C bracket of Hays and Bell (1973) determined in a gas pressure (internally heated) apparatus. Thus, the friction correction applied by Newton (1966a,b) appears to be valid.

Because Newton used natural minerals as starting materials, it is important to address whether his experimental results are complicated by solid solution. Newton's (1966a) experiments on the kyanite-andalusite equilibrium utilized andalusite with < 0.20 wt % Fe_2O_3 and kyanite with 0.10 wt % Fe_2O_3. As discussed in Chapter 4, these impurity levels should have a negligible effect on this equilibrium. Newton (1966a) evaluated the effect of Fe^{3+} in solid solution by another set of experiments using kyanite

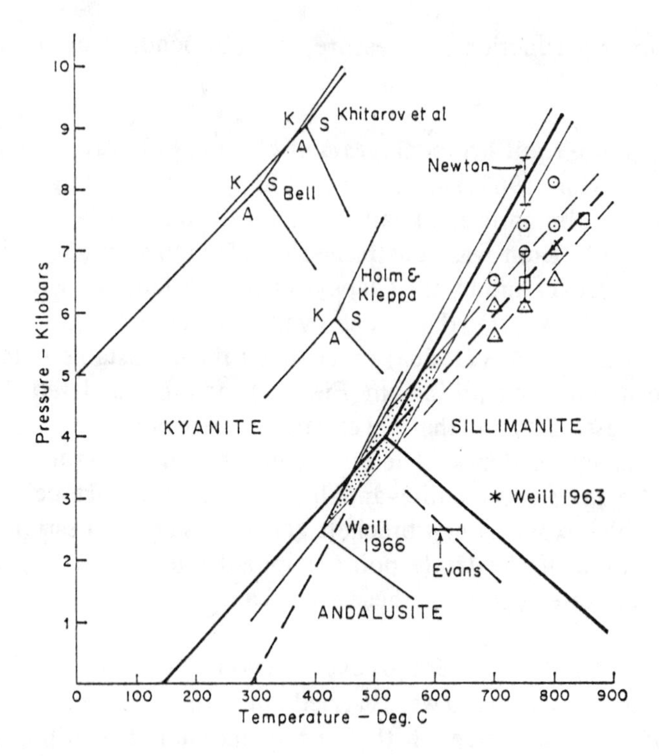

Figure 3.7. Newton's (1966a) determination of the Al_2SiO_5 phase equilibria. The open symbols are his experimental results on the kyanite ⇄ andalusite reaction. Open triangles: kyanite → andalusite; open circles: andalusite → kyanite; open squares: inconclusive. "Newton" refers to Newton's (1966b) 750°C bracket on the kyanite-sillimanite equilibrium. The stippled polygon is the error envelope for the triple point. (From Newton, 1966a, Fig. 2).

with 0.93 wt % Fe_2O_3 and andalusite with 0.3 wt % Fe_2O_3. At 750°C the experimental brackets with the two sets of starting materials are indistinguishable within the pressure uncertainty range of Newton's brackets (i.e., $\Delta P \approx 1$ kbar). Thus, Newton (1966a) provided an empirical confirmation that variations in the Fe_2O_3 contents of the phases in the range < 1 wt % has a negligible effect on the kyanite-andalusite equilibrium. His experiments on the kyanite-sillimanite equilibrium (Newton, 1966b) used kyanite with 0.93 wt % Fe_2O_3 and sillimanite with 0.76 wt % Fe_2O_3 and 0.10 wt % TiO_2. Because of similarities in the Fe^{3+} contents, and because of the negligible partitioning of Fe^{3+} between coexisting kyanite and sillimanite (see Chapter 4), it is concluded that minor element solid solution does not introduce a significant complication in the interpretation of Newton's (1966b) experimental brackets.

<u>Experiments of Richardson et al. (1968) and Richardson et al. (1969)</u>

Kyanite-sillimanite equilibrium. Richardson et al. (1968) experimentally investigated the kyanite-sillimanite equilibrium between 700°C and 1500°C. Starting materials consisted of sillimanite containing some fibrolite from Brandywine Springs, Delaware, and natural kyanite. Electron probe analysis revealed 0.06 wt % Fe_2O_3 and 0.01 wt % CaO in the kyanite, and 0.03 wt % Fe_2O_3 in the "sillimanite". Microscopic examination revealed the presence of quartz intergrown with fibrolite (see also Bell and Nord, 1974). The solid starting mixture was prepared by grinding (with ethyl alcohol) for three hours in an agate mortar. Considerable effort was taken to evaluate the reproducibility of this method by X-ray analyses of several mounts of the starting material. Variation in the [kyanite ($02\bar{1}$)/"sillimanite" (120)] peak height ratio of the starting material (ranging from about 0.30 to 0.60) was attributed to inhomogeneity and to preferred orientation of grains. Only those run products with ratios outside that range were used for determining reaction direction. Figures 3.8 and 3.9 show the experimental results of Richardson et al. (1968) using gas medium and piston cylinder equipment, respectively.

Kyanite-andalusite equilibrium. The kyanite-andalusite equilibrium was studied using internally-heated (gas medium) pressure vessels (Richardson et al., 1969). The starting material was a mixture of natural andalusite and kyanite ground 3 h in an automated agate mortar. Reaction direction was determined by comparing the ratios of peak heights [($02\bar{1}$) kyanite/(220) andalusite] in run products versus starting material. As in the study of Richardson et al. (1968) they determined reaction direction only for those experiments where this peak height ratio differed significantly from that of the starting material. Thus, their analysis provides compelling evidence that reaction direction was demonstrated.

Andalusite-sillimanite equilibrium. Richardson et al. (1969) experimentally investigated the andalusite-sillimanite equilibrium using cold-seal vessels. They determined reaction direction by comparing the ratio of the intensities of selected diffraction peaks [(220) andalusite/(120) sillimanite] in run products versus starting materials. Reaction direction was deduced only for those experiments where the peak height ratio of the run products significantly differed from that of the starting mixture.

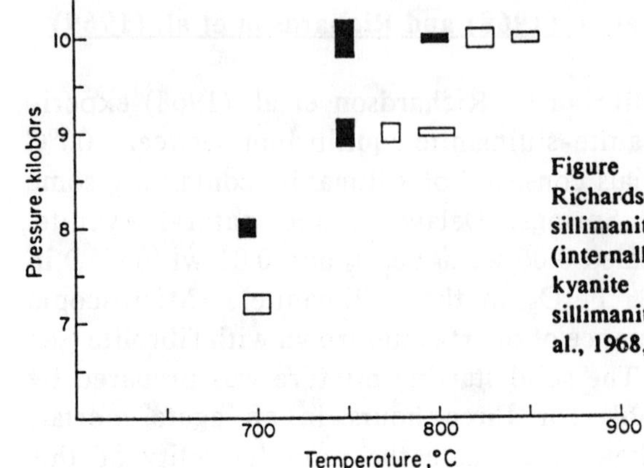

Figure 3.8. Experimental results of Richardson et al. (1968) on the kyanite ⇄ sillimanite reaction using gas medium (internally heated) apparatus. Open symbols: kyanite → sillimanite; closed symbols: sillimanite → kyanite. (From Richardson et al., 1968, Fig. 5).

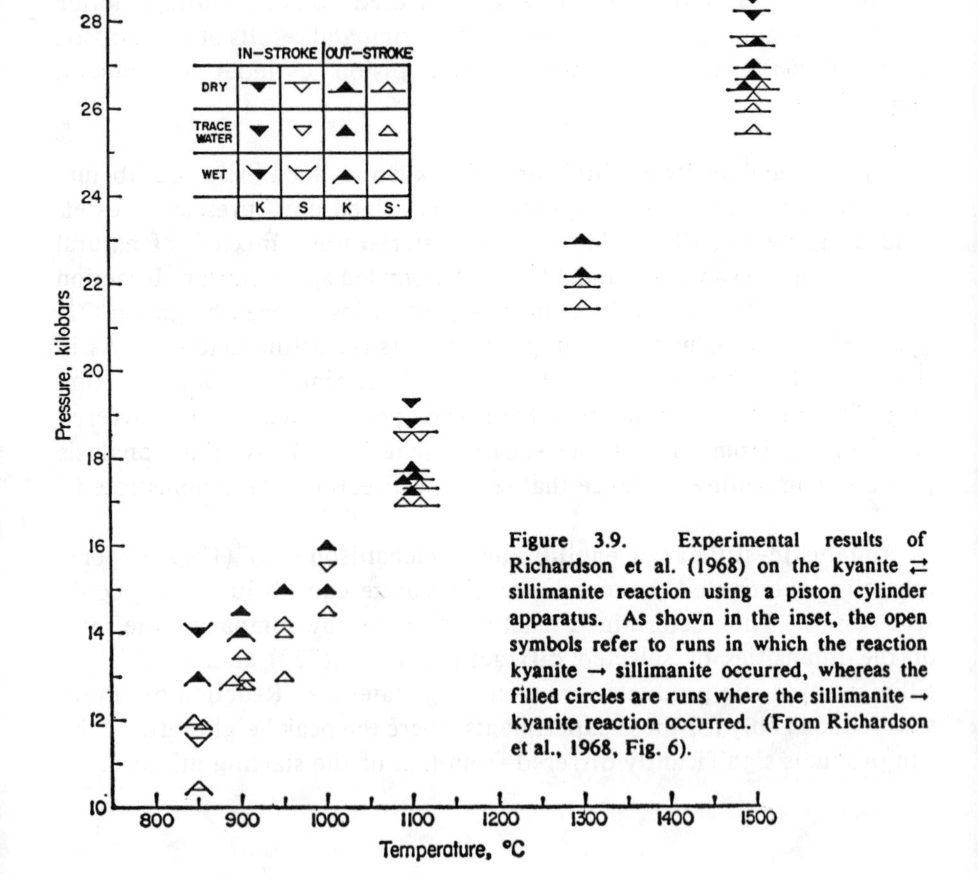

Figure 3.9. Experimental results of Richardson et al. (1968) on the kyanite ⇄ sillimanite reaction using a piston cylinder apparatus. As shown in the inset, the open symbols refer to runs in which the reaction kyanite → sillimanite occurred, whereas the filled circles are runs where the sillimanite → kyanite reaction occurred. (From Richardson et al., 1968, Fig. 6).

Al_2SiO_5 phase equilibrium diagram of Richardson et al. (1969). Richardson et al. (1969) determined the Al_2SiO_5 phase equilibrium diagram using their experimental brackets on the kyanite-andalusite and andalusite-sillimanite equilibria, and the brackets of Richardson et al. (1968) on the kyanite-sillimanite equilibrium. They noted that no straight equilibrium line could be drawn through all of the experimental bracketing data on this equilibrium. As shown in Figure 3.10, Richardson et al. (1968) located the kyanite-sillimanite equilibrium as the line of steepest dP/dT slope allowed by experimental brackets determined from run data of piston cylinder experiments. This location was dictated by their contention that, because of shearing strength of the solid pressure medium, the pressures indicated by the brackets at lower pressures and lower temperatures are falsely high. They assumed that the kyanite-sillimanite equilibrium is a straight line over the 800-1500°C and 8-28 kbar range. Using the Clapeyron equation, this (unjustified) straight line assumption necessitates that the ratio $\Delta S_r/\Delta V_r$ is constant over this P-T range. To derive the location of the triple point shown in Figure 3.11, Richardson et al. (1968) used the "gas medium" experimental data of Richardson et al. (1968) on the kyanite-sillimanite equilibrium coupled with their data on the andalusite-sillimanite and kyanite-andalusite equilibria (Fig. 3.11). The dP/dT slope of the andalusite-sillimanite equilibrium was computed with the Clapeyron equation using available entropy and molar volume data for the Al_2SiO_5 polymorphs.

Critique of the experiments of Richardson et al. (1968) and Richardson et al. (1969). A major source of uncertainty in the interpretation of the experimental results of Richardson et al. (1968) and Richardson et al. (1969) arises from the fact that they intensely ground their starting material. For all of their experiments, starting materials were pulverized with an automated agate mortar for three hours. Intense comminution was particularly significant for their experiments on the andalusite-sillimanite equilibrium. In addition, they interrupted their runs at least twice for 15 to 30 min of regrinding. Newton (1969) clearly demonstrated the deleterious effect of extended grinding of starting materials on experimental bracketing of the kyanite-sillimanite equilibrium. As shown in Figure 3.12, his prolonged grinding of a kyanite + sillimanite mixture resulted in enhanced degradation of kyanite compared to sillimanite. Newton (1969) attributed this phenomenon to the superior cleavage of kyanite. In reviewing previous experimental studies on the kyanite-sillimanite equilibrium, he noted that Althaus (1967) used an intensively pulverized kyanite + sillimanite starting material that was characterized by "remarkable" X-ray line broadening. Thus, Newton

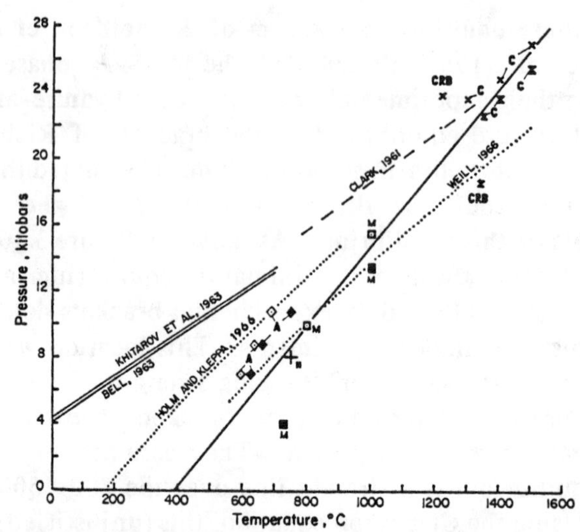

Figure 3.10. Summary of experimental data used by Richardson et al. (1968) to constrain their location of the kyanite- sillimanite equilibrium (heavy line). The letters adjacent to various data points refer to the following experimental studies: A = Althaus (1967), C = Clark (1961), CRB = Clark et al. (1957), M = Matsushima et al. (1967), N = Newton (1966b). Also shown are the locations of the kyanite-sillimanite equilibrium determined from various studies. (From Richardson et al., 1968, Fig. 9).

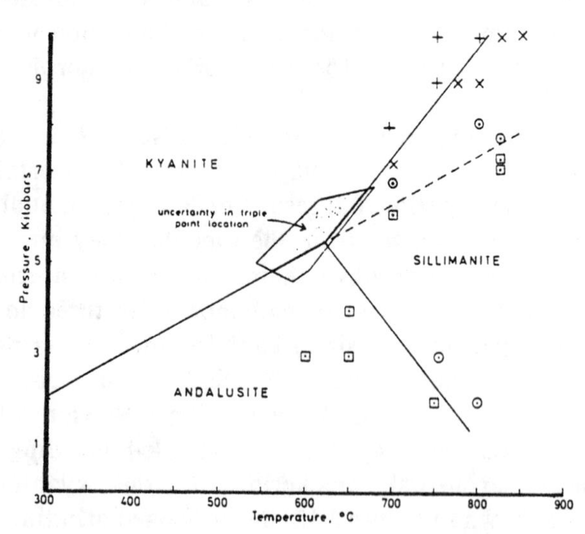

Figure 3.11. Summary of experimental data used to constrain the location of the Al_2SiO_5 equilibria and triple point of Richardson et al. (1969). The crosses and "x" symbols at 8 to 10 kbar are the experimental results of Richardson et al. (1968) on the kyanite ⇄ sillimanite reaction using a gas medium (internally heated) apparatus. The open symbols at 6 to 8 kbar that encompass the dashed metastable extension of the kyanite-andalusite equilibrium are the experimental results of Richardson et al. (1969) on the kyanite ⇄ andalusite reaction. The open symbols at 2 to 4 kbar are the experimental data of Richardson et al. (1969) on the andalusite ⇄ sillimanite reaction. (From Richardson et al., 1969, Fig. 2).

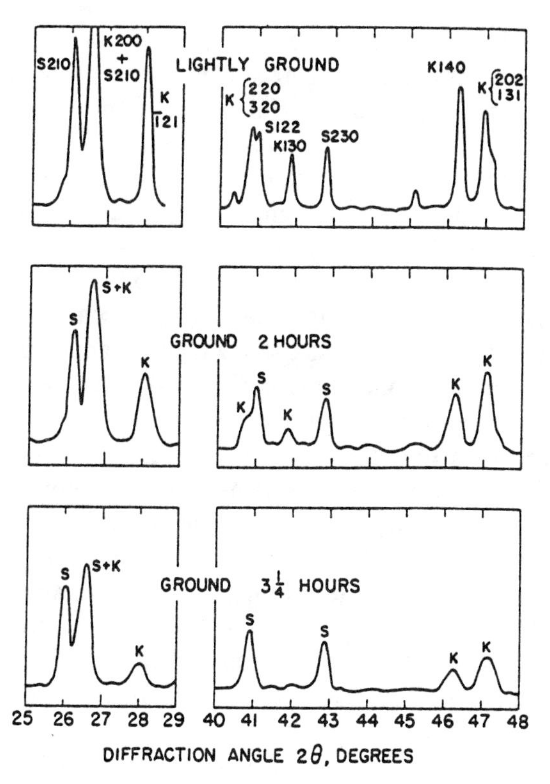

Figure 3.12. X-ray diffractograms of a powdered mixture of kyanite (K) and sillimanite (S) illustrating the effect of the duration of grinding on peak intensities. (From Newton, 1969, Fig. 1).

(1969) concluded that Althaus' (1967) relatively high pressure location of the kyanite-sillimanite equilibrium was affected by a *relatively* enhanced molar surface energy of finely comminuted kyanite, which would have the effect of displacing the kyanite-sillimanite equilibrium to higher pressures.

The effect of intense comminution of powdered materials on the andalusite-sillimanite equilibrium was investigated by A. Nitkiewicz in the author's hydrothermal laboratory. Starting materials consisted of single crystals of andalusite in powdered fibrolite (+ H_2O), and reaction direction was monitored by weight changes of andalusite spheres. Ground fibrolite was sieved into two distinctly different size fractions. In both size fractions, each grain contained numerous, interlocking fibrolite crystals; therefore, as used herein, *grain size* refers to the dimensions of fibrolite *aggregates*, not the individual fibrolite crystals within the aggregates. One fraction had grains that were 38-40 μm across, whereas the other fraction had a mean grain diameter of about 2 μm. Consequently, there were significant differences in the molar surface areas of the two contrasting grain size fractions. Preliminary results were reported by Nitkiewicz and Kerrick (1986). The single crystal data with the fine-grained fibrolite (Fig. 3.13a) appear to demonstrate equilibrium between 700 and 750°C. The andalusite single crystal weight changes in runs with the coarse-grained fibrolite (Fig. 3.13b) *suggest* equilibrium <650°C; i.e., significantly lower in temperature than with the fine-grained fibrolite. Unfortunately, the single crystal weight change data with the coarse-grained fibrolite fraction are inconclusive because the weight changes of the andalusite crystals are within the weighing error of the microbalance used. However, the implied upper temperature limit of the equilibrium (< 650°C at 2 kbar) is in agreement with the experimental data of S.G. Heninger (discussed later in this chapter) who used the same starting materials and grain size as that used in the experiments of A. Nitkiewicz using the coarse-grained fibrolite fraction. It is concluded that at 2 kbar, the *apparent* equilibrium using the fine-grained fibrolite is at least 100-200°C above that using the coarser-grained fraction. The brackets obtained with fine-grained fibrolite (Fig. 3.14) are in good agreement with the experimental data of Richardson et al. (1969) on the andalusite-sillimanite equilibrium.

There are two reasons to suspect that intense grinding of an initially coarse-grained mixture of andalusite + fibrolite, as in the procedure of Richardson et al. (1969), yields marked preferential comminution of fibrolite compared to andalusite. *First*, the cleavage of fibrolite is better developed than that of andalusite. *Second*, preferential degradation of fibrolite over andalusite during grinding may occur by spalling of fibrolite aggregates and fibers by breakage along {*hko*} grain boundaries. Consequently, I suspect that the experiments of Richardson et al. (1969) on the andalusite-sillimanite equilibrium were plagued by a high interfacial surface energy of sillimanite. It is also important to inquire whether ultra-comminution and the resultant high surface free energy effect were important in their experiments on the kyanite-sillimanite (Richardson et al., 1968) and kyanite-andalusite (Richardson et al., 1969) equilibria. In spite of marked differences in the grain size of powdered starting materials, there is good agreement between the experimental brackets of Richardson et al. (1968) and Newton (1966b) on the kyanite-sillimanite equilibrium and the brackets of Richardson et al. (1969) and Newton (1966a) on the kyanite-andalusite equilibrium. Newton's (1966a,b), starting mixtures were prepared by light grinding, which consisted of adding acetone to the powder and hand grinding in an agate mortar for 2 min (R.C. Newton, pers. comm.). In an analysis of the effect

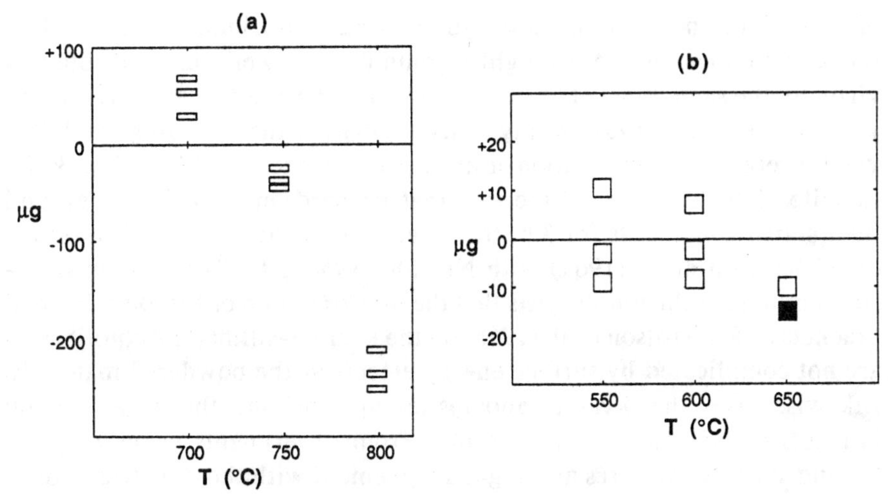

Figure 3.13. Andalusite single crystal weight change data for experiments starting with powdered fibrolite + corundum in the run charges. (a) Results of experiments using fine-grained fibrolite (grain size << 38 µm). (b) Results of experiments using powdered fibrolite with a grain size of 38-44 µm. The open squares are single data points whereas the filled square represents two superimposed data points. (From A. Nitkiewicz, unpublished).

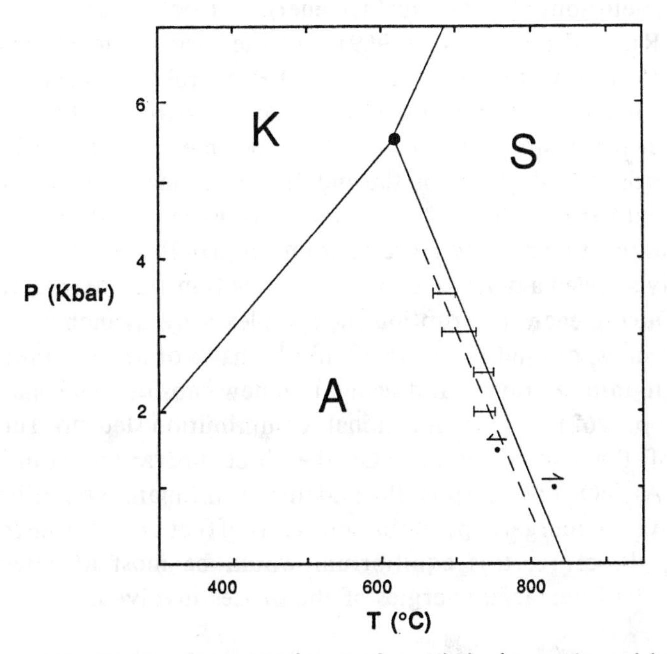

Figure 3.14. Results of the experiments of A. Nitkiewicz on the andalusite-sillimanite equilibrium. The horizontal bars represent reversal brackets whereas the arrows depict runs revealing direction of the andalusite = sillimanite equilibrium. The dashed line is the "best fit" to the experimental results. The Al_2SiO_5 equilibrium diagram of Richardson et al. (1969) is shown for comparison. (From A. Nitkiewicz, unpublished).

of comminution on the kyanite-sillimanite equilibrium, Newton (1969) repeated experiments with a "lightly ground" starting mixture and obtained virtually the same bracket as in his previous experimental study (Newton, 1966b). Newton's (1969) X-ray analysis suggests that grinding for 3.25 h yields preferential degradation of kyanite relative to sillimanite (Fig. 3.12). In spite of this result, and the fact that Richardson et al. (1968) ground their starting materials for 3 h, the agreement of the experimental results of Richardson et al. (1969) with Newton (1966b, 1969) on the kyanite-sillimanite equilibrium suggests that the interpretation of the experimental brackets of Richardson et al. (1968) on the kyanite-sillimanite equilibrium are not complicated by surface energy effects of the powdered materials. Likewise, for the kyanite-andalusite equilibrium, the experimental brackets of Newton (1966a) obtained from experiments using a lightly ground starting mixtures are in good agreement with those of Richardson et al. (1969) that utilized a starting mixture that was ground for 3 h. Thus, as with the kyanite-sillimanite equilibrium, the experimental data of Richardson et al. (1969) on the kyanite-andalusite equilibrium are apparently uncomplicated by surface energy effects of the intensely comminuted powdered starting material. If so, why did comminution, and the apparent deleterious effect of surface energy, affect the experimental brackets of Richardson et al. (1969) on the andalusite-sillimanite equilibrium? In contrast to the experiments of Richardson et al. (1968) on the kyanite-sillimanite equilibrium, and those of Richardson et al. (1969) on the kyanite-andalusite equilibrium, the experimental charges in the study of Richardson et al. (1969) on the andalusite-sillimanite equilibrium were interrupted for regrinding: "The majority of experiments attempted on this sluggish reaction required interruption for grinding. In fact, every run that finally showed a definite direction of reaction was interrupted at least twice. During each interruption the samples were quenched, taken from their capsules, ground for 15 to 30 min by hand or about 5 min in a small agate ball-mill, X-rayed, and sealed into new capsules" (Richardson et at., 1969, p. 262). This additional comminution led to further degradation of the run materials. Of the three univariant equilibria involving the Al_2SiO_5 polymorphs, the andalusite-sillimanite equilibrium is most sensitive to energetic perturbations (this effect is well illustrated in Fig. 1.11); therefore, this equilibrium would be most affected by differences in the interfacial energies of the phases involved.

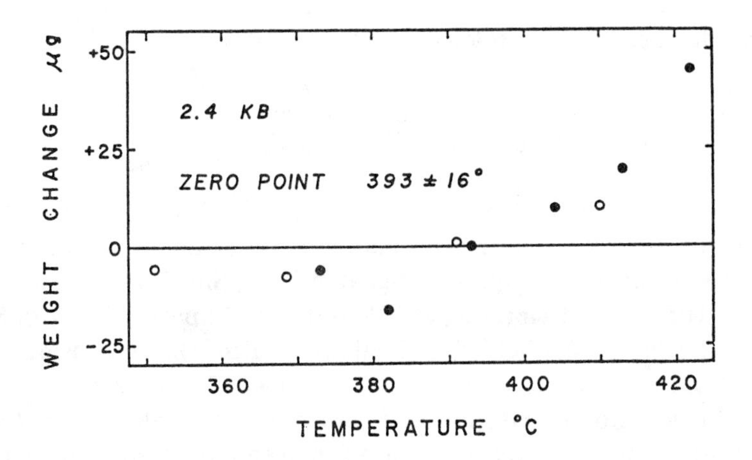

Figure 3.15. Holdaway's (1971) experimental data on the andalusite = kyanite reaction at 2.4 kbar. The ordinate corresponds to weight changes of andalusite single crystals. The filled circles are runs with powdered kyanite whereas open circles represent experiments starting with powdered kyanite + corundum. (From Holdaway, 1971, Fig. 2).

Experiments of Holdaway (1971)

Using cold-seal pressure vessels, Holdaway (1971) studied the kyanite-andalusite and andalusite-sillimanite equilibria. Reaction direction was monitored by measuring weight changes of andalusite single crystals and by optical examination of "seeded" experiments whereby growth or dissolution was determined by textural features.

Kyanite-andalusite equilibrium. At 2.4 kbar, Holdaway monitored the kyanite ⇄ andalusite reaction with weight changes of andalusite crystals with run charges containing kyanite + corundum + water (open circles in Fig. 3.15) and kyanite + water with f_{O_2} buffered by magnetite + hematite (filled circles in Fig. 3.15). The 2.4 kbar bracket was based on the data points at 382°C and 404°C (Fig. 3.15). The *trend* of the data points in Figure 3.15 is consistent with Holdaway's interpretation of the zero point.

Holdaway's bracket at 3.6 kbar was based on optical examination of run products. "In most cases the capsule with andalusite seeds gave the best information, andalusite showing reduction in size and reentrant areas at low temperatures, and thin rims exhibiting crystal faces at high temperatures. ... When kyanite showed a change it was thin rims exhibiting crystal faces at low temperatures and corrosion back from cross fractures at higher temperatures." (Holdaway, 1971, p. 107).

Based on two definitive "seeding" experiments, Holdaway determined an experimental bracket of 574±19°C at 4.8 kbar.

Andalusite-sillimanite equilibrium. Most of Holdaway's (1971) experiments on the andalusite-sillimanite equilibrium were carried out using three sillimanite samples, designated Sill I, Sill II, and Sill III. Sill I and Sill II contained fibrolite; Sill III did not. Experiments with Sill I and Sill II (Fig. 3.16) yielded anomalous results. In addressing these results, Holdaway (1971, p. 112-113) concluded: "It is clear to me that fibrolite reacts to andalusite (and probably to coarsely crystalline sillimanite) during the early stages of a run even though sillimanite is the stable phase." In spite of these complications, Holdaway concluded that the experiments with Sill II (Fig. 3.16) yield an "apparent" reversal of 648±20°C. However, he expressed reservations about the experiments with Sill II because "...the weight loss of the andalusite crystal was not always reproducible near equilibrium temperatures" (p. 109). Accordingly, in Table 6 of Holdaway, which summarized his experiments on the andalusite-sillimanite equilibrium, he referred to the experiments with Sill II as indicating "possible" equilibrium. In his summary analysis (see Table 6 of Holdaway, 1971), he selected experimental brackets on the andalusite-sillimanite equilibrium based on experiments with non-fibrolitic sillimanite.

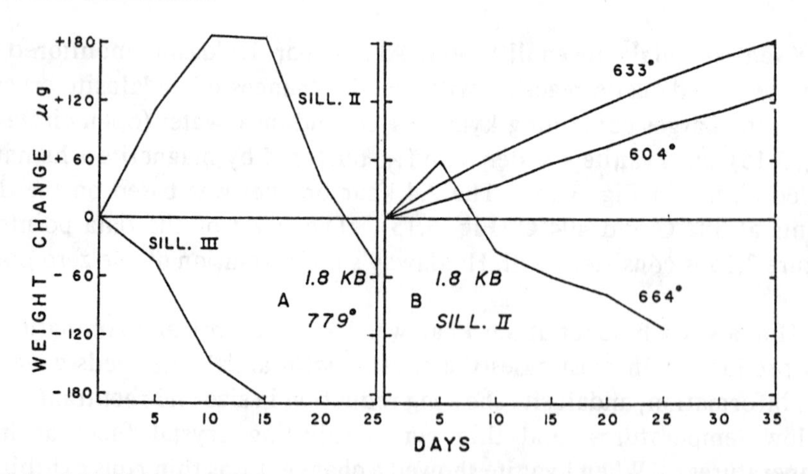

Figure 3.16. Holdaway's (1971) experimental determination of weight changes of andalusite single crystals (in charges with water and powdered "sillimanite") as a function of run duration. Sill II contained fibrolite + sillimanite whereas Sill III contained no fibrolite. In addition to Sill II or Sill III, the powdered starting materials contained corundum. (From Holdaway, 1971, Fig. 3).

At 1.8 kbar, the single crystal weight change data (Fig. 3.17) suggest equilibrium at 617±27°C whereas seeded runs yield 617±46°C. The coincidence of the mid-point temperatures for each set of runs is fortuitous.

Holdaway's bracket at 3.6 kbar is based on andalusite weight change data (Fig. 3.17).

Al_2SiO_5 phase equilibrium diagram of Holdaway (1971). Holdaway's (1971) oft-quoted Al_2SiO_5 phase equilibrium diagram is shown in Figures 3.18 and 3.19. The abstract of his paper gives the triple point as 501°C and 3.76 kbar. Because no uncertainty was given there, Holdaway may have *inadvertently* conveyed an unrealistic accuracy for the location of his triple point. By contrast, in the text Holdaway (1971) carried out a thoughtful error analysis, arriving at 3.76±0.30 kbar and 501 ±20°C (Holdaway, 1971, p. 122). His phase equilibrium diagram was derived from coupling experimental equilibrium bracketing data with Clapeyron slopes calculated with available entropy and molar volume data. He used the entropy data of Todd (1950) and Pankratz and Kelley (1964) and molar volume data of Skinner et al. (1961). Compressibility and thermal expansion data were incorporated in the Clapeyron slope computations. The dP/dT slope of the kyanite-andalusite equilibrium was computed at 100°C intervals, and the P-T location of this equilibrium was constrained by the experimental brackets shown in Figure 3.18. As shown in Figures 3.18 and 3.19, the andalusite-sillimanite and kyanite-sillimanite equilibria were constrained by selected experimental bracketing data. The reader is referred to pp. 114-120 of Holdaway's (1971) paper for justification of his selection of bracketing data for the kyanite-sillimanite equilibrium.

Critique of Holdaway's (1971) experiments. In contrast to the experiments of Richardson et al. (1969) which involved intense comminution of starting materials, Holdaway (1971) removed ultrafine particles in his powdered starting materials by elutriation. Thus, the interpretation of his experiments is uncomplicated by surface energy of powders.

In view of replicate weighings, and standardization of his microbalance against known weights, Holdaway concluded that differences in the weights of crystals before and after runs were reproducible to less than ±4 μg. The single crystal weight change method has been used in numerous

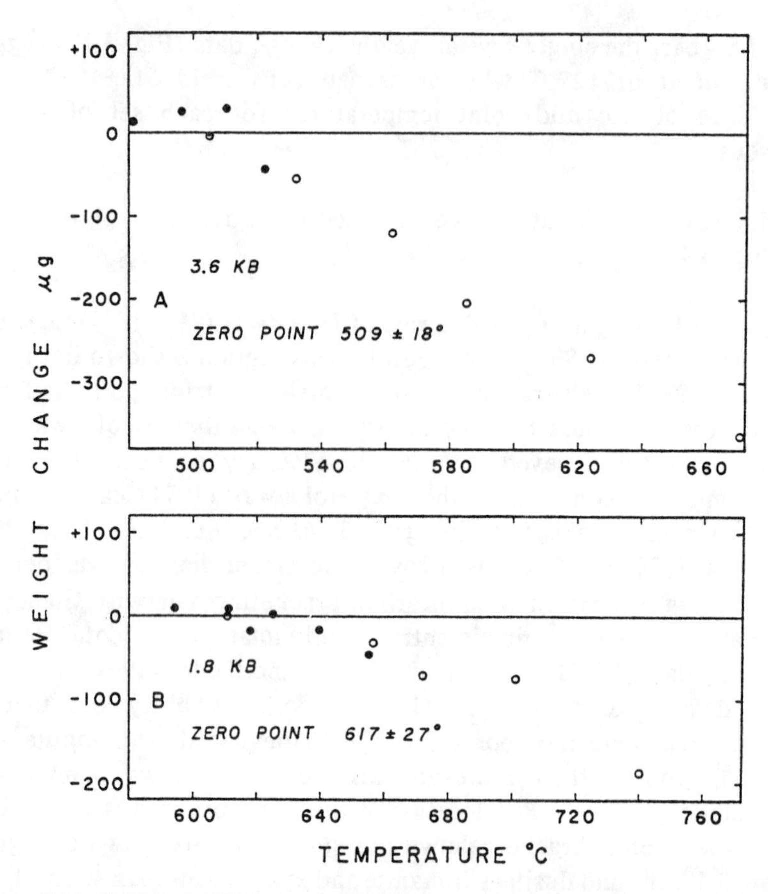

Figure 3.17. Holdaway's (1971) experimental data on the andalusite ⇄ sillimanite reaction at 1.8 kbar (*bottom*) and 3.6 kbar (*top*). The reaction was monitored by weight changes of andalusite crystals in charges containing water and powdered Sill III + corundum. The symbols are differentiated according to run duration: open circles = 15 d, filled circles = 30 d, and half-filled circles = 60 d. The weight changes of the 30 d and 60 d runs were "normalized" to 15 d (see Holdaway, 1971, Fig. 4 caption). (From Holdaway, 1971, Fig. 4).

experimental studies carried out in the author's hydrothermal laboratory (Kerrick, 1972; Slaughter et al., 1975; Hunt and Kerrick, 1977; Kerrick and Ghent, 1979; Eggert and Kerrick, 1981; Jacobs and Kerrick, 1981; Schramke et al., 1987). In all cases we used a regularly-serviced Mettler model M5 microbalance that is virtually identical to that used by Holdaway (1971). Our experience with this microbalance, and our conversations with technical personnel of the Mettler Instrument Corp., suggest that even under the best environmental conditions, single crystal weight changes less than about 15 μg are very difficult to determine with confidence. Consequently, as monitors of the sillimanite → andalusite reaction (Fig. 3.17), I place more credence in the larger weight gains of the

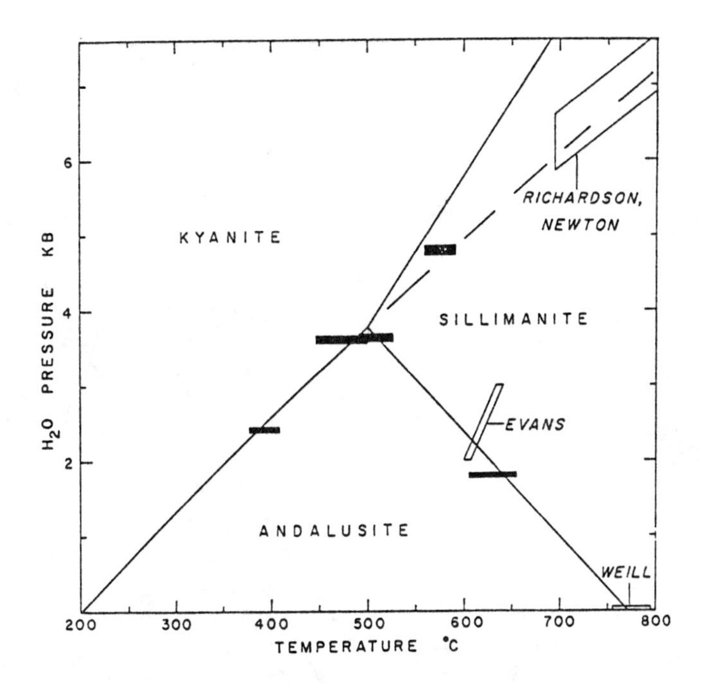

Figure 3.18. Holdaway's (1971) Al_2SiO_5 phase equilibrium diagram showing experimental data used to constrain the location of the kyanite-andalusite and andalusite-sillimanite equilibria. (From Holdaway, 1971, Fig. 5).

single crystals in the 3.6 kbar experiments compared to those of the 1.8 kbar experiments. Interpretation of reaction reversibility and selection of the zero points for the 1.8 and 3.6 kbar data could be questioned in light of the scatter of the data in the vicinity of the zero points (Fig. 3.17). The anomalous weight losses recorded by runs below the selected zero point temperatures *could* be attributed to spalling; however, there is no confirmation of this suggestion. In light of the scatter in the data, and the question regarding the accuracy of weight changes less than 15μg, I conclude that Holdaway's 1.8 kbar data strongly argue for a maximum equilibrium temperature of 655±5°C at 1.8 kbar and 520±5°C at 3.6 kbar. Reversibility *may* have been demonstrated in the 3.6 kbar experiments, whereas reversibility is more questionable for the 1.8 kbar single crystal experiments. At 1.8 kbar, I conclude that the seeded experiment at 576±5°C, where andalusite seeds showed slight growth and sillimanite displayed "definite corrosion" (Holdaway, 1971, p. 109), provides a more reliable lower temperature limit than that derived by Holdaway from his single crystal data. My analysis of Holdaway's single crystal data is biased

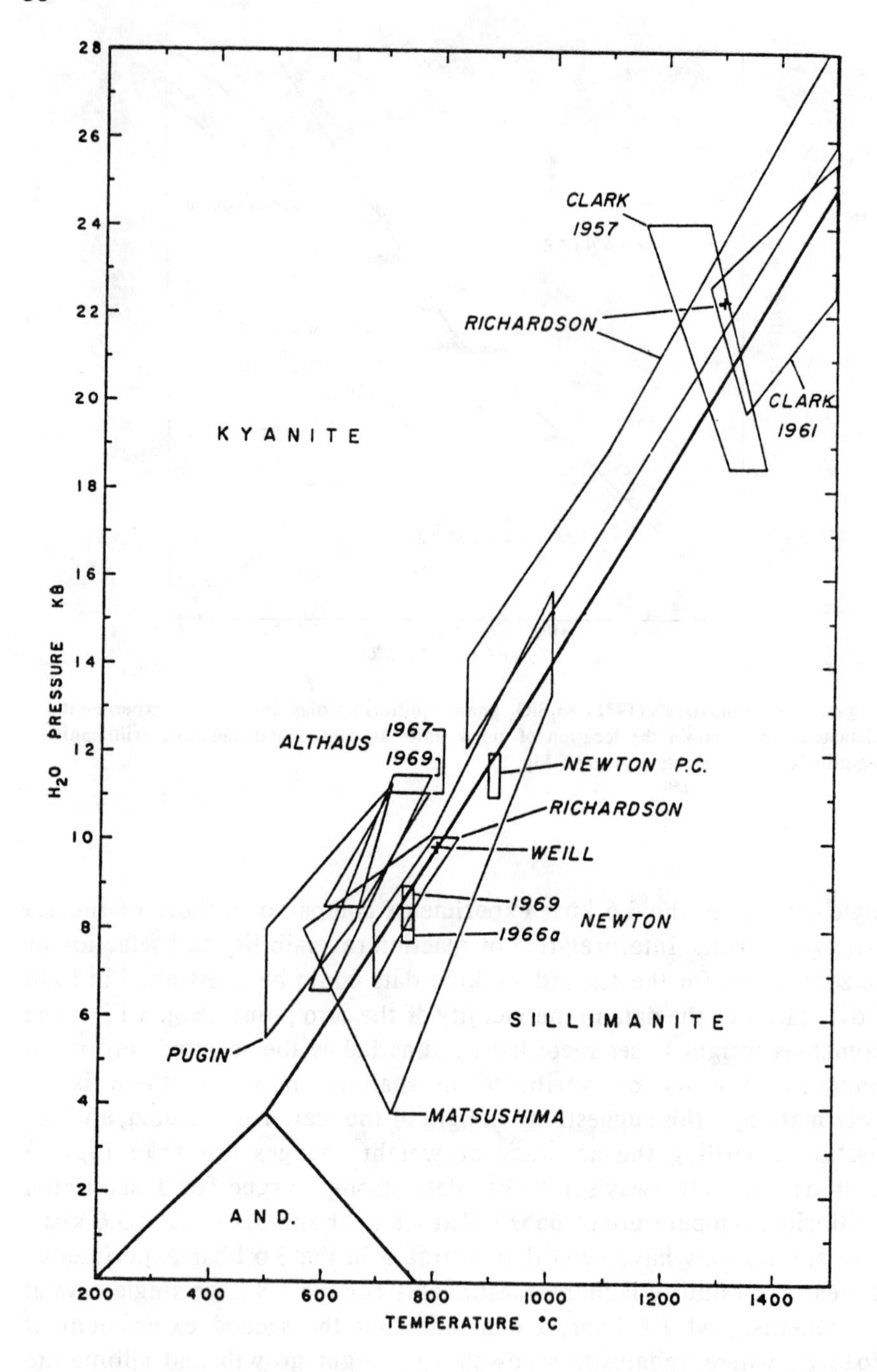

Figure 3.19. Holdaway's (1971) summary of experimental studies on the kyanite–sillimanite equilibrium. (From Holdaway, 1971, Fig. 6).

by our evaluation of accuracy and precision of weight changes measured by a mechanical microbalance. Holdaway's (1971 and pers. comm.) meticulous care in balance calibration and replicate weighings may have resulted in single crystal weight changes that are *significantly* better than the $\pm 15 \mu g$ uncertainty value favored by this author.

Holdaway's experimental brackets on the andalusite-sillimanite equilibrium were determined from experiments starting with andalusite containing 0.38 wt % Fe_2O_3 and Sill III with 1.15 wt % Fe_2O_3. His experimental results must be analyzed in light of the potential complications of solid solution. Compositional data for andalusite-sillimanite pairs in metapelites show that andalusite contains more Fe^{3+} than coexisting sillimanite (see Chapter 4); thus, the andalusite + Sill III starting mixture used in Holdaway's experiments does not represent equilibrium partitioning of Fe^{3+}. Furthermore, there are no data on the compositional changes of andalusite and sillimanite during his experiments. Reaction between andalusite and sillimanite may have occurred to adjust rim compositions toward the equilibrium partitioning of Fe^{3+}. Several alternative reaction mechanisms may be envisioned. Further analysis of the effect of Fe^{3+} solid solution in Holdaway's experiments would necessitate information on the rim compositions of solid run products. A parallel reaction driven by disequilibrium Fe^{3+} partitioning clouds Holdaway's interpretations which implicitly assume polymorphic reactions involving end member compositions. In addressing the effect of solid solution, Holdaway (pers. comm.) noted that the bracket obtained with the Fe-rich Sill III (617±27°C) is in reasonable agreement with that obtained with the Fe-poor Sill II (648±20°C). Thus, he concluded that Fe in solid solution would not have a marked effect on displacing the andalusite-sillimanite equilibrium, although "...it could easily be 20°C" (M.J. Holdaway, pers. comm.).

Experiments of Brown and Fyfe (1971)

Using cold-seal pressure vessels, Brown and Fyfe (1971) determined the kyanite-andalusite equilibrium by measuring the solubilities of andalusite and kyanite in aqueous solutions. They used two different methods to determine solubility: (a) measurements of the concentration of aqueous SiO_2 by molybdate complex spectrophotometry of run products, and (b) measurements of weight losses of single crystals of kyanite and andalusite. To achieve reproducible solubilities they added powdered corundum to the charge. Because of the presence of corundum, Brown and Fyfe (1971) represented the incongruent solution reactions as:

$$Al_2SiO_5 + nH_2O = Al_2O_3 + (SiO_2 \cdot nH_2O)_{aq} \,.$$

The implicit assumption that aqueous silica is present in supercritical aqueous solutions as a monomeric species is supported by numerous studies (e.g., Walther, 1986). Although Brown and Fyfe concluded that the hydration number (n) varies with P and T, more recent analyses of quartz solubility suggest that the hydration number has a constant value of two, reflecting the tetrahedral coordination of silica in silicates (Walther, 1986).

Brown and Fyfe's experimental data are plotted in Figure 3.20, in which the temperature uncertainties for their experimental brackets are determined from the intersection of the error envelopes for the solubility data for kyanite and andalusite.

Brown and Fyfe carried out solubility measurements versus run duration in experiments at 450°C and 2 kbar, and at 500°C and 1 kbar, concluding that equilibrium solubilities were reached in 14 d. Thus, their experimental "equilibrium" solubility measurements (plotted in Fig. 3.20) were based on 14 d experiments. Their 500°C, 1 kbar data on solubility versus run duration (Fig. 3.21) indeed suggest that there was little increase in solubility in runs exceeding 14 d duration; however, for the 450°C, 2 kbar experiments, they did not provide data for runs exceeding 14 d.

Al_2SiO_5 phase equilibrium diagram of Brown and Fyfe (1971). As shown in Figure 3.22, Brown and Fyfe (1971) located the Al_2SiO_5 triple point by the intersection of the kyanite-sillimanite and kyanite-andalusite equilibria. They constrained the kyanite-sillimanite equilibrium to pass through the Al_2SiO_5 triple point of Richardson et al. (1969) and *apparently* extrapolated the equilibrium from this point using Fyfe's (1967) computed Clapeyron slope for this equilibrium. The kyanite-andalusite equilibrium was constrained by the equilibrium temperatures derived from their experimental solubility measurements (see Fig. 3.20). The shaded area in Figure 3.22 was derived by intersection of the estimated error envelopes encompassing the kyanite-sillimanite and kyanite-andalusite equilibria (thin dashed lines in Fig. 3.22). They did not discuss how these error envelopes were derived. However, they *apparently* derived the error band for the kyanite-sillimanite equilibrium from the 750°C experimental data for this equilibrium (Fig. 3.22), whereas the error band for the kyanite-andalusite appears to have been derived from the polygonal error envelope derived by the intersection of the error envelopes encompassing the brackets shown in Figure 3.20. The triple point of Brown and Fyfe (1971) agrees with Weill's (1966) location of the andalusite-sillimanite

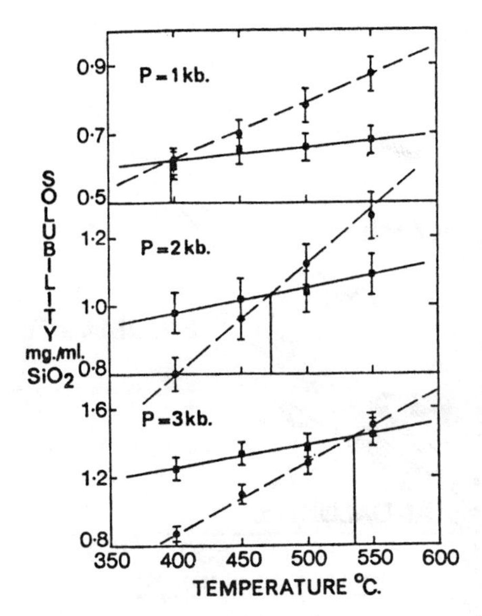

Figure 3.20. Brown and Fyfe's (1971) experimental measurements of silica concentration in aqueous solutions in runs with andalusite (brackets fit with solid lines) and kyanite (dashed lines). For each pressure, intersection of the dashed and solid lines represents Brown and Fyfe's (1971) equilibrium temperature. (From Brown and Fyfe, 1971, Fig. 1).

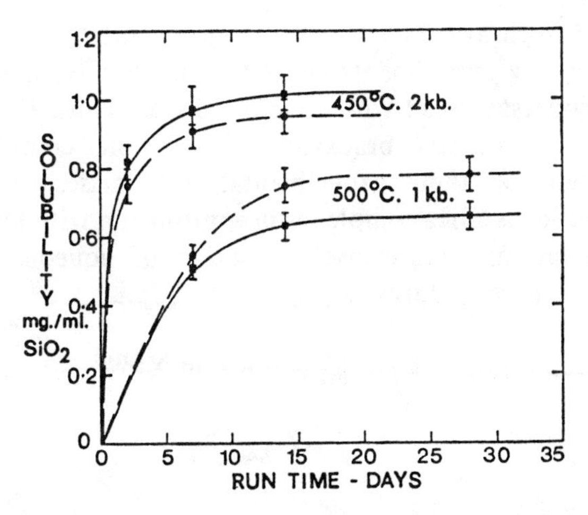

Figure 3.21. Silica solubility in aqueous solutions in runs with andalusite (filled squares and solid lines) and kyanite (filled circles and dashed lines) as a function of run duration. (From Brown and Fyfe, 1971, Fig. 2).

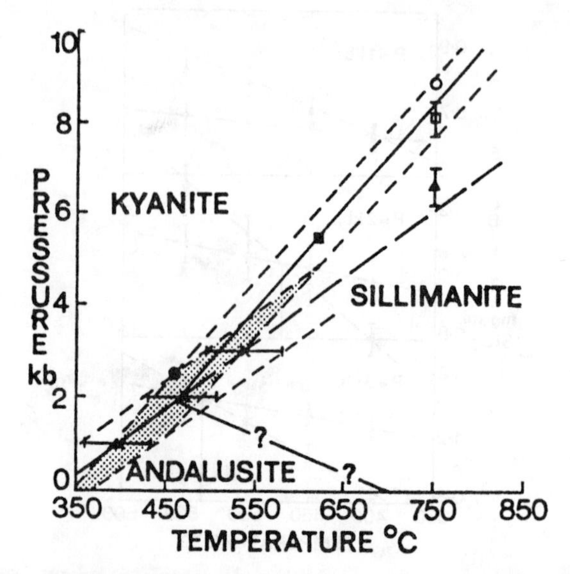

Figure 3.22. Brown and Fyfe's (1971) phase equilibrium diagram for the Al_2SiO_5 polymorphs. The filled circle is the triple point of Fyfe (1967) whereas the filled square is the triple point of Richardson et al. (1969). The brackets on the kyanite-andalusite equilibrium at 1, 2 and 3 kbar were derived from Brown and Fyfe's (1971) experimental solubility measurements at 750°C. The open circle is Weill's (1966) computed kyanite-sillimanite equilibrium, the open square and associated bracket is Newton's (1966b) determination of this equilibrium, and the filled triangle and associated bracket is Newton's (1966a) determination of the kyanite-andalusite equilibrium. The shaded area is Brown and Fyfe's (1971) estimation of the error envelope for the location of the triple point. (From Brown and Fyfe, 1971, Fig. 3).

equilibrium.

Critique of Brown and Fyfe's (1971) experiments. Because Brown and Fyfe's solubility experiments started with pure H_2O, the solutions were initially undersaturated with aqueous silica. Thus, their measured solubilities were not *reversibly* bracketed. An additional complication with the interpretation of their experimental data arises from possible differences in aqueous silica complexes in equilibrium with kyanite versus andalusite. Assuming monomeric speciation of aqueous silica, they expressed their mineral-solution equilibria as

$$\mu_{And} = \mu_{Crn} + \mu^{\circ}_{SiO_2 \cdot nH_2O} + RT \ln X^{And}_{SiO_2 \cdot nH_2O} , \qquad [3.1]$$

and

$$\mu_{Ky} = \mu_{Crn} + \mu^{\circ}_{SiO_2 \cdot nH_2O} + RT \ln X^{Ky}_{SiO_2 \cdot H_2O} . \qquad [3.2]$$

From these equations they derived the following expression for the andalusite-kyanite equilibrium

$$\Delta G_r = RT \ln [X^{Ky}_{SiO_2 \cdot nH_2O}/X^{And}_{SiO_2 \cdot nH_2O}] \ .$$

This treatment assumes that the aqueous silica speciation in equilibrium with kyanite is identical to that in equilibrium with andalusite. Available evidence supports Brown and Fyfe's assumption that monomeric silica is present in aqueous solutions in equilibrium with silicates (Walther, 1986). Furthermore, because silica is present in tetrahedral coordination in both kyanite and andalusite, it is reasonable to conclude that the hydration number (*n*) is the same in equilibria [3.1] and [3.2] (Walther, 1986; pers. comm.). Brown and Fyfe's analysis carries the implicit assumption that the activity coefficients of aqueous silica in equilibrium with kyanite are identical to those in equilibrium with andalusite. Because the assumption of unit activity coefficients for aqueous silica appears to be valid for thermodynamic calculations of quartz solubility (Walther and Helgeson, 1977), it is reasonable to extend this assumption to solutions with lower concentrations of aqueous silica, such as those in the experiments of Brown and Fyfe.

Experiments of Bowman (1975)

Bowman (1975) investigated the andalusite-sillimanite equilibrium at 0.5 kbar and 2 kbar. Starting materials consisted of andalusite (from Minas Gerais, Brazil) and fibrolite-free sillimanite (from Benson Mines, New York). He used two methods for monitoring reaction direction: (a) weight changes of single crystals of andalusite (using a Mettler M5 microbalance) and (b) optical and SEM examination of starting materials and run products. Numerous runs were made with andalusite single crystals that were ground and polished into faceted crystals. Assuming that the growth and dissolution of andalusite occurs primarily along the *c* axis, particular effort was devoted to optical and SEM examination for changes in originally polished surfaces cut perpendicular to *c*.

Experiments at 0.5 kbar. Some of Bowman's (1975) 0.5 kbar single crystal experiments utilized unpolished andalusite crystals, whereas polished crystals were used in other runs. In some cases, the powdered starting materials contained andalusite + sillimanite, whereas sillimanite was the only aluminum silicate present in the powdered starting material of other runs (Table 3.2). The andalusite crystal weight change data are summarized in Figure 3.23. From these data, Bowman assumed

Table 3.2. Bowman's (1975) andalusite single crystal weight change data for the andalusite ⇄ sillimanite reaction at 0.5 kbar. "Q" prefix = runs with powdered quartz in the charge; "C" prefix = runs with powdered corundum in the charge. (From Bowman, 1975, Table 2).

Crystal Monitor (Andalusite)	Run Temperature (T°C)	Run Time (Days)	Pre-Run Weight* (μ grams)	Post-Run Weight* (μ grams)	Weight Change (μ grams)	Relative Weight Change (%)
A1	801	90	10368 ± 4	9921 ± 3	-447	-4.2
A2	801	90	17343 ± 4	16953 ± 4	-390	-2.3
A6	778	70	3769 ± 2	3763 ± 3	-6	-0.2
A9	694	70	12986 ± 7	13001 ± 2	+15	+0.1
A12	674	66	45812 ± 7	45838 ± 4	+26	+0.1

Inferred Reversal @ 752 ± 30°C

*Individual weights represent the mean of ten consecutive weighings of which the ± value is the first standard deviation.

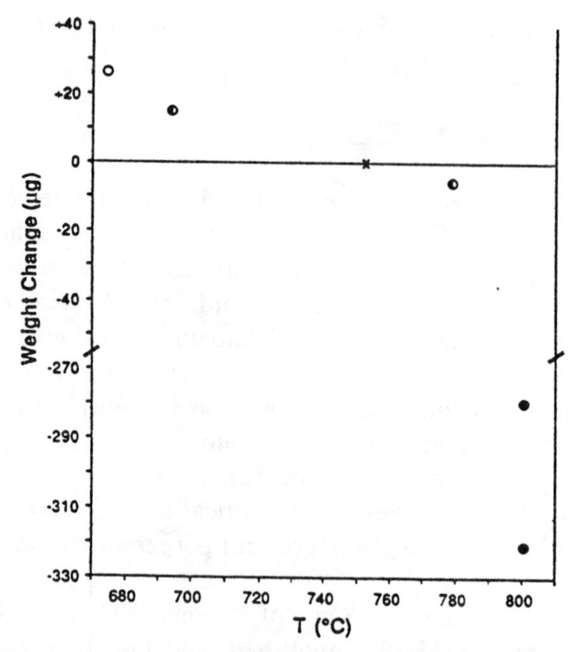

Figure 3.23. Bowman's (1975) experimental data on the andalusite ⇄ sillimanite reaction at 0.5 kbar. The data points are weight changes of andalusite single crystals in experimental charges containing water and powdered sillimanite. The different symbols correspond to different run durations: open circles = 66 d, half-filled circles = 70 d, and filled circles = 90 days. Weight changes for the 90 and 70 d runs were derived by "normalizing" recorded weight changes to 66 d. Accordingly, the recorded weight changes for the 90 and 70 d runs were divided by 1.4 and 1.1, respectively. The "x" corresponds to the equilibrium temperature selected by Bowman. (From Bowman, 1975, Fig. 7).

equilibrium at 752±30°C. The midpoint value (752°C) was obtained by intersection of a straight line joining the data points at 694°C and 778°C. Optical examination of the products of the 801°C run appeared to give conclusive evidence that the andalusite → sillimanite reaction had occurred in that andalusite single crystals appeared corroded whereas sillimanite showed overgrowths. As shown in Figure 6 of Bowman (1975), a polished andalusite crystal from the 778°C run showed a "Localized subhedral phase interpreted as sillimanite growing at the expense of andalusite" (p. 37). The author considers Bowman's interpretation of this feature to be equivocal. In all other 0.5 kbar runs there was no optical or SEM evidence for reaction.

Experiments at 2 kbar. Bowman (1975) used unpolished andalusite crystals in his 2 kbar single crystal experiments. In contrast to the 0.5 kbar runs, powdered starting materials for the 2 kbar runs contained quartz or corundum in addition to sillimanite. His single crystal weight change data are summarized in Figure 3.24 and Table 3.3, from which he inferred equilibrium at 525±15°C. In the 713°C and 665°C runs, optical examination of run products of both the corundum-bearing and quartz-bearing charges showed convincing evidence that the andalusite → sillimanite reaction had occurred. In particular, andalusite displayed corrosion and etch pits whereas sillimanite showed overgrowths. In the 566°C run, the capsule containing corundum showed considerable optical evidence of reaction (i.e., the (001) surface of the andalusite crystal was extensively corroded and displayed numerous etch pits, whereas the sillimanite showed overgrowths). However, products of the runs with quartz-bearing charges displayed no optical evidence of reaction. The products of the experiments at 516°C showed no conclusive optical evidence for reaction.

Al_2SiO_5 phase equilibrium diagram of Bowman (1975). As shown in Figure 3.25, Bowman (1975) located the andalusite-sillimanite equilibrium using his 0.5 and 2 kbar experimental results and the 1 atm "bracket" of Weill (1966). Bowman (1975) noted that the resultant dP/dT slope of -7.4 bars/°C is nearly half the value of the 1 atm Clapeyron slope (-13.6 bars/°C) computed from available entropy and molar volume data. More recent entropy and volume data (Robie and Hemingway, 1984; Hemingway et al., ms.) yield a 1 atm Clapeyron slope very similar to that computed by Bowman (1975). As discussed in Chapter 6, Bowman (1975) attributed the discrepancy between the experimental and "third law" dP/dT slopes to configurational entropy in sillimanite arising from Al/Si disorder. He located the Al_2SiO_5 triple point from the intersection of his andalusite-

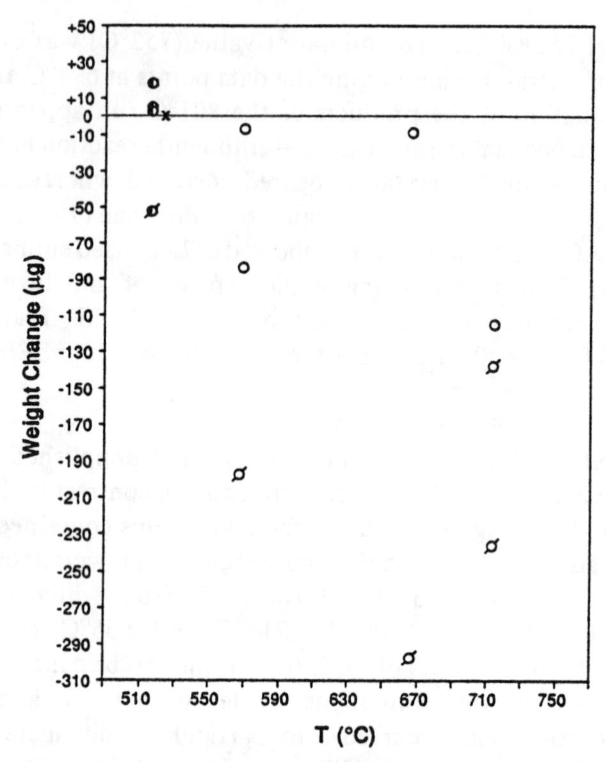

Figure 3.24. Bowman's (1975) experimental data on the andalusite ⇄ sillimanite reaction at 2 kbar. The ordinate represents weight changes of andalusite single crystals. The slash symbol refers to runs with sillimanite + corundum as the powdered starting material, whereas points without the slash are runs with sillimanite + quartz as the powdered starting material. Closed circles = 67 d; half-filled circles = 55 d; open circles = 50 d. The weight changes of the runs of 55 and 67 d duration were "normalized" to 50 d. The run data are summarized in Table 3.3. The "x" corresponds to Bowman's equilibrium temperature. (From Bowman, 1975, Fig. 21).

sillimanite equilibrium and Holdaway's (1971) kyanite-andalusite equilibrium (Fig. 3.25). Bowman's (1975) triple point (425°C and 2.75 kbar) is about 75°C and 1 kbar below that of Holdaway (1971). Bowman's triple point results in a location for the kyanite-sillimanite equilibrium that significantly differs from that of Holdaway (1971). Nevertheless, the kyanite-sillimanite equilibrium emanating from Bowman's (1975) triple point (Fig. 3.25) is within the uncertainty band of the equilibrium brackets defining Holdaway's (1971) kyanite-sillimanite equilibrium (Fig. 3.19). If Bowman's triple point is coupled with Newton's (1966b) 750°C and 8-9 kbar bracket of this equilibrium, the maximum dP/dT slope (18.8 bars/°C) is over 1 bar/°C less than that computed with the Clapeyron equation using entropy and molar volume data (Robie and Hemingway, 1984; Hemingway et al., ms.). However, the uncertainty in the computed Clapeyron slope (Robie and Hemingway, 1984, gave an uncertainty of ±0.4

Table 3.3. Bowman's (1975) andalusite single crystal weight change data for the andalusite ⇄ sillimanite reaction at 2 kbar. (From Bowman, 1975, Table 3).

Crystal Monitor (Andalusite)	Run Temperature (T°C)	Run Time (Days)	Pre-Run Weight* (μ grams)	Post-Run Weight (μ grams)	Weight Change (μ grams)	Relative Weight Change (%)
Q1	713	50	5207 ± 1	5103 ± 3	-104	-2.0
C1	713	50	6029 ± 2	5892 ± 7	-137	-2.3
C2	713	50	2347 ± 1	2111 ± 9	-236	-11.2
Q2	665	67	5135 ± 1	5122 ± 1	-13	-0.3
C3	665	67	3762 ± 3	3365 ± 5	-397	-11.8
Q3	566	67	33238 ± 3	33130 ± 4	-108	-0.3
Q4	566	67	2783 ± 2	2772 ± 8	-11	-0.4
C5	566	67	1575 ± 1	1313 ± 2	-262	-18.0
Q5	516	55	3987 ± 2	3989 ± 6	+2	+0.1
Q6	516	55	1621 ± 3	1644 ± 3	+18	+1.1
Q7	516	55	2035 ± 3	2036 ± 4	+1	+0.1
C6	516	55	3419 ± 6	3361 ± 7	-58	-1.7

Inferred Reversal @ 525 ± 15°C

* ± one standard deviation as in Table 3.2.

"Q" = ∿30 mg qtz included in charge.

"C" = ∿30 mg γ-Alumina included in charge

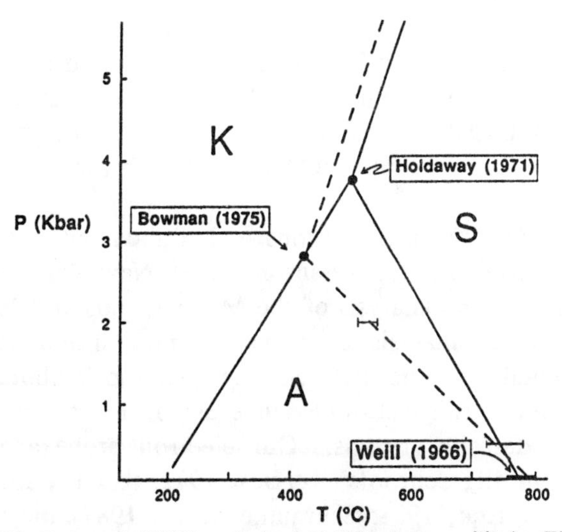

Figure 3.25. Bowman's (1975) analysis of Al_2SiO_5 phase equilibria. The dashed andalusite-sillimanite equilibrium was constrained by his 0.5 and 2 kbar brackets and by Weill's (1966) 1 atm equilibrium. Equilibria depicted with solid lines are from Holdaway (1971). The dashed kyanite-sillimanite equilibrium was extrapolated from Bowman's (1975) triple point using the same dP/dT slope as that determined by Holdaway (1971) for this equilibrium. (Modified from Bowman, 1975, Fig. 26).

bars/°C) is such that there is no compelling reason to discard the location of the kyanite-sillimanite equilibrium based on Bowman's (1975) analysis.

Critique of Bowman's (1975) experiments. Bowman's (1975) interpretation of his andalusite single crystal weight change data must be viewed in the light of the aforementioned uncertainties regarding minimum detectability of weight changes with a Mettler M5 microbalance. In Bowman's (1975) 0.5 kbar runs (Table 3.2), the author considers the -6 μg weight change of the 778°C data point as inconclusive and the +15 μg weight gain at 694°C to be questionable. However, the +26 μg weight gain at 674°C *probably* represents a valid indication of the sillimanite → andalusite reaction. I conclude that Bowman's (1975) 0.5 kbar experiments imply equilibrium between 674°C and 801°C. His 2 kbar data (Fig. 3.24) suggest that the weight losses of andalusite crystals in corundum-bearing charges were significantly larger than in those containing quartz. Bowman based his equilibrium bracket (525±15°C) on the results of his quartz-bearing experiments. He concluded that the equilibrium was bracketed by runs at 516°C and 566°C. Interpretation of the data of the quartz-bearing experiments at 516°C is clouded by relatively small weight changes (+1 to +18 μg). Furthermore, there is significant scatter in data points of the quartz-bearing experiments at higher temperatures (Fig. 3.24). Andalusite crystals in the corundum-bearing charges underwent significant weight losses at all temperatures. In fact, the relatively large weight loss (-58 μg) at 516°C suggests equilibrium below this temperature, thereby contradicting Bowman's (1975) inferred equilibrium based on results of the quartz-bearing charges. Bowman (1975, p. 70) attributed the discrepancy between the data of corundum-bearing versus quartz-bearing charges to differences in the solubility of andalusite in the two mixtures.

Bowman's (1975) starting materials consisted of andalusite from Minas Gerais, Brazil, and sillimanite from Benson Mines, New York. He cited Holdaway's (1971) chemical analysis of the Minas Gerais andalusite and Todd's (1950) chemical analysis of Benson Mines sillimanite. For numerous experimental hydrothermal studies in the author's laboratory, we have purchased (from Ward's Natural Science Establishment) several lots of Minas Gerais andalusite crystals. Our electron probe analyses of selected crystals from different lots (Kerrick, 1972; Kerrick and Ghent, 1979; Jacobs and Kerrick, 1981; Schramke et al., 1987) indicate that Holdaway's (1971) analysis of Minas Gerais andalusite is indeed representative of material from this locality. Todd's (1950) analysis of the Benson Mines sillimanite revealed significant impurity levels (wt % oxides:

FeO = 0.14, Fe_2O_3 = 0.98, MgO = 0.24, P_2O_5 = 0.28). However, there is no guarantee that the chemical analysis of Todd (1950) is applicable to the Benson Mines sillimanite used by Bowman (1975). In fact, electron probe analyses of other samples of Benson Mines sillimanite suggest considerably higher Fe_2O_3 contents than reported by Todd (1950). Grew (1980) reported 1.45 and 1.79 wt % Fe_2O_3 for two samples of sillimanite from this locality, and Kerrick and Speer (1988) obtained a value of 1.54 wt % Fe_2O_3 for a sample of Benson Mines sillimanite. As in the critique of Holdaway's experiments, the high Fe content of the sillimanite casts uncertainty on the interpretation of Bowman's experimental data.

Experiments of Heninger (1984)

In this author's hydrothermal laboratory, Heninger (1984) carried out an experimental study on the andalusite-sillimanite equilibrium at pressures of 1, 2, and 3 kbar. Reaction direction was monitored by weight changes of andalusite spheres. The andalusite spheres were prepared from gem-quality andalusite from Minas Gerais, Brazil. The "sillimanite" sample consisted of nearly phase-pure fibrolite (from Lewiston, Idaho) with crystal diameters ranging from 0.5 to 18 μm. The most significant impurities in the fibrolite are 0.15 wt % Fe_2O_3 and 0.40 wt % K_2O, whereas the andalusite contains 0.04 wt % TiO_2 and 0.25 wt % Fe_2O_3. X-ray diffraction revealed minor amounts of muscovite as the only impurity phase, thereby rationalizing the elevated K_2O content. Following sieving, the powders were washed in cold, concentrated Hcl. Ultra-fine particles were removed by elutriation following ultrasonic agitation in water, then acetone. As in the studies of Holdaway (1971) and Brown and Fyfe (1971), powdered corundum was added to the charge to buffer the activity of Al_2O_3. The andalusite sphere weight change data for these experiments are summarized in Figure 3.26. There is considerable scatter in the weight change data, thereby exemplifying my concern regarding the interpretation of single crystal weight changes $\leq$15 μg. Regression analysis through the 2 kbar and 3 kbar data yields the 95% confidence intervals outlined by the dashed lines in Figure 3.26. For both the 2 kbar and 3 kbar data, these confidence intervals encompass the line of zero weight change. Thus, the sillimanite $\rightarrow$ andalusite reaction was not conclusively demonstrated. It is concluded that the data in Figure 3.26 provide *only* upper temperature limits for the andalusite-sillimanite equilibrium. If we consider only weight changes of 20 μg or more to be meaningful monitors of reaction, the data in Figure 3.26 *suggest* upper temperature limits for the equilibrium of 800°C at 1 kbar, 725°C at 2 kbar, and 600°C at 3 kbar.

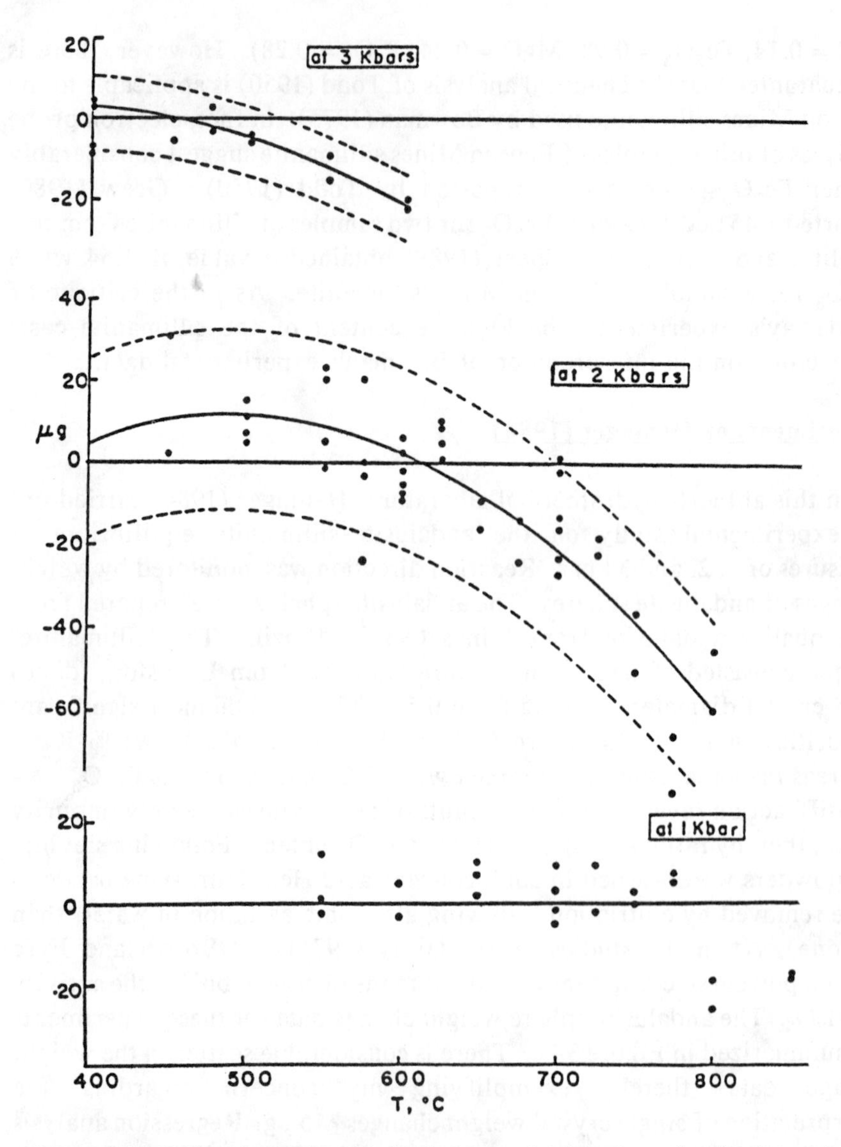

Figure 3.26. Heninger's (1984) andalusite single crystal weight change data for the andalusite ⇄ sillimanite reaction. The solid curves in the 2 and 3 kbar plots were fit by least-squares to the data points. The set of dashed lines at 2 kbar and 3 kbar encompass the 95% confidence interval about the solid curves. (From Heninger, 1984, Fig. 4).

Critique of Heninger's (1984) experiments. Using SEM, Heninger (1984) noted that during his experiments K-feldspar formed on the surfaces of the fibrolite. Because minor amounts of muscovite occur in the fibrolite starting material, K-feldspar undoubtedly formed as a product of the reaction

Muscovite → K-feldspar + Corundum + H_2O .

This reaction probably occurred by growth of corundum on the corundum seed crystals present in the starting materials. Because of the relatively small amount of muscovite in the fibrolite, this reaction is unlikely to have significantly affected weight changes of the andalusite crystals.

Because of Holdaway's (1971) anomalous results in his experiments on the andalusite-sillimanite equilibrium using fibrolite-bearing starting materials, and because of uncertainty regarding the relative thermodynamic properties of fibrolite versus sillimanite (see Chapter 7), there is uncertainty regarding the interpretation of Heninger's (1984) experiments. This question is particularly significant in light of the marked sensitivity of the andalusite-sillimanite equilibrium to relatively minor perturbations in ΔG_r. For example, as discussed in Chapter 7, a few hundred joules of molar grain boundary energy of fibrolite would significantly alter the position of the andalusite-fibrolite equilibrium compared to the andalusite-sillimanite equilibrium.

<u>Experiments of Bohlen et al. (ms.)</u>

In a cooperative investigation with the author, A.L. Montana (UCLA) and S.R. Bohlen (USGS) have experimentally bracketed the kyanite-sillimanite equilibrium using piston cylinder devices equipped with salt cells (Bohlen et al., ms.). Compared to earlier furnace assemblies requiring substantial friction correction to run pressures, salt cells are virtually frictionless (Johannes, 1978; Boettcher et al., 1981; Bohlen, 1984). The starting material in A.L. Montana's experiments (using the UCLA facility) consisted of a powdered mixture of natural, gem-quality kyanite from Brazil and sillimanite from Sri Lanka. In experiments at the USGS laboratory, S.R. Bohlen used gem-quality kyanite from Switzerland and sillimanite from Antarctica. After crushing, the ultrafines in the powdered aluminum silicate samples were removed by washing once in 1N HCl followed by rinsing twice in distilled water. To buffer a_{SiO_2}, a small amount of quartz (1-2 wt %) was added to the kyanite-sillimanite and kyanite-andalusite mixtures. In the experiments at the USGS, NH_4OH was used as a flux because Al and Si are soluble in alkaline solutions. No flux was added to the starting material of the experiments at the UCLA lab. In both laboratories reaction direction was primarily determined by comparing peak areas of kyanite and sillimanite reflections on X-ray diffractograms of starting materials versus run products. Reaction direction determined by X-ray diffractogram peak intensities are consistent with grain morphology features (revealed by optical and SEM examination of run products) that indicated growth of one polymorph and dissolution of the other. As shown in Figure 3.27, there is *excellent* agreement between the experimental results carried out in the UCLA and USGS laboratories. These results provide exceptionally tight control on the location of the kyanite-sillimanite equilibrium at 700-1000°C. The experimental brackets reveal significant curvature of the kyanite-sillimanite equilibrium above 800°C. Figure 3.27 shows the location of the equilibrium computed using the Clapeyron equation with the entropy data of Hemingway et al. (ms.) and molar volume data of Winter and Ghose (1979). In contrast to the experimentally-determined equilibrium, the computed equilibrium is virtually straight in the 700-1000°C range. The increase in the Clapeyron slope of the experimentally-determined equilibrium with increasing temperature above 800°C is compatible with an increase in the configurational entropy of sillimanite due to Al/Si disorder. Evaluation of Al/Si disorder in sillimanite that was experimentally synthesized by A.L. Montana is in progress using NMR.

SOLUBILITY STUDIES AT ATMOSPHERIC PRESSURE

Experimental study of Weill (1966)

Following preliminary notes by Weill and Fyfe (1961, 1964), Weill (1966) published the results of his 1 atm solubility measurements of the Al_2SiO_5 polymorphs in molten cryolite at 800°C and 1010°C (Fig. 3.28). Dissolution of the aluminum silicates in molten cryolite is incongruent, yielding corundum + melt. Accordingly, the isothermal free energy change for the andalusite-sillimanite equilibrium can be calculated from

$$\Delta G_{And=Sil} = RT\ ln\ a^{Sil}_{SiO_2}/a^{And}_{SiO_2} \qquad [3.3]$$

where $a^{Sil}_{SiO_2}$ and $a^{And}_{SiO_2}$ represent the respective activities of silica in melts in equilibrium with sillimanite and andalusite. From his solubility measurements (Fig. 3.28 and Table 3.4), Weill (1966) computed a_{SiO_2} in the cryolite melts using the Temkin model, which assumes: (1) melts are treated as "pseudolattices", (2) independent, random mixing of cations on cation sites and anions on anion sites, and (3) mixing on anion and cation sites is ideal ($\Delta H_{mix} = 0$). Weill's treatment assumed that, at a given temperature, the activity coefficients of silica in cryolite melts in equilibrium with corundum + andalusite are identical to those with corundum + sillimanite. Weill (1966) defended the use of the Tempkin model by the constancy of $a_{Al_2O_3}$ in various liquids in equilibrium with

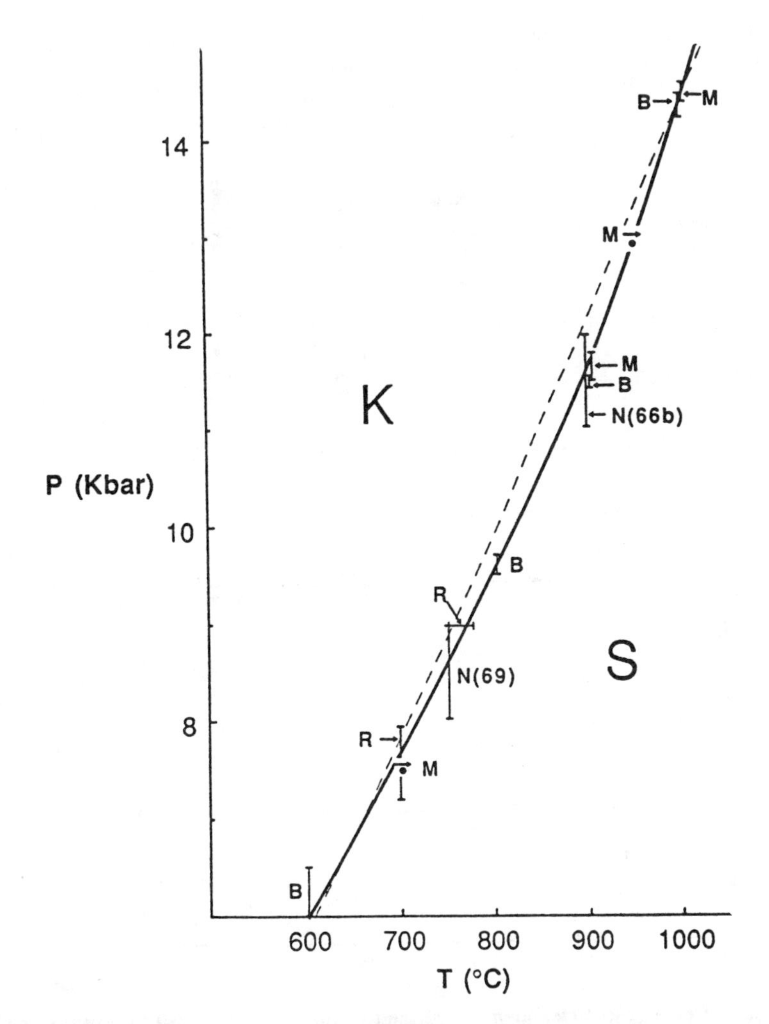

Figure 3.27. Experimental data of Bohlen et al. (ms.) on the kyanite ⇄ sillimanite reaction. The vertical and horizontal bars represent experimental brackets whereas the dots with arrows are runs where direction of the kyanite ⇄ sillimanite reaction was determined. Designations for the experimental studies are: B = S.R. Bohlen (Bohlen et al., ms.); M = A.L. Montana (Bohlen et al., ms.); R = Richardson et al. (1968); N = Newton (1969). The dashed curve was computed by J.A.D. Connolly using the VERTEX program of Connolly and Kerrick (1987). The solid curve represents the graphically-smooth best fit curve to all of the experimental data.

corundum and by the agreement between the change in ΔS_r between 800°C and 1010°C for various reactions, as calculated from free energy data derived from his solubility measurements, compared to the corresponding changes in ΔS_r derived from calorimetric measurements. Using equation [3.3], coupled with entropy data, Weill (1966) computed the equilibrium temperature (T_{eq}) from

Table 3.4. Weill's (1966) determination of the compositions of corundum-saturated cryolite melts in equilibrium with various phases. (From Weill, 1966, Table 1).

Mineral	Liquid designation	mol.%			°C
		SiO_2	Al_2O_3	Na_3AlF_6	
andalusite	*A*	76·2	12·8	11·0	
mullite	*M*	74·9	13·3	11·8	1010
sillimanite	*S*	75·7	13·0	11·3	
tridymite	*T*	83·2	10·8	6·0	
andalusite	*A*	60·6	19·8	19·6	
kyanite	*K*	72·2	14·2	13·6	800
mullite	*M*	62·3	19·0	18·7	
quartz	*Q*	72·7	14·1	13·2	
sillimanite	*S*	60·3	20·0	19·7	

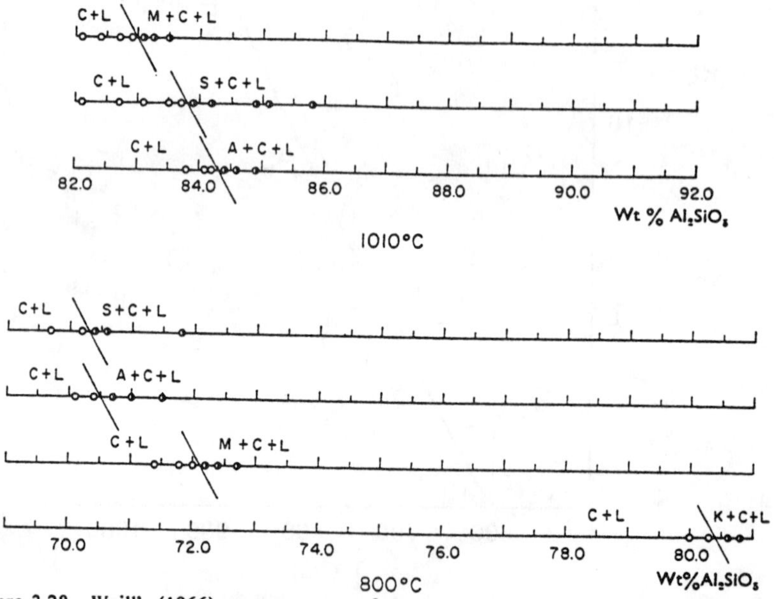

Figure 3.28. Weill's (1966) measurements of the solubility of andalusite (A), kyanite (K), sillimanite (S), and mullite (M), in corundum (C)-saturated melts. The abscissa is the concentration of the Al_2SiO_5 component along the Na_3AlF_6-Al_2SiO_5 join. (From Weill, 1966, Fig. 1).

$$\Delta G_{r,T_{eq}} = 0 = \Delta G_{r,T_{exp}} - \int_{T_{exp}}^{T_{eq}} \Delta S_r dT, \qquad [3.4]$$

where $\Delta G_{r,T_{exp}}$ is Weill's (1966) computed free energy change for the andalusite → sillimanite reaction at the temperatures of his experiments. Using available entropy data, Weill (1966) computed a 1 atm equilibrium temperature of 715°C from the free energy change for the reaction [+100.4 J/(mol·K)] determined from equation [3.3] using silica activities derived from his experimental data at 800°C (D.F. Weill, pers. comm.). Subsequent to Weill's (1966) paper, Holdaway (1971, p. 117) stated: "More recently Weill and his coworkers (pers. comm.) have found that andalusite and sillimanite have equal solubilities in cryolite at 775°." However, contrary to this assertion, there are no additional solubility data for the Al_2SiO_5 polymorphs in molten cryolite beyond those reported by Weill (1966) (D.F. Weill, pers. comm., 1988). Numerous subsequent analyses of Al_2SiO_5 phase equilibria have attributed a 1 atm and 775°C equilibrium bracket to Weill (Day and Kumin, 1980, Robie and Hemingway, 1984; Salje, 1986; Newton, 1987). In fact, this bracket is a key component in the phase equilibrium diagrams of Robie and Hemingway (1984) and Salje (1986). Day and Kumin (1980, p. 265) state: "...a 1 atm bracket of the andalusite-sillimanite reaction (Weill, 1966) is the critical discriminator between triple points based on the experimental data of Holdaway (1971) and Richardson, Gilbert, and Bell (1969)".

Al_2SiO_5 phase equilibrium diagram of Weill (1966). Weill's (1966) phase equilibrium diagram for the Al_2O_3-SiO_2 system is shown in Figure 3.29. It was computed using the ΔG_r data computed from his solubility measurements in molten cryolite (see Table 3.5), coupled with available entropy and molar volume data. Computed ΔV_r values incorporated thermal expansion data; however, because of lack of data, compressibility was not incorporated. Note that Weill's 1 atm temperature for the andalusite-sillimanite equilibrium is 715°C and not 775°C as stated by Holdaway (1971, p. 117). Also note the marked curvature of the andalusite-sillimanite equilibrium. This curvature epitomizes the sensitivity of this equilibrium to changes in ΔS_r and ΔV_r (and, thus, the Clapeyron slope) with changing temperature. Weill's (1966) computed kyanite-sillimanite equilibrium (Fig. 3.29) is also curved. This computed curvature brings into question the assumption of Richardson et al. (1968) that the kyanite-sillimanite equilibrium is a straight line over the P-T range shown in Figure 3.29.

Experiments of Bowman (1975)

In addition to hydrothermal experiments at elevated pressures, Bowman (1975) carried out experiments at atmospheric pressure and 800°C and 1010°C with charges initially containing single crystals of andalusite in powdered sillimanite + quartz + cryolite. Although Bowman's (1975) experiments were apparently intended to provide a re-evaluation of Weill's (1966) solubility measurements, Bowman (1975) actually utilized his measurements of weight losses of andalusite single crystals coupled with SEM examination of run products to determine the direction of the

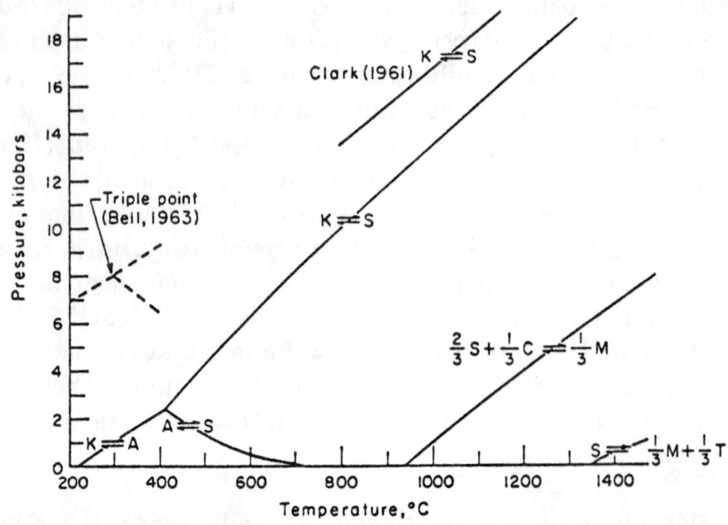

Figure 3.29. Weill's (1966) computed phase equilibrium diagram for the Al_2O_3-SiO_2 system. A = andalusite, S = sillimanite, K = kyanite, C = corundum, M = mullite, and T = tridymite. (From Weill, 1966, Fig. 3).

Table 3.5. Weill's (1966) calculated values of the silica activity in equilibrium with various phases, and the free energy changes of selected reactions. The abbreviations for the minerals listed under *Liquid Designations* are explained in Table 3.4. (From Weill, 1966, Table 2).

°C	Liquid designation (see text)	a_{SiO_2}	ΔG in cal/mol Al_2SiO_5	Reaction
	S	$0{\cdot}280_2$	—	—
1010	*M*	$0{\cdot}267_1$	+345	sillimanite → 1/3 (mullite + tridymite)
	A	$0{\cdot}288_5$	+75	sillimanite → andalusite
	T	$0{\cdot}463_1$	+1273	sillimanite → corundum + tridymite
	S	$0{\cdot}123_0$	—	—
	A	$0{\cdot}124_4$	+24	sillimanite → andalusite
800	*M*	$0{\cdot}135_9$	+598	sillimanite → 1/3 (mullite + quartz)
	K	$0{\cdot}225_8$	+1293	sillimanite → kyanite
	Q	$0{\cdot}233_9$	+1341	sillimanite → corundum + quartz

andalusite ⇄ sillimanite reaction. In Bowman's (1975) experiments at 1010°C, the large weight loss of the andalusite single crystal (-27315 μg), coupled with textural evidence suggesting growth of sillimanite, strongly suggest that the reaction andalusite → sillimanite occurred. In Bowman's 800°C experiments, the andalusite single crystal lost significant weight (42 μg); however, there was no evidence that sillimanite grew. Thus, Bowman's (1975) 800°C experiments yielded inconclusive results regarding reaction direction.

Critique of the experimental results of Weill (1966) and Bowman (1975)

The accuracy of Weill's (1966) values for the free energy change of the andalusite → sillimanite reaction are dependent upon the accuracy of his liquid compositions. Weill's solubility data for andalusite versus sillimanite at 1010°C (Table 3.4) illustrate that his method detected significant differences in solubility between andalusite and sillimanite. However, at 800°C his measured solubilities of andalusite and sillimanite differ by only 0-0.5 wt % (Table 3.4). The contrast between his experimental results at 800°C and 1010°C are exemplified by Weill's (1966) computed values of a_{SiO_2} at these temperatures. Specifically, at 800°C, $a^{And}_{SiO_2}$ = 0.124 and $a^{Sil}_{SiO_2}$ = 0.123, whereas at 1010°C $a^{And}_{SiO_2}$ = 0.288 and $a^{Sil}_{SiO_2}$ = 0.280 (Table 3.5). Consider the total range of uncertainty in Weill's brackets of melt compositions in equilibrium with sillimanite + corundum at 800°C (Fig. 3.28). At one extreme, melt compositions could be identical, thus implying that the andalusite-sillimanite equilibrium is at 800°C (i.e., 85°C above the equilibrium temperature computed by Weill). Although the extremal limits of the uncertainty range may be unlikely, this analysis illustrates that small uncertainties in melt composition yield large errors in the computed equilibrium temperatures.

To derive ΔG_r for the andalusite-sillimanite equilibrium, Weill (1966) utilized a_{SiO_2} values calculated with Temkin's (1945) melt model. The inadequacies of Tempkin's model for silicate melts are now well known (e.g., Bottinga et al., 1981). Weill (1966) recognized the potential inadequacy of the Tempkin model; in particular, he noted the inability of this model to account for polymerization. Weill (1966) conceded that an ideal mixing model is unrealistic for complex silicate melts; however, he considered that, at a given temperature, deviations from ideality would be similar in melts coexisting with andalusite + corundum compared to melts coexisting with sillimanite + corundum. In this context, it is important to emphasize the difference in coordination of Al in sillimanite (Al^{IV}) versus andalusite (Al^{V}). It is possible that such differences in Al coordination would imply corresponding differences in melt speciation. In essence, this possibility is analogous to that of Burnham's (1981) silicate melt model in which melt speciation is *assumed* to reflect the structures of crystalline phases in equilibrium with the melt. However, there is considerable controversy regarding the validity of this assumption (Bottinga et al., 1981). Determination of the speciation of Al in cryolite melts in equilibrium with andalusite versus sillimanite must await further analysis of melt speciation in such systems (e.g., NMR analysis).

An additional problem with Weill's experimental study is the fact that he determined solubility from one direction only. That is, he did not reversibly bracket the equilibrium solubility by starting with both Al-undersaturated and Al-supersaturated melts. It is for this reason that Newton (1987) referred to Weill's 1 atm bracket as "unreversed" (R.C. Newton, pers. comm.).

We reexamine computations of the 1 atm equilibrium based on Weill's (1966) data. Using equation [3.4] with the heat capacity functions of Hemingway et al. (ms.) and Weill's (1966) values of $\Delta G_{r,T_{exp}}$, Weill's 800°C ΔG_r value yields an equilibrium temperature of 764°C, whereas his 1010°C ΔG_r yields equilibrium at 896°C. These significant differences in T_{eq} question Weill's $\Delta G_{r,T_{exp}}$ values. Weill (pers. comm.) calculated the andalusite-sillimanite equilibrium from his ΔG_r data at 800°C rather than 1010°C because he considered 800°C to be closer to the equilibrium temperature and, thus, there would be smaller errors arising from extrapolation. To yield agreement with the equilibrium temperature (764°C) computed from Weill's (1966) ΔG_r value at 800°C, ΔG_r derived from his 1010°C data would have to be +160 J; i.e., twice the value computed by Weill from his data at 1010°C. The discrepancies in T_{eq} computed from Weill's ΔG_r data at 800°C versus 1010°C *may* be attributed to inadequacies in the Tempkin model for computing a_{SiO_2} in cryolite melts. The apparent differences in thermodynamic properties of the melts at 800°C versus 1010°C *may* reflect corresponding differences in viscosity and thus differences in polymerization of the melts at these contrasting temperatures.

An equilibrium temperature of 715°C computed by Weill (1966) markedly differs from the 764°C value computed by this author. These differences result from corresponding differences in ΔS_r, and illustrate the extreme sensitivity of the P-T location of the andalusite-sillimanite equilibrium to relatively small perturbations in ΔG_r.

CALORIMETRIC STUDIES

Calorimetric data provide an important insight into phase equilibria involving the Al_2SiO_5 polymorphs. Heat capacity measurements yield vibrational entropies and solution calorimetry yields enthalpies. Coupled with thermal expansion and compressibility data, calorimetric data provide a means of calculating Al_2SiO_5 phase equilibria that is *independent* of experimental hydrothermal results. However, the most fruitful approach is integration of calorimetrically-derived data with the results from experimental phase equilibrium studies. This combined approach has dominated the analysis of phase equilibria during the last two decades. In fact, combination of experimental phase equilibrium bracketing studies with calorimetrically-derived thermodynamic data has been central to numerous compilations of the thermodynamic data of rock-forming minerals (Helgeson et al., 1978; Robinson et al., 1981; Halbach and Chatterjee, 1984; Holland and Powell, 1985; Berman et al., 1986; Berman, 1988).

Heat capacities and vibrational entropies

Low-temperature heat capacities of the Al_2SiO_5 polymorphs (yielding S_{298} entropies) have been measured in several studies (Todd, 1950; Robie and Hemingway, 1984). Compared to earlier studies, the results of Robie and Hemingway (1984) are strongly preferred in the light of the extended temperature range of measurements. For example, Todd's (1950) measurements of heat capacity covered the temperature range from 54.4 to 296.5K, whereas Robie and Hemingway's (1984) covered the range 10-380K. The ability of modern heat capacity calorimetry to cover the 5-50K temperature range offers a significant advantage over earlier low-temperature calorimeters (Robie, 1987). In addition to a larger temperature range, the results of Robie and Hemingway (1984) are preferred because of the more extensive characterization and description of the preparation of samples used for their calorimetric measurements. The sillimanite used in their study lacked fibrolite. Chemical analyses revealed that Fe_2O_3 was the most significant impurity in all three samples (wt % Fe_2O_3: kyanite = 0.18±0.05; andalusite = 0.38±0.05, sillimanite = 1.03±0.27). Their S_{298} values were not corrected for the Fe_2O_3 contents; however, such a correction would alter the S_{298} values less than 0.1% (Robie and Hemingway, 1984, p. 299).

Using drop calorimetry, Pankratz and Kelley (1964) measured the heat contents (H_T - H_{298}) of andalusite to 1600K, and the heat contents of kyanite and sillimanite to 1500K. The materials were identical to those used for Todd's (1950) low-temperature C_p measurements. Wet chemical analyses reveal minimal impurities in the andalusite and kyanite samples (the major impurities were 0.11 wt % Fe_2O_3 in andalusite and 0.10 wt % Fe_2O_3 in kyanite). However, analysis of the sillimanite sample revealed significant impurities (0.14 wt % FeO, 0.98 wt % Fe_2O_3, 0.24 wt % MgO, and 0.28 wt % P_2O_5). For the past 25 years, the heat capacities derived from Pankratz and Kelley's heat content data have provided the main source for vibrational entropies of the Al_2SiO_5 polymorphs above ambient temperatures.

Salje and Werneke (1982a,b) computed the aluminum silicate phase equilibrium diagram by quantum mechanical treatment of spectroscopic data. Lattice vibrations in crystals propagate as waves. As with photons describing discrete frequencies of electromagnetic radiation, lattice vibrations are quantized, and *phonons* describe the quantized frequencies of lattice vibrations. Using Bose-Einstein statistics to describe the excitation of lattice vibrations with temperature, the total energy is obtained by summing over all frequencies (ω)

$$U = \int_{\omega} h\omega \, [0.5 + n(T)] \, g(\omega) \, d\omega,$$

where $g(\omega)$ is the "phonon density of states" (i.e., the spectral distribution of phonon lines), h = Planck's Constant, and n(T) is the Bose-Einstein term expressing the temperature (T) dependence. Knowledge of $g(\omega)$ allows calculation of heat capacities through

$$C_v = dU/dT$$

and

$$C_p = C_v + TV\alpha^2/\beta$$

where V = molar volume, α = coefficient of thermal expansion, and β = compressibility. The phonon portion of the Gibbs Free energy of the solid phase ($G_{phonon} = H_{phonon} - TS_{phonon}$) is determined by

$$dS_T(P) = -(\delta V/\delta T)_P \, dP$$

and

$$dH_T(P) = [V - T(\delta V/\delta T)_P] \, dP$$

From their Raman and IR measurements of sillimanite and andalusite, Salje and Werneke (1982a,b) assigned the spectral lines of various types of lattice vibrations using the model of Iishi et al. (1979) where SiO_4 tetrahedra are assumed to behave as rigid molecules. Raman and IR spectroscopy measure only the phonons with wave vectors near zero (McMillan and Hofmeister, 1988); however, determination of the phonon density of states requires integration over the entire range of frequencies (mostly representing acoustic phonons). Salje and Werneke (1982a,b) computed the missing acoustic phonon frequencies from the elastic constant data of Vaughan and Weidner (1978). Salje and Werneke (1982a,b) arbitrarily selected a temperature of 800°C for the 1 atm andalusite - sillimanite equilibrium, from which they calculated ΔH_r from

$$\Delta G_r = 0 = \Delta H_r - T\Delta S_r \ .$$

From the resultant ΔH_r value, they calculated ΔG_o, the change in the potential energy of the reaction with the atoms of andalusite and sillimanite at rest, from the relation

$$\Delta G_r = \Delta G_o + \Delta G_{phonon} \ .$$

As shown in Figure 3.30, Salje and Werneke (1982a,b) computed the andalusite-sillimanite equilibrium from two arbitrary 1 atm equilibrium temperatures: 800°C and 900°C. They referenced both Richardson et al. (1969) and Brown and Fyfe (1971) for the kyanite-andalusite equilibrium. However, the kyanite-andalusite equilibrium shown in Figure 3.30 is compatible with Richardson et al. (1969) and not Brown and Fyfe (1971).

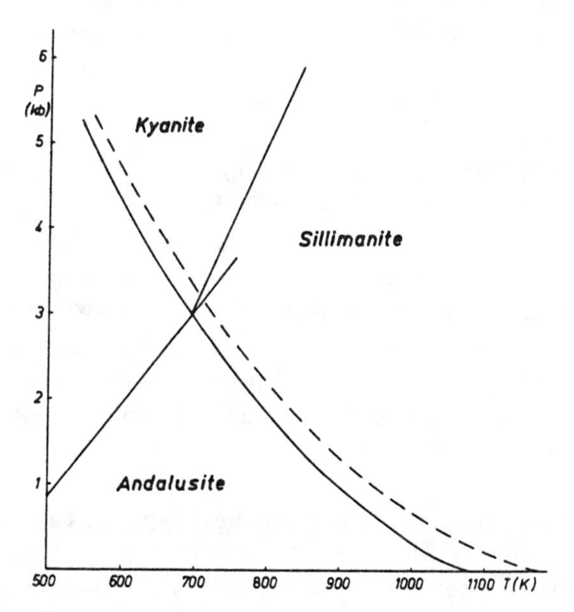

Figure 3.30. Andalusite-sillimanite equilibrium (curved lines) computed by Salje and Werneke (1982a,b) from heat capacities derived from measurements of phonon spectra. The solid curve was computed assuming a zero pressure equilibrium point of 800°C, whereas the dashed curve assumes zero-pressure equilibrium at 900°C. (From Salje and Werneke, 1982, Fig. 12).

Using a differential scanning calorimeter, Salje (1986) measured the heat capacities of andalusite and sillimanite at 380-485 K. Electron probe analyses of his samples suggest relatively low impurity levels. He derived high temperature entropies from

$$S = \int_{298}^{T} (C_p/T)dT + S_{298} \,.$$

For andalusite, Salje (1986) used the S_{298} entropy data of Robie and Hemingway (1984). In contrast, the S_{298} value for sillimanite was computed by quantum theory, i.e., "... the phonon densities of states were fitted to give a best fit with the experimental heat capacities at high temperatures ... The heat capacities at low temperatures were then calculated using Debye temperatures determined from the elastic constant data of Vaughan and Weidner (1978) and integrated to obtain standard entropies" (Salje, 1986, p. 1368). As shown in Figure 3.31, Salje (1986) chose a 1 atm equilibrium temperature of 775°C (i.e., Weill's bracket) and computed the andalusite-sillimanite equilibrium from this point. There are significant differences in the Clapeyron slopes of the equilibrium curves computed from his high-temperature data for the fibrolite-free Sri Lanka sillimanite (curve **a** in Fig. 3.31) compared to the P-T slopes for the equilibrium with two other fibrolite-free sillimanite samples (curves **b** and **c** in Fig. 3.31). Salje (1986) gave no explanation for these differences in P-T slopes. It should be noted that the convergence of the andalusite-sillimanite curves at 1 atm (Fig. 3.31) implies that $\Delta G_r = 0$ for the equilibrium with fibrolite and all three fibrolite-free sillimanites samples studied by Salje (**a**, **b**, and **c** in Fig. 3.31). Such convergence necessitates that the Gibbs free energies of the fibrolite and three sillimanite samples are identical at 1 atm. This equivalence would only be possible with the unlikely situation that the differences in entropies of the fibrolite and three sillimanites are compensated by the differences in enthalpies. Nevertheless, Figure 3.31 correctly illustrates that differences in the vibrational entropies of different sillimanite samples result in different Clapeyron slopes.

Using the low-temperature adiabatic heat capacity calorimeter at the U.S. Geological Survey (described by Robie, 1987), B.S. Hemingway measured the low-temperature C_p of a sample of phase-pure, gem-quality sillimanite from Sri Lanka (Hemingway et al., ms.). This sample has low minor element contents (Fe_2O_3 = 0.26 wt %). The computed value of S_{298} = 95.40±0.52 J/(mol·K) for this specimen is in excellent agreement with S_{298} = 95.79±0.29 J/(mol·K) obtained by applying a correction for Fe^{3+} in

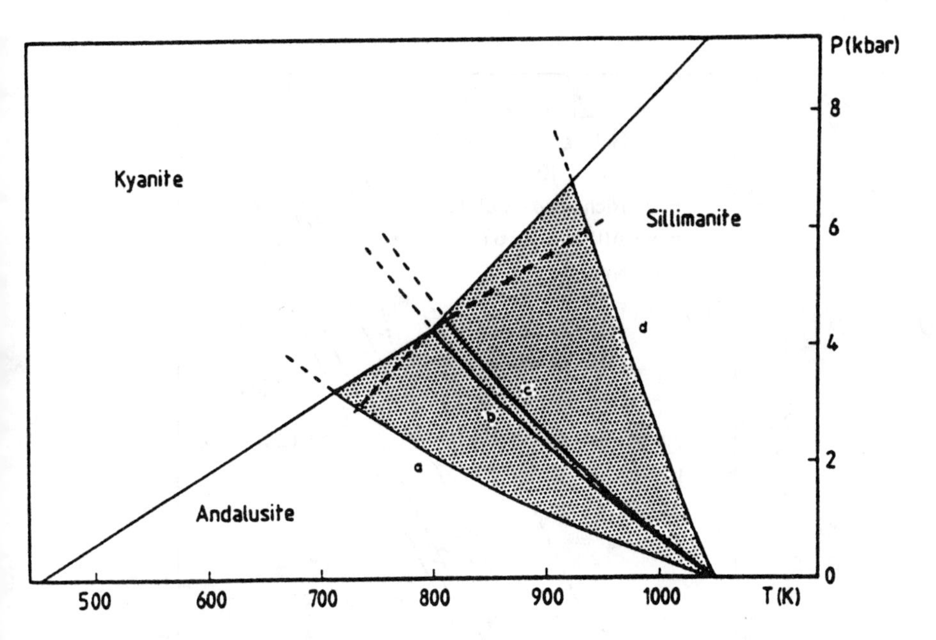

Figure 3.31. Phase equilibrium diagram calculated from heat capacities computed and measured by Salje (1986). Curves a, b and c represent the andalusite-sillimanite equilibrium for three different sillimanite samples (a = sillimanite from Sri Lanka; b = sillimanite from Waldeck, Germany, and c = sillimanite from Träskbole, Finland). Curve d is the equilibrium between andalusite and fibrolite (from Harcujuela, Spain). (From Salje, 1986, Fig. 3).

solid solution to Robie and Hemingway's (1984) S_{298} value derived from their C_p measurements of a specimen of sillimanite from Antarctica. Thus, the author considers 95.40±0.52 J/(mol·K) to be a reliable, accurate value for the vibrational entropy of sillimanite at 298.15K.

Solution calorimetry

The refractory nature of the Al_2SiO_5 polymorphs precludes determination of the enthalpies of solution with acid calorimetry. In a pioneering effort representing the first application of oxide melt solution calorimetry to silicate systems, Holm and Kleppa (1966) determined the heats-of-solution of the Al_2SiO_5 polymorphs in lead-cadmium-borate at 695°C. Recognizing the uncertainties of this early work, molten oxide enthalpies of solution of kyanite and sillimanite were redetermined by Anderson and Kleppa (1969), and andalusite was redetermined by Anderson et al. (1977). Coupling their value of $\Delta H_{Ky=Sil}$ determined from molten salt calorimetry at 700°C with available C_p and molar volume data, Anderson and Kleppa's (1969) calculated kyanite-sillimanite equilibrium curve for non-fibrolitic sillimanite [which they designated as Si(II)] did

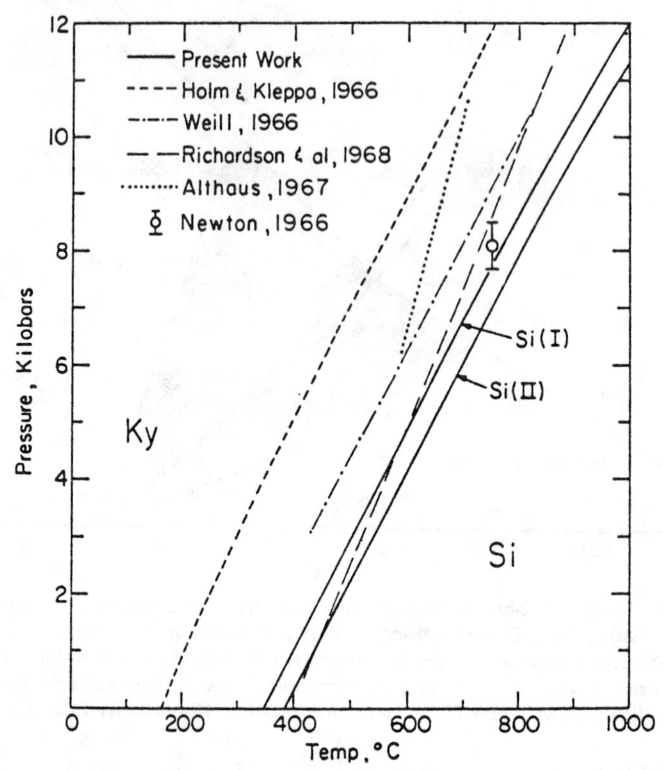

Figure 3.32. Comparison of experimentally-determined kyanite-sillimanite equilibria with those computed by Anderson and Kleppa (1969) from molten oxide heat-of-solution measurements. The equilibrium labeled Si(I) was derived from $\Delta H_{solution}$ measurements of a fibrolite sample (Custer, South Dakota), whereas that labeled Si(II) corresponds to a sample of sillimanite (Benson Mines, New York). (From Anderson and Kleppa, 1969, Fig. 1).

not pass through the experimental brackets of Newton (1966b) and Richardson et al. (1968) (Fig. 3.32). Anderson and Kleppa (1969) attributed this discrepancy to Al/Si disorder in sillimanite. Anderson et al. (1977) concluded that the andalusite-sillimanite equilibrium computed from their heat-of-solution data was in agreement with Holdaway's (1971) experimental results but inconsistent with the experimental data of Richardson et al. (1969). In addition to favoring Holdaway's (1971) experimental results, they concluded that their calorimetric data, and Holdaway's (1971) experimental results, suggest some Al/Si disorder in sillimanite. The conclusions of Anderson and Kleppa (1969) and Anderson et al. (1977) regarding Al/Si disorder are critiqued in Chapter 6.

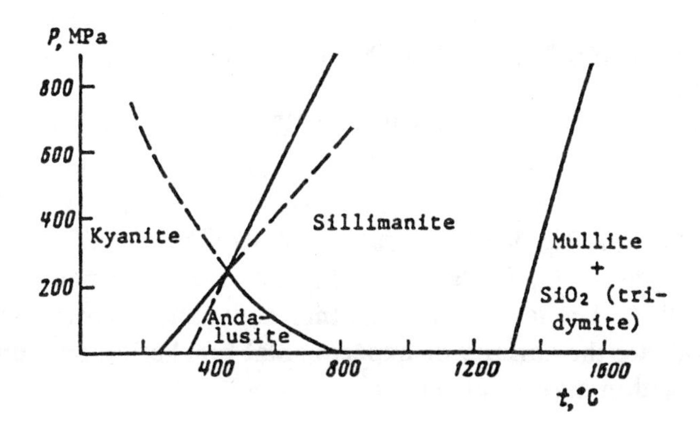

Figure 3.33. Al_2SiO_5 phase equilibrium diagram computed by Kiseleva et al. (1983) using calorimetric data. (From Kiseleva et al., 1983, Fig. 2).

Using a Calvet-type microcalorimeter with $2PbO{\cdot}B_2O_3$ as a solvent, Kiseleva et al. (1983) measured the enthalpies of solution of kyanite, sillimanite, andalusite and synthetic mullite at 700°C. Chemical analyses suggest relatively low impurity levels in the Al_2SiO_5 samples. Their enthalpies of solution are in good agreement (i.e., within 300 J/mol) with the values of Anderson and Kleppa (1969) for kyanite and Anderson et al. (1977) for andalusite. However, for sillimanite there are significant differences (1.34 kJ/mol) between the enthalpy of solution values determined by Anderson and Kleppa (1977) and Kiseleva et al. (1983). In referring to the molten oxide enthalpy of solution calorimetry of Navrotsky et al. (1973), Kiseleva et al. (1983) considered that the discrepancies in the values of the heat of solution of sillimanite are attributed to Al/Si disorder. Kiseleva et al. (1983) adopted the heat-of-solution value of 29.75±1.21 KJ/mol from Navrotsky et al. (1977) as representative of the most ordered sillimanite. Kiseleva et al. (1983) computed the phase equilibrium diagram shown in Figure 3.33 using their computed enthalpies of formation, data on entropies, specific heats and molar volumes from Ostapenko et al. (1978), and the thermal expansion data of Skinner et al. (1961). It is important to emphasize that there are large uncertainties in their computed phase equilibria. For example, their error in $\Delta H_{r,298}$ of the andalusite-sillimanite equilibrium is ±1.38 kJ/mol. Translating this error into ΔG_r, and considering the expression: $(d\Delta G_r)_P = \Delta S_r dT$ with the entropy data of Hemingway et al. (ms.), an enthalpy error of ±1.38 kJ/mol yields an uncertainty in the 1 atm equilibrium temperature of ±476°C! This large error epitomizes the large uncertainties in calculating the Al_2SiO_5 phase equilibria from ΔH_f data derived from heat-of-solution measurements.

ANALYSIS OF Al_2SiO_5 PHASE EQUILIBRIA USING THERMODYNAMICALLY CONSISTENT DATA SETS

Within the past two decades there have been several studies utilizing hydrothermal experimental equilibrium bracketing data to derive an internally consistent thermodynamic data base for minerals (Fisher and Zen, 1971; Zen, 1972; Helgeson et al., 1978; Robinson et al., 1982; Halbach and Chatterjee, 1984; Holland and Powell, 1985; Berman et al., 1986; Berman, 1988). The test for internal thermodynamic consistency offers a way to evaluate the validity of experimental bracketing data on multiple equilibria within a given chemical system.

Gordon (1973) outlined a linear programming technique to evaluate the thermodynamic consistency of experimental data. Day and Kumin (1980) used this technique to evaluate the thermodynamic consistency of experimental data on Al_2SiO_5 phase equilibria. Newton (1987) provides a succinct, clear review of Day and Kumin's (1980) paper. However, for completeness, the following is a brief summary of Day and Kumin's (1980) contribution. Experimental studies reveal the *direction* of a reaction (and, thus, the sign of ΔG_r) for selected P-T conditions. At each P-T condition where reaction was detected, the Gibbs free energy change for the reaction is either negative or positive in sign. Accordingly, Day and Kumin (1980) derived the expression

$$\Delta H_f{}^{o}{}_{,298} \ \overset{\geq}{\underset{\leq}{\text{or}}} \ T\Delta S_f{}^{o}{}_{,298} - G'(T,P) \ ,$$

where G'(T,P) was computed from heat capacities, molar volumes, compressibility and thermal expansion data. On a plot of $\Delta H_f{}^{o}{}_{,298}$ versus $\Delta S_f{}^{o}{}_{,298}$ (Fig. 3.34), each data point delimits an area of possible values for *pairs* of $\Delta H_f{}^{o}{}_{,298}$ and $\Delta S_f{}^{o}{}_{,298}$ values. Multiple data points for a reaction outline a polygonal-shaped *feasible solution space* for the equilibrium (Fig. 3.34). Within the constraints of the feasible solution space, Day and Kumin (1980) derived values of $\Delta H_f{}^{o}{}_{,298}$ and $\Delta S_f{}^{o}{}_{,298}$ that provide the best agreement between the enthalpies and entropies of formation computed from linear parametric analysis of experimental data and those derived from calorimetric measurements. Day and Kumin (1980) noted that, taken separately, the experimental data of Richardson et al. (1969) versus Holdaway (1971) and Weill (1966) are internally consistent (Figs. 3.35 and 3.36, respectively). They concluded that the critical discriminator is the 1 atm. "bracket" of Weill (1966). No thermodynamically consistent Al_2SiO_5

phase equilibrium diagram is possible if Weill's bracket is included with the data of Richardson et al. (1969). In contrast, thermodynamic consistency results when the data of Holdaway (1971) is coupled with that of Weill (1966). Because Day and Kumin (1980) found no reason to doubt the validity of Weill's bracket, they favored the phase equilibrium diagram of Holdaway (1971). Nevertheless, they considered that their analysis did not yield a "non-controversial" choice between the phase equilibrium diagrams of Richardson et al. (1969) versus Holdaway (1971).

Robie and Hemingway (1984) coupled experimental equilibrium brackets with the dP/dT slopes calculated with their computed high-temperature entropy data (giving ΔS_r) and molar volume data (giving ΔV_r) from several sources (Skinner et al., 1961; Schneider, 1979a; Winter and Ghose, 1979). From compressibility data they showed that the pressure effect on ΔV_r is negligible below 10 kbar; thus, their calculations utilized 1 atm. ΔV_r for the equilibria (i.e., ΔV_r^o). The dP/dT slopes given at low pressures in Figure 3.37 were derived from 1 atm. ΔS_r values. However, their computed slope of -19.5 bar/K shown in Figure 3.37 for the

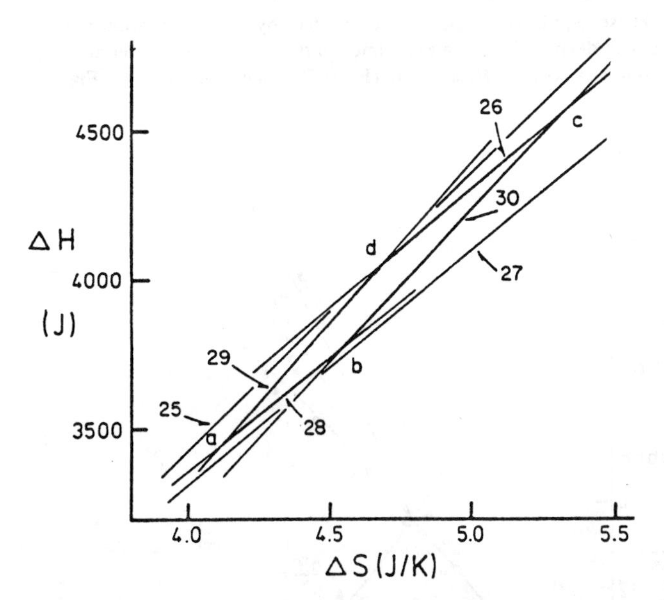

Figure 3.34. Day and Kumin's (1980) plot of enthalpies and entropies of formation (from the elements) for the andalusite ⇄ sillimanite reaction at 298K and 1 bar. The numbered lines correspond to experimental data points for the andalusite ⇄ sillimanite reaction (25 to 28 are from Holdaway, 1971, whereas 29 and 30 are from Weill, 1966). The polygon bounded by the apices: a, b, c and d, represents the *feasible solution space* that is compatible with all of the experimental constraints. (From Day and Kumin, 1980, Fig. 2).

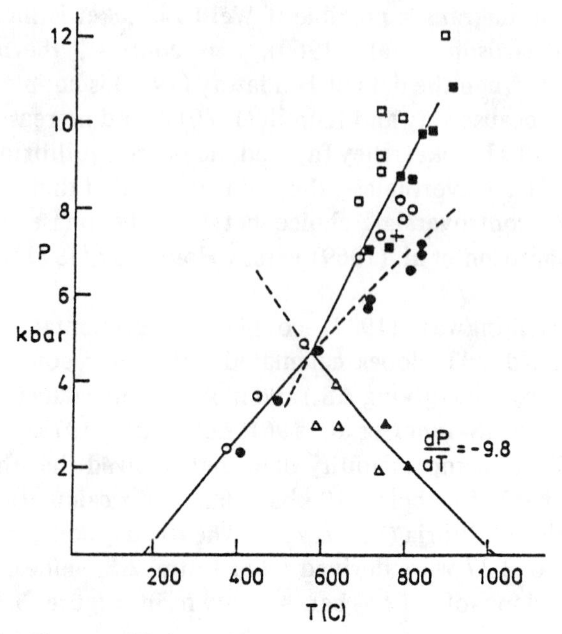

Figure 3.35. Phase equilibrium diagram computed by Day and Kumin (1980) from thermodynamic data derived from the experimental data of Richardson et al. (1969) on the andalusite ⇄ sillimanite reaction (triangles). (From Day and Kumin, 1980, Fig. 4).

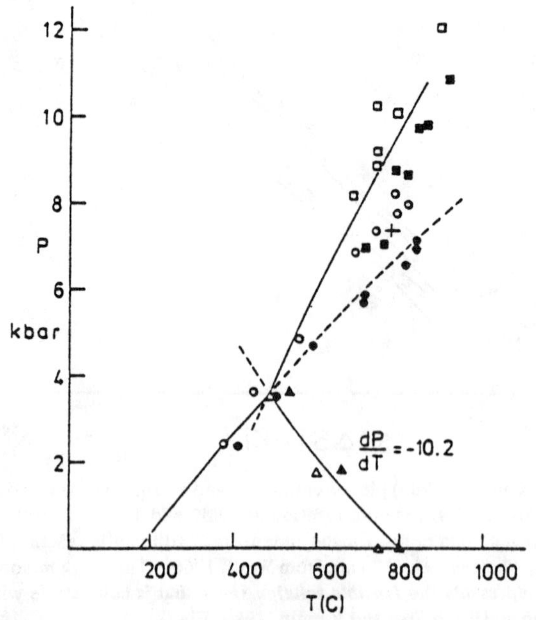

Figure 3.36. Phase equilibrium diagram computed by Day and Kumin (1980) from thermodynamic data derived from the experimental data of Holdaway (1971) and Weill (1966) for the andalusite ⇄ sillimanite reaction (triangles). (From Day and Kumin, 1980, Fig. 3).

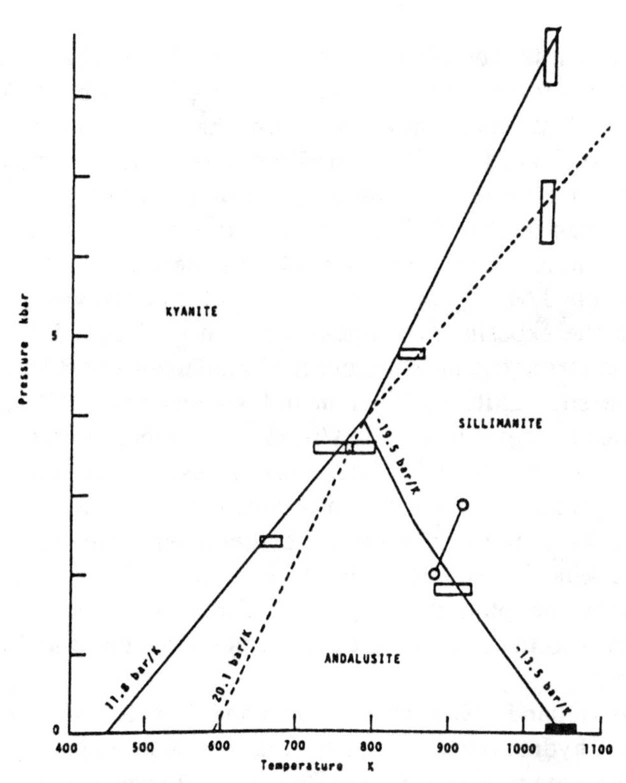

Figure 3.37. Phase equilibrium diagram showing experimental brackets for the Al_2SiO_5 equilibria (rectangles) and the univariant equilibria computed with the Clapeyron equation using entropies derived from the heat capacity measurements of Robie and Hemingway (1984). (From Robie and Hemingway, 1984, Fig. 5).

Table 3.6. Comparison of thermodynamic data derived by Robie and Hemingway (1984) with that derived from molten salt calorimetry (Anderson and Kleppa, 1969, and Anderson et al., 1977). (From Robie and Hemingway, 1984, Table 8).

Property		Kyanite	Andalusite	Sillimanite
$\Delta H^\circ_{f,298}$	kJ	-2596.01 ±1.70	-2591.90 ±1.72	-2587.77 ±1.73
$\Delta G^\circ_{f,298}$	kJ	-2445.12 ±1.69	-2443.72 ±1.71	-2440.90 ±1.72
$\Delta S^\circ_{f,298}$	J/(mol·K)	- 506.08 ±0.24	- 497.00 ±0.25	- 492.60 ±0.25
S°_{298}	J/(mol·K)	82.30 ±0.13	91.39 ±0.14	95.79 ±0.14
$\Delta H^\circ_{r,970}$	kJ	0	3.46 (3.34)[a,b]	
		0		6.52 (5.94)[a] (6.36)[b]
			0	3.06 (2.80)[b]

a. Anderson and Kleppa (1969)

b. Anderson, Newton, and Kleppa (1977)

andalusite-sillimanite equilibrium at Holdaway's (1971) triple point (501°C, 3.76 kbar) incorporates the change in ΔS_r from 1 bar to 3.76 kbar [= $\int \Delta(\alpha V)\, dP$]. It is important to emphasize that Robie and Hemingway (1984) accepted Weill's (1966) and Holdaway's (1971) experimental brackets on the andalusite-sillimanite equilibrium. A P-T location of the andalusite-sillimanite equilibrium that differs from these experimental studies would necessitate alteration of the derived thermodynamic parameters (Table 3.6). Their values for ΔH_r (Table 3.6) were computed by combining the experimental brackets shown in Figure 3.37 with the 700°C, 1 atm calorimetric measurements of Anderson and Kleppa (1969) for the kyanite-sillimanite equilibrium and Anderson et al. (1977) for the kyanite-andalusite equilibrium. Within the uncertainties in these experimental and calorimetric data, they adjusted ΔH_f for all three polymorphs to give a thermodynamically consistent set of ΔH_f values (B.S. Hemingway, pers. comm.). Their phase diagram may *inadvertently* imply that there is a kink in the andalusite-sillimanite equilibrium boundary. Rather, the two slopes plotted in Figure 3.37 for this equilibrium represent limiting *tangents* to the univariant curve at the triple point and at 1 bar.

Powell and Holland (1985) and Holland and Powell (1985) integrated experimental hydrothermal equilibrium bracketing data with thermochemical data to derive a thermodynamically consistent dataset for silicate minerals. Their strategy was to ignore calorimetrically-determined ΔH_r data; thus, they treated enthalpies as unknowns. For an isobaric experimental bracket, they computed ΔH_r at the upper and lower limits of the bracket with the equation

$$\Delta H_r = T\Delta S_r - P[\Delta V_r + \Delta(\alpha V)(T - 298) - \Delta(\beta V)P/2]$$

$$- \int_{298}^{T} \Delta C_p\, dT + T \int_{298}^{T} (\Delta C_p/T)dT - \sum r_i\, RT\, ln\, f_i - RT\, ln\, K\ .$$

It is instructive to review their treatment of the data used to compute the terms on the right side of the above equation. They computed ΔS_r from $\Delta S_{vib} + \Delta S_{cfg}$, where ΔS_{vib} is the vibrational entropy (derived from heat capacity calorimetry), and ΔS_{cfg} is the configurational entropy arising from Al-Si disorder. For each of the 60 experimentally determined equilibria they computed ΔS_{cfg} only if the third law entropies yielded Clapeyron slopes (dP/dT) that are inconsistent with experimental brackets. In such cases, ΔS_{cfg} was computed as the *minimal* value yielding an equilibrium that passes through all of the experimental brackets. For phases lacking measured coefficients of thermal expansion (α) and/or coefficients of isothermal compressibility (β), they estimated these parameters using measured values for phases of similar crystal structure (i.e., sheet silicates, chain silicates, etc.). For pure H_2O and CO_2 they fitted the quadratic: RT *ln* f = a + bt + cT^2 to available P-V-T data. For H_2O-CO_2 mixtures the activity data of Kerrick and Jacobs (1981) was fitted to a subregular (asymmetric) solution model. For a given system, their computational strategy involved defining vectors containing ΔH_f of the various phases involved (**h**), vectors of the enthalpy changes of the reactions (**b**), and a matrix of reaction coefficients (**R**). Their computational strategy was to compute **h** by the least squares minimization of: $||\mathbf{b} - \mathbf{Rh}||^2$. Figure 3.38a shows Powell and Holland's (1985) least-squares solution to the enthalpies of paragonite and albite. As input, this solution involved five equilibria involving albite and/or paragonite. Further refinement involved inclusion of a weighting factor for location of equilibria within brackets - they arbitrarily selected a temperature equivalent to 1/4 of the temperature interval of each bracket. The centroid of this ellipse (Fig. 3.38b) yielded their reported ΔH_f values for albite and paragonite. Figure 3.39 summarizes Holland and Powell's (1985) analysis of the Al_2SiO_5 phase equilibria. Because the experimental brackets of Richardson et al. (1969) on the andalusite-sillimanite equilibrium are thermodynamically inconsistent with those of Holdaway (1971), Powell and Holland (1985) did not include experimental data on the andalusite-sillimanite equilibrium in their computation of the enthalpies of the polymorphs. Rather, their input included only experimental data for the kyanite-sillimanite and kyanite-andalusite equilibria. Their computed triple point and andalusite-sillimanite equilibrium are between those of Richardson et al. (1969) and Holdaway (1971). In light of the critique of the experimental data of Richardson et al. (1969) and Holdaway (1971) on the andalusite-sillimanite equilibrium, presented earlier in this chapter, the author supports Powell and Holland's (1988) exclusion of the data of Richardson et al. (1969). However, I contend that Heninger's (1984) experiments provide *upper* temperature limits on the andalusite-sillimanite equilibrium. Because Powell and Holland's (1988) location of the andalusite-sillimanite equilibrium is at significantly higher temperatures than Heninger's (1984) *apparently* definitive andalusite → sillimanite data points (Fig. 3.39), I suggest that Holland and Powell's (1988) calibration of the Al_2SiO_5 P-T phase equilibrium diagram (Fig. 3.39) is in error.

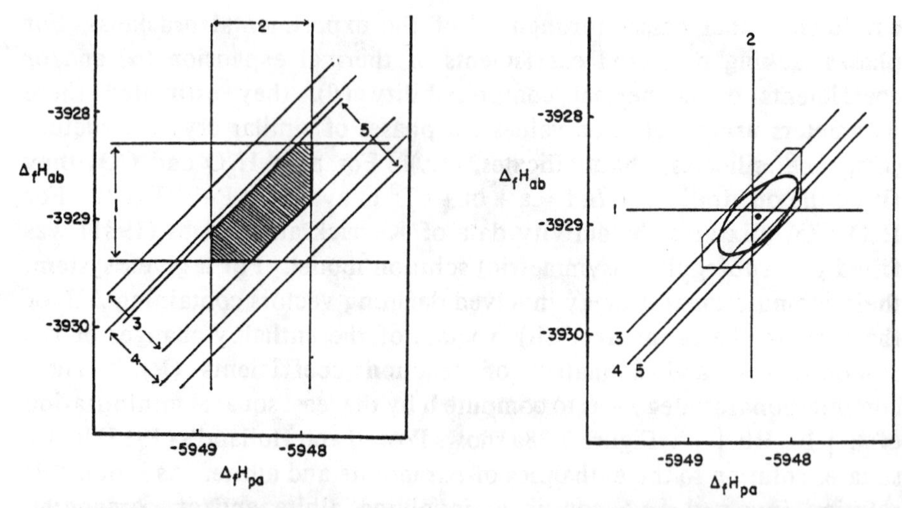

Figure 3.38. (a) Powell and Holland's (1985) least-squares solution of the "feasible solution space" (hachured area) for the enthalpies of formation of albite ($\Delta_f H$) and paragonite ($\Delta_f H_{pa}$) derived by experimental brackets on five equilibria involving albite and/or paragonite. (From Powell and Holland, 1985, Fig. 5). (b) Covariance matrix error ellipse and centroid (dot) within the feasible solution space (polygonal area corresponding to the hachured region in Fig. 3.38a). (From Powell and Holland, 1985, Fig. 6).

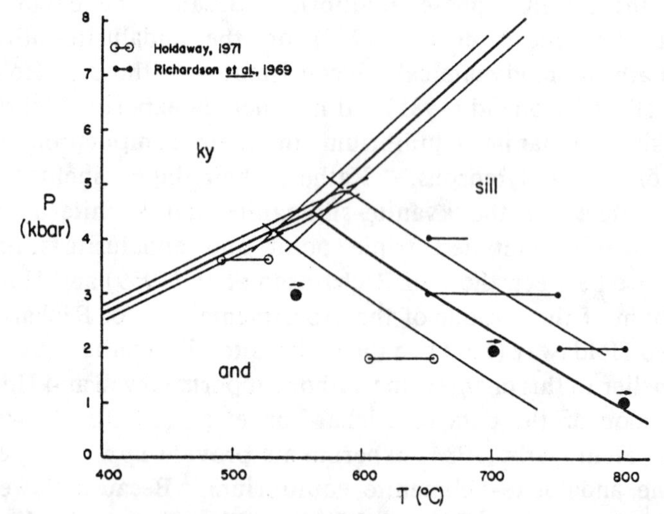

Figure 3.39. Holland and Powell's (1985) calculated Al_2SiO_5 phase equilibrium diagram. The horizontal lines bounded by filled circles are the experimental brackets of Richardson et al. (1969) for the andalusite-sillimanite equilibrium. The horizontal lines terminated by open circles are Holdaway's (1971) experimental brackets for this equilibrium. The thin horizontal lines are 2σ error estimates for experimental brackets on the andalusite-sillimanite equilibrium. Intersection of the estimated error envelopes about the three univariant equilibria yields the oval-shaped error envelope encompassing the triple point. The large filled circles with arrows are Heninger's (1984) experimental determination of the andalusite → sillimanite reaction. (From Holland and Powell, 1985, Fig. 5).

Using an extension of the linear programming method outlined by Gordon (1973), Berman (1988) derived an internally consistent thermodynamic data base for 67 minerals. Input for the solutions involved hydrothermal experimental equilibrium brackets, and measured heat capacities and molar volumes. Compared to other methods of extracting thermodynamic data from phase equilibrium data, the mathematical programming technique used by Berman (1988) explicitly accounts for uncertainties in experimental bracketing data. Final solutions were obtained by optimizing the nonlinear objective function

$$\sum_{i}^{n}(X_i - M_i)^2/W_i ,$$

where M_i are measured H, S and V data, W_i are associated variances of this data, and X_i are calculated values of H, S and V. In his analysis of the Al_2SiO_5 polymorphs, Berman (1988) cited Kerrick and Heninger (1984) and Salje (1986) in justifying his selection of Holdaway's (1971) location of the andalusite-sillimanite equilibrium. Accordingly, the andalusite-sillimanite equilibrium computed with Berman's data base is constrained by Holdaway's 1.8 kbar and 3.6 kbar experimental brackets and Weill's (1966) 1 atm bracket (Fig. 3.40), and Berman's computed triple point (3.73 kbar and 506°C) is nearly identical to that of Holdaway (i.e., 3.76 kbar and 501°C).

Coupling entropies computed from heat capacity measurements of the aluminum silicates with experimental hydrothermal data, Hemingway et al. (ms.) reevaluated the Al_2SiO_5 phase equilibria. From the Gibbs function

$$\frac{(G_T - H_{298})}{T} = \int_{298}^{T} C_p \, dT - T \int_{298}^{T} (C_p/T)dT - S_{298} ,$$

the Clapeyron slopes of the univariant equilibria were computed from

$$- \Delta H_{r,298} = T\Delta \left[(G_T - H_{298})/T\right] + \int_{1}^{P} \Delta V_{r,T} \, dP . \qquad [3.5]$$

Using equation [3.5], $\Delta H_{r,298}$ was adjusted to yield the best agreement with available experimental data. This approach yielded good agreement

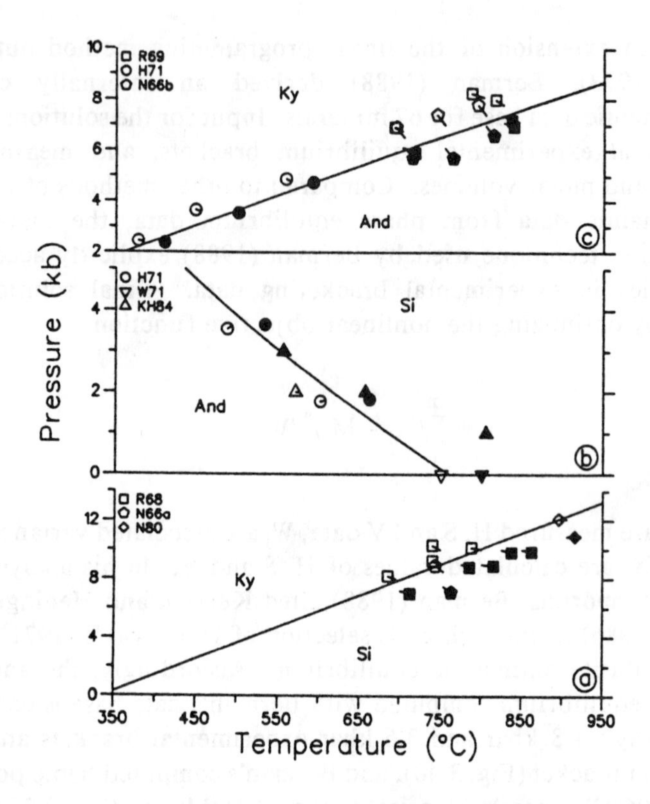

Figure 3.40. Comparison between univariant equilibria calculated with Berman's (1988) thermodynamic data base and experimental hydrothermal bracketing data. R68, R69 = Richardson et al. (1968, 1969); H71 = Holdaway (1971); N66a = Newton (1966b), N66b = Newton (1966a); W71 = Weill (as reported by Holdaway, 1971); KH84 = Kerrick and Heninger (1984); N80 = Newton (reported by Day and Kumin, 1980). In each diagram the filled symbol represents runs in which the high temperature polymorph formed whereas the open circles represent the reverse reaction. (From Berman, 1988, Fig. 5).

with the kyanite-andalusite and andalusite-sillimanite equilibria; however, the computed kyanite-sillimanite equilibrium has a steeper slope than the equilibrium constrained by experimental brackets. In order to yield agreement with experimental data it was necessary to change the 298.15K entropy of kyanite by +0.5 J/(mol·K) and that of sillimanite by -0.5 J/(mol·K). These adjustments are within the uncertainties in 298.15K entropies derived from calorimetry. The computed Al_2SiO_5 triple point is at 511°C and 3.87 kbar (i.e., very close to that of Holdaway, 1971). However, if we consider the P-T uncertainties of the experimental brackets, the triple point computed by Hemingway et al. (ms.) is subject to an *approximate* error of ±500 bars and ±25°C. The ΔH_f data of Topor et al. (1989) is inconsistent with the thermodynamic analysis of Hemingway et al. (ms.).

Using a piston-cylinder apparatus with NaCl pressure medium, Peterson and Newton (1990) bracketed the kyanite = corundum + quartz equilibrium at 8.5±5°C and 7.75±0.25 kbar. Coupling their experimental data with the entropy and volume data of Robie et al. (1978) and Robie and Hemingway (1984) they derived a value of $\Delta H^{o}_{f,298}$ for kyanite that is in good agreement with that of Berman (1988).

CALIBRATION OF THE Al_2SiO_5 PHASE EQUILIBRIA WITH MINERAL PARAGENETIC DATA

There have been several attempts to calibrate the Al_2SiO_5 P-T equilibrium relations by estimating pressure and temperature from mineral assemblages and thermobarometry.

Schuiling (1957) provided one of the earliest attempts to quantify the aluminum silicate phase equilibria by geologic thermobarometry. He based his calibration of the Al_2SiO_5 phase equilibria on estimates of the P-T conditions of various isograds, metamorphic facies, and mineral assemblages, combining Heim's (1952) estimates of the P-T conditions of selected isograds and metamorphic facies with Bowen's (1940) P-T diagram for equilibria in metamorphosed siliceous dolomites (Table 3.7). For each sample plotted in Figure 3.41, temperature was chosen from geothermometric evidence, whereas pressure was determined by the intersection of the isotherm with an estimated geothermal gradient. Schuiling (1957) also considered pressures computed from estimates of the total thickness of overburden during metamorphism. In referring to Figure 3.41, Schuiling (1957, p. 226) concluded: "The constructed diagram may be used with confidence as an aid in determining the conditions of metamorphism in metamorphic rocks".

Hietanen (1956, 1961, 1962) carried out an extensive analysis of metamorphosed rocks in northern Idaho. All three polymorphs are common in metapelitic rocks of the Precambrian Prichard Formation. Hietanen (1956, p. 1) concluded: "In some thin sections all three [Al_2SiO_5] modifications occur side by side, suggesting that they were crystallized close to the physical-chemical conditions in which all three may exist together". Accordingly, Hietanen (1961) used independent thermometry and barometry of other mineral assemblages in rocks of this area to estimate the Al_2SiO_5 triple point at 500°C and 5000 atmospheres. Because

Table 3.7. Schuiling's (1957) estimates of the depth of burial and P-T conditions for selected isograds and metamorphic facies. (From Schuiling, 1957, Table 1).

	km	°C	atm.
Highest part of region greenschist facies	9.5	400	2610
Biotite isograde, top of epid.-amph.facies	11.5	480	3165
Staurolite isograde	12.5	520	3435
K-feldspar isograde, top of amph.-facies	13	540	3575
Sillimanite isograde	13.5	560	3715
Corderierite compatible with K-feldspar	14.5	600	3985
Anatexis becomes general in gneisess	16	660	4400
Top of hornfels facies	17.5	725	4815

this estimate was based on relatively limited thermobarometric data it is subject to considerable uncertainty. Her awareness of this uncertainty is indicated by the following quotes from Hietanen (1961, p. 94- 95): "For *illustrative* purposes the triple point was *assumed* to be near 500°C"..."It was *assumed*, for *illustrative* purposes, that the pressure of this point was about 5000 atm"[italics added by D.M. Kerrick].

Aluminum silicates in Precambrian metapelites of the Truchas Range, New Mexico, suggest that metamorphism occurred at P-T conditions essentially coincident with the Al_2SiO_5 triple point (Grambling, 1981). Grambling (1981) used experimental dehydration equilibria involving ferromagnesian silicates to estimate that metamorphism occurred at 500-575°C and 3.5-4.5 kbar (Fig. 3.42). As shown in Figure 3.42, this P-T estimate is in considerably better agreement with Holdaway's (1971) location of the triple point than with that of Richardson et al. (1969). In using this as an argument favoring Holdaway's (1971) triple point over that of Richardson et al. (1969) we must temper this conclusion with possible uncertainties in Grambling's (1981) thermobarometry. In particular, Grambling derived activities of components in ferromagnesian phases by assuming ideal multisite mixing. Significant non-ideal mixing in these phases would markedly alter the P-T estimated from equilibria involving these phases.

Grambling (1984) considered paragonite-bearing metapelitic rocks in New Mexico as independent evidence regarding the P-T location of the triple point. The assemblage paragonite + quartz + sillimanite (in textural equilibrium) occurs in four samples studied by Grambling (1984). From Chatterjee's (1972) experimental determination of the paragonite + quartz = albite + Al_2SiO_5 + H_2O equilibrium, Grambling (1985) argued that the assemblage paragonite + quartz + sillimanite is stable with Holdaway's (1971) Al_2SiO_5 phase equilibrium diagram. However, as shown in Figure

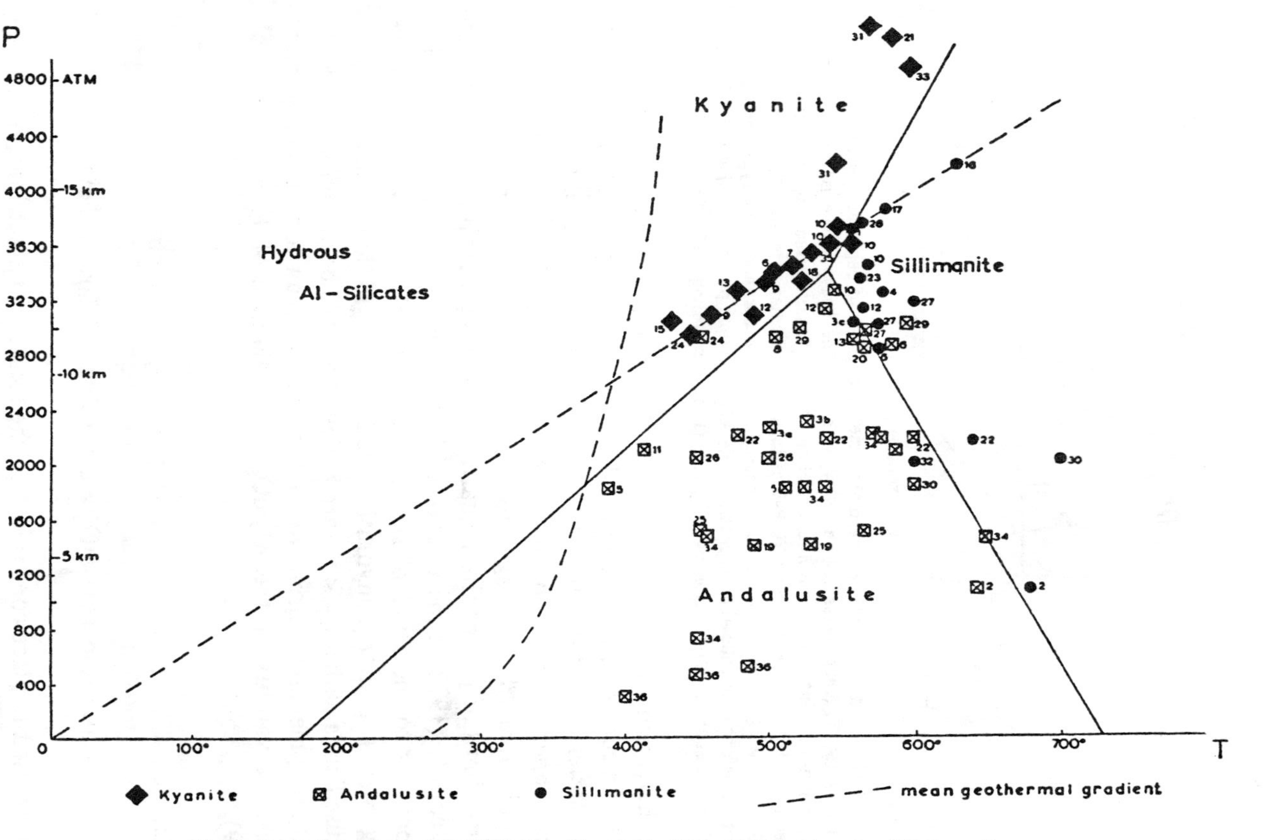

Figure 3.41. Schuiling's (1957) "field calibration" of the Al_2SiO_5 phase equilibrium diagram. The ordinate gives pressure (in atm) and depth (in km). (From Schuiling, 1957, Fig. 1).

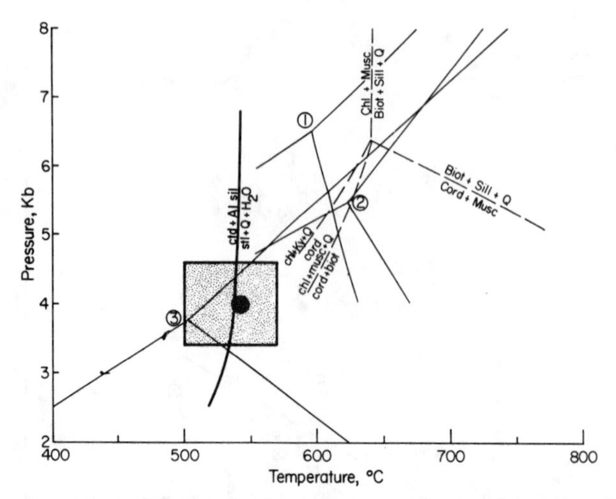

Figure 3.42. Grambling's (1981) thermobarometric analysis of regionally metamorphosed pelites in the Truchas Peaks region, New Mexico. The heavy curve is the calculated equilibrium relevant to the staurolite-chloritoid-Al_2SiO_5-quartz assemblage in a quartzite, whereas the filled circle represents his P-T estimate for a cordierite schist. The stippled area is Grambling's (1981) estimate of the uncertainty in the thermobarometry. The circled numbers refer to the following estimates of the Al_2SiO_5 triple point: (1) Althaus (1967), (2) Richardson et al. (1969), (3) Holdaway (1971). (From Grambling, 1981, Fig. 15).

3.43, this assemblage is not stable with the triple point of Richardson et al. (1969). In considering the chemical compositions of paragonites, coupled with Chatterjee and Froese's (1975) analysis of the effect of solid solution on the upper stability of paragonite, Grambling (1984) concluded that non-stoichiometry in paragonite would not significantly alter the upper stability limit of paragonite + quartz. Grambling's analysis was based on Chatterjee's (1972) experiments with a_{H_2O} = 1.0. Addition of other components to the fluid phase (e.g., CO_2, NaCl, etc.) would shift the paragonite + quartz dehydration equilibrium to lower temperatures. Because fluids with $a_{H_2O} < 1.0$ are believed to have been common in rocks undergoing metamorphism, we are provided with an additional argument against the validity of the Al_2SiO_5 phase diagram of Richardson et al. (1969).

Coupled with assemblages in metapelites, the equilibrium

$$\text{Chloritoid} + Al_2SiO_5 = \text{Staurolite} + \text{Quartz} + \text{Water}$$

has been used to place constraints on the location of the triple point. The assemblage chloritoid + sillimanite has been noted in several localities (Holdaway, 1978; Grambling, 1981, 1983; Milton, 1986). This assemblage is stable with Holdaway's (1971) Al_2SiO_5 phase equilibrium diagram but

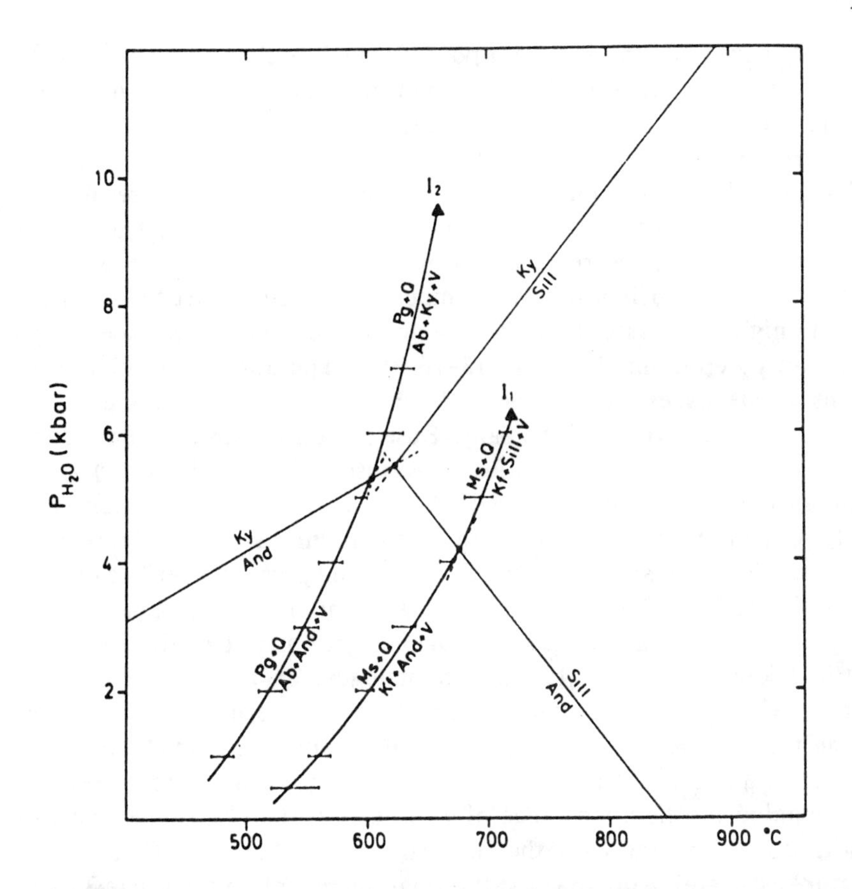

Figure 3.43. Experimental locations of the paragonite + quartz and muscovite + quartz dehydration equilibria. (From Chatterjee and Froese, 1975, Fig. 1). Note that the Al_2SiO_5 phase diagram of Richardson et al. (1969) does not yield a stability field for the assemblage paragonite + sillimanite + quartz that occurs in some metapelites (Grambling, 1984).

is precluded with the Al_2SiO_5 phase diagram of Richardson et al. (1969). Milton (1986) argued that these conclusions are unaltered by consideration of the effect of solid solution in chloritoid and staurolite.

The equilibrium

$$\text{Muscovite} + \text{Quartz} = Al_2SiO_5 + \text{K-feldspar} + H_2O$$

is also useful in analyzing the Al_2SiO_5 equilibria. Numerous experimental studies (Schramke et al., 1987) support the validity of Chatterjee and Johannes' (1974) experimental equilibrium curve for the upper stability limit of muscovite + quartz. In numerous contact aureoles the prograde index mineral sequence: andalusite → sillimanite → K-feldspar, suggests pressures exceeding the invariant point formed by the intersection of the andalusite-sillimanite equilibrium with the muscovite + quartz dehydration equilibrium. For this invariant point, Holdaway's (1971) andalusite-sillimanite equilibrium yields a pressure of about 2 kbar, whereas the andalusite-sillimanite equilibrium of Richardson et al. (1969) results in a pressure above 4 kbar (Fig. 3.44). In the likely case that the fluid phase was not pure H_2O, the resultant displacement of the muscovite+quartz dehydration equilibrium to lower temperatures would shift the invariant point to higher pressures and lower temperatures. Pressures exceeding 4 kbar imply depths in the range 13-16 km. The author considers such depths to be excessive for many "high-level" contact aureoles. For example, in the aureole of the Santa Rosa pluton in Nevada, breakdown of muscovite + quartz to sillimanite + K-feldspar (i.e., equilibrium) occurs simultaneously with the andalusite → sillimanite transformation, thus implying conditions close to the invariant point shown in Figure 3.44. During metamorphism the depth of cover at the present level of exposure was estimated by Compton (1960) as 3-8 km (P = 1-2.5 kbar). This analysis favors Holdaway's (1971) andalusite-sillimanite equilibrium over that of Richardson et al. (1969). An analogous example is provided by Elan's (1985) study of the contact aureole of the Kernville pluton in the southern Sierra Nevada. As in the Santa Rosa aureole, Elan (1985) found that the transformation from the andalusite to sillimanite zones was accompanied by the appearance of K-feldspar, thereby suggesting P-T conditions coincident with the invariant point shown in Figure 3.44. Pressure estimates using the GASP geobarometer fall in the range 3.0±0.8 kbar. As with the Santa Rosa aureole, these pressures support Holdaway's (1971) andalusite-sillimanite equilibrium, but are incompatible with the andalusite-sillimanite equilibrium of Richardson et al. (1969). Rubenach and Bell (1988) contended that the K-feldspar-muscovite-andalusite-sillimanite assemblages in high-grade gneisses of the aureole of the Tinarroo Batholith (North Queensland, Australia) formed at the invariant point shown in Figure 3.44. With Holdaway's (1971) Al_2SiO_5 phase equilibrium diagram, and assuming a pressure of 2.5 kbar, garnet-biotite thermometry of two lower grade (andalusite zone) rocks yielded temperatures just below the invariant point (Fig. 3.45). "The Richardson, Gilbert and Bell (1969) triple point is less satisfactory...as it gives a temperature for the gneisses of 650°C, which is 50°C higher than the garnet-biotite temperatures in the adjacent schists" (Rubenach and Bell, 1988, p. 657-658).

The foregoing analysis assumes that the muscovite + quartz dehydration equilibrium corresponds to that of Chatterjee and Johannes (1974) who

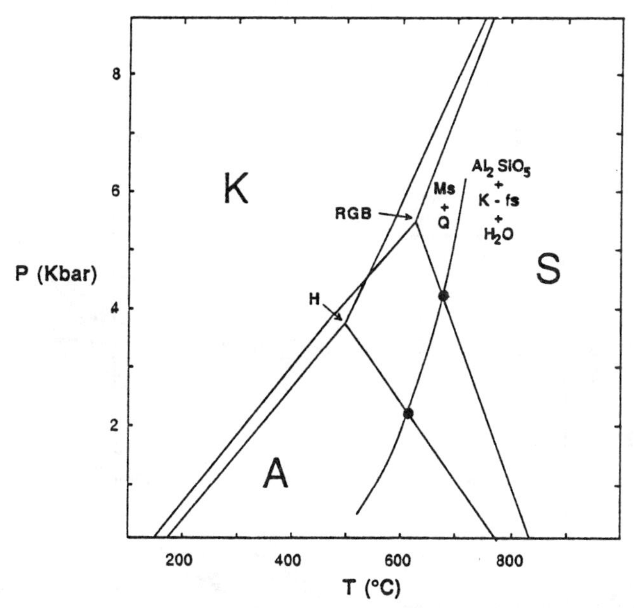

Figure 3.44. Phase equilibrium diagram showing the invariant point (filled circles) formed by the intersection of the equilibrium muscovite + quartz = Al_2SiO_5 + sanidine + water (from Chatterjee and Johannes, 1974) and the andalusite-sillimanite equilibrium determined by Holdaway (H) versus Richardson et al. (RGB).

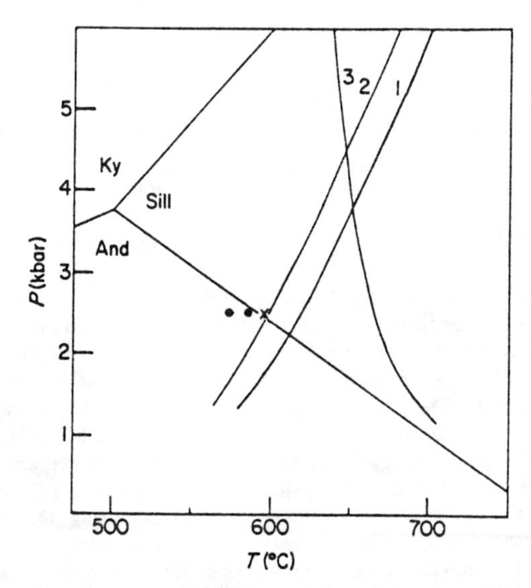

Figure 3.45. Rubenach and Bell's (1988) thermobarometric analysis of metapelites in the aureole of the Tinaroo Batholith, north Queensland, Australia. The "x" corresponds to the high-grade andalusite-sillimanite-K-feldspar-muscovite gneiss, whereas the filled circles were derived from garnet-biotite thermometry (assuming P = 2.5 kbar) of two andalusite-zone rocks. Curves 1 and 2 represent Kerrick's (1972) determination of the muscovite + quartz = Al_2SiO_5 + K-feldspar + H_2O equilibrium for X_{H_2O} values of 1.0 and 0.8, respectively. Curve 3 is the water-saturated solidus for granodiorite. (From Rubenach and Bell, 1988, Fig. 7).

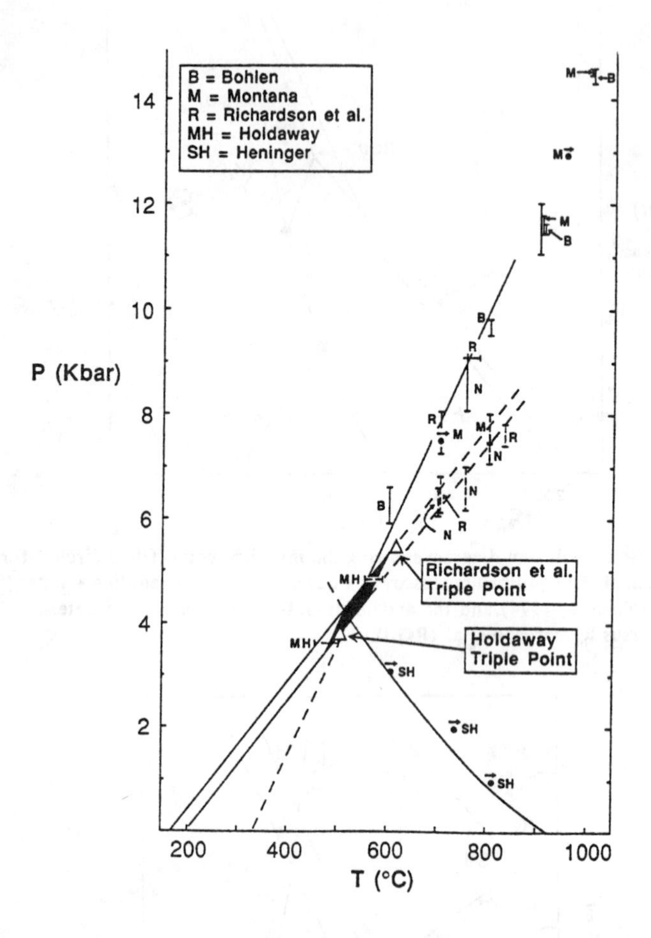

Figure 3.46. Summary of experimental data providing constraints on locating the Al_2SiO_5 univariant equilibria and the triple point. The dots with arrows, and the solid brackets, refer to the following experimental studies on the kyanite ⇄ sillimanite reaction: B = S.R. Bohlen (Bohlen et al., ms.); M = A.L. Montana (Bohlen et al., ms.); N = Newton (1969); R = Richardson et al. (1968). The dashed brackets correspond to the following experimental studies on the kyanite ⇄ andalusite reaction: M = A.L. Montana; MH = Holdaway (1971); N = Newton (1966a); R = Richardson et al. (1969). The dots labeled "SH" are Heninger's (1984) determination of the reaction: andalusite → sillimanite. The kyanite-sillimanite equilibrium, and the parallel lines delimiting error envelopes encompassing the kyanite-andalusite equilibrium, are constrained by the experimental data and were calculated by J.A.D. Connolly using the VERTEX program of Connolly and Kerrick (1987). The black polygon locating the triple point is formed by the intersection of the error bands encompassing the kyanite-sillimanite and kyanite-andalusite equilibria (see text). The andalusite-sillimanite equilibrium is constrained by Heninger's (1984) 3 kbar data point.

bracketed this equilibrium using synthetic, stoichiometric muscovite and K-feldspar as starting materials. Thus, their curve for this equilibrium corresponds to that with K_{eq} = 1.0. However, we must consider the effect of solid solution in natural muscovite and K-feldspar when applying Chatterjee and Johannes'(1974) equilibrium to metapelites. Elan's (1985) electron probe analyses of muscovite and K-feldspar in most samples of metapelites of the Kernville aureole reveal similar K/(K + Na) ratios for these phases. Therefore, it is reasonable to assume that the reduced activity in muscovite is compensated by that in K-feldspar, such that the equilibrium constant is not significantly different from unity.

SUMMARY AND CONCLUSIONS

The preceding review highlights the difficulties in the quantification of the P-T phase equilibria involving the Al_2SiO_5 polymorphs. In spite of such difficulties, a judicious evaluation of existing experimental and thermodynamic data provides reasonable constraints on the location of the Al_2SiO_5 univariant phase equilibria and the triple point.

The triple point can be located by considering recent experimental data on the kyanite-sillimanite and kyanite-andalusite equilibria, coupled with P-T slopes calculated with the Clapeyron equation (computed from the entropy data of Hemingway et al., ms., and the molar volume data of Winter and Ghose, 1979). The error band for the kyanite-andalusite equilibrium shown in Figure 3.46 was derived from the constraints of the experimental brackets coupled with the calculated Clapeyron slope. The kyanite-sillimanite equilibrium shown in Figure 3.46 represents the best fit of the computed equilibrium with the experimental brackets at 600-800°C. The resultant location of the triple point (black polygon in Figure 3.46) was derived by *arbitrarily* assuming a ±20°C uncertainty on the location of the kyanite-sillimanite equilibrium, and intersection of the resultant error band with that of the kyanite-andalusite equilibrium. This error polygon does not consider the precision uncertainties of C_p measurements, errors in molar volumes, or the expansion of experimental brackets arising from uncertainties in the measurement and control of run pressures and temperatures. Addition of such uncertainties could expand the error envelope by 40°C and 400 bars. This expanded error envelope encompasses the triple points of Richardson et al. (1969) and Holdaway (1971); however, the triple point of Brown and Fyfe (1971) is precluded.

In view of the earliest efforts to calibrate the Al_2SiO_5 phase equilibrium diagram (e.g., Schuiling, 1957), and the considerable subsequent research to improve this calibration by experimental and thermodynamic methods, it is ironic that this chapter concludes by turning to field-oriented studies. The extensive petrologic analyses by J.A. Grambling and colleagues of the triple point parageneses in several mountain ranges in northern New Mexico (Grambling, 1981; Grambling and Williams, 1985; Grambling et al., 1989) serves as an excellent field calibration of the triple point. Their thermobarometry of rocks with triple point assemblages yields: Picuris Range: 505°C, 3.8 kbar; southern Manzano Range: 520°C, 4.5 kbar; Truchas Range: 530°C, 4.3 kbar; Rio Mora Range: 540°C, 4.5 kbar (Grambling et al., 1989). These estimates are generally compatible with Holdaway's triple point. However, these conclusions should be tempered by potentially significant uncertainties in thermobarometers. For example. Mössbauer spectroscopy reveals significant Fe^{3+} in biotites of metamorphosed pelites from northwestern Maine (Dyar, 1990). As shown by Dyar et al. (ms.), correction of the Ferry and Spear (1978) garnet-biotite geothermometer for Fe^{3+} in biotite reduces the computed temperatures by 25-35°C. Assuming that the Pre-Cambrian rocks in north-central New Mexico have significant Fe^{3+} contents in biotite (which is likely in light of the analyses of Dyar, 1990, Dyar et al., ms., and Guidotti and Dyar, in press), the computed temperatures would be notably lower than those of Holdaway's triple point. Nevertheless, this analysis would argue against a triple point higher in temperature than that of Holdaway (e.g., the black polygon in Fig. 3.46). In addition to errors in garnet-biotite geothermometry, the uncertainty in "field" estimates of the triple point pressure by geobarometry must also be considered. McKenna and Hodges (1988) carried out an error analysis of the garnet-plagioclase-Al_2SiO_5-quartz geobarometer. They concluded that even with the addition of the "tight" experimental brackets of Koziol and Newton (1988), there is a ±2.5 kbar pressure uncertainty with this geobarometer. In light of Figure 3.46, a pressure uncertainty of this magnitude would render this geobarometer of little use in the P-T location of the triple point.

Brown and Fyfe's (1971) triple point (Fig. 3.22) yields a very confined P-T stability field for andalusite. I consider this restricted stability field for andalusite to be unlikely in view of the widespread abundance of andalusite in contact aureoles and in regimes of low-pressure regional metamorphism. Furthermore, such a stability field for andalusite results in petrologically unreasonable implications when combined with other metamorphic equilibria. For example, considering the muscovite + quartz dehydration equilibrium, and the andalusite-sillimanite equilibrium deduced using Brown and Fyfe's (1971) experimental results on the kyanite-andalusite equilibrium, the invariant point shown in Figure 3.44

would be at very low pressures (P < 1 kbar). The implied depths of burial are unreasonably shallow for parageneses displaying prograde index mineral sequences suggesting either intersection with the invariant point (e.g., Rubenach and Bell, 1988) or pressures below this invariant point (e.g., the Onawa aureole, Maine; G. Symmes, unpub.). An additional petrologic argument against Brown and Fyfe's (1971) phase equilibrium diagram is that it does not provide a P-T stability field for andalusite-bearing peraluminous melts (see Chapter 11).

OTHER EQUILIBRIA IN THE Al_2SiO_5 SYSTEM

The equilibria

$$3Al_2SiO_5 = \underset{(3/2\ \text{Mullite})}{3Al_2O_3{\cdot}2SiO_2} + \text{Quartz}\ , \qquad [3.6]$$

and

$$Al_2SiO_5 = \text{Corundum} + \text{Quartz}\ , \qquad [3.7]$$

are relevant for the upper temperature stability limits of the Al_2SiO_5 polymorphs. In light of *synthesis* experiments, coupled with available entropy and molar volume data, Beger (1979) suggested two possible phase diagrams involving the above equilibria (Fig. 3.47). It is important to emphasize that these diagrams apply to stoichiometric sillimanite and 3/2mullite. As discussed in Chapter 4, variations in the Al/Si ratios of mullite and sillimanite will introduce additional variancy. Within the P-T stability field of sillimanite, Peterson and Newton (1990) bracketed the metastable kyanite = corundum + quartz equilibrium (Fig. 3.48). Their conclusion that corundum + quartz is always metastable in relation to the Al_2SiO_5 can be evaluated by *coupling* the kyanite = sillimanite and kyanite = corundum + quartz equilibria. From Figure 3.48, we assume the kyanite-sillimanite equilibrium to be at 9.75 kbars and 815°C, and the kyanite-corundum-quartz equilibrium at 7.75 kbars and 815°C. From the relation

$$\Delta G_r = \int_{9.75}^{7.75} \Delta V_r \, dP\ ,$$

ΔG_r = -1408 J for the kyanite-sillimanite reaction at 7.75 kbars and 815°C. Accordingly, ΔG_r = +1408 J for the sillimanite = corundum +

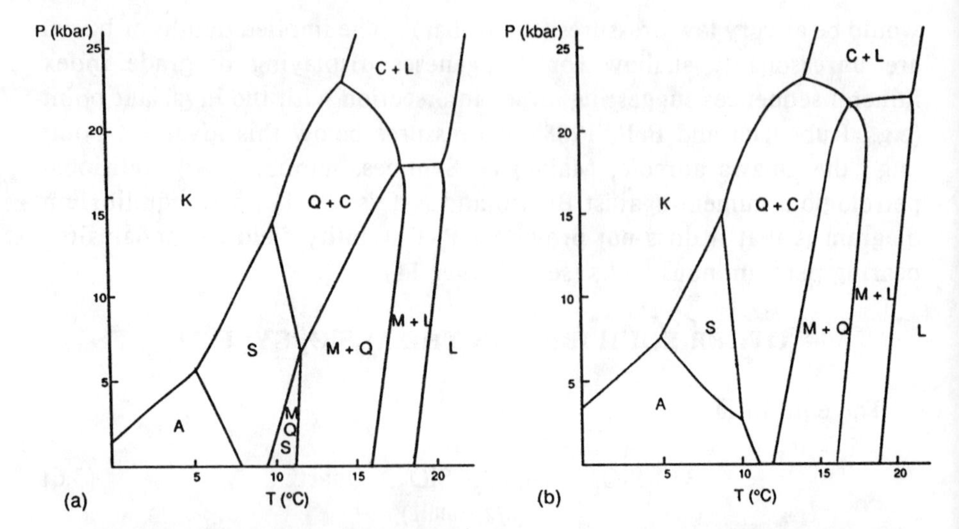

Figure 3.47. Beger's (1979) "possible" phase equilibrium diagrams for the Al_2SiO_5 system. (a) corundum + quartz field does not extend to zero pressure, (b) corundum + quartz field extends to zero pressure. Symbols for phases: A = andalusite, C = corundum, L = liquid, K = kyanite, M = mullite, Q = quartz; S = sillimanite. (Redrawn from Beger, 1979, Figs. 2.22 and 2.23).

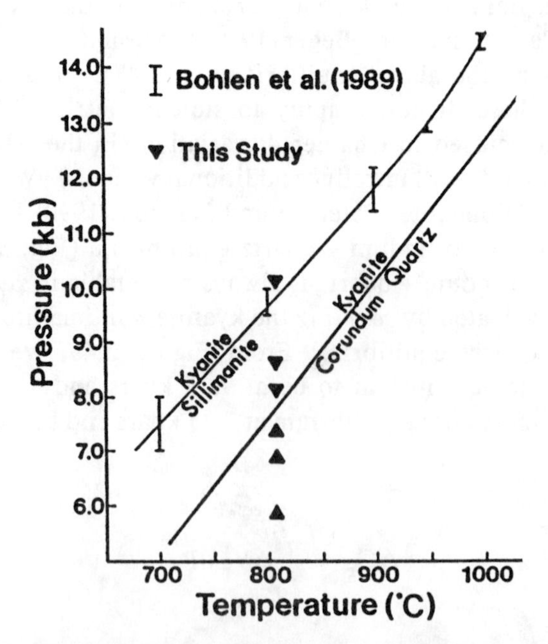

Figure 3.48. Peterson and Newton's (1990) experimental results on the kyanite ⇄ corundum + quartz reaction. The filled triangles represent their data points (▲ = kyanite → corundum + quartz; ▼ = corundum + quartz → kyanite). (From Peterson and Newton, 1990).

quartz reaction at this P-T condition. The entropy data of Hemingway et al. (ms.) and the molar volume data of Winter and Ghose (1979), coupled with the relation

$$\left[\Delta G_r\right]_{T_2}^{P_2} - \left[\Delta G_r\right]_{T_1}^{P_1} = \int_{P_1}^{P_2} \Delta V_T \, dP - \int_{T_1}^{T_2} \Delta S_r dT ,$$

confirms that sillimanite is more stable than corundum + quartz over the P-T range of geologic interest. Nevertheless, the corundum + quartz assemblage (in textural equilibrium) has been reported in several high-grade parageneses (Tracy and McLellan, 1985; Powers and Bohlen, 1985; Motoyoshi et al., 1990). These apparent metastable corundum + quartz parageneses serve as an example of the sluggish kinetics of a solid-solid reaction with small ΔG_r (see Chapter 9).

EPILOGUE

The preceding review is intended to provide a *critical* analysis of experimental, thermodynamic, and field data relevant to the calibration of the Al_2SiO_5 phase equilibria. I have provided alternatives to some of the interpretations given in many outstanding publications. My critique is *not* intended to downplay the quality or impact of these contributions. Rather, I wish to stimulate further research to shed light on controversial issues regarding the stability relations of the Al_2SiO_5 polymorphs.

This author is of the opinion that little further quantification of the Al_2SiO_5 phase equilibrium diagram is possible with additional calorimetric measurements using existing instrumentation. We await development of higher precision solution calorimeters for measuring the heats-of-solution of the refractory aluminum silicates. My recommended priority for future research on this problem is to obtain *tight* experimental brackets on the andalusite-sillimanite equilibrium using thoroughly characterized starting materials that are close to end member composition, meticulous characterization of run products, and incontrovertible monitors of reaction direction. SEM examination of starting materials and run products holds considerable promise in monitoring small amounts of this sluggish reaction. In so doing, careful attention must be given to the possibility of anisotropies in the morphology of growth and dissolution features.

CHAPTER 4

NON-STOICHIOMETRY

Chemical analyses of the Al_2SiO_5 polymorphs show that most are fairly close to end-member stoichiometry. However, because the stability relations are affected by relatively small perturbations in the chemical potentials of these phases (Fig. 1.11), minor deviations from end-member stoichiometry can have relatively large effects on the phase equilibrium relations. In addition to affecting the P-T stabilities of the polymorphs, non-stoichiometry can significantly affect the physical properties of these phases.

MAJOR ELEMENT NON-STOICHIOMETRY

In this section we examine the stoichiometry of Al, Si and O in the aluminum silicates. Because minor element solid solution is discussed in the following section, we exclude here major element non-stoichiometry produced by substitution of minor elements.

It has traditionally been assumed that all three polymorphs have essentially Al_2SiO_5 stoichiometry. In view of crystal structure refinements coupled with chemical analyses, this assumption is undoubtedly valid for kyanite and for andalusite with low concentrations of minor elements. However, because of solid solution with mullite, there are questions concerning the major element stoichiometry of sillimanite. The following discussion addresses this problem.

Crystal chemistry of sillimanite-mullite solid solution

Before reviewing geologic evidence bearing on sillimanite-mullite solid solutions, it is instructive to compare the crystal structures of sillimanite and mullite and to examine the mechanism by which sillimanite is transformed into mullite. From a comparison of the crystal structures of sillimanite and mullite, Burnham (1964b) concluded that there is no structural reason why there could not be complete solid solution between sillimanite and mullite. Sillimanite and mullite have similar crystal structures: both have chains of AlO_6 octahedra that are cross-linked by double chains of tetrahedrally coordinated Si and Al (Fig. 4.1). However, in comparison with sillimanite, mullite has oxygen vacancies, and the excess Al occupies a unique tetrahedral site. The transformation of sillimanite into mullite ("mullitization") may be visualized with the aid of Figure 4.2. It involves replacement of Si^{IV} by Al^{IV}, creation of vacancies

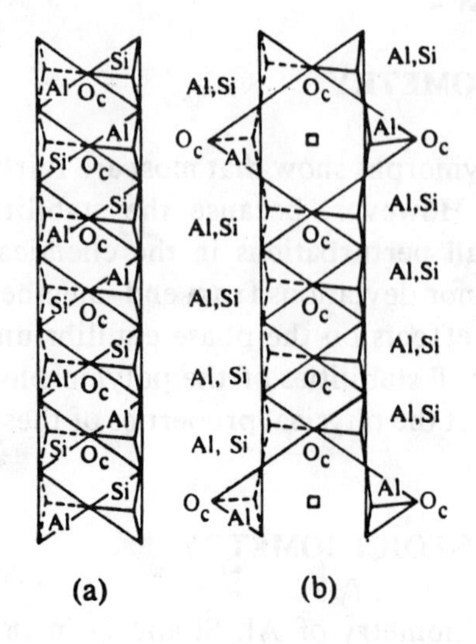

Figure 4.1. Comparison of the tetrahedral chains in sillimanite (a) and mullite (b). (From Deer et al., 1982, Fig. 309; copyright Longman Group Ltd.).

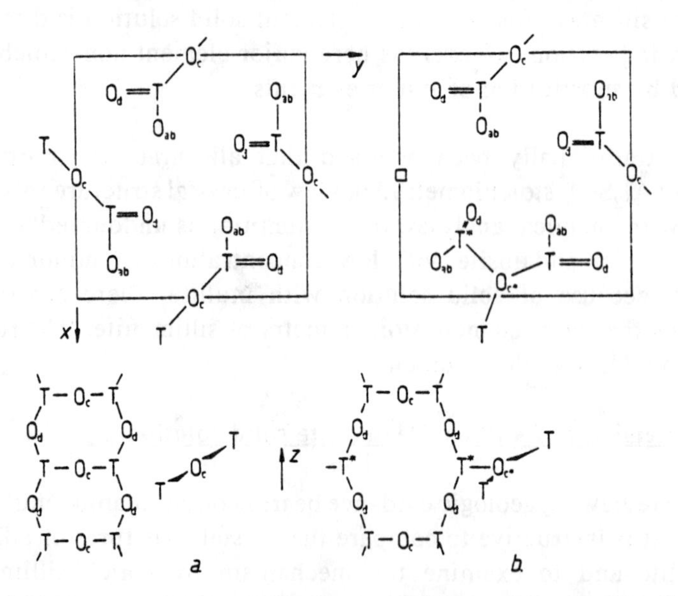

Figure 4.2. *Top*: (001) projection illustrating the mechanism of the transformation from sillimanite (left) to mullite (right). The open square in the upper right diagram represents an O_c oxygen vacancy. *Bottom*: Schematic illustration of the linkages between oxygens and tetrahedral cations in sillimanite (*a*) and mullite (*b*). (From Angel and Prewitt, 1986, Fig. 1).

by removal of O_c oxygens (i.e., oxygens bridging the double tetrahedral chains), and concomitant rearrangement of the tetrahedra (T^* in Fig. 4.2). This process is represented by the reaction $2Si^{4+} + O^{2-} = 2Al^{3+} + \square$, where $\square$ represents an oxygen vacancy. In expressing the mullite formula as $Al_2{}^{VI}[Al_{2+2x}Si_{2-2x}]^{IV}O_{10-x}$, which is supported by density determinations of mullites (Fig. 4.3), Cameron (1977a) considered mullite to have x values between 0.085 and 0.295. Although it is generally believed that the T^* site contains mostly Al, there is controversy regarding the proportion of Si on this site. Some workers (Sadanaga et al., 1962; Saalfeld and Guse, 1981) concluded that the T^* site contains only Al, whereas others (Burnham, 1964a; Durovic and Fejdi, 1976) considered that this site contains a small proportion of Si. Angel and Prewitt's (1986) X-ray structure refinement of mullite failed to resolve the question regarding the Al/Si occupancy of the T^* site.

The mullitization of sillimanite was studied by Guse et al. (1979) by X-ray examination of sillimanite heated at 1600°C (1 atm). After 5 h the X-ray reflections with $l = 2n + 1$ were weakened. After 16 h these reflections were absent and reflections for mullite appeared. Accordingly, the space group of the sillimanite changed from *Pbnm* to *Pbam*. The disappearance of the l-odd reflections was accompanied by a slight increase in the a cell dimension. Guse et al. (1979) concluded that the topotaxic sillimanite → mullite reaction involves thermally-induced disorder of Al and Si in the tetrahedral chains and little or no change in the AlO_6 octahedral chains. Sillimanite was completely transformed to mullite between 25 h and 30 h. Amorphous silica was produced by the incongruent transformation.

Evidence for sillimanite-mullite solid solution

From electron probe analyses of sillimanite and fibrolite from a variety of localities, Kwak (1971a) concluded that all of the sillimanite and fibrolite analyzed in his study had essentially stoichiometric Al/Si ratios, but he gave an analytical uncertainty of ±3 wt %. Deviations from non-stoichiometry within ±3 wt % uncertainty could have a significant effect on the stability relations of sillimanite (especially the andalusite-sillimanite equilibrium). Thus, further analysis of the stoichiometry of sillimanite is warranted.

Based on solvus relations, solid solution between sillimanite and mullite is expected to be more extensive with increasing temperature. Thus, Al enrichment of sillimanite should be most prevalent in the highest grades of metamorphism (i.e., the sanidinite facies). Indeed, Aramaki (1961) described fibrolite in pelitic xenoliths from the Asama volcano in Japan with a composition intermediate between sillimanite and mullite. There have been numerous subsequent papers suggesting that sillimanite may be enriched in Al.

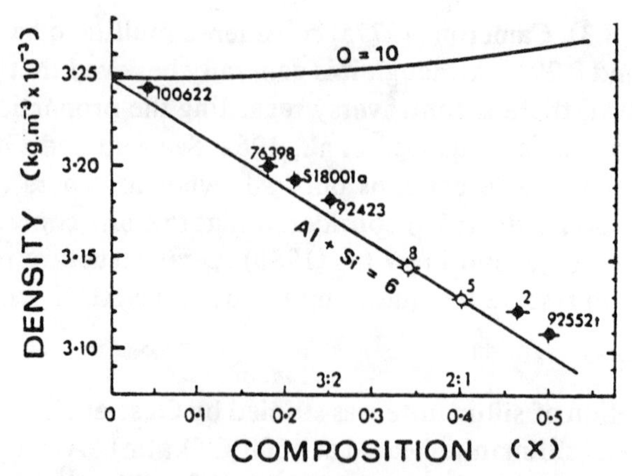

Figure 4.3. Densities of the sillimanite-mullite solid solution series. The circles represent experimentally measured values: filled circles = Fe, Ti-enriched; open circles = $Fe_2O_3 + TiO_2$ < 0.5 wt %. The straight line was calculated assuming a fixed number of cations per unit cell (Al + Si = 6), whereas the curved line was calculated on the basis of a fixed number of oxygens (O = 10). The composition axis corresponds to sillimanite-mullite solid solutions expressed as $Al_2^{VI}[Al_{2+2x}Si_{2-2x}]^{IV}O_{10-x}$. (From Cameron, 1977b, Fig. 1).

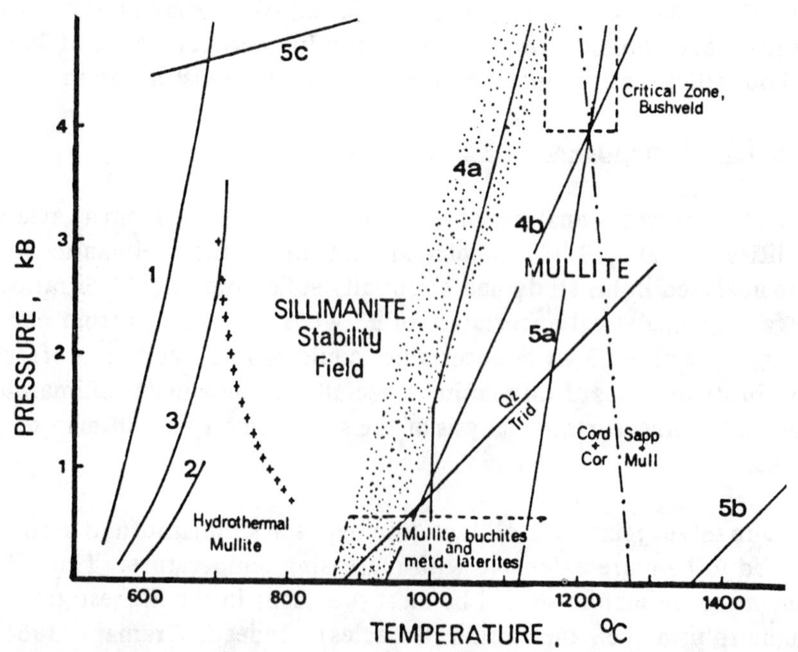

Figure 4.4. Cameron's (1976a) phase equilibrium diagram illustrating possible P-T stability fields of sillimanite and mullite. The numbered curves relevant to mullite/sillimanite stability relations are: (4) sillimanite + corundum = mullite, (5) sillimanite = mullite + "silica". The lower case letters refer to various experimental determinations of equilibria (4) and (5): (a) Holm and Kleppa (1966), (b) Weill (1966), and (c) Khitarov et al. (1963). The stippled area separates the stability fields of sillimanite and mullite. (From Cameron, 1976a, Fig. 2).

In contrast to the Asama fibrolite, Kwak's (1971a) electron probe analyses of fibrolite in xenoliths (buchites) in basic volcanics from Cerro del Hoyazo, Spain, and Finkenberg, Federal Republic of Germany, revealed no measurable Al/Si non-stoichiometry. Thus, Kwak (1971a, p. 1758) concluded that "...possible mullite solid solutions...need not worry the petrologist". However, in a critique of Kwak's paper, Zeck (1973) concluded that fibrolite in the buchites analyzed by Kwak (1971a) crystallized during *medium grade* metamorphism prior to incorporation of the xenoliths into the magmas. Thus, Zeck (1971) contended that Kwak (1971a) did not address the question of Al/Si stoichiometry of "sillimanite" that crystallized during sanidinite facies metamorphic P-T conditions (i.e., the P-T field of buchites in Fig. 4.4). In a reply to Zeck's (1973) critique, Kwak (1973, p. 559) argued that early-formed "sillimanite" subjected to later sanidinite facies P-T conditions "...could be expected to equilibrate, at least partly, to the volcanic conditions if the sillimanite-mullite solid solution does occur".

Cameron and Ashworth (1972) questioned Aramaki's (1961) conclusion of significant Al enrichment of fibrolite from the Asama volcano, Japan. Based on their electron probe analysis of the Asama fibrolite, Cameron and Ashworth (1972) concluded that only 4 % of the tetrahedral Si is replaced by Al.

Cameron (1976a) documented a sillimanite-mullite miscibility gap through his description of coexisting sillimanite + mullite from a variety of high-grade metapelitic lithologies. In a subsequent study (Cameron, 1977a) he showed that complete solid solution between mullite and sillimanite occurs only in Fe-rich, Ti-poor crystals (Fig. 4.5). He concluded (p. 270), "It would therefore seem likely that a solvus exists in the binary system Al_2O_3-SiO_2 which is depressed by several hundred degrees when 2% Fe_2O_3 is added". Using the compositions of coexisting mullite and sillimanite from two natural parageneses, coupled with the composition of mullite synthesized from melts, Cameron (1976b) presented a "possible" phase diagram for the pseudobinary Al_2SiO_5-Al_2O_3 system (Fig. 4.6). The sillimanite limb of the solvus shown in Figure 4.6 suggests that sillimanite is virtually stoichiometric for $T < 800°C$.

Recent support for the existence of sillimanite-mullite solid solution is provided by Grant and Frost's (ms.) description of sillimanite + corundum intergrowths in high grade metapelitic rocks in the aureole of the Laramie Anorthosite Complex, Morton Pass, Wyoming. Because sillimanite has "...the mosaic texture of pseudomorphs...", and "...crystals of corundum are partially controlled by the structure of sillimanite..."), they concluded that

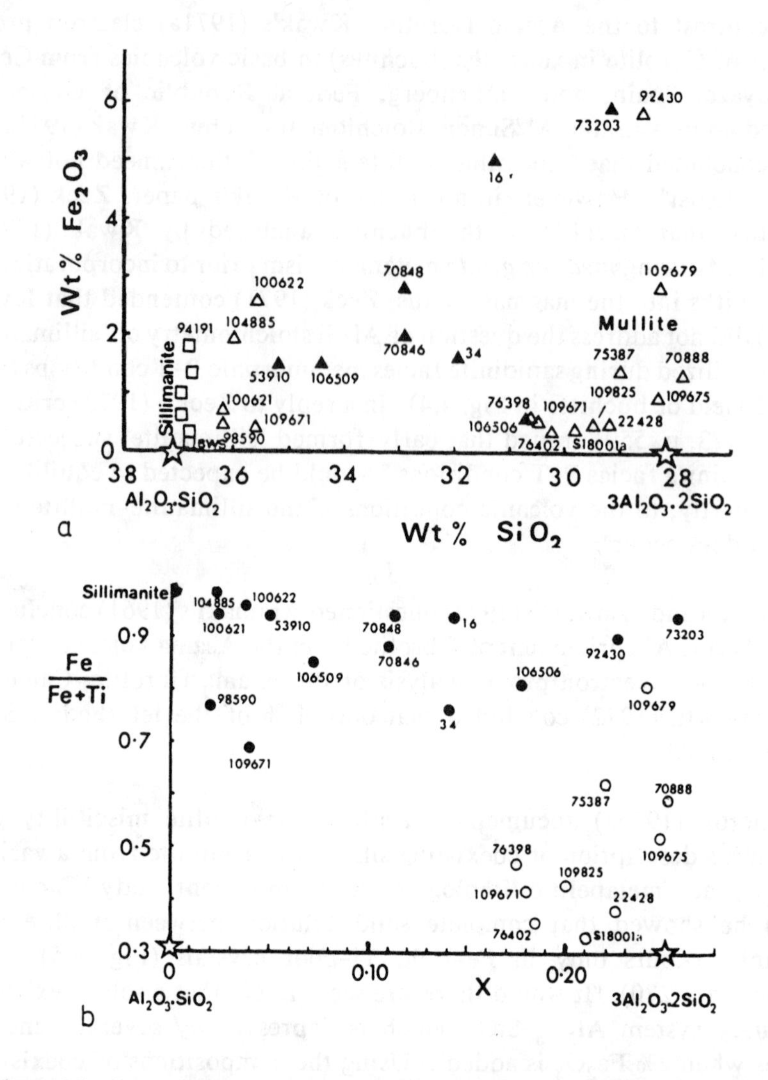

Figure 4.5. Compositions of naturally-occurring members of the sillimanite-mullite solid solution series. *Top* (a): Relationship between concentrations of Fe_2O_3 and SiO_2. The open squares are Cameron's (1977a) "stoichiometric sillimanite", open triangles represent a phase that is interpreted to have crystallized from a melt, and the filled triangles are a phase that formed from dehydration of an aluminous gel. *Bottom* (b): Correlation between the Fe/(Fe + Ti) atomic ratio versus $X_{Al_2O_3}$ in the system Al_2SiO_5-Al_2O_3. Filled circles are crystals with the sillimanite-type superstructure whereas the open circles are crystals with the mullite-type superstructure. (From Cameron, 1977a, Fig. 1a,b).

the sillimanite + corundum intergrowths formed from pseudomorphic replacement of mullite. Their estimates of *minimum* temperatures of metamorphism (650-800°C) are compatible with the contention that mullite initially formed.

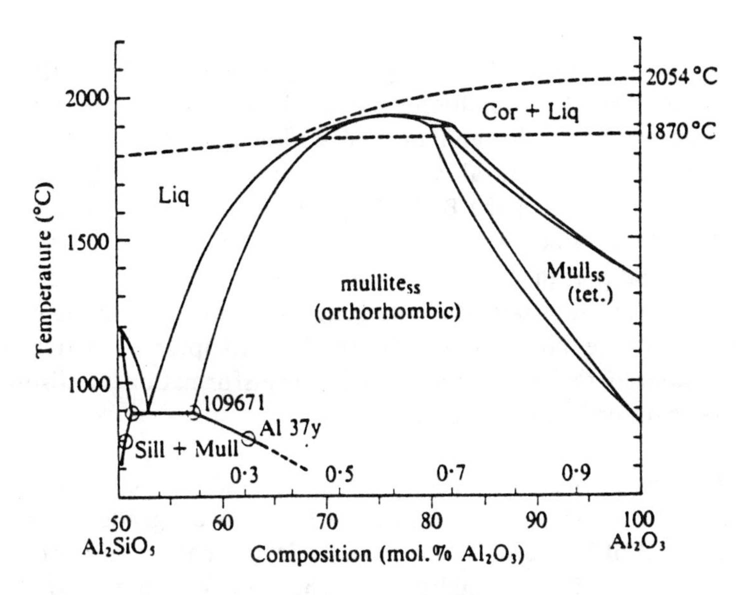

Figure 4.6. Cameron's (1977b) "possible" phase equilibrium diagram for the Al_2SiO_5-Al_2O_3 system. The field of mullite + corundum is omitted. The circled points represent the compositions of phases in two samples containing mullite + sillimanite. (From Deer et al., 1982, Fig. 312; copyright Longman Group Ltd.).

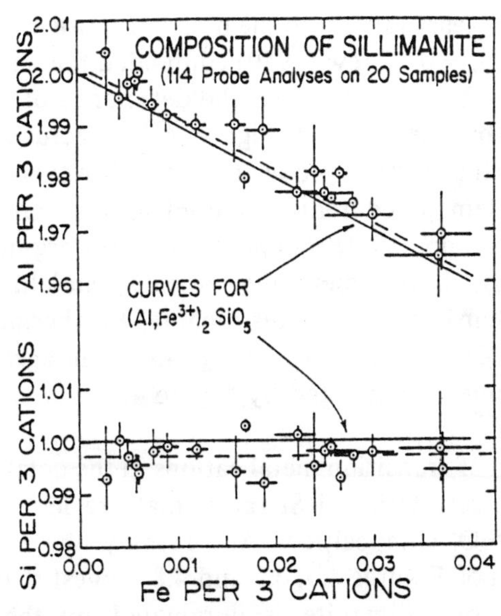

Figure 4.7. Stoichiometry of Al, Si and Fe in sillimanite. The solid lines correspond to stoichiometric sillimanite [i.e., $(Al_xFe^{3+}_{1-x})_2SiO_5$]. The dashed lines were fit by linear least-squares regression. (Modified from Grew, 1980, Fig. 1).

The study of Wenk (1983) brings into question the implication that sillimanite is virtually stoichiometric at T < 800°C. He described intergrowths of sillimanite and mullite within pelitic inclusions in the Bergell Tonalite in the Central Alps of Switzerland. Thermobarometry of the high grade portions of the Bergell contact aureole yields temperatures that are considerably below those shown in Figure 4.4 for mullite. Wenk (1983) concluded that the sillimanite-mullite intergrowths were produced by exsolution from an originally homogeneous phase. Sillimanite contains abundant antiphase boundaries (discussed in Chapter 5) which Wenk (1983) considered to have formed during transformation of disordered sillimanite to an ordered structure.

In addition to Wenk's (1983) study, reexamination of published analytical data suggests some excess Al in lower grade sillimanite. Specifically, all of Cameron's "stoichiometric sillimanite" analyses (open squares in Fig. 4.5) are slightly enriched in Al compared to the stoichiometric end member (Al_2SiO_5). Further evidence for Al-enrichment in sillimanite is provided by Figure 4.7. In this diagram the bulk of the Al values lie above the line for stoichiometric Al_2SiO_5. In concert with the mullitization reaction which involves replacement of tetrahedral Al by Si, most data points for the Si analyses in Figure 4.7 lie below the line for the stoichiometric end member. Furthermore, there is no apparent approach to the stoichiometric Al and Si contents as $X_{Fe} \rightarrow 0$. Lines fit with linear least squares to the data in Figure 4.7 yield Al = 2.002 and Si = 0.998 at $X_{Fe} = 0$. Grew (1980) contended that sillimanite has the stoichiometry $(Al_{2-x}M_x)SiO_5$ and that the deficiency of Si and excess Al + M^{3+} suggested by the data in Figure 4.7 is actually the result of systematic error in the electron probe analyses. However, because kyanite with low minor element content was used as the standard in Grew's analyses, and there is no *apparent* crystal chemistry argument why such a kyanite should not have essentially Al_2SiO_5 stoichiometry, Grew's suggestion of a systematic error is questionable. Furthermore, more recent analytical data by Evers and Wevers (Fig. 4.8) is in agreement with the Al/Si non-stoichiometry suggested by Figure 4.7.

The substitution of transition metal cations in the octahedral site of the aluminum silicates results in Al/Si ratios that are less than that of the stoichiometric Al_2SiO_5 composition. Accordingly, the excess Al/Si ratios implied by the data of Figures 4.5, 4.7 and 4.8 suggest that major element non-stoichiometry of sillimanite is dominated by the "mullitization" reaction, not by transition element solid solution.

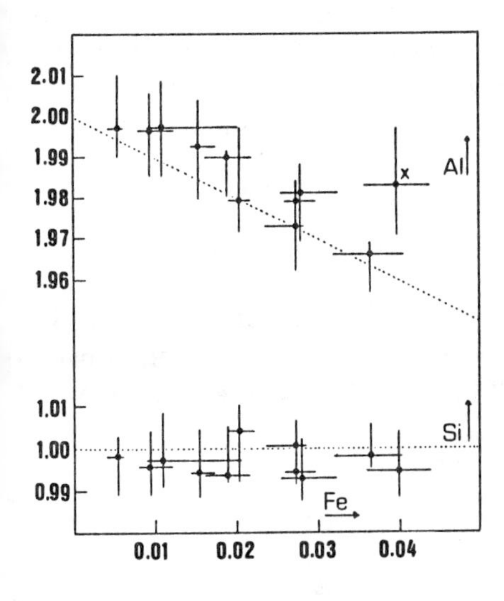

Figure 4.8. Stoichiometry of Al, Si and Fe in sillimanite. The axes are identical to those in Figure 4.7. (From Evers and Wevers, 1984, Fig. 3).

In evaluating aluminum silicate stoichiometry determined from electron probe analytical data it is important to consider that analyses are typically made using an Al_2SiO_5 standard (kyanite is typically used). In some cases, the standard has been *assumed* to be of stoichiometric Al_2SiO_5 composition (e.g., Grew, 1980). In such cases, any deviation from Al/Si stoichiometry of the standard will introduce corresponding error in the major element stoichiometry of the unknowns. Because of the possibility of solid solution of sillimanite and mullite, this problem is particularly significant if sillimanite is used as a standard. Most electron probe analyses of the aluminum silicates reported in the literature are not accompanied by the analytical methods used to determine the stoichiometry of the standards, and many do not report the standards used. Because of the importance of Al/Si stoichiometry in the aluminum silicates, it is hereby recommended that future analytical studies on the Al_2SiO_5 polymorphs discuss the standards, and methods used to determine the compositions of the standards.

An example of questionable accuracy of Al_2O_3 and SiO_2 compositions of the aluminum silicates by electron probe analysis is provided by the analytical data reported by Hemingway et al. (ms.) for gem-quality sillimanite from Sri Lanka. Electron probe analysis yielded $Al_2O_3 = 63.08$ wt % and $SiO_2 = 36.68$ wt %, whereas emission spectroscopy yielded $Al_2O_3 = 62.41$ wt % and $SiO_2 = 37.10$ wt %. Thus, electron probe analysis suggests excess Al whereas emission spectroscopy reveals essentially stoichiometric Al_2O_3 and SiO_2 (stoichiometric Al_2SiO_5 has Al_2O_3 = 62.89 wt % and SiO_2 = 37.11 wt %). Accordingly, the author questions the validity of the excess Al in sillimanite implied by the data in Figures 4.5, 4.7 and 4.8.

Thermodynamic analysis of sillimanite-mullite solid solution

Using the compositions of Fe-free sillimanite suggested in Figure 4.7, we can compute the displacement of the andalusite-sillimanite equilibrium from that with pure stoichiometric sillimanite. Sillimanite with excess Al can be formulated as

$$Al_2^{IV}\,[Al_{2+2x}\,Si_{2-2x}]^{IV}\,O_{10-x}\,.$$

Assuming ideal multisite mixing, the activity of the Al_2SiO_5 component in sillimanite can be computed from

$$a^{Sil}_{Al_2SiO_5} = 16\,(X^{IV}_{Al})^2\,(X^{IV}_{Si})^2\,(X_{O_c})^2.$$

The first term on the right side of this equation is the "normalization factor" for tetrahedral Al and Si (see Nordstrom and Munoz, 1985, p. 157). The term involving fractional site occupancy of oxygen was derived from the fact that only the O_c oxygens are involved in the mullitization reaction and that there are two O_c oxygens for every 10 total oxygens. From Figure 4.7, X^{IV}_{Al} = 0.501; thus, X^{IV}_{Si} = 0.499. Considering the stoichiometric formula: $Al_2^{IV}\,[Al_{2+2x}\,Si_{2-2x}]^{IV}\,O_{10-x}$, these tetrahedral site occupancies yield x = 0.002. Using the above activity expression, the fractional site occupancies yield $a^{Sil}_{Al_2SiO_5}$ = 0.998. If we *arbitrarily* select 800°C as the 1 atm andalusite-sillimanite equilibrium with stoichiometric sillimanite (x = 1.0), and adopt the standard state as unit activity of the pure stoichiometric phase (Al_2SiO_5) at P and T, an equilibrium temperature (T_2) of 794°C for sillimanite with $a^{Sil}_{Al_2SiO_5}$ = 0.998 is computed from

$$0 = -\int_{1073}^{T_2} \Delta S_r\,dT + RT_2 ln(0.998)\,.$$

For petrologic purposes, it is important to note that because natural sillimanite *may* contain a slight excess of Al, and because natural sillimanite has been used in hydrothermal experimental studies and in calorimetric investigations, experimental studies using natural sillimanite as starting material relate to naturally-occurring sillimanite. Thus, the Al_2SiO_5 equilibria with strictly stoichiometric sillimanite are petrologically irrelevant.

MINOR ELEMENT NON-STOICHIOMETRY

Transition elements

By far the largest amount of data on minor elements in the aluminum silicates exists for transition elements (especially Ti, V, Cr and Fe) belonging to the fourth period of the periodic table. Because electronic configurations in the outer two orbitals are similar, absorption of electromagnetic energy, and the resultant changes in the energy levels of the outer **d** electron orbitals, is common in fourth period transition elements. Absorption of electromagnetic radiation in the visible spectrum by transition metal ions is a primary cause for colors in minerals (Burns, 1970).

Numerous spectroscopic studies have been carried out on transition elements in the aluminum silicates. Aside from revealing the fundamental cause of colors, such studies provide insight into the atomic sites in which the transition elements reside. Knowledge of site distribution is important in the thermodynamic modeling of crystalline solutions.

Mössbauer spectroscopy, electron paramagnetic resonance (EPR) spectroscopy, and optical absorption spectroscopy (OAS) have been the principal experimental methods for investigating transition elements in the aluminum silicates. Hawthorne (1988) provides a recent, detailed review of these and other spectroscopic techniques in mineralogy. The following discussion summarizes spectroscopic studies on the aluminum silicate polymorphs.

Kyanite. Troup and Hutton (1964a,b) carried out EPR spectroscopic studies on Fe^{3+} and Cr^{3+} in kyanite. Using perturbation calculations (see Troup and Hutton, 1964b, p. 1498-1499) the spectra were fit with a spin Hamiltonian assuming that certain crystal field terms are dominant. They concluded that Fe^{3+} and Cr^{3+} are concentrated in three of the four octahedral sites. Using X-ray diffraction, Langer and Seifert (1971) deduced that Cr^{3+} is randomly distributed in the M1 and M2 octahedra that form chains parallel to the *c* axis. In his diffuse reflectance spectroscopic analysis of synthetic Cr-rich kyanite, Langer (1976) concluded that the width of the most prominent spectral band for Cr^{3+} (which he assigned to the 10Dq transition) is attributed to splitting that is compatible with the (approximate) tetragonal (D_{4h}) symmetry of the M1 and M2 octahedral sites. Langer and Abu-Eid (1977) studied synthetic Cr^{3+}-bearing kyanite with polarized absorption spectroscopy. They assigned spectral bands to spin-allowed and spin-forbidden transitions involving electrons of the 3d orbital of octahedrally-coordinated Cr^{3+}. However, their band assignments do not provide independent confirmation of the site occupancies of Cr^{3+}. From the bulk of the available data it is most likely that transition metal ions dominantly substitute into the M1 and M2 sites of kyanite.

Kyanite typically displays blue or green colors. Spectroscopic studies have been oriented toward determining the mechanisms responsible for color in kyanite. Several contrasting theories have been suggested to explain the color of blue kyanite. Correlation of intensity of coloration with Ti content suggests that Ti^{3+} is responsible for the color of blue kyanite. Coupling this observation with optical spectroscopy, White and White (1967) and Rost and Simon (1972) attributed the blue color to a spin-allowed crystal field transition in Ti^{3+}. Smith and Strens (1976) attributed the blue color to a $Fe^{2+} \rightarrow Ti^{4+}$ charge transfer mechanism. However, Faye and Nickel (1969) questioned the correlation between color and Ti content. They instead suggested that the blue coloration arises from $Fe^{2+} \rightarrow Fe^{3+}$ charge transfer. Based on Mössbauer spectroscopy, Parkin et al. (1977) proposed that the blue color of kyanite results from a combination of $Fe^{2+} \rightarrow Ti^{4+}$ and $Fe^{2+} \rightarrow Fe^{3+}$ charge transfer mechanisms and crystal field transitions of Fe^{2+} and Fe^{3+} ions. However, they presented no definitive arguments defending the assumption that their Mössbauer spectra suggest the presence of significant amounts of Fe^{2+} in blue kyanite. Ghera et al. (1986) also studied the mechanism responsible for the blue color of kyanite. In contrast to earlier studies, they utilized wet chemical analysis to establish the presence of significant quantities of Fe^{2+} and Fe^{3+}. Heat treatment of blue kyanite at 1200°C produced bleaching and nearly complete disappearance of the 16,500 cm^{-1} absorption band. Assuming that heat treatment results in oxidation of Fe^{2+} to Fe^{3+}, such data support the argument that Fe^{2+} is important in producing the blue coloration. This suggests that the blue color is attributable to a charge transfer mechanism ($Fe^{2+} \rightarrow Fe^{3+}$ and $Fe^{2+} \rightarrow Ti^{4+}$) and/or crystal field transitions involving Fe^{2+}. Based on electron probe analyses of blue kyanites, Cooper (1980) noted that there is a general correlation between the intensity of the absorption color and the Cr and Fe contents. However, there are notable exceptions to this correlation. For example, he described a strongly colored kyanite containing low concentrations of Cr and Fe. "Clearly factors other than the content of Cr and Fe are involved in the production of the blue color"

(Cooper, 1980, p. 156). Altherr et al. (1982) described Cr-rich, Fe-poor blue kyanite. In light of all of the above studies, questions remain regarding the transition element(s) causing the color of blue kyanites.

In their Mössbauer spectroscopic study of *green* kyanite, Parkin et al. (1977) suggested that $Fe^{2+} \rightarrow Fe^{3+}$ charge transfer is responsible for the color. However, as noted above for *blue* kyanite, they presented no discussion defending their interpretation of the spectra. It is notable, however, that electron probe analysis of this sample revealed virtually no Cr. Thus, the emerald-green color of natural Cr-rich kyanites may not necessarily be attributed to absorption phenomena involving Cr^{3+}.

In kyanite, Ti, V, Cr and Fe are the only transition elements present in quantities exceeding 0.1 wt % (Chinner et al., 1969).

In metasedimentary rocks, and in veins hosted in metasediments, Fe^{3+} is the most abundant transition element in kyanite. Fe_2O_3 is typically in the 0.25-1 wt % range; however, Fe_2O_3 contents up to 1.6 wt % have been recorded in kyanite from oxidized (hematite-bearing) metapelites (Chinner et al., 1969).

Significant contents of Cr occur only in kyanite within blueschists and eclogites. The maximum Cr_2O_3 content of kyanite is 18 mol % of the Cr_2SiO_5 component in a grospydite xenolith within a kimberlite (Sobolev et al., 1968). Delor and Leyreloup (1986) described kyanite with $\leq$ 6.99 wt % Cr_2O_3 in eclogites from the Massif Central (France), and Enami and Zang (1988) found 1.18 wt % Cr_2O_3 in kyanite within eclogites from the Jiangsu province, east China. The highest Cr contents occur in close proximity to inclusions of Cr-bearing rutile. However, most kyanites in eclogites (Carswell et al., 1981; Smyth et al., 1984; Miller, 1986; Klemd, 1989) and kyanite inclusions in diamonds (Meyer, 1987; Jaques et al., 1989) have relatively low Cr_2O_3 contents. Kyanites in eclogite-facies metapelites also have negligible Cr contents (Ballevre et al., 1989). Thus, elevated Cr_2O_3 contents are atypical of kyanites in high-pressure parageneses.

Seifert and Langer's (1970) experimental synthesis study of the Al_2SiO_5-Cr_2SiO_5 system at 20 kbar and 30 kbar reveals extensive solid solution of Cr in kyanite (Fig. 4.9). Their study shows that increasing pressure is far more important than increasing temperature in enhancing the solubility of Cr in kyanite. The relatively small ionic radius of Cr^{3+} is apparently conducive to the enhanced concentration of Cr in kyanite at high pressure (Fig. 4.10). However, the increase in Cr content with

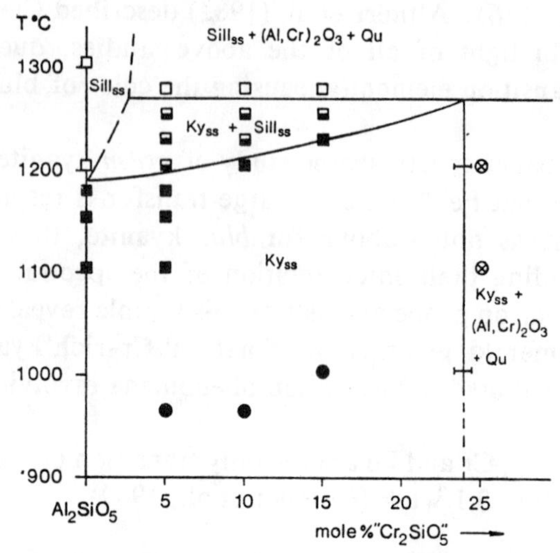

Figure 4.9. T-X section of the aluminous portion of the pseudobinary Al_2SiO_5-Cr_2SiO_5 system at 20 kbar. The vertical line represents the maximum solubility of Cr in kyanite. Note the bivariant field of kyanite + sillimanite at 1190-1270°C. The circular and square symbols represent phases or phase assemblages (e.g., the half-filled squares are runs in which kyanite$_{ss}$ + sillimanite$_{ss}$ formed). The horizontal brackets define the maximum Cr_2SiO_5 content of kyanite as determined from rate studies (see Seifert and Langer, 1970, Fig. 3). Note that this diagram is based on *synthesis* experiments; thus, the phase boundaries were not bracketed by reaction reversal. (From Seifert and Langer, 1970, Fig. 4).

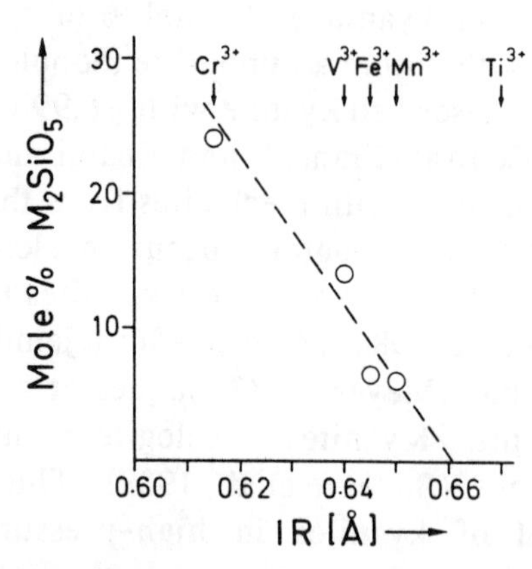

Figure 4.10. Maximum solubility of transition metal cations (M) in kyanite at 20 kbar and about 1000°C as a function of effective ionic radii (IR). (From Langer, 1976, Fig. 3).

increasing pressure may also result from favorable crystal field effects due to tetragonal distortion of Cr^{3+}-bearing octahedra with increasing pressure (Langer, 1976).

<u>Andalusite</u>. In comparison to kyanite and sillimanite, some andalusite contains large amounts of Fe^{3+} and Mn^{3+}. To the author's knowledge, the maximum Fe content of andalusite reported is 3.4 wt % Fe_2O_3 (Grapes,

1987). Up to 35.5 wt % Mn_2O_3 (40 mol % Mn_2SiO_5) has been found in kanonaite (Kramm, 1979b).

Holuj et al. (1966) measured the ESR spectra of Fe in andalusite. With no *apparent* justification, they assigned major spectral lines to Fe^{3+} substituting for Al^{3+} in octahedral and 5-coordinated sites. However, they concluded that much of the Fe^{3+} occurs in octahedral sites. Holuj et al. (1966) suggested that the hyperfine splitting of the lines assigned to octahedrally-coordinated ions results from the presence of Cr^{3+}.

Hålenius (1978) studied two Mn-rich andalusites using Mössbauer and polarized absorption spectroscopy. As shown in Figure 4.11, he considered the Mössbauer spectrum to result from three doublets. Based on previous studies showing the correlation between Mössbauer parameters (i.e., isomer shift and quadrupole splitting) and site occupancies of Fe^{2+} and Fe^{3+} in other silicates, Hålenius assigned two of the doublets to Fe^{3+} in octahedral and 5-coordinated sites, and a third to Fe^{2+} in octahedral coordination. Hålenius found three major absorption bands in the polarized absorption spectra of manganian andalusite samples. As shown in Figure 4.12, he correlated the three bands with spin-allowed d-d transitions in distorted octahedra having D_{4h} symmetry. From the energy levels of the split ^5D states (Fig. 4.12), Hålenius computed a crystal field stabilization energy (CSFE) of 47.6 kcal. He concluded that this CFSE is comparable with that of Mn^{3+} in distorted octahedra in other silicate minerals.

Abs-Wurmbach et al. (1981) studied a variety of natural and synthetic Mn-rich andalusite samples using X-ray diffraction and polarized absorption and Mössbauer spectroscopy. Their X-ray diffraction data were refined with the assumption that Fe occupies octahedral sites. More specific information on Fe^{2+} and Fe^{3+}, and site assignments of these ions, were derived from Mössbauer spectra. Their conclusions regarding the site occupancies of Fe^{2+} and Fe^{3+} were based on correlation of their results with the interpretations of previous Mössbauer studies on a variety of Fe-bearing silicate minerals. They concluded that the ^{57}Fe spectra of viridines are best fit with two doublets (Fig. 4.13). The isomer shift of the largest doublet was assigned to Fe^{3+} in octahedral coordination, whereas the isomer shift of the smaller doublet was considered to represent Fe^{3+} in the 5-coordinated site. From the relative intensities of the two doublets they concluded that 85-95% of the Fe^{3+} resides in the octahedral site whereas the remaining 10-15% of the Fe^{3+} occurs in the 5-coordinated site.

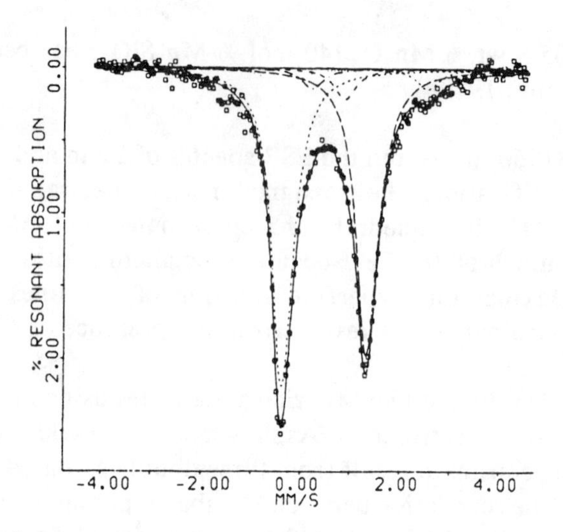

Figure 4.11. Mössbauer spectrum of a sample of Mn-andalusite (at 77 K). (From Hålenius, 1978, Fig. 3).

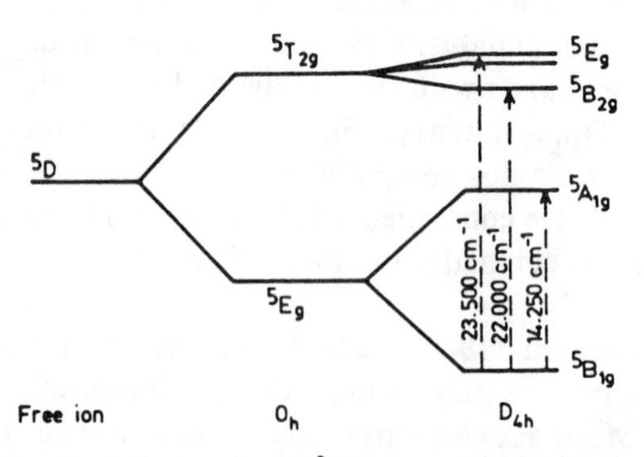

Figure 4.12. Energy level diagram for Mn^{3+} in the octahedral site in Mn-andalusite as deduced from polarized absorption spectra. (From Hålenius, 1978, Fig. 6).

Abs-Wurmbach et al. (1981) utilized polarized absorption spectroscopy to interpret the color and pleochroism of manganian andalusite. They considered that the greenish-yellow color observed with the electrical polarization vector parallel to the X and Z optical indicatrix directions is attributed to an intense absorption band at about 22,000 cm^{-1}, whereas the emerald-green color with polarization parallel to Y was attributed to an absorption minimum at 19,000 cm^{-1} between two adjacent strong polarization bands. Abs-Wurmbach et al. (1981) presented an excellent discussion of the crystal chemistry of manganian andalusite. The following is a summary of their discussion. Their single-crystal X-ray structure refinement suggests that Mn and Fe are confined to the octahedral site. As shown in Figure 4.14a, the most prominent change in

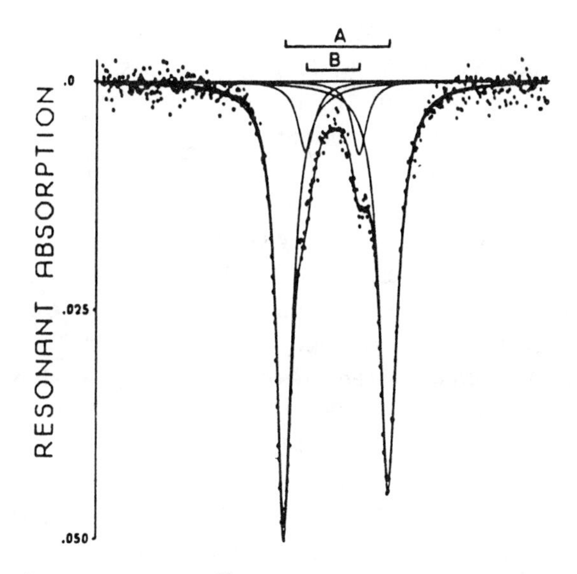

Figure 4.13. Mössbauer spectrum of ^{57}Fe in a viridine from Yakutia, U.S.S.R. The spectrum was deconvoluted into two doublets. (From Abs-Wurmbach et al., 1981, Fig. 3).

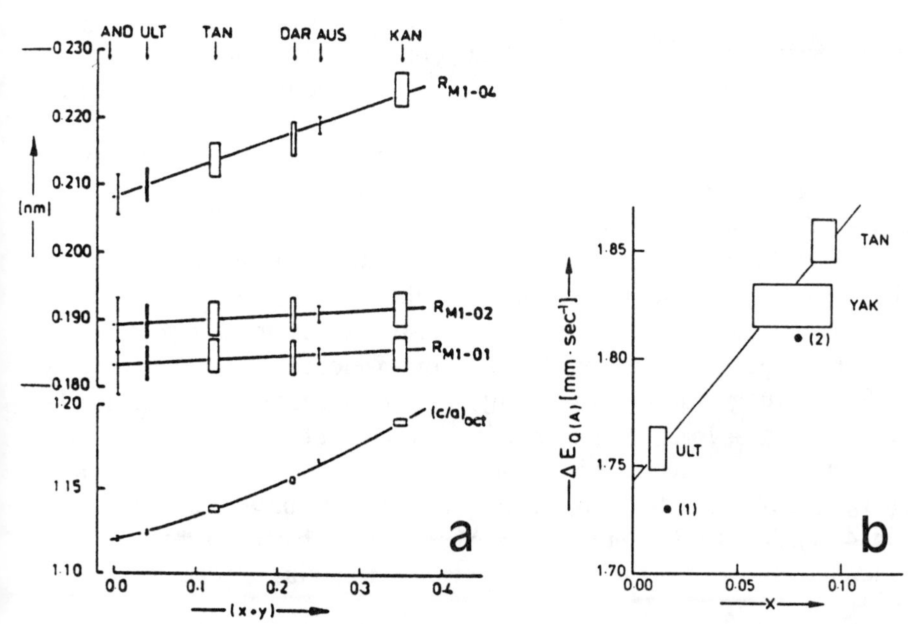

Figure 4.14. (a) Cation-oxygen bond distances of the octahedral site in andalusite as a function of the fractional site occupancy of Mn^{3+} (x) + Fe^{3+} (y). The designations for the oxygen atoms are as follows: O1 = O_A, O2 = O_B, and O4 = O_D. The rectangles are estimates of uncertainties in the data for various samples. (From Abs-Wurmbach et al., 1981, Fig. 7). (b) Correlation between quadrupole spitting ($\Delta E_{Q(A)}$) of $^{57}Fe^{3+}$ and fractional occupancy (x) of Mn^{3+} in the octahedral site in Mn-andalusites. The rectangles are estimates of uncertainties in the data. (From Abs-Wurmbach et al., 1981, Fig. 8).

the octahedra with increasing Mn + Fe content is distortion arising from an increase in the M1-O4 bond length. Increased distortion of the octahedra produces rotation of the trigonal bipyramids and tetrahedra. Because of the progressive distortion of the octahedra with increasing Mn + Fe substitution, the trigonal bipyramids undergo an increase in the M2-O1 and M2-O3 distances and a corresponding decrease in the M2-O4 distances. Overall, there is a relatively small increase in the mean M2-O bond distances. Presumably using ionic radii arguments, Abs-Wurmbach et al. (1981) concluded that this small change in the average cation-oxygen bond distances contributes to the very limited substitution of Mn^{3+} and Fe^{3+} for Al^{3+} in the 5-coordinated sites of andalusite. As shown in Figure 4.14b, Mössbauer studies show that there is an increase in quadrupole splitting of $^{57}Fe^{3+}$ with increasing Mn content. In this diagram, it is notable that extrapolation to $x = 0$ reveals relatively significant quadrupole splitting (ΔE_Q = 1.74) for Mn-free andalusite. Accordingly, the octahedral site of Mn-free andalusite is relatively distorted. Although the site symmetry of the distorted octahedra in andalusite is C_2 (space group *Pnnm*), the polyhedron in the plane perpendicular to the axis of elongation is close to square symmetry (Fig. 4.15). For this reason, Hålenius (1978) and Abs-Wurmbach et al. (1981) assumed D_{4h} symmetry (i.e., an elongated octahedron) for the correlation of the optical absorption bands with *d-d* transitions of Mn^{3+} and Fe^{3+}. Abs-Wurmbach et al. (1981) assigned the two strongest bands in the polarized absorption spectra to transitions I and II (Fig. 4.16). They argued that the absorption band at 23,300 cm^{-1} is produced by transition III (Fig. 4.16) and a spin-forbidden *d-d* transition of Fe^{3+}. Abs-Wurmbach et al. (1981) assigned several other spectral bands to spin-forbidden *d-d* transitions of Fe^{3+}. Using energies of the bands in the polarized absorption spectra, they calculated the energies of various crystal field parameters. In Figure 4.17, the parameter δ_2 represents the energy separation of *d*-level orbitals in an elongated octahedron (Fig. 4.18), and $10Dq$ represents the energy difference between the ^5E_g and $^5T_{2g}$ orbitals (Fig. 4.16). With increasing Mn content, the increase in δ_2 is compensated by a decrease in $10Dq$ such that the crystal field stabilization energy (CFSE) is virtually independent of the Mn content (bottom plot in Fig. 4.17). The CFSE computed by Abs-Wurmbach et al. (1981) for Mn^{3+} in andalusite is similar to that computed by Hålenius (1978).

Weiss et al. (1981) carried out an X-ray refinement of the crystal structure of kanonaite. This refinement was initiated with three different assumptions regarding the percentage occupancy of Mn^{3+} in the octahedral site: (1) 100%, (2) 95%, and (3) 60%. Only model (2) yielded convergence. Their refinement, which was made by reiteration starting with this model, yielded $Mn_{0.74}$ $Al_{0.26}$ for the octahedral site and $Mn_{0.12}$ $Al_{0.88}$ for the 5-

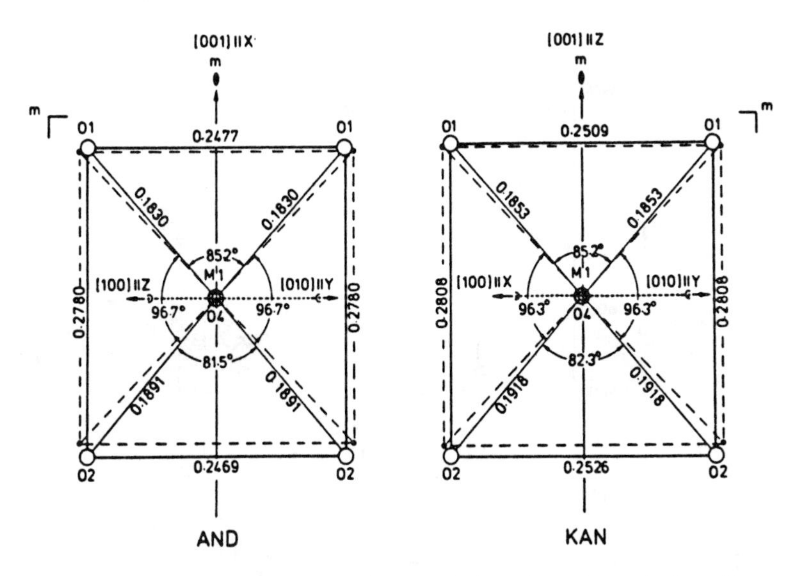

Figure 4.15. Projection of the octahedra of andalusite (AND) and kanonaite (KAN) along the M1-O4 (= M1-O_D) axes. The solid lines correspond to the actual octahedra whereas the dashed lines represent undistorted octahedra (with point symmetry D_{4h}) with the same area as the actual octahedra. Note the relatively small distortion of the actual octahedra from D_{4h} symmetry. (From Abs-Wurmbach et al., 1981, Fig. 9).

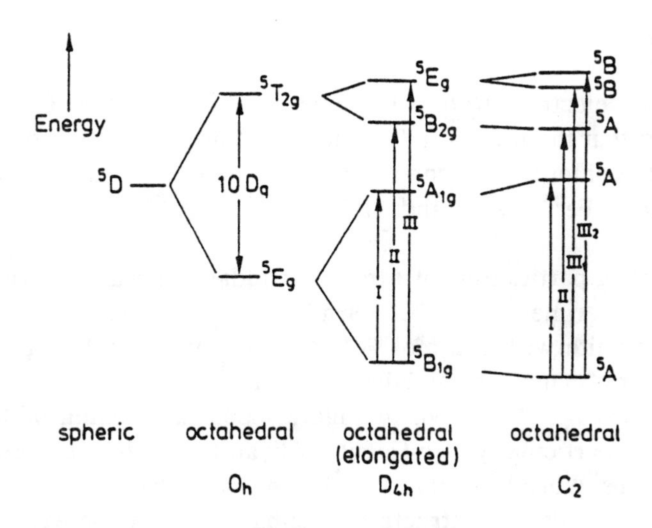

Figure 4.16. Schematic energy level diagram for splitting of the ^5D ground state of M^{3+} ions in three different octahedral site symmetries. O_h is "perfect" octahedral symmetry, D_{4h} represents tetragonal distortion from octahedral symmetry, and C_2 is the actual symmetry (monoclinic) of the M^{3+} site in andalusite. (From Abs-Wurmbach, 1981, Fig. 10).

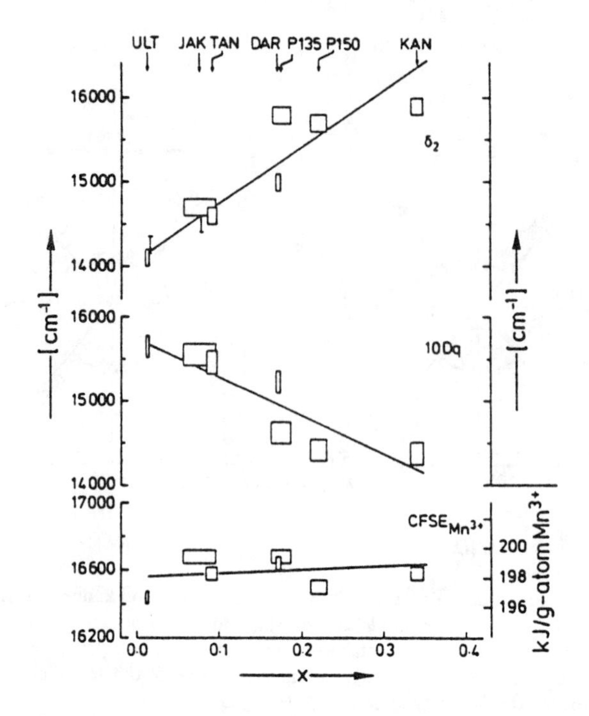

Figure 4.17. Crystal field parameters for the Mn^{3+} ion in andalusite. CFSE = crystal field stabilization energy, $10D_q$ = crystal field splitting for octahedral symmetry (see Fig. 4.16), and δ_2 = ^5E_g ground state splitting. (From Abs-Wurmbach et al., 1981, Fig. 12).

coordinated site.

From an electron probe study of a zoned andalusite, Grapes (1987) concluded that the MgO + TiO_2 concentration is directly proportional to the Fe_2O_3 content. He concluded that these correlations suggest the substitutions: $Fe^{3+} = Al^{3+}$ and $Ti^{4+} + Mg^{2+} = 2Al^{3+}$

The strong partitioning of Mn^{3+} into andalusite was confirmed by the synthesis experiments of Abs-Wurmbach et al. (1983). At 900°C (Fig. 4.19) sillimanite with 1-2 mol % Mn_2SiO_5 coexists with "viridine" (now referred to as "manganian andalusite"; Gunter and Bloss, 1982) having 8-9 mol % Mn_2SiO_5. As shown in Figure 4.20, partitioning of Mn^{3+} into "viridine" is particularly significant for kyanite-viridine assemblages at T < 700°C. Abs-Wurmbach et al. (1983) confirmed that the concentration of Mn^{3+} in the andalusite structure is enhanced with increasing f_{O_2}. In agreement with the presence of hematite in viridine-bearing rocks, their experimental results suggest that viridine is stable only in the P-T-f_{O_2} stability field of hematite.

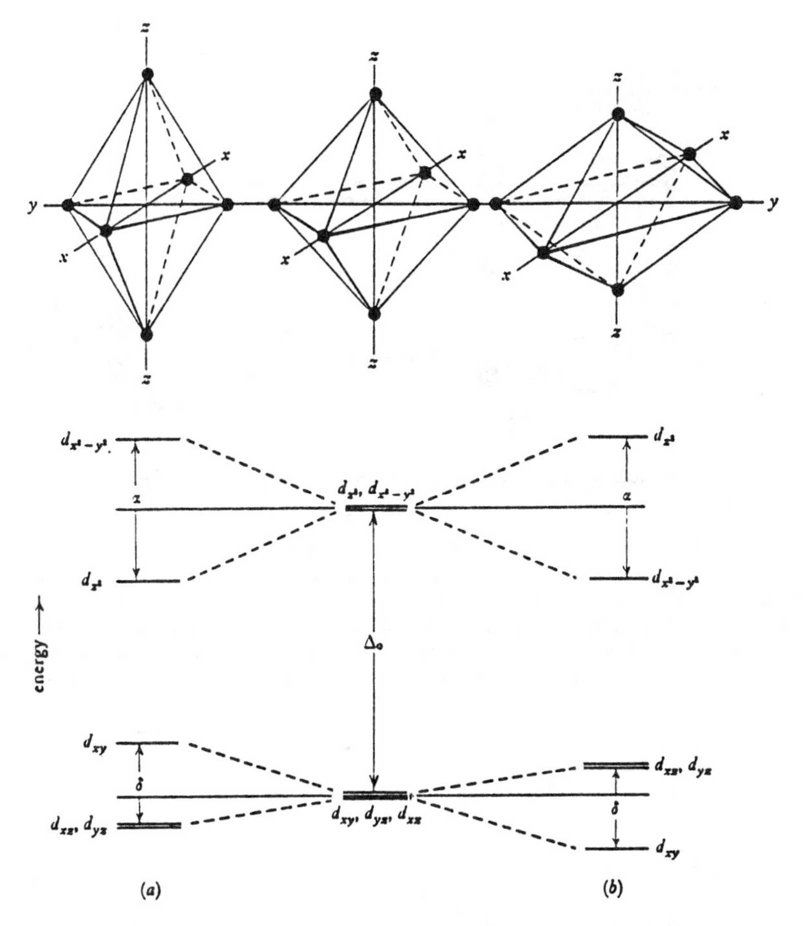

Figure 4.18. Top diagrams: regular octahedron (center), octahedron distorted by elongation (left), and compression (right) along the z axis. The energy level diagrams illustrate the difference in energy levels ("splitting") of the *d* orbitals of transition metal ions in the distorted octahedra. (From Burns, 1970, Fig. 2.7).

Acharyya et al. (1990) described manganian andalusite (containing 14.8-21.4 mol % Mn_2SiO_5) in manganiferous metasediments subjected to contact metamorphism. The presence of coexisting rutile + hematite provides another example of the oxidizing environment during crystallization of manganian andalusite.

Partitioning of Mn into andalusite can be rationalized with crystal field theory. In the high-spin $3d^4$ Mn^{3+} ion, the sole electron of the e_g orbital group can occupy either the $d_{X^2-Y^2}$ or d_{Z^2} orbitals. The d_{Z^2} orbital, which is elongated in the Z direction, yields lower repulsion with adjacent oxygen ions than the $d_{X^2-Y^2}$ orbitals that form elongated lobes parallel to

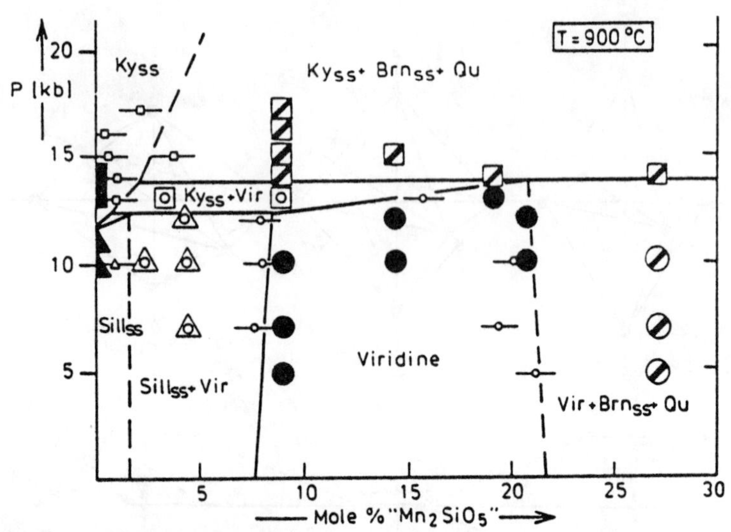

Figure 4.19. P-X section of the aluminous portion of the Al_2SiO_5-Mn_2SiO_5 system at 900°C. Oxygen fugacity was controlled by the MnO_2/Mn_2O_3 buffer. The large symbols represent phases or phase assemblages (e.g., the filled circles are runs in which viridine formed). The small open symbols with horizontal bars are solid solution compositions determined by X-ray methods. Note that this diagram is based on *synthesis* experiments; thus, the phase boundaries were not bracketed by reaction reversal. (From Abs-Wurmbach et al., 1983, Fig. 4).

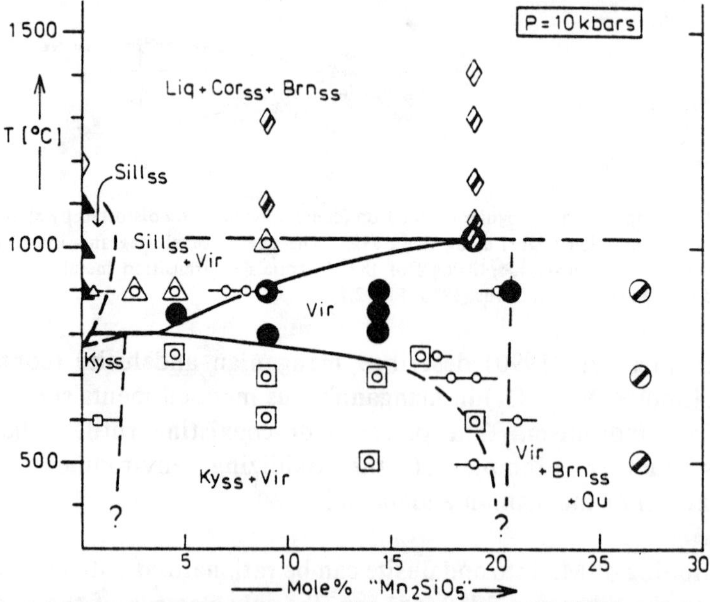

Figure 4.20. T-X section of the aluminous portion of the pseudobinary Al_2SiO_5-Mn_2SiO_5 system at 10 kbar. Oxygen fugacity was controlled by the MnO_2/Mn_2O_3 buffer. Symbols as in Figure 4.19. Note that this diagram is based on *synthesis* experiments; thus, the phase boundaries were not bracketed by reaction reversal. (From Abs-Wurmbach et al., 1983, Fig. 5).

X and Y (Fig. 4.18). Thus, for energetic reasons, Mn^{3+} prefers the distorted (elongated) octahedron over an undistorted octahedron (Fig. 4.18). The preference of Mn^{3+} and other transition elements for distorted polyhedra is referred to as the "Jahn-Teller effect". The amount of distortion of a coordination polyhedron may be expressed by the *longitudinal strain* parameter (Ghose and Tsang, 1973):

$$|\alpha| = \sum_i |ln\ (l_i/l_o)|$$

where l_i is an individual Al-O bond length, and l_o is the bond length of an undistorted polyhedron having the same volume as the distorted polyhedron. For the octahedral site of the aluminum silicates, the $|\alpha|$ values are: sillimanite = 0.068, kyanite = 0.149, andalusite = 0.300 (Ghose and Tsang, 1973). Thus, the octahedra in andalusite are considerably more elongated (distorted) than those of kyanite and sillimanite. Accordingly, the Jahn Teller effect predicts that, in comparison to kyanite and sillimanite, Mn^{3+} is preferentially concentrated in the octahedral site of andalusite.

Aside from Fe and Mn, most other transition elements are typically present in trace quantities in andalusite. However, elevated Cr and V have been reported in a limited number of samples (Deer et al., 1982). Particularly notable is Carlson and Rossman's (1988) description of andalusite with V_2O_3 = 0.54-0.81 wt % and Cr_2O_3 = 0.09-0.12 wt %. Based on data from other vanadian minerals, Carlson and Rossman (1988) assigned most of the absorption maxima in the polarized absorption spectra of their andalusite to V^{3+} and Cr^{3+} in octahedral sites. They attributed the unusual yellow-to-colorless pleochroism of this andalusite to a polarization-dependent transition from the $^3T_{1g}$ state of V^{3+}.

Blue andalusite is a rare but notable variety. Langer et al. (1984) studied blue andalusite from the Venn-Stavelot Massif in Belgium. Polarized absorption spectra of this andalusite revealed a strong, broad absorption band centered around 13,000 cm^{-1}, which they assigned to $Fe^{2+} \rightarrow Fe^{3+}$ charge transfer. Electron probe analysis showed that there is a correlation between the blue coloration and elevated Fe and P contents. Because of the inverse correlation between Si and P contents, Langer et al. (1984) concluded that P substitutes for tetrahedral Si in the andalusite structure. In light of these correlations, and the evidence from polarized absorption spectra that both Fe^{2+} and Fe^{3+} are present, they concluded that the introduction of P into the andalusite structure occurs by

$$[M^{3+}]^{VI} + [Si^{4+}]^{IV} = [Fe^{2+}]^{VI} + [P^{5+}]^{IV}\ .$$

Sillimanite. Using EPR spectroscopy, LeMarshall et al. (1971) concluded that Fe^{3+} occupies both octahedral and tetrahedral sites in sillimanite.

Hålenius (1979) analyzed Fe in sillimanite with Mössbauer and optical absorption spectroscopy. As shown in Figure 4.21, he concluded that the Mössbauer spectra were resolved into three doublets. The most intense doublet (Fig. 4.21) was assigned to Fe^{3+} in octahedral sites, another doublet was assigned to Fe^{2+} in octahedral coordination, and the third doublet was attributed to a small amount of included hematite. From the relative intensities of the Mössbauer doublets assigned to Fe^{3+} and Fe^{2+}, he concluded that for Sil1 and Sil2 (Fig. 4.21) Fe^{2+} respectively occupies 26% and 14% of the total iron in the octahedral sites. While admitting that no *conclusive* assignments can be made, he suggested that many of the polarized absorption bands result from spin-forbidden transitions in Fe^{3+}. Two of the bands were attributed to cation-cation charge transfer.

Using optical and Mössbauer spectroscopy, Rossman et al. (1982) investigated the three colored varieties of sillimanite (yellow, brown and blue). From the lack of optical absorption bands expected from Fe^{2+} in the 800-1200 nm region, they concluded that most of the iron in yellow sillimanite is Fe^{3+}. Comparing the absorption spectra of yellow sillimanite with that of kyanite, they concluded that tetrahedrally-coordinated Fe^{3+} is responsible for the most prominent absorption bands in yellow sillimanite. As shown in Figure 4.22, they fit their Mössbauer spectrum of yellow sillimanite with two doublets. They assigned the more intense doublet to Fe^{3+} in octahedral coordination and the less intense doublet to tetrahedral Fe^{3+}. From the relative areas of the two doublets (Fig. 4.22) they concluded that this sillimanite contains 80% of the Fe^{3+} in octahedral coordination and 20% in tetrahedral coordination. Using Mössbauer spectroscopy, Rossman et al. (1982) reexamined a sample of yellow sillimanite that had been analyzed by Hålenius (1979). In contrast to the study of Hålenius (1979), Rossman et al. (1982) found no evidence for Fe^{2+} in solid solution. Their optical absorption study suggested that the coloration of brown sillimanite results from inclusions of Fe-Ti oxides. Rossman et al. (1982) found that the optical spectra of blue sillimanite are similar to that of blue kyanite. Based on the explanation of the blue coloration of kyanite proposed by other investigators, Rossman et al. (1982) attributed the blue color in sillimanite to either $Fe^{2+} \rightarrow Fe^{3+}$ or $Fe^{2+} \rightarrow Ti^{4+}$ charge transfer.

Using neutron diffraction, Peterson and McMullan (1986) carried out a single-crystal refinement of sillimanite. From neutron scattering

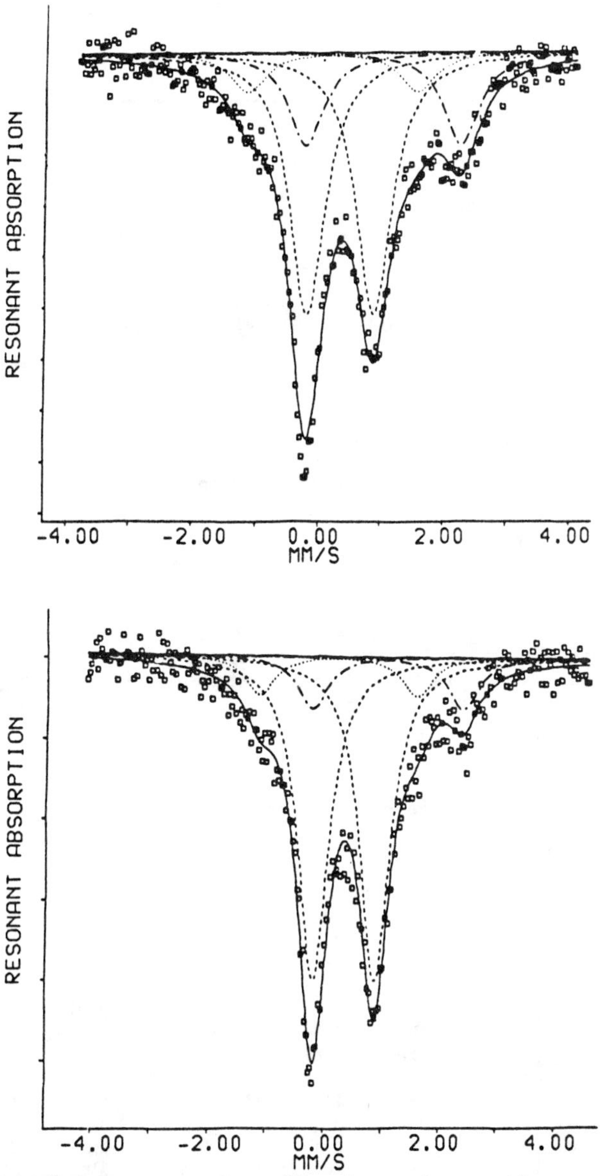

Figure 4.21. Mössbauer spectra of two sillimanite samples at ambient temperature. The dashed lines correspond to Fe^{3+} in octahedral coordination, the dot-dash lines represent Fe^{2+} in octahedral coordination, and the dotted lines are hematite (α-Fe_2O_3). (From Hålenius, 1979, Fig. 1).

amplitudes, they concluded that the fractional site occupancies of Fe were 0.013 in the octahedral site and 0.007 in the tetrahedral site.

In addition to Fe, Cr and V are the only transition elements consistently detected in sillimanite (Chinner et al., 1969; Albee and Chodos, 1969;

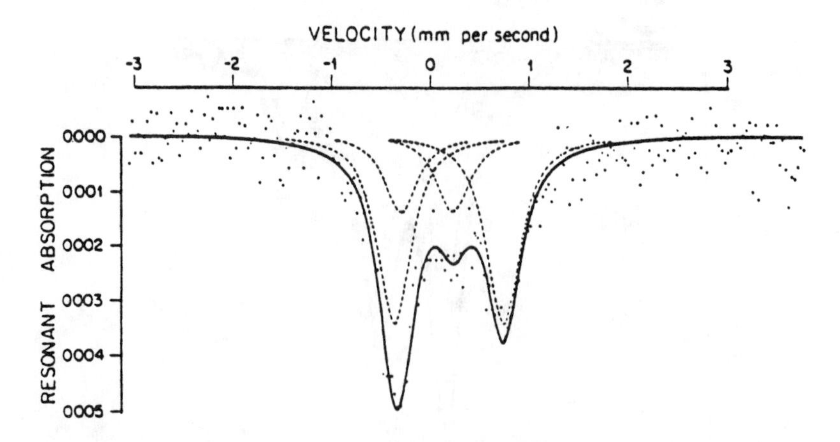

Figure 4.22. Mössbauer spectrum of yellow sillimanite (from Benson Mines, New York) at ambient temperature. The dashed lines were derived by spectral "deconvolution": the intense doublet was assigned to Fe^{3+} in octahedral coordination whereas the less intense doublet is probably Fe^{3+} in tetrahedral coordination. (From Rossman et al., 1982, Fig. 4).

Table 4.1. Summary of Strens' (1968) computations of the solid solution of Fe^{3+} and Mn^{3+} in the Al_2SiO_5 polymorphs and in viridine. For each phase, L2 represents the maximum transition element content, and M is the minimum on a plot of ΔG versus composition (see Fig. 4.23). (From Strens, 1968, Table 1).

Andalusite	0·035Fe^{3+} @ 773° K	ΔH = 6·7 kcal/mole $FeAlSiO_5$
Sillimanite	0·015	8·0
Kyanite	0·04	6·5
1:1 viridine	0·20$(Fe,Mn)^{3+}$	4·9
Mn-viridine	0·15Mn^{3+}	4·3

Limiting (L(2)) and most stable (M) compositions

		327° C	427°	527°	627°	727°	827°
Kyanite	L(2)	0·013	0·026	0·045	0·070	0·100	0·130
	M	—	0·010	0·019	0·029	0·039	0·050
	$-\Delta G_M$	2	13	27	47	75	111
Sillimanite	L(2)	—	—	0·018	0·031	0·047	0·069
	M	—	—	0·007	0·012	0·019	0·026
	$-\Delta G_M$	1	3	9	20	35	56
Andalusite	L(2)	0·016	0·028	0·046	0·068	0·094	0·135
	M	0·005	0·010	0·016	0·023	0·034	0·045
	$-\Delta G_M$	2	11	23	43	68	101
1:1 viridine	L(2)	0·09	0·15	0·22	0·30	0·37	0·44
	M	0·03	0·06	0·09	0·13	0·15	0·18
	$-\Delta G_M$	35	80	140	215	310	410
Mn-viridine	L(2)	0·076	0·116	0·162	0·213	0·260	0·314
	M	0·035	0·048	0·064	0·080	0·100	0·120
	$-\Delta G_M$	26	60	103	152	213	280

Calculated maximum widths of divariant zones (in kilobars)

	327° C	427°	527°	627°	727°	827°
$A+K$	0·00	0·01	(0·02)	(0·03)	(0·04)	(0·06)
$A+S$	(0·05)	(0·23)	0·36	0·59	0·85	1·15
$S+K$	(0·01)	(0·08)	0·13	0·19	0·29	0·40
$K+V(1:1)$	0·19	0·38				
$S+V(1:1)$	(0·85)	(1·92)	3·3*	5·0*	7·3*	9·3*
$K+V(1:1)$	—	—	0·55	0·35	0·25	0·15

* exceeds width of sillimanite field; ( ) metastable reaction

Okrusch and Evans, 1970; Dodge, 1971). The concentration of Cr_2O_3 and V_2O_3 in most sillimanites does not exceed 0.2 wt %; however, larger concentrations of these oxides have been reported. Particularly notable are 1.4 wt % V_2O_3 and 0.8 wt % Cr_2O_3 (Speer, 1982), 0.16-0.49 wt % Cr_2O_3 (Grew et al., 1987), and 0.46-1.20 wt % Cr_2O_3 (Dawson and Smith, 1987).

Thermodynamic analysis of transition element solid solution. There have been limited rigorous analyses of the thermodynamics of mixing of transition elements in the aluminum silicates.

Strens (1968) carried out the first detailed thermodynamic analysis of the effect of solid solution of Fe^{3+} and Mn^{3+} on the thermodynamic properties and P-T stability relations of the Al_2SiO_5 polymorphs. He considered the *unbalanced* equilibrium

$$Al_2SiO_5 + Fe_2O_3 + SiO_2 = (Al,Fe)AlSiO_5 \quad . \qquad [4.1]$$

Strens (1968, p. 842) utilized the equation: $\Delta G = \Delta H \cdot x - TS_c$, which he defined as: "The free energy change of the reaction, i.e. the stabilization of andalusite solid solution relative to pure andalusite or sillimanite...". Strens (1968) defined ΔH as the enthalpy change of reaction [4.1] and ΔS_c as the configurational entropy of mixing of Fe and Al [$\Delta S_c = -R\sum_1(x_i ln x_i)$]. He contended that saturation in the $FeAlSiO_5$ component is expected in parageneses containing aluminum silicates in equilibrium with hematite + quartz. In such assemblages, he concluded that for reaction [4.1] $\Delta G = 0$ and, thus, $\Delta H \cdot x = TS_c$. Strens (Table 4.1) estimated T and x from "paragenetic data"; unfortunately, he presented no discussion of the source(s) for such data. With T-x data, Strens (1968) computed ΔG at selected P-T conditions. Figure 4.23 shows four isotherms, each computed at "...a pressure in the sillimanite stability field such that (pure) kyanite is destabilized relative to (pure) sillimanite by 150 cal/mole" (Strens, 1968, p. 843, Fig. 2 caption). For each temperature, the position of the free energy minimum gave ΔG_m; for example, ΔG_m is about -200 cal for the 1300 K isotherm (Fig. 4.23). For each isotherm, the difference between ΔG_m for the polymorphs yielded ΔG_r values from which the widths of the divariant intervals were computed from: $\Delta P = \Delta G_r/\Delta V_r$. As shown in Figure 4.24, the computed divariant zone is largest for the andalusite-sillimanite equilibrium, less for the kyanite-sillimanite equilibrium, and negligible for the kyanite-andalusite equilibrium. Strens' (1968) thermodynamic treatment requires some clarification. Assuming ideal mixing (as in Strens' treatment), the free energy change for reaction [4.1] is

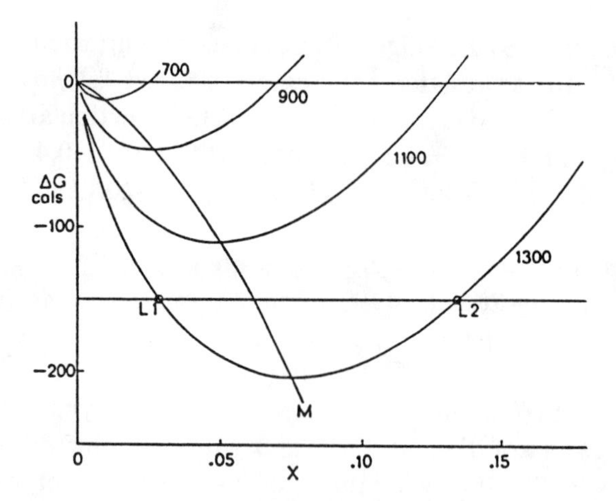

Figure 4.23. Isotherms (°K) of ΔG of Strens' (1968) *unbalanced* reaction: $Al_2SiO_5 + Fe_2O_3 + SiO_2 = (Al,Fe)AlSiO_5$, versus mole fraction of the $FeAlSiO_5$ component in kyanite. The curve labeled "M" connects minima of the isothermal ΔG versus x curves. Points L1 and L2 represent limiting compositions at 1300K and at a pressure in the stability field of stoichiometric sillimanite where the stoichiometric kyanite → sillimanite reaction has a ΔG value of 628 J/mol (= 150 cal/mol). The stable assemblages are: x < L1: kyanite + sillimanite; L1 < x < L2: kyanite; x > L2: kyanite + hematite + quartz. (From Strens, 1968, Fig. 2).

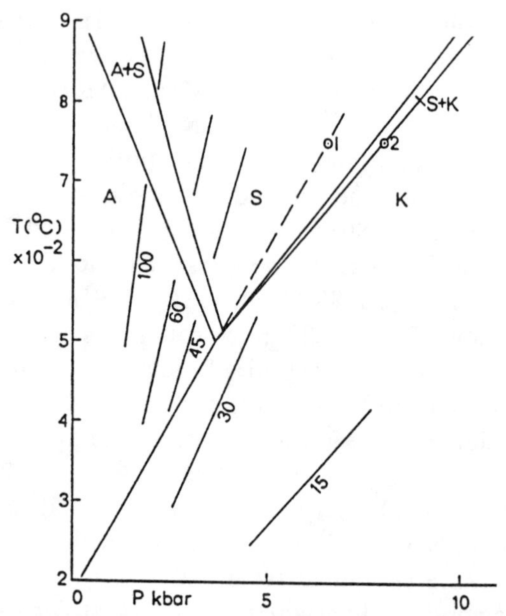

Figure 4.24. Strens' (1968) "suggested" phase equilibrium diagram illustrating the effect of solid solution of Fe^{3+} on the stability relations of the Al_2SiO_5 polymorphs in equilibrium with hematite + quartz. The bivariant field for andalusite + sillimanite is denoted as "A + S" whereas that of sillimanite + kyanite is marked "S + K". Also illustrated are selected geothermal gradients between 15°C/km and 100°C/km. Points 1 and 2 are Newton's (1966b,a) respective experimental determinations of the kyanite-sillimanite and kyanite-andalusite equilibria at 750°C. (From Strens, 1968, Fig. 3).

$$\Delta G_r = \Delta G_r^o + RT ln[(1-x) ln(1-x) + x ln x] \qquad [4.2]$$

or

$$\Delta G_r = \Delta H_r^o - T\Delta S_r^o + RT ln[(1-x) ln(1-x) + x ln x] \ .$$

Considering the configurational entropy,

$$\Delta S_c = -R\sum_i (x_i ln x_i),$$

equation [4.2] can be written as

$$\Delta G_r = \Delta H_r^o - T\Delta S_r^o - T\Delta S_c \ .$$

In contrast, Strens (1968) expressed the free energy change for the reaction as

$$\Delta G = \Delta H \cdot x - T\Delta S_c \ . \qquad [4.3]$$

As outlined by Kerrick and Speer (1988), this equation can be derived by starting with the fundamental expression for a binary mixture of the components $FeAlSiO_5$ and Al_2SiO_5:

$$\Delta G_m = X_{Al_2SiO_5}\Delta\overline{H}_{Al_2SiO_5} + X_{FeAlSiO_5}\Delta\overline{H}_{FeAlSiO_5}$$

$$+RT[X_{Al_2SiO_5}\ ln\ X_{Al_2SiO_5} + X_{FeAlSiO_5}\ ln\ X_{FeAlSiO_5}]$$

$$- T[X_{Al_2SiO_5}\Delta\bar{S}^{ex}_{Al_2SiO_5} + X_{FeAlSiO_5}\Delta\bar{S}^{ex}_{FeAlSiO_5}] \ . \qquad [4.4]$$

From this expression, equation [4.3] can be derived by assuming that the excess entropy terms and $\Delta\overline{H}_{Al_2SiO_5}$ are zero. However, it should be noted that ΔG in equation [4.3] is the free energy of mixing (ΔG_m) of the $FeAlSiO_5$ and Al_2SiO_5 components and not ΔG_r of equation [4.2] as stated by Strens (1968). He evaluated equilibrium by assuming that ΔG_r is zero such that $\Delta H \cdot x = T\Delta S_c$. As stated, this approach is misleading. From reaction [4.1], Strens' derivation erroneously implies the equilibrium coexistence of two aluminum silicates, i.e., a pure (Fe-free) Al_2SiO_5 phase coexisting with an Fe-bearing phase $(Al_{1-x}Fe_x)AlSiO_5$. Instead, equation [4.3] can be derived by differentiating equation [4.4] (Kerrick and Speer, 1988, p. 155):

$$d\Delta G_m/dX_{FeAlSiO_5} = 0 = \Delta\overline{H}_{FeAlSiO_5} + RT[ln(X_{FeAlSiO_5}/X_{Al_2SiO_5})] \ .$$

Thus, Strens' (1968) ΔH parameter represents $X_{FeAlSiO_5}\Delta H_{FeAlSiO_5}$ rather than the enthalpy change for reaction [4.1], and his ΔG value (e.g., Fig. 4.23) is actually ΔG_m. In spite of such errors in definition of the thermodynamic parameters, Strens (Table 4.1) correctly evaluated ΔG_m and, thus, correctly computed the divariant intervals shown in Figure 4.24. The major uncertainty regarding his computations is the lack of discussion of the source(s) for the T-x parameters (Table 4.1).

Using an approach similar to that of Strens (1968), Holdaway (1971) analyzed the thermodynamics of Fe^{3+} substitution in the Al_2SiO_5 polymorphs. Assuming one-site substitution of Fe^{3+} for Al, he expressed (p. 99) the "Solid solution free energy..." as:

$$G_s = xH_s - TS_s = xH_s - RT[x\ ln\ x\ - (1-x)ln(1-x)]\ .$$

In this equation, x is the fractional site occupancy of Fe^{3+}. As illustrated in Figure 4.25, Holdaway computed the position of the minima of the G_s curves from the criterion of horizontal tangency:

$$dG_s/dx = 0 = H_s + RT\ ln\ (x_{max} /1 - x_{max})$$

or

$$H_s = -RT\ ln\ (x_{max} /1 - x_{max})\ . \qquad [4.5]$$

Holdaway (1971) assumed that the x_{max} values correspond to the compositions of the aluminum silicates in equilibrium with hematite + quartz. He derived H_s values by considering the compositions of the most Fe-rich, Mn-free andalusite then reported, i.e., that from the Steinach aureole in Bavaria. Coupling an estimated metamorphic temperature of 700°C for this locality with available K_D data on Fe partitioning between Al_2SiO_5 polymorphs from Glen Clova, Scotland, Holdaway (1971) used equation [4.5] to compute H_s, and thus derived G_s versus $X_{FeAlSiO_5}$ plots for 700°C (Fig. 4.25). The difference in free energy between the minima of the G_s versus $X_{FeAlSiO_5}$ curves permitted computation of the pressure displacement (ΔP) of the Al_2SiO_5 equilibria compared to that with the pure phases: $\Delta P = \Delta G_{s,min} / \Delta V_r$. Using this equation, Holdaway (1971) computed the following displacements of the equilibrium curves:

	500°C	700°C
And = Sil	+284 bars	+497 bars
And = Ky	+100 bars	———
Ky = Sil	+31 bars	+62 bars

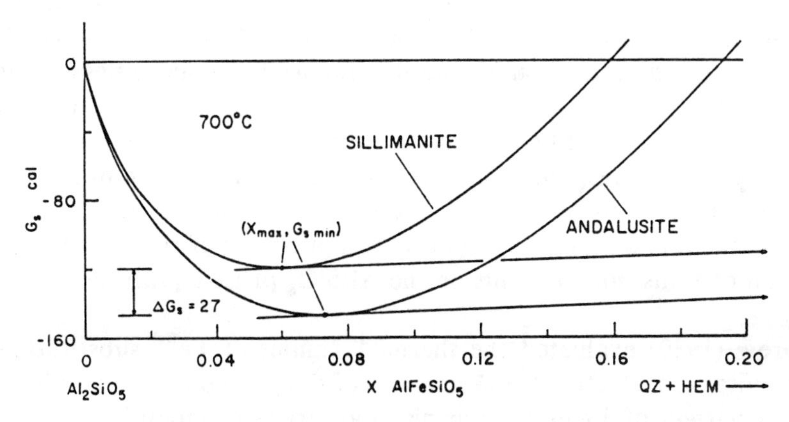

Figure 4.25. Holdaway's (1971) plot of the "solid solution free energy" (G_s) of sillimanite and andalusite as a function of composition. The horizontal lines with arrows are tangents to the free energy minima ($G_{s,min}$). (From Holdaway, 1971, Fig. 1).

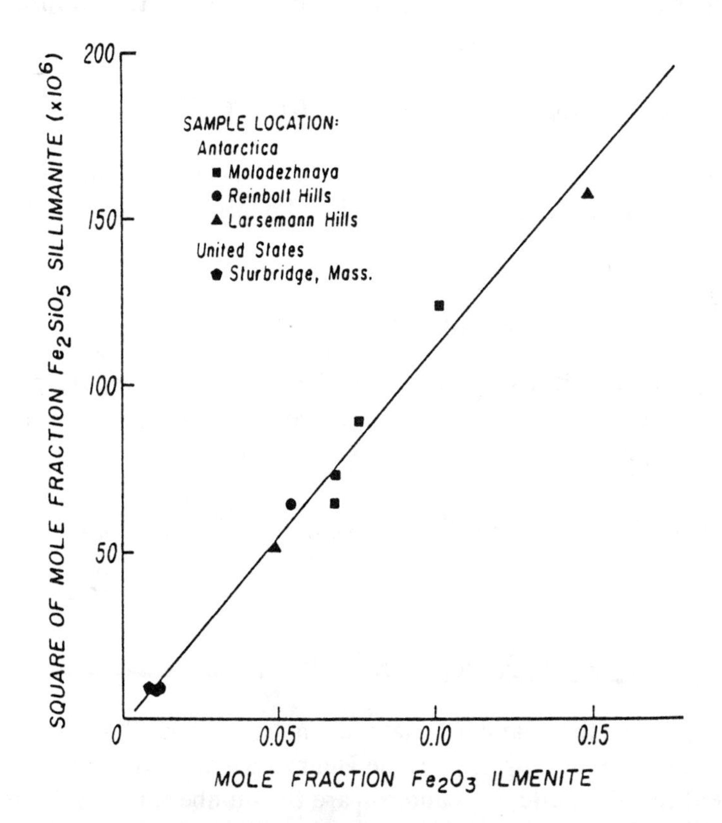

Figure 4.26. Iron contents of coexisting sillimanite and ilmenite. The ordinate assumes that Fe^{3+} substitutes in two sites in sillimanite. (From Grew, 1980, Fig. 6).

Holdaway (1971, p. 100) concluded that, with the exception of aluminum silicates in the relatively rare Mn-rich lithologies, "...solid solution of Fe^{3+} is not an important factor affecting aluminum silicate stabilities except, to a small extent, in the andalusite-sillimanite boundary". Kerrick and Speer (1988, Fig. 1 and p. 155) noted that the minima in G_s curves (Fig. 4.25) do not necessarily correspond with the compositions of the aluminum silicates in equilibrium with hematite + quartz. Thus, there is uncertainty in Strens' (1968) and Holdaway's (1971) computations of the effect of solid solution of transition elements on the Al_2SiO_5 phase equilibria.

Grew (1980) evaluated the thermodynamics of Fe^{3+} substitution in sillimanite from electron probe analyses of the iron contents of sillimanite from a variety of ilmenite-bearing, high grade metapelitic rocks. His analytical data show a correlation between the Fe content of sillimanite and coexisting ilmenite (Fig. 4.26), thereby suggesting equilibrium between these phases. As shown in Figure 4.26, Grew (1980) concluded that the partitioning of Fe between coexisting sillimanite and ilmenite is best fit by the empirical relation

$$[(X^{Sil}_{Fe_2SiO_5})^2 \ / \ X^{Ilm}_{Fe_2O_3}] = 1.110 \times 10^{-3}. \qquad [4.6]$$

He showed that this equation could be derived starting with the fundamental equation for the sillimanite-ilmenite-quartz assemblage at equilibrium

$$\mu^{Sil}_{Fe_2SiO_5} = \mu^{Ilm}_{Fe_2O_3} + \mu^{Qtz}_{SiO_2} \ .$$

Assuming that ilmenite can be represented as a symmetric regular solution, and Henry's Law behavior for the Fe_2SiO_5 component in sillimanite, Grew derived the equations

$$\mu^{Ilm}_{Fe_2O_3} = \mu^{\circ}_{Fe_2O_3} + \alpha_{Ilm} RT \ ln \ X_{Fe_2O_3} + W^{Ilm}_G X^2_{FeTiO_3} \ , \qquad [4.7]$$

and

$$\mu^{Sil}_{Fe_2SiO_5} = \mu^{*}_{Fe_2SiO_5} + \alpha_{Sil} RT \ ln \ X^{Sil}_{Fe_2SiO_5} \ , \qquad [4.8]$$

where $\mu^{\circ}_{Fe_2O_3}$ is the standard state chemical potential of the Fe_2O_3 component in ilmenite, $\mu^{*}_{Fe_2SiO_5}$ is the Henry Law constant of the Fe_2SiO_5 component in sillimanite, α_{Ilm} and α_{Sil} are the number of sites for mixing ("site multiplicities") of Fe^{3+} in ilmenite and sillimanite, respectively, and W^{Ilm}_G is the regular solution interaction parameter for ilmenite. He noted that the empirical equation [4.6] could be derived from equations [4.7] and

[4.8] with the following parameters: $\alpha_{Sil} = 2$, $\alpha_{Ilm} = 1$, and $W_G^{Ilm} = 0$. For sillimanite, Grew defended two-site mixing on the basis of the optical absorption data of Grew and Rossman (1976a,b) and W.A. Dollase's Mössbauer data (subsequently published in Rossman et al., 1982). He concluded that in sillimanite Fe^{3+} mixes ideally in the dilute compositional range $X_{Fe_2SiO_5} < 0.013$ (i.e., up to 0.64 wt % Fe_2O_3). As discussed in the preceding section reviewing spectroscopic studies on transition elements in sillimanite, the study of Rossman et al. (1982) indeed suggests mixing of Fe^{3+} on both octahedral and tetrahedral sites. However, in contrast to Grew's assumption of equal amounts of Fe^{3+} in these two sites, Fe^{3+} in sillimanite appears to be significantly fractionated into the octahedral site (Rossman et al., 1982). Grew (1980) defended his conclusion that $\alpha_{Ilm} = 1$ on the basis of local electrostatic neutrality (equivalent to Kerrick and Darken's (1975) model 4 for plagioclase solid solutions). However, Spencer and Lindsley's (1981) thermodynamic analysis of solution models for Fe-Ti oxides suggests that two-site mixing for ilmenite ($\alpha = 2$; mixing of Fe^{2+} and Fe^{3+} on "A" sites and Fe^{3+} and Ti on "B" sites) yields the best fit to their phase equilibrium data for the Fe-Ti oxides. The results of Rossman et al. (1982) and Spencer and Lindsley (1981) thus question the validity of Grew's (1980) interpretation of ideal mixing of Fe^{3+} in sillimanite for $X_{Fe_2SiO_5} < 0.013$.

Brothers (1986, 1987) used the Gibbs method of algebraic analysis to derive mathematical relations between pressure, temperature and composition for the rutile-ilmenite(hematite)-Al_2SiO_5-quartz assemblage. This equilibrium was calibrated against geothermometry of regionally metamorphosed rocks in northern New Mexico. She evaluated activities of Fe^{3+}-bearing sillimanite using site multiplicities of 1 and 2 for Fe^{3+}-Al^{3+} substitution. A multiplicity value of 1 yielded the best agreement with independent thermobarometric estimates for rocks with this assemblage in other areas. Thus, Brothers' (1986, 1987) study supports Grew's (1980) assumption that Fe substitutes for Al in one site in sillimanite.

<u>Partitioning of transition elements between coexisting polymorphs</u>. Pearson and Shaw (1960) carried out the first detailed field-oriented study of the partitioning of trace elements between the Al_2SiO_5 polymorphs. Using spectrographic analysis they determined the compositions of aluminum silicates from many different localities and concluded (p. 808) that: "No indisputable differences in trace-element contents were found, but andalusite probably contains less Cr, whereas B, Be and Ba may be concentrated in sillimanite. It is unlikely that trace elements are factors in the polymorphic relations between these minerals".

From electron probe analysis, Chinner et al. (1969) determined the transition element contents of coexisting aluminum silicates from numerous localities. Their study excluded manganiferous lithologies. Ti, V, Cr and Fe were the only transition elements present in quantities exceeding 0.01 wt %. With the exception of Cr-rich kyanites in eclogites and blueschists, Fe is the only transition element consistently present in amounts greater than 0.2 wt %. Their analysis of Fe in kyanite-andalusite, kyanite-sillimanite, and andalusite-sillimanite pairs suggested no marked partitioning of Fe. They stressed the correlation between f_{O_2} and Fe^{3+} in solid solution. Chinner et al. (1969, p. 111) concluded that partitioning of Fe^{3+} may yield divariant coexistence of polymorphs in oxidizing environments (characterized by the presence of hematite); in contrast, "...for aluminum silicates crystallized under the relatively low oxygen partial pressures characterized by graphite and magnetite equilibria, partitioning of iron between two polymorphs is small and divariant bands of coexistence negligible". They suggested that, in general, disequilibrium is a more important factor than minor element partitioning in the coexistence of polymorphs.

Albee and Chodos (1969) used the electron probe to analyze the compositions of coexisting aluminum silicates in samples from the area studied by Hietanen (1956) in Idaho and that studied by Woodland (1963) in Vermont They found no significant differences in the Fe/Al ratios between coexisting polymorphs (Fig. 4.27), thus concluding that "Solid solution of minor elements does not appear to account for the coexistence of the polymorphs in these two areas..." (Albee and Chodos, 1969, p. 310).

Okrusch and Evans (1970) analyzed the compositions of coexisting andalusite + sillimanite in metapelites from several contact aureoles. As shown in Figure 4.28, their data suggest that the ratio of the wt % of Fe_2O_3 in sillimanite to that in andalusite ranges from 1:1 to 2:1. From calculations assuming ideal mixing they concluded: "The resultant stabilization of andalusite relative to sillimanite is geologically rather insignificant" (Okrusch and Evans, 1970, p. 261).

Dodge (1971) reported spectrographic analyses of andalusite and sillimanite from samples of metapelites from the Sierra Nevada and Inyo Mountains of California. He concluded that the concentration of Cr is consistently higher in sillimanite compared to andalusite. Other transition elements analyzed by Dodge (Fe, Ti, Mn, Ni) show no consistent partitioning between andalusite and sillimanite.

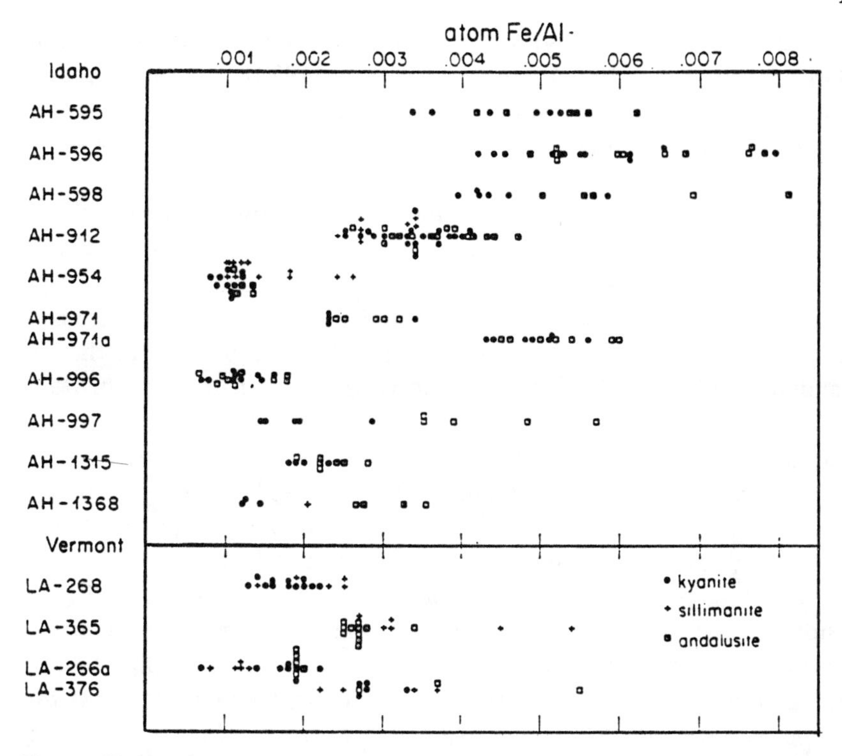

Figure 4.27. Atomic Fe/Al ratio for aluminum silicates in samples from the Boehls Butte area (central Idaho) and the Burke Mountain area (northern Vermont). (From Albee and Chodos, 1969, Fig. 2).

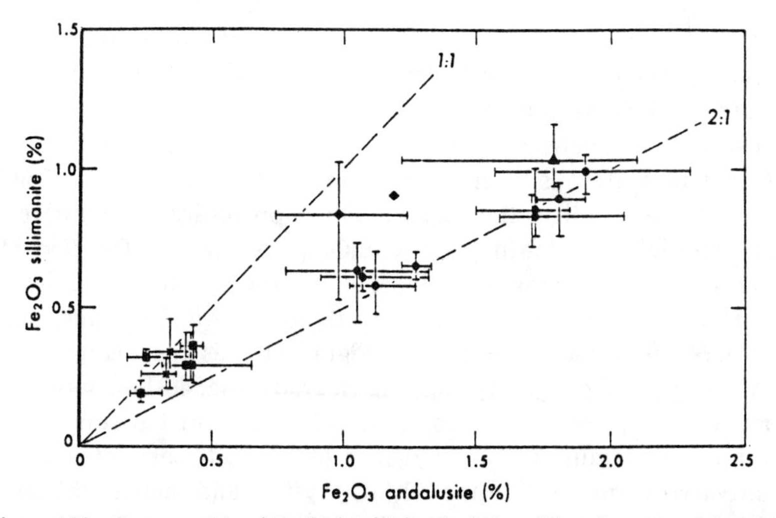

Figure 4.28. Concentration of Fe_2O_3 (wt %) in coexisting sillimanite and andalusite from various localities (dots = Steinach aureole, Bavaria; squares = Fanad aureole, Ireland; triangles = May Lake, Yosemite National Park, California; crosses = Ross of Mull, Invernesshire, Scotland). (From Okrusch and Evans, 1970, Fig. 1).

In an analysis of regionally metamorphosed pelites containing andalusite, kyanite and sillimanite, Rumble (1973) concluded that the partitioning of Fe is insufficient to significantly affect the P-T stability relations of the polymorphs.

Yokoi (1983) studied the role of Fe in solid solution in andalusite and sillimanite in samples collected from low-pressure regionally metamorphosed pelites of the Ryoke metamorphic belt in central Japan. Electron probe analyses revealed that andalusite and sillimanite are zoned in Fe; the cores of the crystals are enriched in Fe compared to the rims. As shown in Figure 4.29, the cores and rims of coexisting andalusite and sillimanite show a fairly systematic variation in Fe content with metamorphic grade. From thermodynamic calculations assuming ideal mixing in andalusite and sillimanite, Yokoi (1983) concluded that for sample AS1 (Fig. 4.29) partitioning of Fe^{3+} into andalusite resulted in a 10°C displacement of the andalusite-sillimanite equilibrium compared to that involving pure (Fe-free) andalusite and sillimanite. Yokoi (1983) considered that the temperature range (grade) over which andalusite and sillimanite coexist in the central part of the Ryoke metamorphic belt may have been considerably larger than 10°C. Thus, he suggested that there may be significant deviations from ideal mixing in andalusite and sillimanite.

Grambling and Williams (1985) studied in detail the effects of Fe^{3+} and Mn^{3+} on the aluminum silicate phase relations of regionally metamorphosed pelitic rocks in three fault-bounded uplifts (Picuris, Truchas, and Rio Mora) in north-central New Mexico. Two of these areas (Rio Mora and Picuris) are advantageous in that aluminum silicates occur in typical Mn-poor pelitic lithologies as well as in a distinctive Mn-rich layer (the manganiferous layer is absent in the Truchas uplift). The Rio Mora and Picuris uplifts thus afford the opportunity to compare and contrast Mn-rich versus Mn-poor Al_2SiO_5 parageneses. The Rio Mora uplift (Fig. 4.30) contains a well-defined kyanite-sillimanite isograd; andalusite is confined to the Mn-rich lithology. All three polymorphs occur where the kyanite-sillimanite isograd crosses the manganiferous horizon. Because of the gently-dipping isograds (Fig. 4.31), coupled with significant vertical relief, the southern Truchas uplift offers the rare opportunity to examine the convergence of isograds correlated with all three univariant equilibria in the Al_2SiO_5 system and, hence, this area is a "triple point" locality. In the Picuris uplift the isograds are similar to those of the Truchas uplift; i.e., subhorizontal isograds with the kyanite-sillimanite isograd underlying the andalusite-sillimanite isograd. Grambling and Williams (1985) obtained electron probe analyses of the

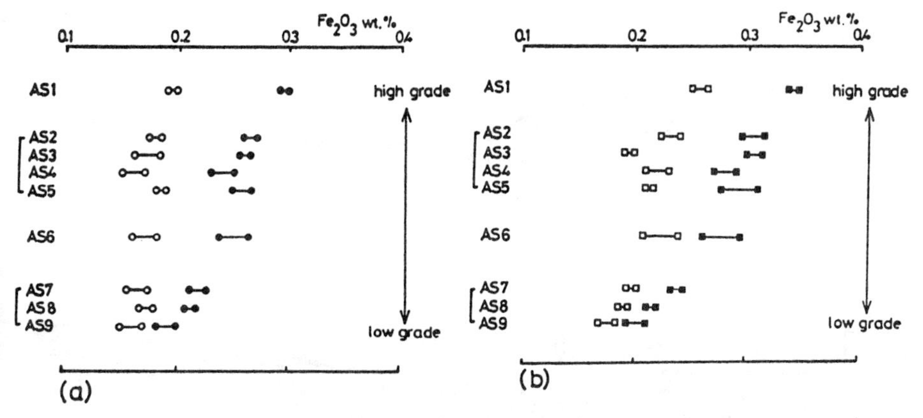

Figure 4.29. Fe_2O_3 content of coexisting sillimanite (open circles and squares) and andalusite (filled circles and squares) in metapelites from the central Ryoke metamorphic belt, Japan. The compositions of crystal *cores* are given in (a) whereas the *rim* compositions are displayed in (b). (From Yokoi, 1983, Fig. 9).

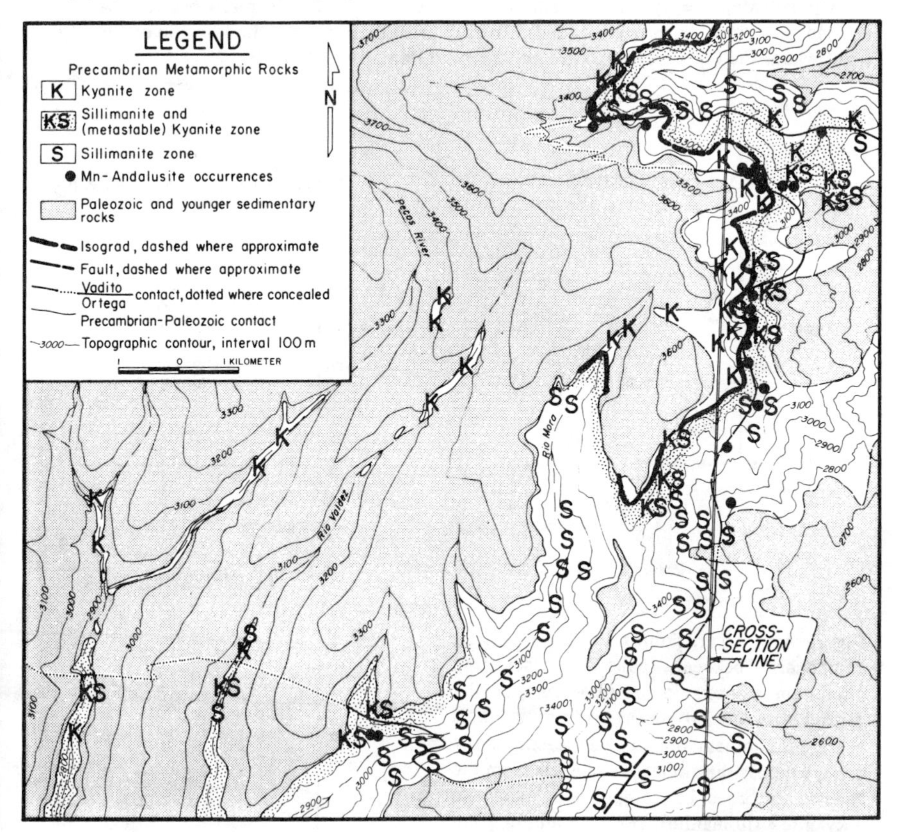

Figure 4.30. Geologic map showing the distribution of kyanite, sillimanite and Mn-andalusite in the Rio Mora uplift, north-central New Mexico. Note that Mn-andalusite is confined to a manganiferous horizon adjacent to the Vadito-Ortega contact. (From Grambling and Williams, 1985, Fig. 2).

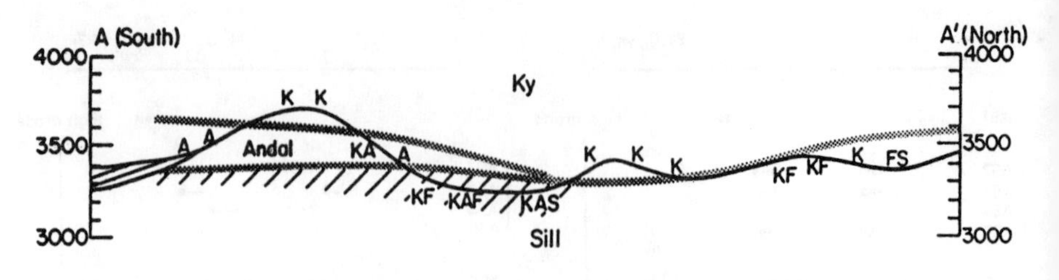

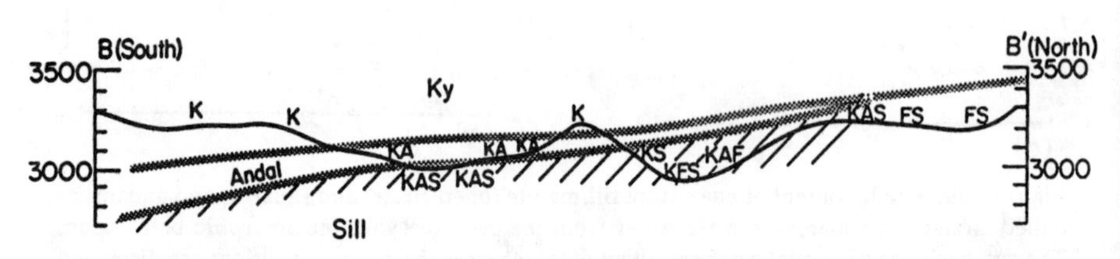

Figure 4.31. Cross sections through the Truchas Range showing aluminum silicate localities (A = andalusite, K = kyanite, S = sillimanite, F = fibrolite) and isograds based on Al_2SiO_5 polymorphic transformations (shaded). In each cross section, note the location of the triple point where the three isograds intersect. (From Grambling, 1981, Fig. 7).

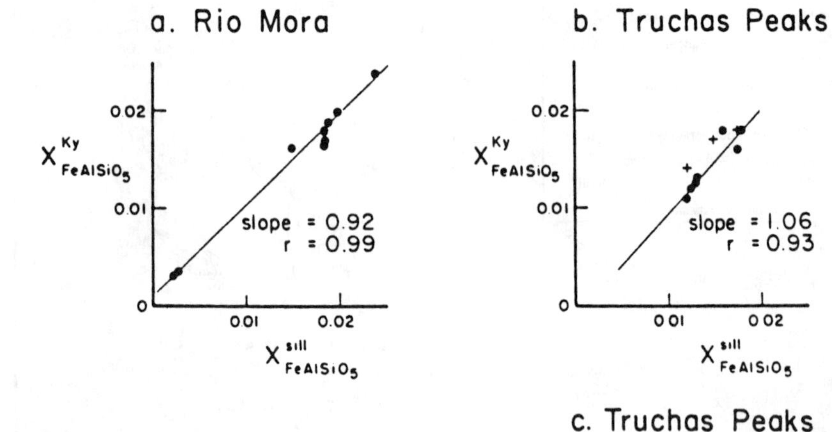

Figure 4.32. Partitioning of the $FeAlSiO_5$ component between pairs of coexisting aluminum silicates in samples from north-central New Mexico. In (b) the "plus" symbols represent samples from the kyanite-sillimanite zone, whereas the circles are samples from the "triple point" zone (i.e., samples with andalusite + kyanite + sillimanite). (From Grambling and Williams, 1985, Fig. 10).

aluminum silicates in these areas. Textural relationships coupled with the distribution of Fe between kyanite and sillimanite (Fig. 4.32a) suggest equilibrium. Grambling and Williams (1985) contended that equilibrium between andalusite and the other polymorphs is evidenced by the consistent partitioning of Fe and Mn (Fig. 4.32 and tie lines in Fig. 4.33). In the Rio Mora (Fig. 4.33) and Picuris (Fig. 4.34) uplifts, andalusite collected from the manganiferous horizon has an essentially constant value of the $FeAlSiO_5$ component. In contrast to the conclusions of Abs-Wurmbach et al. (1981, 1983), Figures 4.33 and 4.34 strongly suggest that the Fe content of iron-saturated andalusite does not increase with increasing Mn content. In the three uplifts studied by Grambling and Williams (1985) there is an excellent correlation between the compositions of the Fe-Ti oxides and both the transition element contents and assemblages of the aluminum silicates. In the Rio Mora uplift the Fe content of kyanite and sillimanite is lowest in hematite-free rocks containing rutile or ilmenite, intermediate in ilmenite-free rocks containing titaniferous hematite, and highest in rocks containing virtually pure hematite. Mn-rich andalusite is confined to rocks containing pure hematite. Grambling and Williams (1985) attributed the virtually constant concentration of the $FeAlSiO_5$ component of Mn-andalusite (Figs. 4.33 and 4.34) to the equilibrium

$$\mu_{Fe_2O_3} + \mu_{SiO_2} + \mu_{Al_2SiO_5} = 2\mu_{FeAlSiO_5}$$

(the stoichiometric reaction coefficients are corrected from those of Grambling and Williams). At constant P and T, buffering of $\mu_{FeAlSiO_5}$ results from the presence of quartz, virtually pure hematite, and aluminum silicates with $\mu_{Al_2SiO_5}$ close to that of the stoichiometric end member. In the Truchas Range, rocks with titaniferous hematite (i.e. hematite with relatively low $X_{Fe_2O_3}$) contain kyanite + sillimanite and no andalusite. In contrast, samples with pure hematite (and implicitly higher f_{O_2}) contain only andalusite. In the Picuris uplift, rocks with hematite contain only andalusite. Hematite-free rocks with ilmenite contain various combinations of the three aluminum silicates. The minor element content of andalusite correlates with the opaque oxide assemblage: andalusite in hematite-bearing assemblages is considerably richer in Fe and Mn than andalusite in rocks containing ilmenite. Assuming that all Fe^{3+} and Mn^{3+} is contained within octahedral sites in all three polymorphs, and assuming ideal mixing, Grambling and Williams (1985) computed the kyanite-andalusite and andalusite-sillimanite P-T equilibria in the three areas. Figure 4.35 suggests that the triple point assemblage found in the Rio Mora uplift (i.e. where the manganiferous horizon crosses the kyanite-sillimanite isograd) crystallized at about 4.6 kbar and 540°C. The

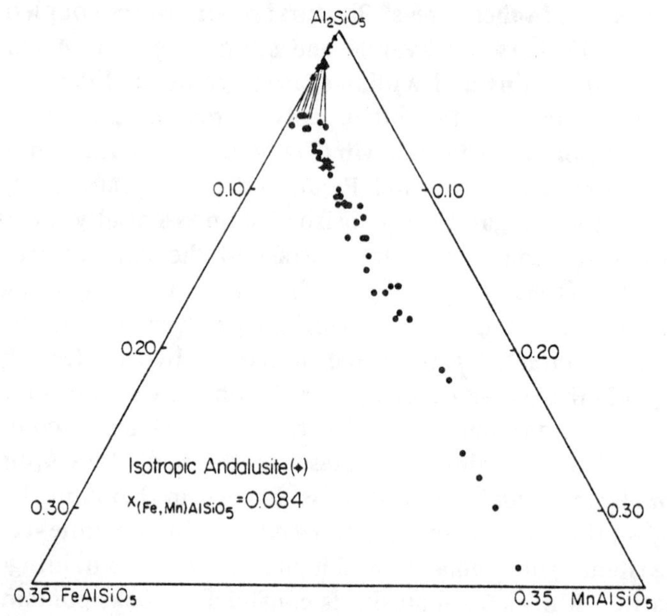

Figure 4.33. Compositions (atomic proportions) of aluminum silicates from the Rio Mora uplift, New Mexico. Triangles = kyanite and sillimanite, circles = andalusite, and crosses = isotropic andalusite ($X_{(MnFe)AlSiO_5}$ = 0.084). Tie lines connect the compositions of coexisting phases. (From Grambling and Williams, 1985, Fig. 9).

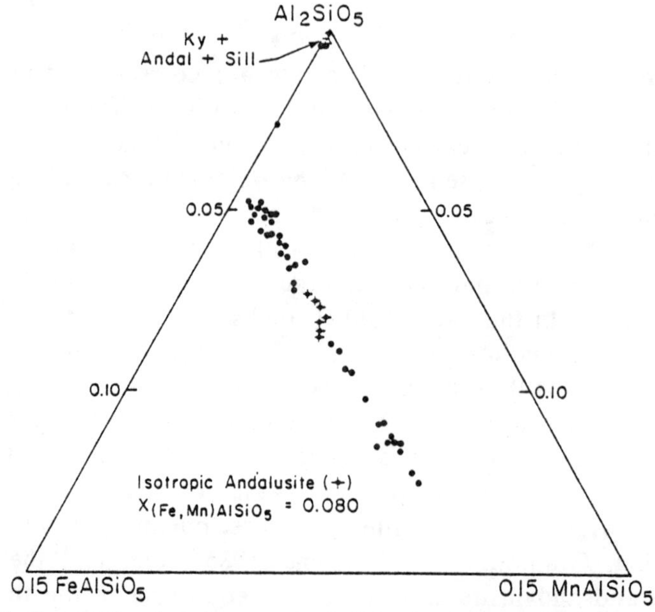

Figure 4.34. Compositions (atomic proportions) of aluminum silicates from the Pilar-Copper Hill area, Picuris Range, New Mexico. Open triangles = compositions of coexisting andalusite + sillimanite + kyanite. Other symbols as in Figure 4.33. (From Grambling and Williams, 1985, Fig. 11).

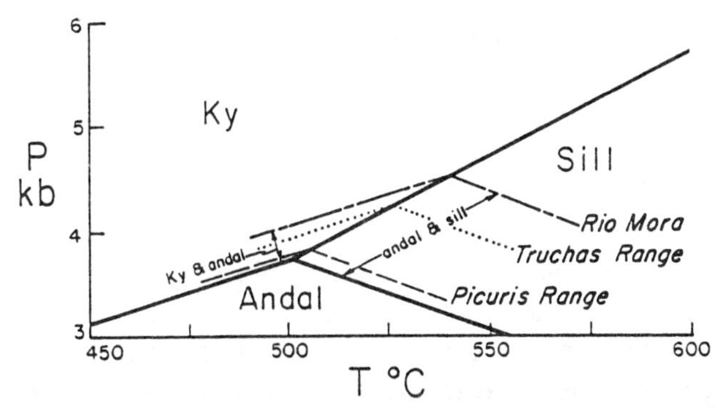

Figure 4.35. The effect of solid solution of Fe^{3+} and Mn^{3+} on phase equilibria of aluminum silicate assemblages in three areas in north-central New Mexico. The arrows delimit bivariant fields for kyanite + andalusite and andalusite + sillimanite. (From Grambling and Williams, 1985, Fig. 12).

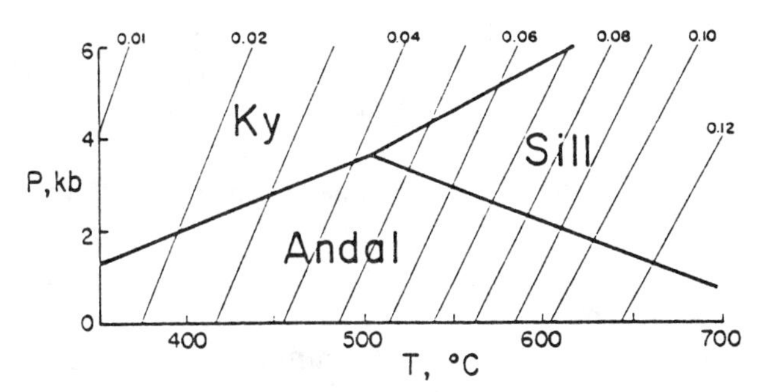

Figure 4.36. Calculated isopleths of $X_{FeAlSiO_5}$ in iron-saturated andalusite (i.e., andalusite coexisting with hematite + quartz). Univariant equilibria for the Fe^{3+}-free Al_2SiO_5 system are from Holdaway (1971). (From Grambling and Williams, 1985, Fig. 15).

computed triple point for this locality is at a P-T condition significantly different from that of the stoichiometrically pure phases [which Grambling and Williams assumed to be that determined by Holdaway (1971)]. Similar calculations for the andalusite-kyanite-sillimanite assemblage of the Truchas Peak uplift yield P-T conditions that are about 400 bars and 20°C higher that of Holdaway's (1971) triple point. Only the Picuris Range has aluminum silicates suggesting P-T conditions close to Holdaway's (1971) triple point (Fig. 4.35).

Following an approach similar to that of Grew (1980), Grambling and Williams (1985) defined five Gibbs-Duhem equations for the Al_2SiO_5-hematite-quartz subassemblage. Simultaneous solution yielded a single equation for $(dP/dX_{FeAlSiO_5})_T$. From an integrated version of this equation, Grambling and Williams (1985) computed isopleths of $X_{FeAlSiO_5}$

in Fe-saturated andalusite (Fig. 4.36). These results show that the Fe content of andalusite in hematite-bearing assemblages systematically varies with pressure and temperature.

The study of Grambling and Williams (1985) illustrates that Fe^{3+} and Mn^{3+} in the aluminum silicates can have important effects on the stability relations of these phases. In particular, the fractionation of Fe^{3+} and Mn^{3+} into andalusite yields an enlarged P-T stability field of this phase. It is notable that in the Truchas uplift, extensive solid solution of Fe^{3+} into andalusite occurs in hematite-bearing, Mn-poor lithologies. Thus, in Mn-poor metapelites metamorphosed under high oxygen fugacities, Fe in solid solution can have significant effects on the aluminum silicate phase equilibrium relations.

Grambling and Williams' (1985) calculations of the displacement of the Al_2SiO_5 equilibria due to solid solution of Fe^{3+} and Mn^{3+} assumed single site mixing for all three polymorphs. However, spectroscopic studies suggest that mixing occurs on multiple sites. From the results of Abs-Wurmbach et al. (1981) and Weiss et al. (1981), let us assume that in andalusite, 85% of the Fe^{3+} + Mn^{3+} resides in the octahedral site, whereas the remaining 15% occurs in the 5-coordinated site. For sillimanite, the Fe^{3+} site occupancies of Rossman et al. (1982) for yellow sillimanite are assumed to apply to Fe^{3+} + Mn^{3+}; i.e., 80% of the Fe^{3+} + Mn^{3+} occurs in octahedral coordination whereas the remaining 20% resides in tetrahedral sites. With ideal multisite mixing, the isobaric displacement of the andalusite-sillimanite equilibrium from that with stoichiometric end members may be calculated from

$$\Delta P = (RT/\Delta V_r) ln(X_{Al}^{VI}\, X_{Al}^{IV})^{Sil}/(X_{Al}^{VI}\, X_{Al}^{V})^{And} \; .$$

Using the molar volume data of Gunter and Bloss (1982) for andalusite, and Grew (1980) for sillimanite, the calculated Al_2SiO_5 phase equilibria for multisite mixing for the Truchas and Rio Mora uplifts are such that at the scale of Figure 4.35 there is no significant difference in the positions of the Al_2SiO_5 equilibria for single site versus multisite mixing. The largest difference in the equilibria computed with single-site versus multisite mixing is with the Rio Mora uplift; however, the differences are petrologically insignificant (ΔP = 200 bars and ΔT = 10°C for the andalusite-sillimanite equilibrium). Thus, it is concluded that the assumption of single-site mixing is adequate for computing the displacement of the Al_2SiO_5 equilibria due to partitioning of Fe and Mn amongst the polymorphs.

As shown in Figure 4.35, solid solution of Fe^{3+} and Mn^{3+} can significantly affect the kyanite-andalusite and andalusite-sillimanite equilibria. However, the kyanite-sillimanite equilibrium is essentially univariant. The univariant behavior of this equilibrium reflects minimal partitioning of Fe^{3+} between coexisting kyanite and sillimanite (Fig. 4.32), and the fact that the kyanite-sillimanite equilibrium is less affected by perturbations in ΔG_r compared to the other two Al_2SiO_5 equilibria (see Fig. 1.11). The insignificance of solid solution of Fe^{3+} on the kyanite-sillimanite equilibrium is bolstered by the study of Ghent et al. (1980) on the kyanite-sillimanite isograd in the Mica Creek area, British Columbia. Electron probe analysis revealed very similar Fe contents in coexisting sillimanite (FeO = 0.22 wt %) and kyanite (FeO = 0.14-0.18 wt %). In the vicinity of the kyanite-sillimanite isograd in the Truchas Range, New Mexico, Grambling (1981) showed that there is virtually no partitioning of Fe^{3+} between coexisting kyanite and sillimanite.

Kerrick and Speer (1988) carried out a field-oriented study of minor element solid solution on the andalusite-sillimanite equilibrium. In several locales they focused on the role of solid solution on the andalusite → sillimanite isograd reaction. The Ardara aureole (Ireland) and the Waterville-Vassalboro area (Maine) contain sharp isograds with little or no zone of coexisting andalusite + sillimanite. The low minor element contents of andalusite and sillimanite in these areas are compatible with field evidence suggesting virtually univariant behavior of the andalusite → sillimanite isograd reaction. The Kiglapait aureole (Labrador) exhibits different behavior of the andalusite → sillimanite isograd between two contrasting lithologies, i.e., graphite-free metaquartzites and metapelites versus graphitic metasiltstones. In non-graphitic rocks (Fig. 4.37), andalusite + sillimanite coexist over a zone 300 m wide (measured perpendicular to the isograd surfaces). Throughout this zone, modal analyses of andalusite and sillimanite (Fig. 4.38) suggest a prograde increase in the progress of the andalusite → sillimanite reaction. Compositional data of coexisting andalusite + sillimanite in this zone (Fig. 4.39) suggest prograde evolution along a $T\text{-}X_{MAlSiO_5}$ partitioning "loop" (M = transition element). In contrast to graphite-free assemblages, andalusite and sillimanite in graphitic metasiltstones have very low concentrations of minor elements such that a zone of coexisting andalusite + sillimanite is virtually absent in graphitic metasiltstones. The contrast in minor element contents of the graphite-free versus graphite-bearing rocks is compatible with the implicit differences in f_{O_2}. Specifically, low concentrations of Fe^{3+} in the aluminum silicates are expected in the relatively reducing conditions of graphitic rocks, whereas higher Fe^{3+} contents are expected with the relatively higher f_{O_2} conditions of non-graphitic lithologies. As

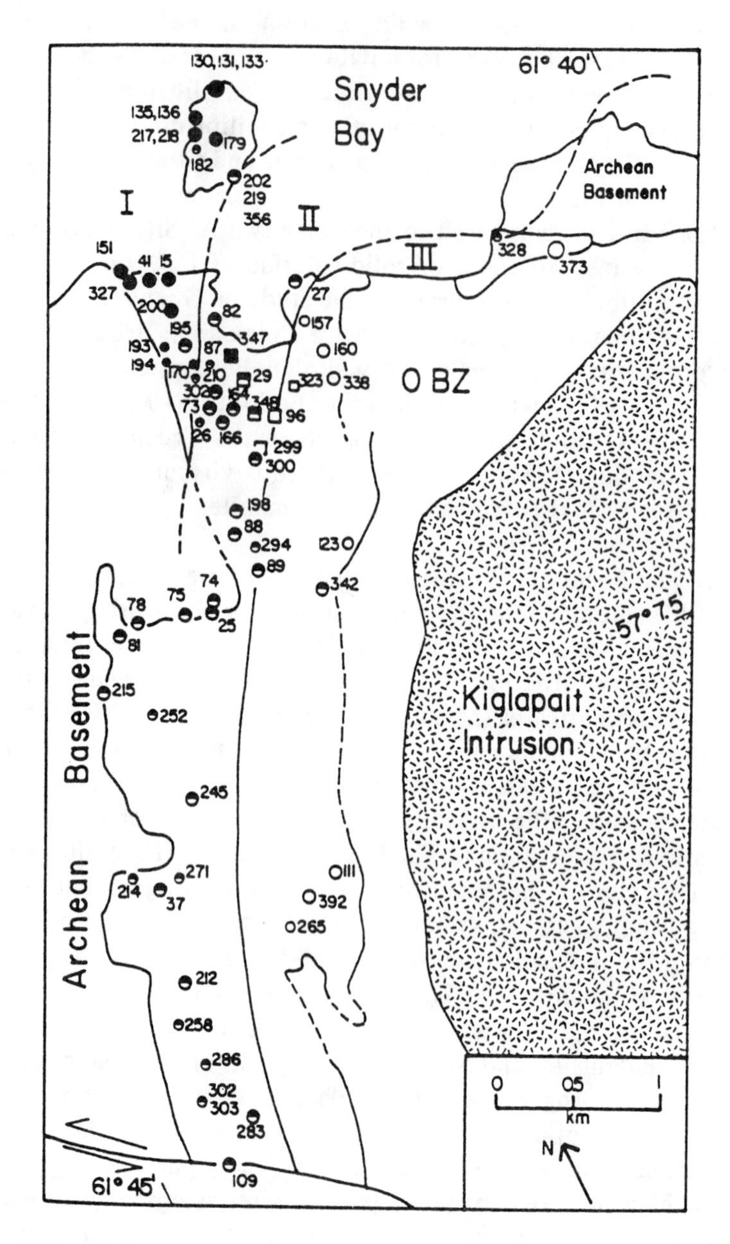

Figure 4.37. Kerrick and Speer's (1988) sample locality map of the Kiglapait aureole, Labrador. The squares represent graphite-sulfide metasiltstones, whereas the circles are metaquartzites. The aluminum silicates are denoted by: filled symbols = andalusite; half-filled symbols = andalusite + sillimanite; open symbols = sillimanite only. (From Kerrick and Speer, 1988, Fig. 9).

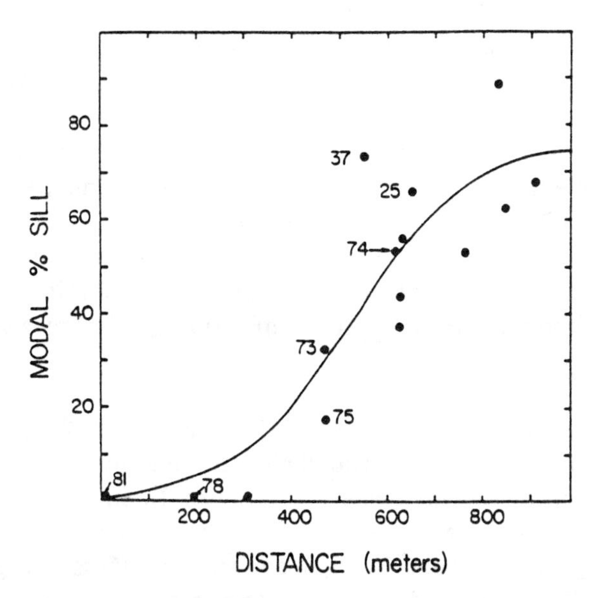

Figure 4.38. Modal percent sillimanite [Sill/(And + Sill)] as a function of map distance in metamorphic zone II of the Kiglapait aureole, Labrador (see Fig. 4.37). On the abscissa, the zero value corresponds to the boundary between metamorphic zones I and II, whereas the boundary between zones II and III is at 1 km. The sigmoidal curve was obtained by least-squares regression. Sample numbers are next to each data point. (From Kerrick and Speer, 1988, Fig. 10).

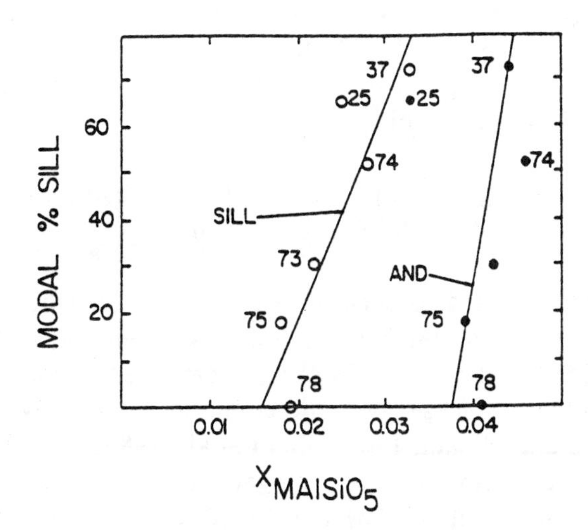

Figure 4.39. Modal percent sillimanite [Sill/(And + Sill)] as a function of composition of andalusite-sillimanite pairs from zone II of the Kiglapait aureole, Labrador (see Fig. 4.37). Open circles: sillimanite, filled circles: andalusite. Sample numbers are next to each data point. (From Kerrick and Speer, 1988, Fig. 11).

in the study of Grambling and Williams (1985), the Kiglapait aureole demonstrates the correlation between f_{O_2} and minor element contents of the aluminum silicates.

In addition to metapelites, Kerrick and Speer (1988) also evaluated the role of minor element solid solution in Al_2SiO_5-bearing peraluminous granitoids. Their electron probe analyses of andalusite + sillimanite and andalusite + fibrolite in selected samples of peraluminous granitoids show negligible minor element contents and partitioning of minor elements. The petrologic implications of these compositional data are discussed in Chapter 11.

Recent studies have provided additional confirmation of the low minor element concentration of the aluminum silicates in graphitic assemblages. In a study of graphitic metapelites in west-central Maine, Holdaway et al. (1988) analyzed sillimanite in six samples. The sillimanites have Fe_2O_3 contents ranging from 0.19 to 0.35 wt %, and trace amounts of MgO, TiO_2, and Mn_2O_3. They concluded that "...this amount of Fe is of no consequence in calculating reaction boundaries from experimental equilibria" (Holdaway et al., 1988, p. 32). Kerrick and Woodsworth (1989) analyzed the compositions of andalusite and sillimanite in metapelites from the Mt. Raleigh pendant, British Columbia. Most of the aluminum silicates in their samples of graphitic metapelites have Fe_2O_3 contents in the range 0.15-0.35 wt %. K_D values for andalusite-sillimanite (= $X^{Sil}_{Al_2SiO_5}/X^{And}_{Al_2SiO_5}$) range from 0.997 to 1.007. According to Figure 4.40 minor element partitioning has an insignificant effect on the stability of andalusite and sillimanite in this pendant.

Zoning of transition elements. Zonation of transition elements has been described in all of the Al_2SiO_5 polymorphs.

There have been several descriptions of uneven distribution of blue color in kyanite. Faye and Nickel (1969) and Rost and Simon (Figs. 4.41 and 4.42) showed that the blue coloration is arranged in tabular-shaped zones parallel to {100}. These zones are coincident with elevated Fe and Ti contents (Faye and Nickel, 1969; Ghera et al., 1986). In addition, the blue regions are marked by relatively longer *a* and *b* unit cell dimensions (Fig. 4.43). Ghera et al. (1986) concluded that, in contrast to the inhomogeneous distribution of Fe and Ti, Cr is homogeneously distributed. They argued that because of their relatively large ionic radii, Fe^{2+} and Ti^{3+} do not readily substitute into the kyanite structure. Thus, these cations cluster into regions where "...possible distortion of the structure is expected to result into local point defects...". In contrast,

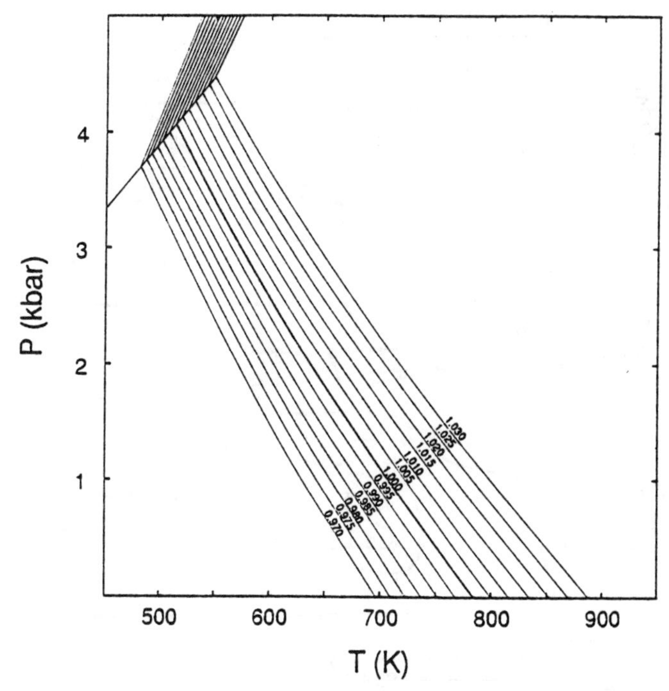

Figure 4.40. Isopleths of selected values of the equilibrium constant for the andalusite-sillimanite equilibrium. The heavy line (K = 1.0) corresponds to the equilibrium with stoichiometrically pure phases. This diagram was calculated by J.A.D. Connolly using the VERTEX program of Connolly and Kerrick (1987).

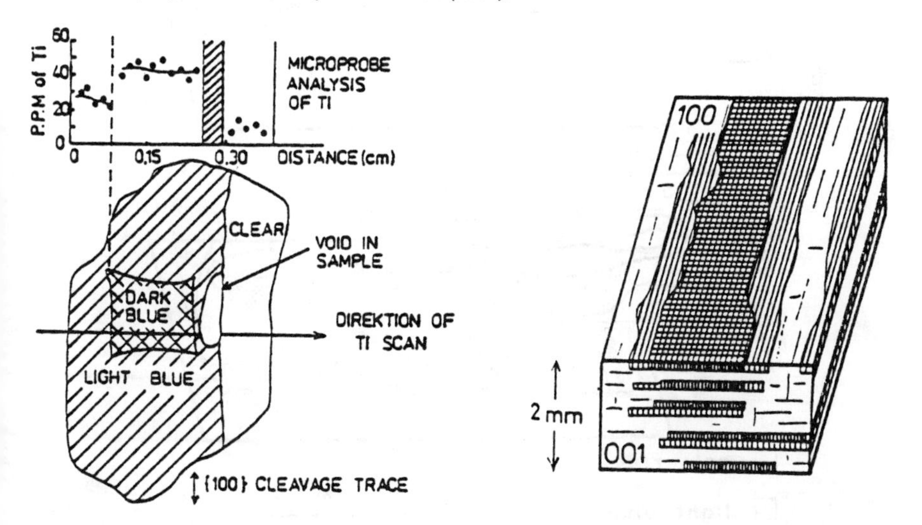

Figure 4.41. (*Left*). Patchy color zonation in a kyanite crystal (bottom) showing location of an electron probe traverse (zoning profile is shown in top diagram). (From Rost and Simon, 1972, Fig. 1).

Figure 4.42. (*Right*). Tabular-shaped regions of dark blue coloration (corresponding to elevated Ti content) in a kyanite crystal. (From Rost and Simon, 1972, Fig. 2).

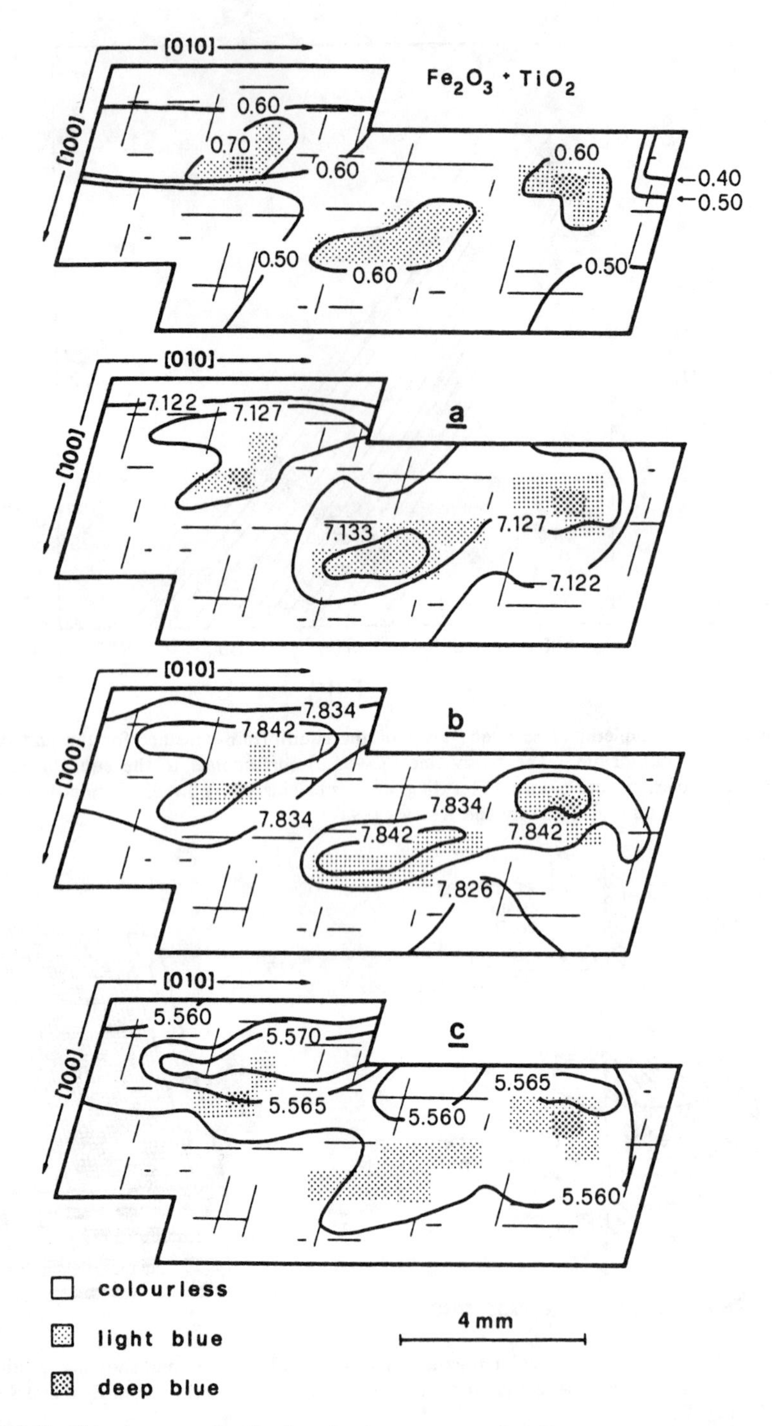

Figure 4.43. Heterogeneous distribution of color, contours of wt % Fe_2O_3 + TiO_2 (top), and *a*, *b*, and *c* unit cell parameters, in a {110} section of a kyanite crystal. (From Ghera et al., 1986, Fig. 8).

because of a relatively smaller ionic radius, Cr^{3+} is able to more readily substitute in the kyanite structure and thus Cr displays relatively even spatial distribution.

Andalusite displays two types of zoning: *concentric* zoning with rims concentrically disposed about cores, and *sector* zoning characterized by "patchy" distribution of zones of differing composition.

Several papers have described *concentric* zoning of andalusite with pink cores and colorless rims (Okrusch and Evans, 1970; Plummer, 1980; Grapes, 1987; Kerrick and Speer, 1988). Okrush and Evans (1970) showed that the pink cores are enriched in Ti, Fe, and Mg relative to the rims; however, there are no significant differences in the concentration of Cr and V between cores and rims. Using back-scattered electron imaging, Grapes (1987) showed that the central Fe-rich core of an andalusite crystal actually consists of micron-scale lamellar and patchy zones of Fe-poor areas intergrown with Fe-rich domains. Figure 4.44 shows a distinct parallelism of such lamellae thereby suggesting that the orientation of the lamellae is crystallographically controlled. Grapes (1987) concluded that the cores formed during the thermal maximum of regional metamorphism whereas the rims formed during retrogression.

Sector zoning in andalusite has been described by Macdonald and Merriam (1938), Hollister and Bence (1967), and Burt and Stump (1983). Hollister and Bence (1967) described sector zoning parallel to {001}, {100} and {010}. Dowty (1976b, p. 466) developed a model for sector zoning based on differences in the ease of adsorption of impurity cations on different faces of growing crystals. The model involves locating favorable "protosites" within which the adsorbed impurity cations can be incorporated into the structure of the growing crystal. Dowty (1976b) noted that his model yields ambiguous results for the {110} faces of andalusite in that these faces lack favorable octahedral Al_1 protosites. However, he contended that impurities should be readily incorporated within octahedral sites exposed on {001} sectors.

Regardless of the actual mechanism(s) producing sector zoning in andalusite, this type of zoning complicates the petrologic interpretation of solid solution in andalusite. This dilemma is epitomized by Hollister's (1969, p. 365) statement: "...which part of the crystal can be considered the stable andalusite and how did the different regions behave with time or external conditions?". Burt and Stump (1983) found inclusions of sector-zoned andalusite within a single crystal of andalusite with lower birefringence. In this case the sector zoned andalusite may not be

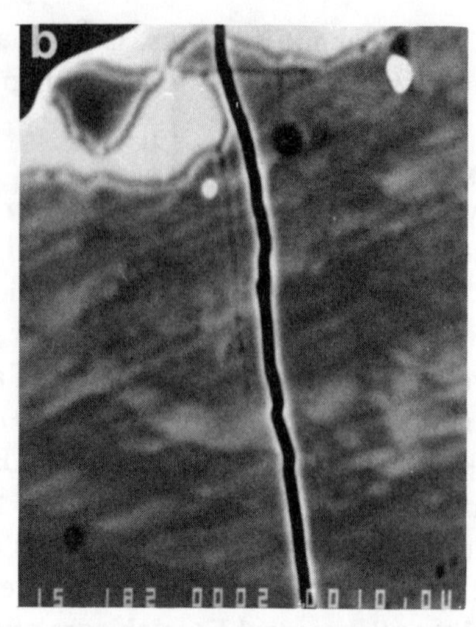

Figure 4.44. Backscattered electron image of a portion of an andalusite crystal. The light-colored lamellae and patches are enriched in Fe_2O_3, MgO and TiO_2. Scale bar = 10μm. (From Grapes, 1987, Fig. 1b).

problematical in that the low birefringent andalusite was presumably in equilibrium with the surrounding matrix. Grambling and Williams (1985) found kyanite and sillimanite within metapelites of north-central New Mexico to be unzoned. However, andalusite displays "erratic zoning patterns" of Fe_2O_3 and Mn_2O_3. "The zoning is not concentric. In many samples, one edge of a porphyroblast has the highest Mn_2O_3 and the opposite edge has the lowest Mn_2O_3. In other samples, the zone of highest Mn_2O_3 traces relict bedding planes which were transposed then overgrown with andalusite. The erratic zoning patterns are interpreted as representing chemical heterogeneities in the pre-metamorphic rock, overgrown and incorporated into andalusite during metamorphism" (Grambling and Williams, 1985, p. 332). As with the above query by Hollister (1969), andalusite crystals with irregular zoning throughout present difficulties in the petrologic interpretation using an equilibrium model.

In contrast to the abrupt compositional discontinuity between cores and rims of andalusite as discussed in the preceding paragraph, Yokoi (1983) and Shiba (1988) described andalusite and sillimanite without discontinuities in minor element content across crystals. Figures 4.45 and 4.46 show "parabolic" zoning profiles of Fe_2O_3 concentrations across crystals of andalusite and sillimanite. However, the zoning profiles of the larger andalusite and sillimanite crystals analyzed by Yokoi (Fig. 4.45) suggest that much of the central portions of the crystals are either unzoned or have much gentler zoning profiles than the rims of the crystals. There

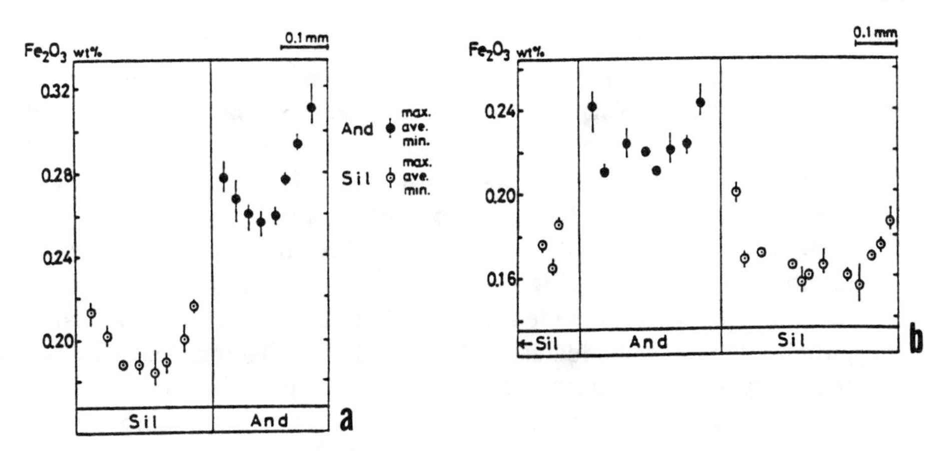

Figure 4.45. Fe_2O_3 zoning profiles of adjacent andalusite and sillimanite in a metapelite from the central Ryoke belt, Japan. (From Yokoi, 1983, Fig. 8a).

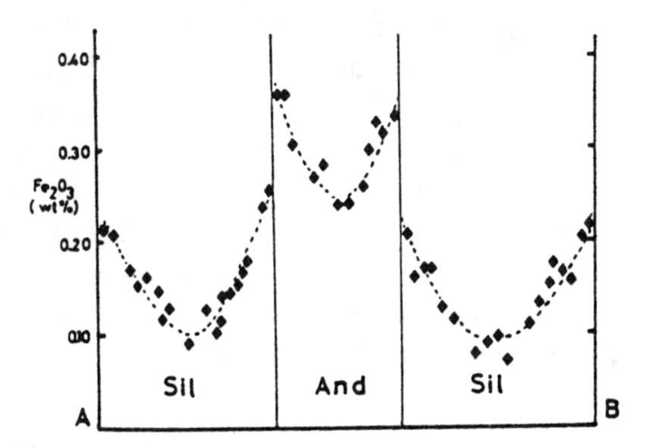

Figure 4.46. Fe_2O_3 zoning profiles of intergrown andalusite + sillimanite in a sample of metapelite from the southern part of the Hidaka metamorphic belt, Hokkaido, Japan. (From Shiba, 1988, Fig. 7).

are several possible explanations for this zoning trend: (a) a heterogeneous reaction such as: $Fe_2O_3^{Ilm} + SiO_2^{Qtz} = Fe_2SiO_5^{Sil}$ (Grew, 1980), (b) late-stage diffusion of Fe^{3+} into the rims of the crystals, (c) increase in f_{O_2} during growth of the crystals, or (d) prograde evolution along a bivariant T-$X_{Fe_2SiO_5}$ "loop". Kerrick and Speer (1988) concluded that the prograde compositional changes of coexisting andalusite and sillimanite in the central Ryoke metamorphic belt of Japan were controlled by zoning mechanism (d). However, with a model of *progressive* prograde metamorphism where all rocks in a given area were subjected to the same P-T-f_{O_2} path, it is difficult to reconcile this conclusion with the prograde compositional changes of the *cores* of coexisting andalusite and sillimanite (Fig. 4.29). Furthermore, Yokoi's (1983) analysis of the spatial distribution of Fe concentration in juxtaposed andalusite and sillimanite (Fig. 4.47) is incompatible with the hypothesis of partitioning of Fe between andalusite and sillimanite. In particular, with partitioning exchange of Fe, the compositional isopleths in andalusite and sillimanite (Fig. 4.47) should conform to the contact between the andalusite and sillimanite crystals. However, as shown in Figure 4.47, the spatial distribution of Fe zoning contours within the andalusite and sillimanite crystals is not related to the grain boundary between andalusite and sillimanite. Truncation of the isopleths in sillimanite by the andalusite-sillimanite grain boundary (Figure 4.47) suggests that andalusite replaced sillimanite. This textural evidence contradicts Yokoi's (1983) conclusion that sillimanite replaced andalusite. In light of this dilemma, and with the heterogeneity of minor element zoning in andalusite (Grapes, 1987), considerable additional research will be necessary to unravel the significance of minor element zoning in andalusite and sillimanite.

W.A. Dollase (pers. comm.) found zoned vanadian sillimanite in a small roof pendant in the Mt. Bancroft pluton, White Mountains, eastern California. Back scattered electron imaging showed patches of V-rich regions that are inhomogeneously distributed. V_2O_3 contents range from below detection to 4.0 wt %, with an approximate average concentration of 0.5 wt %.. To my knowledge, the maximum V_2O_3 content of this sillimanite far exceeds that of any sillimanite. In this area, sillimanite-free rocks at lower metamorphic grades contain magnetite with approximately 0.25 wt % V_2O_3. Thus, Dollase concluded that V in the sillimanite was derived from the decomposition of magnetite.

Boron

Pearson and Shaw (1960) and Dodge (1971) suggested that sillimanite *may* be enriched in B in comparison with the other Al_2SiO_5 polymorphs. D.R. Wones (see Ribbe, 1980, p. 208) believed that the fractionation of B into sillimanite could have a significant effect on the Al_2SiO_5 equilibria. There is considerable variation in the B concentrations of sillimanite. Pearson and Shaw (1960) and Evers and Wevers (1984) report negligible amounts of B (< 200 ppm) in sillimanite. However, in four sillimanite samples Dodge (1971) found B concentrations ranging from 0.014-0.036 wt %. In sillimanite from six granulite-facies, kornerupine-bearing rocks, Grew and Hinthorne (1983) reported B_2O_3 values ranging from 0.035 to 0.43 wt %. Grew and Rossman (1985) argued that B is present in solid solution rather than inclusions of a B-rich phase. Analytical data (Fig. 4.48) reveal an inverse correlation between the concentrations of B and Si, and a direct correlation between the Mg and B contents. Grew and Rossman (1985) concluded that there is a coupled substitution of B for Si and Mg for Al, and concomitant loss of oxygen (to maintain charge

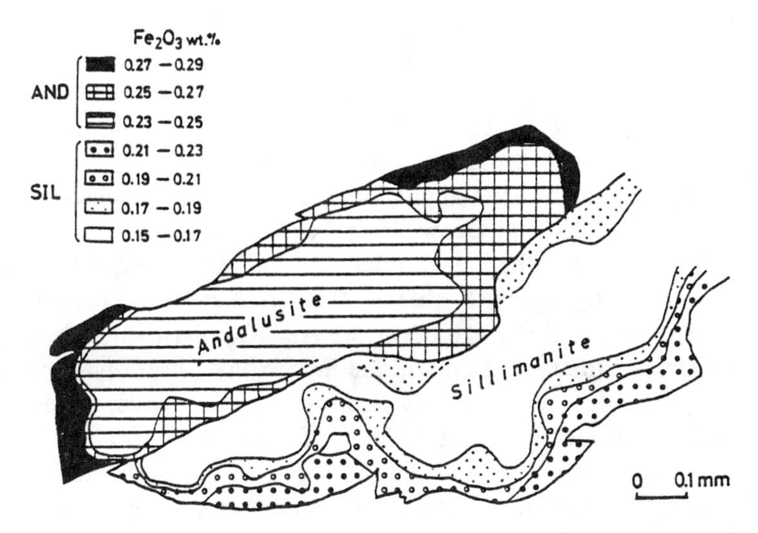

Figure 4.47. Contours of Fe_2O_3 concentration in adjacent andalusite and sillimanite in a sample of metapelite from the central Ryoke belt, Japan. (From Yokoi, 1983, Fig. 7).

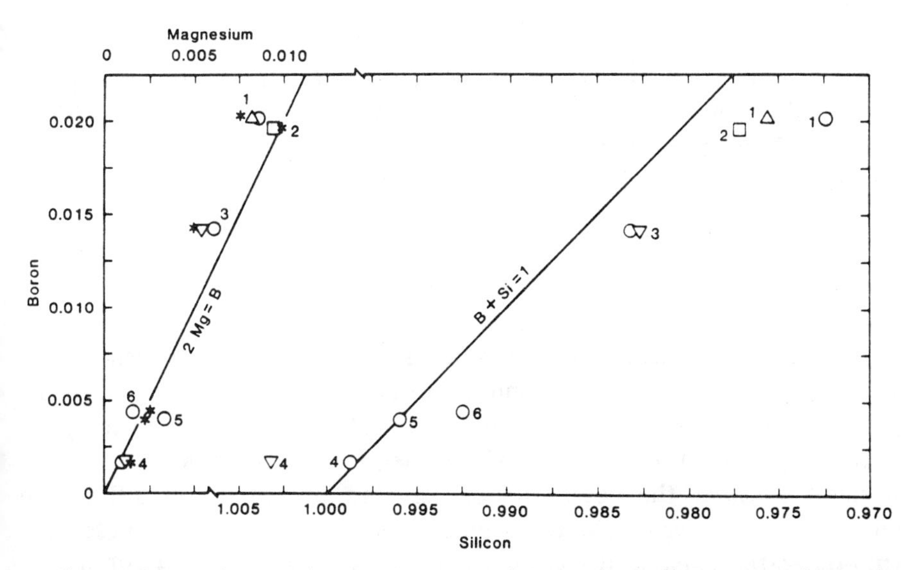

Figure 4.48. Atomic proportions (based on a total of three cations) of boron, magnesium, and silicon in sillimanite. The asterisks are ion microprobe analyses, whereas the other points are electron probe analyses. The numbers correspond to analyses listed in Table 1 of Grew and Hinthorne (1983). (From Grew and Hinthorne, 1983, Fig. 1).

balance): 2(B +xMg) = 2(Si + xAl) + (1+x)O, where x = 0.5 (Grew and Rossman, 1985). On the basis of similarities in the crystal structures of sillimanite and grandidierite $[(Mg,Fe^{2+})^{V}Al^{V}Al_2^{VI}Si^{IV}B^{III}O_9]$, Grew and Rossman (1985) considered the boron component to be represented by grandidierite. Accordingly, Grew and Hinthorne's (1983) sillimanite sample with the largest B content (0.43 wt % B_2O_3) would be considered as a binary solid solution with 2.0 mol % of the grandidierite component. Assuming ideal mixing, and expressing the stoichiometry of B-bearing sillimanite according to the scheme of Grew and Hinthorne (1983), $a_{Al_2SiO_5}$ = 0.947 for sillimanite with 0.43 wt % B_2O_3. Assuming that kyanite and andalusite have negligible B contents (as suggested by Grew and Hinthorne, 1983, and Grew, 1986), extrapolation of the isopleths in Figure 4.40 suggests that the andalusite-sillimanite equilibrium with $a_{Al_2SiO_5}$ = 0.947 significantly differs from that with B-free sillimanite. Grew (1986) analyzed the effect on the Al_2SiO_5 phase equilibria of 0.32 wt % B_2O_3 in sillimanite from Waldheim, Saxony. *Presumably* using the same stoichiometry and activity expressions as above, Grew (1986) computed a displacement (at 800°C) of +200 bars for the kyanite-sillimanite equilibrium and -600 bars for the andalusite-sillimanite equilibrium. Lithologies with kornerupine and/or grandidierite are relatively rare. Lonker (1988) noted that kornerupine has been found in about 40 localities, whereas grandidierite has been found in about 20 localities.

Grew et al. (in press) carried out a study of the compositional controls of B in sillimanite from high-grade rocks containing kornerupine. They concluded that enhanced B contents are favored by elevated temperature (B contents exceed 0.4 wt % with T > 900°C) and extremely low water activities. Correlation with a_{H_2O} is manifest by the *inverse* correlation between the B content of sillimanite and the H_2O concentration of coexisting cordierite (see Grew et al., in press, Fig. 9).

Tourmaline, a phase with B contents intermediate between kornerupine and grandidierite, is a very common phase in metapelitic rocks. Most metapelites have accessory amounts of tourmaline; however, the author has observed significant quantities of tourmaline in metapelites. Thus, we need to consider the possibility of elevated B contents of sillimanite coexisting with tourmaline. Using an ion microprobe, Kerrick and Houser (in preparation) analyzed B contents of coexisting andalusite + sillimanite in metapelites from several locales. Although these samples lacked kornerupine and/or grandidierite, tourmaline is present in most of their samples. The B contents of andalusite and sillimanite analyzed in this study are quite small (B_2O_3 < 500 ppm), and there is no apparent fractionation of B into sillimanite compared to andalusite. Such low B contents, coupled with the lack of fractionation of B into sillimanite, suggest that the solid solution of B is insignificant in altering the phase equilibria of the Al_2SiO_5 polymorphs in most metapelites.

Hydroxyl

Minor amounts of H_2O have been reported in all three polymorphs. In fact, several of the analyses of the aluminum silicates given by Deer et al. (1982) reveal significant H_2O^+ contents. Of the three polymorphs, the maximum H_2O^+ given by Deer et al. (1982) is 1.31 wt % for a viridine. However, interpretation of the water contents of the aluminum silicates has been a matter of contention. Petrographic studies have shown that the aluminum silicates commonly contain inclusions of hydrous minerals (e.g., muscovite, margarite). In their description of a sample of andalusite containing *submicroscopic* intergrowths of chlorite, Ahn and Buseck (1988) suggest caution in interpreting the analyzed H_2O contents. It is also possible that a significant amount of the H_2O is contained in fluid inclusions (Beran and Götzinger, 1987). In their analysis of the water content of some nominally anhydrous silicates, Wilkins and Sabine (1973, p. 512) concluded that for all three polymorphs "...the high water contents that have been reported doubtless are due to foreign material". Wilkins and Sabine (1973) noted that there are several mechanisms for the incorporation of structural OH in silicates: (1) substitution of SiO_4 groups by $(OH)_4$ (i.e., the hydrogarnet substitution), (2) the H_3O^+ (oxonium) ion, (3) interstitial water, (4) OH ions associated with dislocations, and (5) isolated "substitutional" OH ions. Beran et al. (1983) considered that OH was incorporated by removal of an Al^{3+}-$3O^{2-}$ pair resulting in a vacancy:

$$Al^{3+} + 3O^{2-} = 3(OH)^- + \square . \qquad [4.9]$$

In contrast, Hålenius (1979) suggested that OH is incorporated into sillimanite by replacement of Al^{3+} by Fe^{2+}:

$$Al^{3+} + O^{2-} = Fe^{2+} + OH^- .$$

Infrared absorption spectroscopy offers a definitive tool for addressing the nature of H_2O^+ in the aluminum silicates. H_2O from fluid inclusions is readily identified by the broad absorption band at 3450 cm^{-1}, which is attributed to two overlapping stretching modes (Wilkins and Sabine, 1973). Furthermore, due to differences in crystal structures, different hydrous silicates will have different absorption spectra. Thus, IR spectroscopy offers a way to determine whether H_2O^+ occurs in inclusions of hydrous phases. For example, from absorption bands at about 3600 cm^{-1}, Beran and Götzinger (1987) concluded that the H_2O in a "nongemmy" kyanite was attributed to intergrowths of muscovite.

Beran and Götzinger (1987) used IR spectroscopy to analyze the H_2O^+ contents of kyanite from ten localities. From petrographic examination they chose portions of the kyanite crystals that contained no visible inclusions. The key IR signature for structural OH is the strong absorption bands at 3270 cm^{-1} and 3380 cm^{-1} (Fig. 4.49). To confirm that these bands are attributable to stretching vibrations of OH groups within the kyanite structure, they performed "deuteration experiments" as described by Wilkins and Sabine (1973). This method consists of subjecting a kyanite sample to hydrothermal treatment with D_2O as the fluid medium. The diffusion of D into the crystal structure produced a shift in the frequencies of the absorption bands by an amount that is equivalent to that of the OH-OD exchange (Wilkins and Sabine, 1973, p. 509). Thus, they attributed the two major absorption bands observed in the kyanite spectrum (Fig. 4.49) to structural OH.

Beran and Götzinger (1987) considered that the incorporation of OH into the kyanite structure occurs by replacement of oxygen atoms not bonded to Si. Using Beer's Law to express absorption as a function of sample thickness, they computed H_2O contents ranging from less than 0.005 wt % to 0.180 wt %. They noted that there is a general correlation between the H_2O^+ content of kyanite and the paragenesis. Specifically, kyanites in granulites (with implicit low a_{H_2O}) are virtually H_2O-free. In contrast, they considered the higher H_2O^+ contents of kyanites from other parageneses (e.g., eclogites from the Swiss Alps) to reflect higher water activities.

Beran et al. (1989) used IR spectroscopy to examine the H_2O contents of 30 sillimanite samples from a variety of parageneses. They found the maximum amount of structural OH to be 0.02 wt % H_2O equivalent. As with the study of Beran and Götzinger (1987), Beran et al. (1989) found a correlation between H_2O content and water activity during metamorphism (e.g., sillimanites from granulites have the lowest H_2O content). Beran et al. (1989) concluded that the apparent correlation of sillimanite water content with water activity supports reaction [4.9] for the incorporation of OH into the sillimanite structure. However, they noted that there were some marked discrepancies in the correlation of H_2O content and apparent water activity during metamorphism. For example, sillimanite from two "hydrated" parageneses lack absorption bands in the range 2500-4000 cm^{-1}. They postulated that these samples may have undergone alteration at the unit cell scale forming lamellae of hydrated aluminosilicate. Because of the small size of such lamellae, they suggested that the OH vibration would be controlled by the host sillimanite and thus the absorption spectrum would be that of sillimanite rather than a hydrous

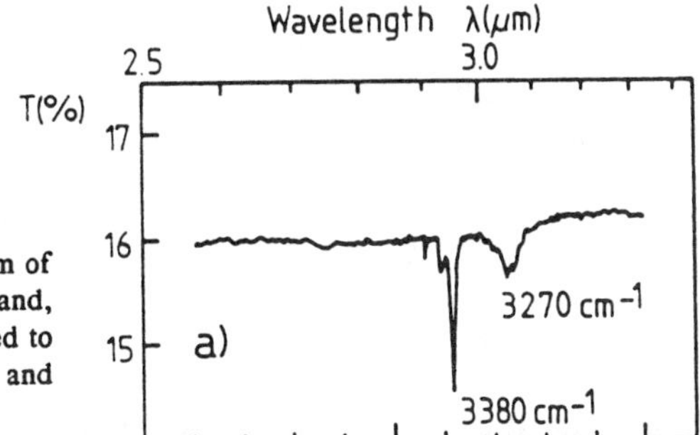

Figure 4.49. Infrared absorption spectrum of kyanite from Pizzo Forno, Switzerland, showing the two absorption bands assigned to OH stretching frequencies. (From Beran and Götzinger, 1987, Fig. 2a).

phase. Further evaluation of the validity of this hypothesis awaits detailed HRTEM studies.

Beran et al. (1989) suggested that the fractionation of OH between coexisting aluminum silicates could have a significant effect on the stability relations of the Al_2SiO_5 polymorphs. To evaluate their suggestion, consider the H_2O concentrations of the aluminum silicates that have been determined by IR studies, and assume that the H_2O reflects structural OH. The activity of the Al_2SiO_5 component may be calculated with the assumption of ideal mixing. Consider reaction [4.9] in the light of the assumption of Beran et al. (1983) that the Al deficiency occurs on a single site. The fractional site occupancy of Al would be

$$X_{Al}^{VI} = 1 - (1/3\ X_{OH})\ . \qquad [4.10]$$

The data of Beran and Götzinger (1987) and Beran et al. (1989) suggest *maximum* H_2O contents of about 0.02 wt % in kyanite and sillimanite. Using equation [4.10], a structural water content of 0.02 wt. % corresponds to $X_{Al}^{VI} = 0.998$. To maximize the effect of this OH content on the andalusite-sillimanite equilibrium, consider the unlikely case where OH-bearing sillimanite is in equilibrium with OH-free andalusite. Constraining the stoichiometry by [4.10], $a_{Al_2SiO_5} = 0.998$ for sillimanite with 0.02 wt % H_2O. According to Figure 4.40 there would be little displacement of the andalusite-sillimanite equilibrium with $K_{eq} = 0.998$ compared to that with pure (OH-free) sillimanite. Considering that this calculation was based on the *maximum* H_2O content of the 30 sillimanite samples analyzed by Beran et al. (1989), it would appear that structural H_2O has relatively little effect on the Al_2SiO_5 phase equilibria. In contrast, if we consider the larger H_2O^+ contents of andalusite, kyanite and sillimanite given in Deer et al. (1982) to represent structural OH, such

H_2O contents could significantly displace the Al_2SiO_5 equilibria. Of particular importance is the 1.34 wt % H_2O reported for a viridine. However, without detailed phase characterization of such samples (especially IR), there is little reason to seriously consider structural H_2O as a significant factor in perturbing the Al_2SiO_5 phase equilibria.

Other elements

Other than the transition elements, B, and OH, concentrations exceeding trace element levels (i.e., > 0.1 wt %) are common for a number of other elements. Because of the possibility of small amounts of included impurity phases, the interpretation of the elevated concentrations of some elements is questionable. For example, the correlation between Ca and P contents has been attributed to inclusions of apatite (Chinner et al., 1969). Some workers concluded that elevated concentrations of K and Na result from fine inclusions of alkali-bearing phases (Henriques, 1957; Chinner et al., 1969; Deer et al., 1982). However, others (Pearson and Shaw, 1960; Dodge, 1971) concluded that the amounts of inclusions are insufficient to account for the elevated K and Na contents: "...concealment in lattice voids seems the best explanation at present" (Pearson and Shaw, 1960, p. 816).

Based on ionic radii arguments, Pearson and Shaw (1960) evaluated the substitution of minor and trace elements in the aluminum silicate polymorphs. Accordingly, they suggested that Be substituted for Al in sillimanite. However, such interpretations are questionable without a detailed microscopic *and* submicroscopic search for impurity phases. The presence of submicroscopic intergrowths of chlorite in kyanite (Wenk, 1980) and andalusite (Ahn and Buseck, 1988), and the suggestion by Beran et al. (1989) that sillimanite may contain unit cell scale intergrowths of a hydrous phase, illustrate the importance of TEM examination in the interpretation of the minor element and trace element chemistry of the Al_2SiO_5 polymorphs. Grew and Rossman's (1985) study serves as a model for future studies on this problem. By TEM examination they confirmed that there were no submicroscopic inclusions of B-bearing phases. Thus, they were assured that B indeed resides within the sillimanite lattice rather than within inclusions. Future studies should also evaluate minor elements by mass balance. For example, if elevated K contents are attributed to intergrowths of muscovite, an amount of H_2O^+ correlative with muscovite stoichiometry must be confirmed.

CHAPTER 5

LATTICE DEFECTS

"In these minerals with similar free energies, energy contributions from defects such as dislocations and stacking faults and in addition nonstoichiometry could considerably affect their stability. This may explain differences in reaction conditions established in various laboratories. In fact a thermodynamic interpretation of reaction equilibria remains tenuous unless microstructures of starting materials and run products are specified."

H.R. Wenk (1983, p. 13)

Lattice defects represent a potentially significant perturbation of the molar free energies of crystals. In fact, Helgeson et al. (1978) concluded that lattice defects (specifically vacancy point defects and dislocations) could contribute many kilojoules of excess molar energy to rock-forming minerals. Excess energies of this magnitude would significantly alter the Al_2SiO_5 phase equilibrium relations. The following discussion follows the three-fold classification of defects in crystals, i.e., point defects, line defects, and planar defects.

POINT DEFECTS

Point defects represent deviations in the locations and types of atoms in a crystal compared to a perfectly crystalline, defect-free crystal. *Intrinsic* point defects include *vacancy* defects produced by the absence atoms on lattice sites, and *interstitial* defects corresponding to atoms in interstitial lattice positions (i.e., atoms not located at the atomic positions in a geometrically perfect crystal structure). *Extrinsic* point defects result from the presence of impurity atoms. Certain minerals (e.g., mullite, wüstite and pyrrhotite) have significant concentrations of point defects. In such cases, consideration of the consequences of point defects on the thermodynamic properties and stabilities of these phases is vital. For many other rock-forming minerals it has been traditionally assumed that point defects have a negligible effect on the chemical potentials; however, this contention was challenged by Helgeson et al. (1978). The following discussion focuses on the effects of point defects on the thermodynamic properties and P-T stability relations of the Al_2SiO_5 polymorphs.

Intrinsic point defects

To the author's knowledge, there have been no studies on intrinsic point defects in the aluminum silicate polymorphs. An empirical approach is available for estimating the abundance and energetics of vacancy (Schottky) defects from the linear correlation between melting points and enthalpies of formation of defects for the alkali halides (Lasaga, 1981, p. 284):

$$\Delta H_s = 2.14 \times 10^{-3}\, T_m \; . \qquad [5.1]$$

In this equation ΔH_s is the enthalpy of formation of a Schottky defect (in electron volts = e.v.), and T_m is the 1 atm melting temperature in kelvins. Taking the 1 atm melting temperature of sillimanite as 1473 K (Cameron, 1977b), equation [5.1] yields ΔH_s = 3.15 e.v. per vacancy. Let us assume that we form a vacancy defect by removing Si^{4+} from a structural atomic site. For charge balance, removal of a Si^{4+} requires removal of two O^{2-} ions. Using Kröger's notation (Lasaga, 1981, p. 264), the formation of such a defect would be written:

$$Si^{x}_{Si} + 2\, O^{x}_{O} = V^{''''}_{Si} + 2\, V^{''}_{O} + SiO_2 \; .$$

In treating vacancies as quasi-chemical species (Lasaga, 1981), we may express an equilibrium constant for the vacancy forming reaction:

$$K_{eq} = \frac{X_{Si,v}\; X^2_{O,v}}{X_{Si,Si}\; X^2_{O,O}} \; .$$

Because most of the atomic sites are occupied, it is reasonable to assume that $X_{Si,Si}$ and $X_{O,O}$ are both unity (Lasaga, 1981); thus,

$$X_{Si,v} = 2X_{O,v} = \exp\,(-\Delta G_f/2RT) \; . \qquad [5.2]$$

Assuming $\Delta H_s = \Delta G_f$ (Lasaga, 1981, p. 267), where ΔG_f is the Gibbs free energy of formation of a vacancy defect, we have ΔG_f = 3.15 e.v. per vacancy. Equation [5.2] illustrates that the concentration of Schottky defects increases with increasing temperature. Accordingly, the maximum concentration of Schottky defects will be at the melting temperature. At 1473 K equation [5.2] yields $X_{Si,V} = 2X_{O,V} = 4.09 \times 10^{-6}$/mol. Multiplying this site defect concentration by Avogadro's number and by the energy per defect (5.05 x 10^{-19} J) yields 1.24 J/mol for vacancy defects in sillimanite at 1473 K. This excess energy would have a negligible effect on the P-T stability of sillimanite. For example, if we assume a temperature of 800°C for the andalusite-sillimanite equilibrium at 1 atm, and compute ΔS_r from the data of Hemingway et al. (ms.), an energy perturbation of 1.24 J/mol would shift this equilibrium less than 1°C. Using a similar approach, the author computed negligible values for molar energies of vacancy defects in other rock-forming minerals, such as periclase and olivine (Kerrick, unpublished). Thus, in contrast to the contention of Helgeson et al. (1978), I suggest that vacancy defects do not introduce significant perturbations in the molar Gibbs free energies of most rock-forming minerals. However, this conclusion hinges on a non-rigorous theoretical treatment of the energetics of vacancy defects. Advances in our knowledge of the nature, concentration, and thermodynamic properties of point defects in the aluminum silicates awaits sensitive measurements of defect concentrations (e.g., Huebner and Voigt, 1988) coupled with improved modeling (e.g., Anderson, 1985; Catlow and Parker, 1985).

Extrinsic point defects

Extrinsic point defects correspond to departures from the stoichiometric Al_2SiO_5 composition. Strictly speaking, all solid solutions can be interpreted with extrinsic point defect theory. However, in the mineralogical literature, most studies of extrinsic point defects are concerned with those involving vacancy defects.

Considering the aluminum silicates, the most significant type of extrinsic point defect involving vacancies arises from the mullitization of sillimanite. To preserve charge balance, oxygen vacancies are formed from the replacement of Si^{4+} by Al^{3+}. Because the sillimanite → mullite transformation is covered in Chapter 4, the reader is referred to the "MAJOR ELEMENT NON-STOICHIOMETRY" section of that chapter.

Extrinsic vacancy-forming defect reactions can be written for a variety of impurities present at minor and trace concentration levels. For example, Beran et al. (1983, 1989) hypothesized that oxygen vacancies are created by the incorporation of OH into the sillimanite structure (see Chapter 4). However, there is no *direct* evidence supporting this mechanism, and alternative reaction mechanisms involving OH have been proposed (Hålenius, 1979). Experimental measurements of the correlation between the concentrations of vacancies and impurity elements are needed to further address this problem.

LINE DEFECTS

This category of lattice defects involves a disturbance of the crystal structure in the vicinity of a line in a crystal. Dislocations are a vital aspect of the intracrystalline deformation of crystals. Compared to a dislocation-free structure, distortion of atomic structures near dislocations adds strain energy. Thus, a crystal with dislocations is less stable than a dislocation-free crystal.

The amount of slip along a dislocation line is measured by the *Burgers vector* **b**. Three types of dislocations are classified according to the angular relationship between the dislocation line and **b**: *edge* dislocation: **b** is perpendicular to the dislocation line; *screw* dislocation: **b** is parallel to the dislocation line; and *mixed* dislocation: **b** is oblique to the dislocation line. In general, dislocations in the Al_2SiO_5 polymorphs are straight (Fig. 5.1). Dislocations with a Burgers vector corresponding to a structural translation vector are referred to as *perfect* or *unit* dislocations. Perfect (= unit) dislocations may undergo dissociation (= splitting) into two collinear *partial* dislocations with an intervening *stacking fault* (Figs. 5.2 and 5.3). The abundance of dislocations within a crystal is measured by the *dislocation density* (ρ), which is the total length of dislocation per unit volume (typically in cm^3). Dislocation density is typically measured with transmission electron microscopy by determining the number of dislocations within a unit area. This method yields the true "volume" dislocation density for parallel dislocations. However, for totally random

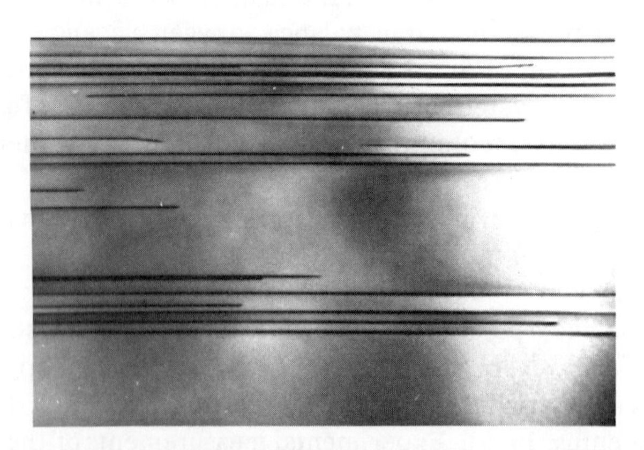

Figure 5.1. Parallel [001] screw dislocations in andalusite experimentally deformed at 900°C and 1 kbar. Width of photo corresponds to about 7.0 μm. (From Doukhan et al., 1985, Fig. 4a).

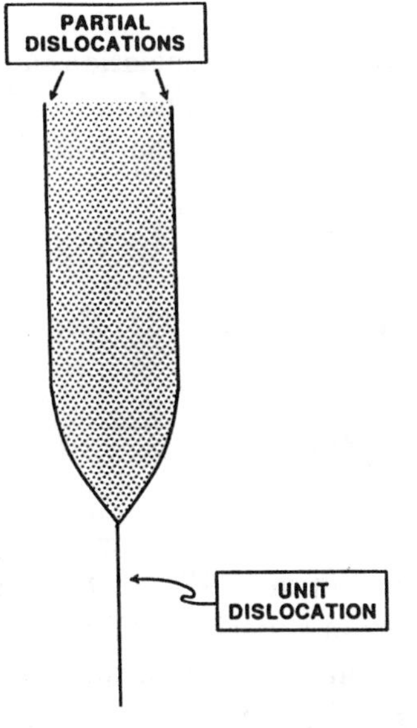

Figure 5.2 Schematic illustration of *dissociation* of a perfect (unit) dislocation into two collinear *partial* dislocations with an intervening stacking fault (stippled).

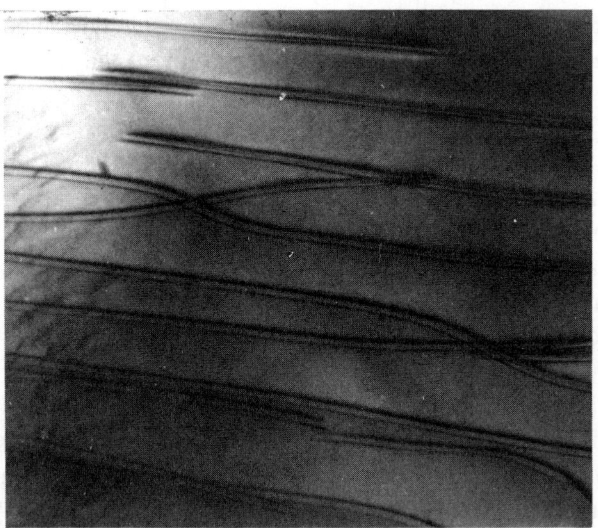

Figure 5.3. Dissociated [001] screw dislocations in fibrolite. The width of the stacking faults between each set of dissociated dislocations is about 300 Å. Width of photo corresponds to about 2.2 μm. (From Menard and Doukhan, 1978, Fig. 3).

dislocations, dislocation density determined in this manner yields a dislocation density approximately one-half the volume density (Hull and Bacon, 1984, p. 22). Dislocation densities range from about $10^2/cm^3$ to $10^{12}/cm^3$.

[The following five paragraphs reviewing dislocations in the Al_2SiO_5 polymorphs are from Kerrick (1986, p. 221-222).]

TEM studies reveal that the only slip system in kyanite is (100)[001] (i.e., slip plane = (100), Burgers vector (**b**) = [001]). Dissociation into two collinear partials is common. From the studies of Boland et al. (1977), Menard et al. (1977), and Lefebvre and Menard (1981), the dissociation

$$[001] \rightarrow 1/2[001] + 1/2[001] \quad (1)$$

appears to be typical; however, Lefebvre and Menard (1981) also suggest dissociation via

$$[001] \rightarrow 1/2[001] + 1/2[0\bar{1}1] \,. \quad (2)$$

The sole slip system in experimentally deformed kyanite is (100)[001] (Raleigh, 1965; Boland et al., 1977; Menard et al., 1979). Dissociation (according to the above schemes) is common in experimentally deformed kyanite. Doukhan et al. (1985) concluded that dissociation scheme (1) is prevalent in kyanite experimentally deformed at room temperature and without confining pressure, and that dissociation scheme (2) becomes more significant as P and T are increased above ambient values.

Natural (undeformed) sillimanite contains predominantly [001] screw dislocations. In some cases (Lefebvre and Paquet, 1983; Doukhan et al., 1985) these dislocations are dissociated according to scheme (1). However, Wenk (1983) reported [001] screw dislocations in sillimanite that have unit Burgers vector (i.e., perfect dislocations). The dominant slip systems in experimentally deformed sillimanite are (010)[100] and (100)[010] (Doukhan and Christie, 1982; Doukhan et al., 1985).

Very little work has been devoted to characterizing dislocations in natural andalusite (i.e., andalusite not subjected to experimental deformation). In a study of andalusite from nodules in a metamorphic series at Vendée, France, Doukhan et al. (1985) found that [001] screw dislocations are most common. The fringe contrast of these dislocations suggests that there may be some dissociation into collinear partials with **b** = 1/2[001]. In a study of andalusite subjected to uniaxial compression, Doukhan and Paquet (1982) found that the only glide systems are (110)[001] and ($0\bar{1}1$)[001]. They observed long, straight [001] screw dislocations that are dissociated out of the {110} glide planes. Dissociation is believed to occur by (1) (above). However, Doukhan et al. (1985) observed evidence of dissociation of dislocations in andalusite subjected to uniaxial compression at 900°C and 1 kbar, and at 450°C and 3 kbar. In a study of andalusite deformed by indentation, Lefebvre (1982a) found mainly undissociated [001] screw dislocations.

In summary, [001] screw dislocations predominate in the Al_2SiO_5 polymorphs. The preponderance of [001] slip in part reflects the fact that the *c* unit cell dimension is shorter than either *a* or *b*. Because the energy (per unit length) is proportional to $\mathbf{b}^2$ (see Kerrick, 1986), it is expected that dislocations with the shortest lattice translation vector will be most stable. In addition, [001] slip is energetically favored because the chains of AlO_6 octahedra are not distorted or cut (Menard and Doukhan, 1978; Doukhan and Christie, 1982).

Because no *direct* measurements have been made on the strain energies of dislocations in silicates, an alternative is to compute strain energy in the vicinity of dislocations using elastic continuum theory. This theory is based on Hooke's law of elasticity (i.e., stress is proportional to strain). In applying it to dislocations it is assumed that deformation is frictionless and that the work done in producing dislocations is stored as elastic strain energy. Strain energy calculations near dislocations are made by considering strain within a cylindrical region that is coaxial with the dislocation line. Because structural displacements are largest near the dislocation line, this region is excluded because strain may be high enough that Hooke's law is not applicable. Thus, "non-core" computations of elastic strain energy exclude the cylindrical "core" region coaxial with the dislocation line. There is uncertainty regarding the choice of values for the inner and outer radii of the hollow cylinder within which the strain energies are calculated. Typically, the magnitude of the inner radius is equated to that of the Burgers vector, whereas the outer radius is given as $1/2\sqrt{\rho}$ (Kerrick, 1986).

Screw dislocations with a Burgers vector parallel to [001] predominate in the Al_2SiO_5 polymorphs. For andalusite and sillimanite, Figure 5.4 shows the calculated molar elastic strain energies of [001] screw dislocations as a function of dislocation density.

The computed alteration in the Al_2SiO_5 phase diagram for selected values of excess molar energy in sillimanite and andalusite is shown in Figures 5.5a and 5.5b, respectively. Note that *minimum* strain energies of 100 J/mole are required to significantly perturb the P-T stability relations. TEM measurements show that dislocation densities of the Al_2SiO_5 polymorphs are typically $\lesssim 10^8/cm^2$ (Kerrick, 1986). According to Figure 5.4, such dislocation densities would yield negligible molar strain energies.

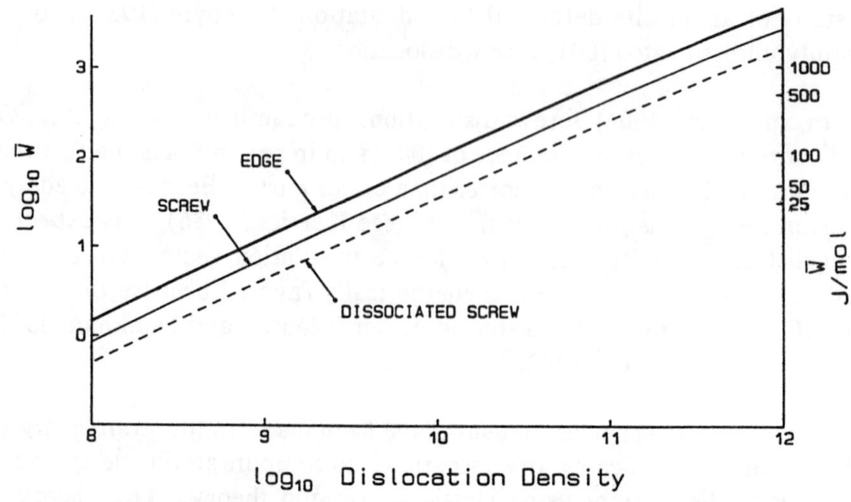

Figure 5.4. Calculated molar strain energy ($\bar{W}$) versus dislocation density. Solid lines represent perfect (unit) [001] dislocations in andalusite and sillimanite. The dashed line represents dissociated screw dislocations. (From Kerrick, 1986, Fig. 1).

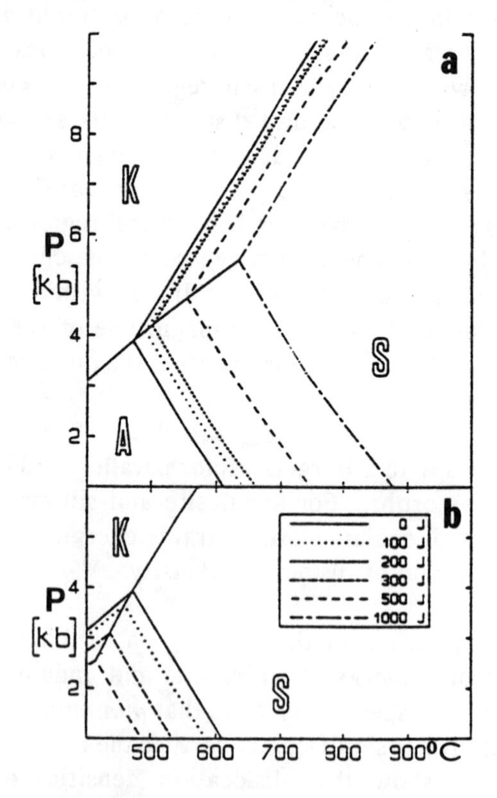

Figure 5.5. Alteration of the P-T stability field of sillimanite (a) and andalusite (b) for selected values of molar strain energy. (From Kerrick, 1986, Fig. 2).

Thus, in my 1986 paper I concluded that dislocation strain energies have a negligible effect on the P-T stability relations of the Al_2SiO_5 polymorphs in nature. At that time there was uncertainty regarding dislocation densities in fibrolite. The author and Doukhan et al. (1985) observed low dislocation densities ($\lesssim 5 \times 10^8/cm^2$). However, Wenk (1983) found high dislocation densities (ca. $10^{10}/cm^2$) in a fibrolite sample. Medrano, Wenk and Kerrick (unpub.) recently examined several additional fibrolite samples and found low *bulk* dislocation densities. Thus, it is concluded that in general dislocation densities are such that the computed strain energies will have a negligible effect on the P-T stability of fibrolite.

It is important to note that dislocation densities in aluminum silicates do not necessarily correspond to those that existed during metamorphism. It is possible that they were higher during metamorphism but that subsequent annealing (recovery) yielded lower dislocation densities. The presence of planar dislocation arrays forming subgrain boundaries (discussed in the section on "PLANAR DEFECTS"), which are observed in sillimanite, fibrolite and andalusite, attest to recovery via the migration of dislocations.

The above conclusions regarding dislocation strain energies assume that the excluded core energy is not substantial. For metals, atomistic calculations suggest that the core energy contributes no more than 5-10% of the total dislocation strain energy (Kerrick, 1986). However, to the author's knowledge, there are no data on core energies of dislocations in silicates. In light of covalent bonding, dislocation core strain energies in silicates could be considerably larger than those in metals. If so, the total dislocation strain energies (i.e., including the core) could be significantly larger than those computed with the non-core model. Accordingly, it is conceivable that significant molar strain energies could arise for dislocation densities considerably less than the "cut off" value of $10^{10}/cm^2$ implied by Figure 5.4. A definitive statement regarding this problem awaits theoretical modeling and/or measurements of core energies. Measurements of heats of solution as a function of dislocation density (e.g., Gross, 1965) offers an intriguing way to evaluate the total strain enthalpy and thus a potential way to extract core energies through comparison with non-core strain energy calculations.

PLANAR DEFECTS

Planar defects in crystals include: stacking faults, twin planes, antiphase boundaries, subgrain boundaries, and grain boundaries.

Stacking faults

Stacking faults perturb the otherwise regular arrangement of the atomic structure. Formation of *intrinsic* stacking faults may be envisioned by removing a portion of an atomic layer (Fig. 5.6a) thereby creating a localized layer of vacancies. In contrast, *extrinsic* stacking faults represent an extra layer of atoms (Fig. 5.6b). Both types of stacking faults are bounded by *partial* dislocations. Partial dislocations are mutually repelled by elastic energy, which tends to increase the distance of separation of the two dislocations. However, this repulsive interaction is offset by the stacking fault energy, which increases with increasing distance of separation of the partial dislocations. The counterbalance between the repulsive interaction of the collinear partials and the stacking fault energy produces an equilibrium distance of separation of the partial dislocations and thus an equilibrium width of the intervening stacking fault. In general, widely dissociated collinear partials, separated by a stacking fault, are common in kyanite and sillimanite. In contrast, dissociation of dislocations is not marked in andalusite. Where observed, partials in andalusite are narrowly separated. Doukhan and Paquet (1982) and Lefebvre (1982a) suggested that the limited dissociation of dislocations in andalusite reflects the fact that the stacking fault has relatively high energy (i.e., is relatively unstable) because it necessitates modification of the cationic polyhedra from that of the double chains of alternating SiO_4 and AlO_6. In contrast, they suggested that the wide dissociation of dislocations in kyanite and sillimanite is attributed to stacking faults that do not modify the cationic polyhedra and are thus of lower energy than those in andalusite. It is notable that, for a given dislocation density, dissociated screw dislocations have lower strain energies than perfect (unit) edge or screw dislocations (Fig. 5.4), thereby rationalizing the preponderance of dissociated screw dislocations in kyanite and sillimanite.

In kyanite Wenk (1980) found abundant (100) stacking faults concentrated near an interface between kyanite and staurolite. This interface is virtually coherent, such that (100) of kyanite parallels (010) of staurolite. Wenk (1980) implied that kyanite replaced staurolite, and that the abundant stacking faults near the interface were produced because of

(a)

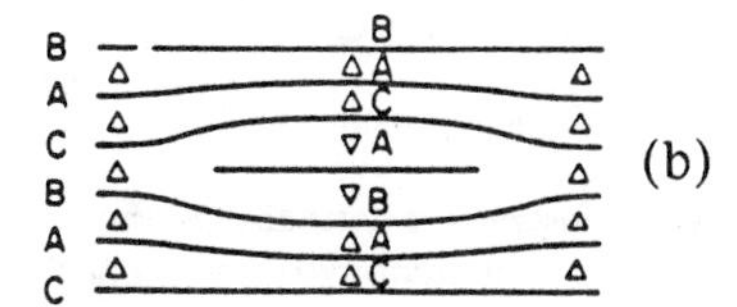

Figure 5.6. Schematic illustration of the perturbations produced by (a) intrinsic and (b) extrinsic stacking faults. Atomic planes in the normal stacking relationship are separated by triangles with the same orientation. (From Hull and Bacon, 1984, Fig. 1.13; copyright Pergamon Press, Oxford, England).

differences in the stacking sequences of staurolite and kyanite.

Dislocation energies for dissociated screw dislocations in andalusite and sillimanite (Fig. 5.4) were calculated by summing the computed elastic strain energies of the two partial dislocations, the forces between the partials, and the stacking fault energy (for details see Kerrick, 1986, p. 223-224). Because dissociated dislocations have lower strain energies (per unit length) than perfect dislocations, I suggest that, for dislocation densities observed in the Al_2SiO_5 polymorphs (i.e., $\lesssim 5 \times 10^8/cm^2$), stacking faults will have an insignificant effect on the P-T stability relations of the Al_2SiO_5 polymorphs.

Antiphase boundaries

Antiphase boundaries (APB's) are planar structural imperfections that occur in the Al_2SiO_5 polymorphs. As shown in Figure 5.7, APB's correspond to discontinuities in the atomic ordering sequence. As shown in this diagram, *conservative* APB's have a displacement vector (R) that is parallel to the APB; in contrast, R is not parallel to the APB with *non-conservative* APB's. No compositional change occurs across a conservative APB, whereas non-conservative APB's are characterized by local compositional changes (note, for example, the concentration of the black atoms along the non-conservative APB in Fig. 5.7). Because of repulsive forces resulting from the juxtaposition of identical atomic species, non-conservative APB's should have higher energies than conservative APB's. Thus, conservative APB's are expected to be more common than non-conservative APB's. Because there is no change in lattice orientation across an APB, antiphase domains cannot be imaged with a light optical microscope; rather, imaging of APB's requires examination by TEM.

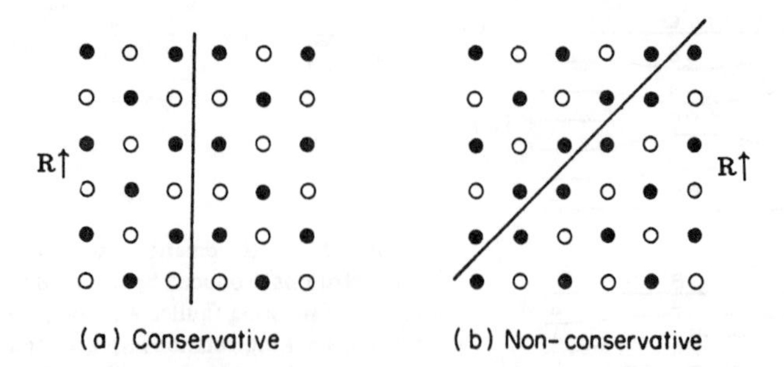

Figure 5.7. Schematic illustration of conservative and non-conservative antiphase boundaries (APB's). The arrows labeled "R" are the displacement vectors, representing the offset of the atomic structures on opposite sides of the APB. (From Putnis and McConnell, 1980, Fig. 5.35; copyright Blackwell Scientific Publications, Oxford, England).

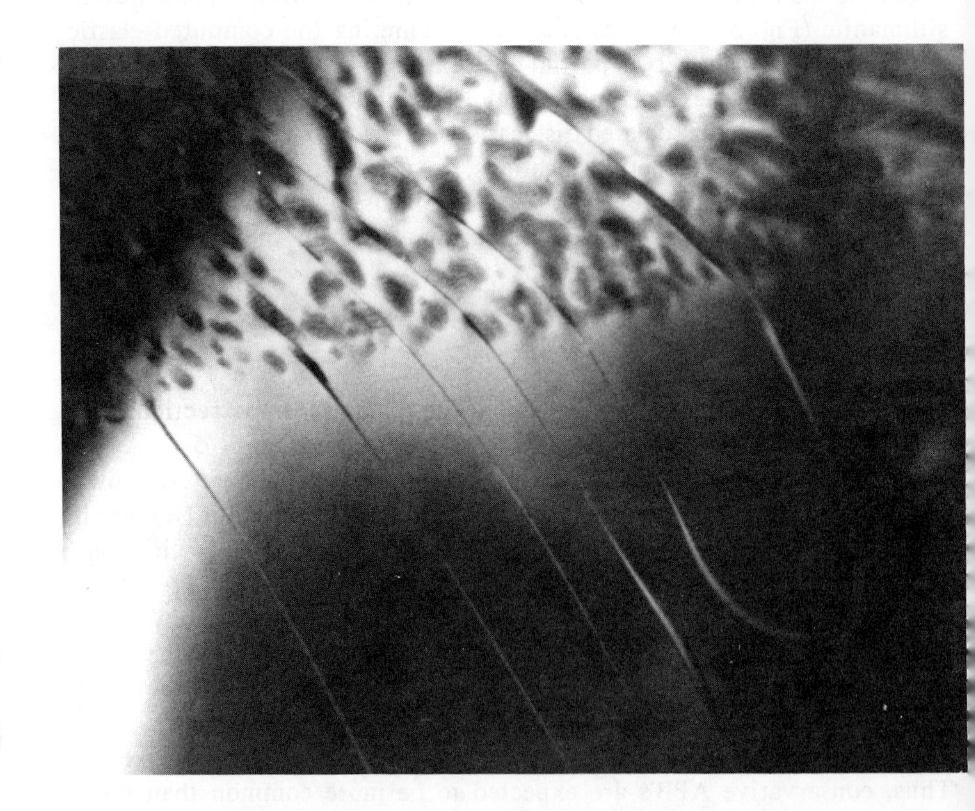

Figure 5.8. Antiphase boundaries in sillimanite (thin, curved lines). The mottled area in the upper part of the photo is a mixture of sillimanite + mullite. Width of photo corresponds to about 1.87 mm. (From Wenk, 1983, Fig. 3c).

Boland et al. (1977) described two types of APB's in natural and experimentally deformed kyanite. One type occurs parallel to (110) and may be non-conservative, thus explaining the relative rarity of this type of APB. A more common APB in kyanite is parallel to (100) with R = 1/2 [001]; thus, Boland et al. (1977) suggested that this APB is conservative.

In sillimanite, Wenk (1983) described APB's which he interpreted to be parallel to (010) (Fig. 5.8). The displacement vector R = 1/2[001]; thus, he considered that these APB's are conservative and have relatively low energy. He suggested that the APB's may have formed during the transformation from disordered, non-stoichiometric aluminum silicate to mullite + ordered sillimanite. If so, this could have important petrologic implications because the thermodynamic properties and, thus, the P-T stability relations of the early-formed (disordered and non-stoichiometric) phase could significantly differ from that of the ordered sillimanite now present.

Using HRTEM Hamid Rahman (1987) observed inhomogeneities in the structure of andalusite. As shown in Figure 5.9 he suggested that the disorder involves 1/2[001] displacement of the upper and lower halves of the unit cell. Ordered domains were hypothesized to be separated by disordered domains that trend parallel to [110] or [$1\bar{1}0$]. Although not specifically designated by Hamid Rahman (1987), the domain boundaries are correlative with conservative APB's in that the anionic structure is preserved and the domain boundaries do not differ chemically from the ordered regions within domains.

Twinning and kink bands in kyanite

Kink bands and twins are a notable feature in natural and experimentally-deformed kyanite. Mügge (1883) noted that kink bands and twins in kyanite are indistinguishable upon cursory optical examination.

Simple and lamellar twinning occurs in natural kyanites. Multiple twinning with the twin plane = (100) and twin axis normal to (100) is the most common twin law.

Lefebvre and Menard (1981) used electron microscopy to study twinning in a sample of natural kyanite. They found two types of twins both with twin planes parallel to (100). One twin is generated by rotation along [001] whereas the other is formed by *b*-glide across (100). In concert

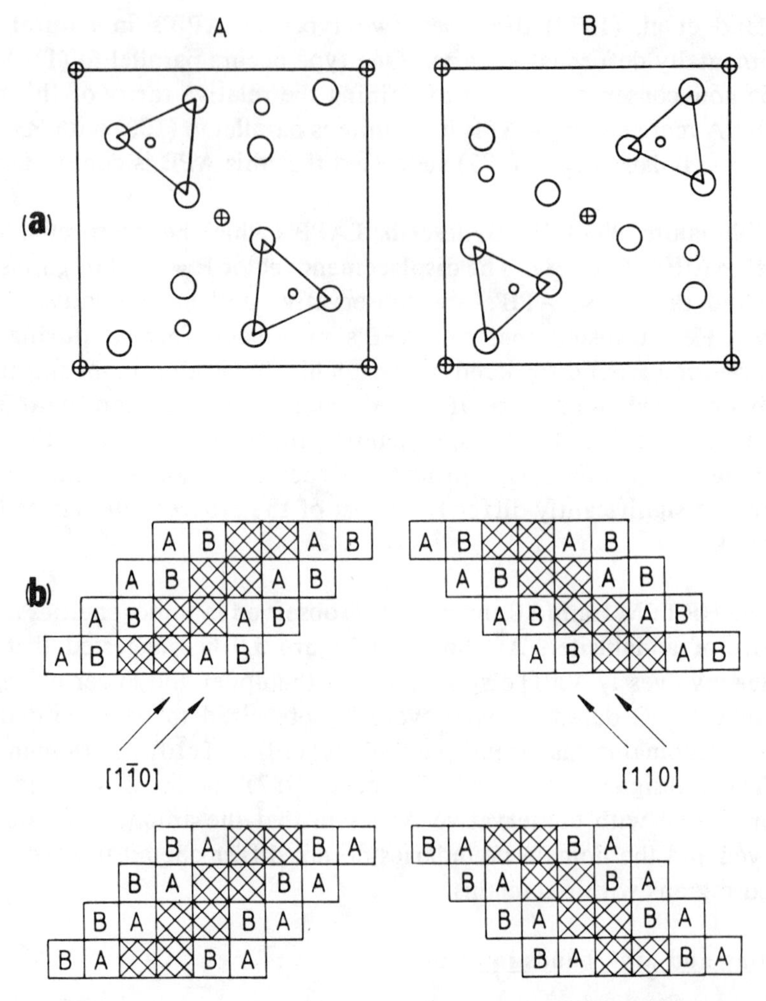

Figure 5.9. Model for the structural inhomogeneities observed in andalusite by HRTEM. (a) upper half (A) and lower half (B) of the andalusite unit cell, (b) structural disorder scheme; the half unit cells A and B are marked accordingly whereas the half unit cells depicted by cross hachuring are disordered regions parallel to [110] or [1$\bar{1}$0]. (From Hamid Rahman, 1987, Fig. 4).

with the (100)[001] slip system in kyanite, they modeled both types of twins by displacing the kyanite structure with periodic shears parallel to (100).

Kink bands are a notable feature of deformed kyanite (Fig. 5.10). Raleigh (1965) experimentally induced kink banding by straining kyanite at confining pressures of 5 to 7 kbar and temperatures of 700C to 800°C (Fig. 5.11). From measurements of optical indicatrix directions with the universal stage, he determined that kinking occurs by (100)[001] slip. The

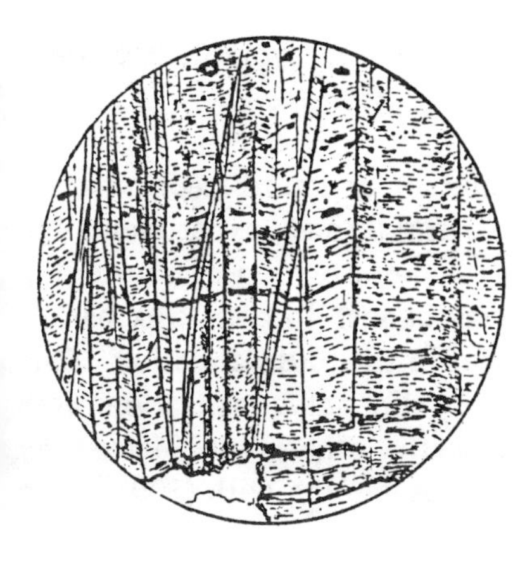

Figure 5.10. Kink bands in kyanite from the Moine Schist, Ross of Mull, Argyll, Scotland. (From Nockolds et al., 1978, Fig. 33.1C; copyright Cambridge University Press, Cambridge, England).

Figure 5.11. Experimentally deformed kyanite. The prominent banded structure represents (100) slip planes. Kink bands are oriented at a high angle to the slip planes. (From Raleigh, 1965, Fig. 1).

boundaries between kink bands are nearly perpendicular to slip directions. A slip plane parallel to (100) is compatible with the excellent (100) cleavage and with the fact that slip parallel to (100) plane would involve no breakage of strong Si-O bonds. Because the Burgers vector is shortest in the [001] direction, slip parallel to this direction is favorable because it minimizes the amount of unit translation.

Boland et al. (1977) examined the intragranular strain mechanism of kyanite through TEM examination of natural and experimentally-deformed kyanite. Samples of kyanite with the direction of shortening parallel to (100) revealed two conjugate kink bands. In both kink bands, slip occurred parallel to the (100) plane and in the [001] direction. One type of kink band primarily deformed by "slip polygonalization" (see Nicolas and Poirier, 1976, p. 47), whereas the other type formed by dislocation glide, microkinking, and by microtwinning with (100) as the twin plane.

Grain boundaries

The external boundary of a crystal may be considered as a planar defect. In rocks we must consider two types of grain boundary energetics. First, if a grain boundary lacks a fluid phase (or a thin, intergranular film of adsorbed fluid), then the grain boundary energy is dependent upon the atomic structures of the adjacent crystals, the angular mismatch of the adjacent lattices, and the nature of the bonding of the atoms at the interface. Second, a grain boundary may be represented by the contact between a crystal and an adjacent fluid phase. The interfacial energetics between a crystal and an adjacent fluid phase depends upon the interaction (i.e., chemical and physical forces) between the crystal and the adjacent fluid. Chemisorption and physisorption will be important in dictating the interfacial energy between a crystal and the adjacent fluid. Recent work suggests that there may not be a pervasive intergranular fluid in rocks undergoing metamorphism; rather, the fluid phase may instead reside in intergranular "pockets" or at triple junctions (White and White, 1981; Watson and Brenan, 1987). Thus, the energetics of grain boundaries in rocks undergoing metamorphism may be dominated by solid-solid interfacial energies.

The molar grain boundary energy is directly proportional to the molar grain boundary area. Accordingly, grain boundary energy will become larger as grain size becomes smaller. Based on measurements of surface energies of silicate minerals, it is generally considered that grain sizes ≤ 1 μm are necessary to yield grain boundary energies large enough to significantly perturb the molar Gibbs Free energy compared to that of coarse-grained crystals (where the molar surface energy is effectively zero).

For the aluminum silicate polymorphs, the most likely candidate for significant molar grain boundary energy is fibrolite, which typically occurs as aggregates of acicular crystals with individual crystals commonly having micron- to sub-micron diameters. Aside from the possibility of significant molar grain boundary energy, there are a number of questions regarding the chemistry, structure, thermodynamic properties, and P-T stability of fibrolite. The "fibrolite problem" (including the question of interfacial energy) is separately addressed in Chapter 7.

Subgrains are polygonized aggregates of a single phase. Subgrain boundaries are composed of dislocation walls that may be imaged with TEM, and they have been observed in andalusite (Doukhan et al., 1985), sillimanite and fibrolite (Wenk, 1983; Doukhan et al., 1985). However, I am not aware that any subgrain boundaries have been reported in kyanite. The lattice planes of crystals on either side of a subgrain boundary are related by low angles of mismatch (<10-15°); hence, subgrain boundaries are also referred to as *low-angle* grain boundaries. Subgrain boundaries are indicative of recovery from strain (Poirier, 1985). In naturally deformed sillimanite, Lambregts and van Roermund (1990) found abundant subgrain boundaries subparallel to (001) or (010). As shown in Figure 5.12, they modeled the (001) boundaries as tiltwalls, the (010) boundaries as twistwalls, and they hypothesized that both orientations formed by rotation about the *b* axis. The nature and energetics of low-angle grain boundaries in fibrolite aggregates are discussed in Chapter 7.

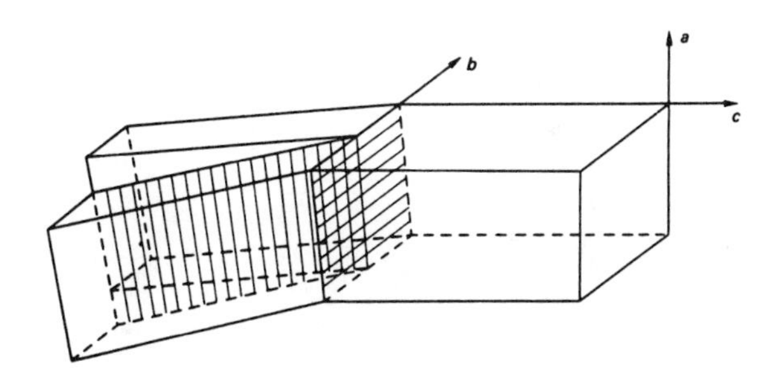

Figure 5.12. Model for the formation of (001) and (010) subgrain boundaries in naturally deformed sillimanite. Rotation about the *b* axis is accomodated by (001) tiltwalls and (010) twistwalls. (From Lambregts and van Roermund, 1990, Fig. 8).

CHAPTER 6

Al-Si DISORDER IN SILLIMANITE

Sillimanite is the most likely Al_2SiO_5 polymorph to have Al-Si disorder because it contains both Al and Si in tetrahedral coordination.

While noting that tetrahedral disorder in sillimanite would lead to localized charge imbalance between "nearest neighbor" tetrahedra, Zen (1969) considered that it could be an important factor in perturbing the Al_2SiO_5 phase equilibria. He also noted that Al-Si disorder would violate the *aluminum avoidance principle* (Lowenstein, 1954) in that Al^{IV}-O-Al^{IV} linkages would be produced. However, because such linkages occur in some minerals (e.g., scapolite), he implicitly questioned the validity of this principle.

THERMODYNAMIC MODELING

Greenwood (1972) carried out a theoretical analysis of Al-Si disorder in sillimanite using two models: (A) random mixing of Al and Si on tetrahedral sites, and (B) considering the Al^{IV}-O_c-Si^{IV} linkage as a unit. Because Al^{IV}-O_c-Al^{IV} linkages are precluded, model B adheres to the aluminum avoidance principle. For fully ordered sillimanite, Greenwood (1972) defined the Al^{IV} and Si^{IV} sites as α and β, respectively. For model A, the configurational entropy of Al-Si mixing was given as

$$\Delta S = R[2 \ln 2 - \{(1 + s) \ln (1 + s) + (1 - s) \ln (1 - s)\}] , \qquad [6.1]$$

where the "Bragg-Williams order parameter" s is a function of the probabilities of finding Al and Si on a particular site:

$$s = (p - p_{cd}) / (p_{po} - p_{cd}) .$$

In this equation, p_{cd} is the probability of complete disorder and p_{po} is the probability of perfect order. The limiting values of s are 0.0 for complete disorder and 1.0 for complete Al-Si order. For complete disorder, ΔS = 11.53 J/mol·K. Although not noted by Greenwood (1972), the validity of equation [6.1] is confirmed by the fact that the same value of ΔS is obtained with the more familiar multisite configurational entropy formulation:

$$\Delta S = -2R(X^{\alpha}_{Al} \ln X^{\alpha}_{Al} + X^{\beta}_{Si} \ln X^{\beta}_{Si}) ,$$

with complete disorder given by $X^{\alpha}_{Al} = X^{\beta}_{Si} = 0.5$. With model B each Al^{IV}-O_c-Si^{IV} linkage involves two cations; thus, the configurational entropy of model B is one half that of model A. For a given cation, only the potential energies of the nearest neighbor and next nearest neighbor atoms were considered. Figure 6.1 shows the linkages considered by Greenwood (1972) for model A. For the α and β sites, the energy of interaction was computed for Al and Si by summing the energy terms for various linkages; each term was weighted for the probability of that linkage. Greenwood then expressed the energies of the four sites ($U_{Al,\alpha}$, $U_{Al,\beta}$, $U_{Si,\alpha}$, $U_{Si,\beta}$) as a function of the energies of the individual linkages (i.e., ϕ_1, ϕ_2, see Fig. 6.1) weighted by their respective probabilities (p). Adding the four individual site energies to yield U_{total}, followed by algebraic simplification, Greenwood derived the energy difference between the totally ordered and disordered sillimanite

$$(U_{total,\, s=1} - U_{Total,\, s<1}) = \phi_2(1-s^2) \ .$$

In noting that H = U + PV, and assuming the PV term to be negligible: $\Delta H_{disorder} = \Delta H°(1 - s^2)$. From the relationship: $\Delta G = \Delta H - T\Delta S$, Greenwood derived for model A:

$$\Delta G = \Delta H°(1 - s^2) + RT[(1 + s)ln(1 + s) + (1 - s)ln(1 - s) - 2ln2]. \quad [6.2]$$

Noting that the most stable degree of disorder occurs where ΔG is minimum, Greenwood set the first derivative of equation [6.2] equal to zero (corresponding to a horizontal tangent to the curve of ΔG versus s) and thus arrived at:

$$\Delta H° = RT\left[\frac{ln(1 + s) - ln(1 - s)}{2s}\right] \ . \quad [6.3]$$

He then evaluated $\Delta H°$ values from equation [6.3] by selecting "reasonable" ranges of temperature (500-700°C) and s values (0.9 to 1.0). Greenwood (1972) noted that single-crystal X-ray refinements of sillimanite show no disorder, although up to 5% disorder would be undetected by this technique. Thus, he chose s values corresponding to 0 to 5% disorder. For model A these temperature and s values yield $\Delta H°$ ranging from 12.5 to 23 kJ/mol. By arbitrarily assuming a "midrange" value of $\Delta H°$ = 19.25 kJ/mol (= 4 kcal) and noting that because of the small $\Delta V_{disorder}$ the degree of order will be virtually independent of pressure, Greenwood contoured the sillimanite P-T stability field with isopleths of selected s values (Fig. 6.2). As expected, the degree of disorder increases (i.e., s

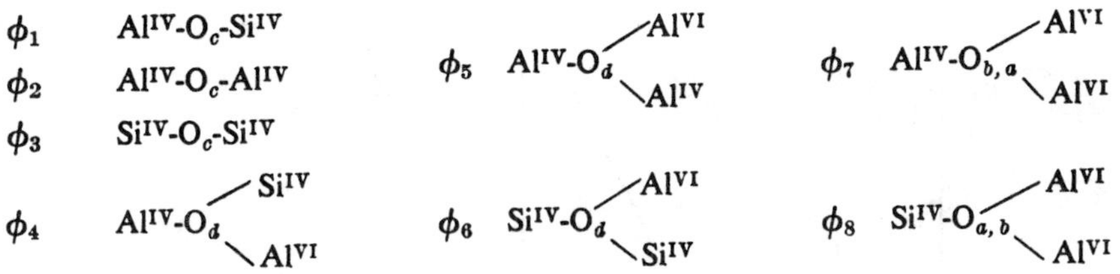

ϕ_1 Al^{IV}-O_c-Si^{IV}

ϕ_2 Al^{IV}-O_c-Al^{IV}

ϕ_3 Si^{IV}-O_c-Si^{IV}

ϕ_4 Al^{IV}-O_d (–Si^{IV}, –Al^{VI})

ϕ_5 Al^{IV}-O_d (–Al^{VI}, –Al^{IV})

ϕ_6 Si^{IV}-O_d (–Al^{VI}, –Si^{IV})

ϕ_7 Al^{IV}-$O_{b,a}$ (–Al^{VI}, –Al^{VI})

ϕ_8 Si^{IV}-$O_{a,b}$ (–Al^{VI}, –Al^{VI})

Figure 6.1. Coordination linkages considered in Greenwood's (1972) model of Al-Si disorder in sillimanite. (From Greenwood, 1972, p. 558).

decreases) with increasing temperature. Figure 6.3 provides graphical confirmation of Greenwood's (1972) conclusion that his $G_{disorder}$ has a significant effect on the stability of sillimanite. He suggested that differing degrees of disorder in sillimanite may explain differences in the experimental phase equilibrium bracketing studies on the Al_2SiO_5 equilibria. In addition, he considered that disorder may have an important effect on the Al_2SiO_5 polymorphic transformations in the field, suggesting that coexisting kyanite + sillimanite may be attributed to disorder in sillimanite.

Jones et al. (ms.) used a static, minimum lattice energy, ionic model to compute the change in lattice energy of sillimanite in the transformation from full tetrahedral Al-Si disorder to complete Al-Si order. From comparison with estimates of changes in lattice energy for Al-Si ordering of other silicate minerals, their computed Al-Si ordering energy value of 122.4 kJ/mol is "...certainly in the range expected for this reaction". However, they emphasized that their computations are for *lattice energy*, not *free energy*. Because their computations did not include temperature-dependent effects (e.g., Boltzmann averaging), they concluded that their computed value of Al-Si ordering energy may be excessive.

PHASE EQUILIBRIUM EXPERIMENTS

Anderson and Kleppa (1969) observed that the dP/dT slope of the kyanite-sillimanite equilibrium determined by Richardson et al. (1968) was steeper than that computed from the Clapeyron equation using third-law entropies, and suggested that this discrepancy may be attributed to configurational entropy arising from Al-Si disorder in the sillimanite formed in the experimental run charges of Richardson et al. (1968).

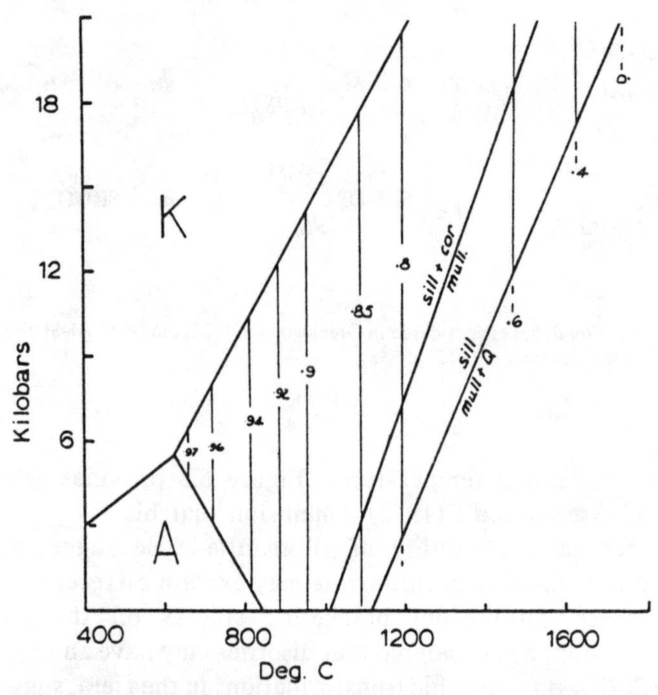

Figure 6.2. Greenwood's (1972) phase equilibrium diagram showing the stability relations of the Al_2SiO_5 polymorphs and mullite. The vertical lines correspond to Greenwood's (1972) Al/Si disorder parameter *s*. The limiting values of *s* are 0.00 for complete disorder and 1.0 for complete order. (From Greenwood, 1972, Fig. 6).

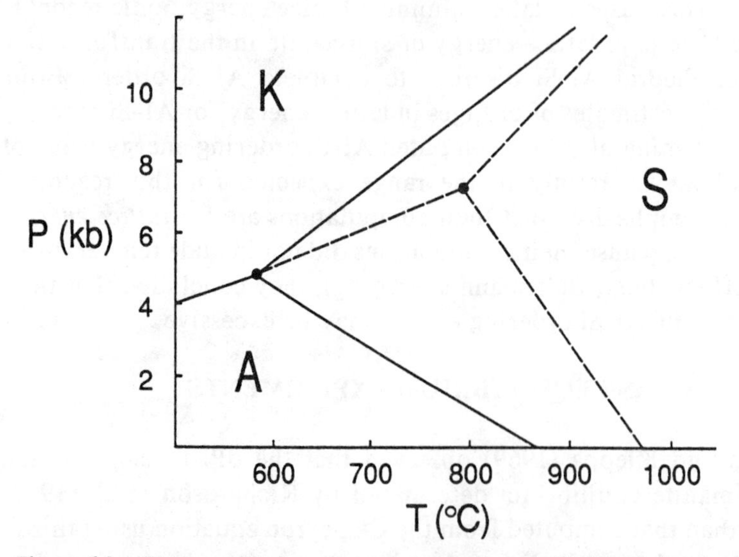

Figure 6.3. Phase equilibrium diagram comparing equilibria calculated with ordered sillimanite (solid curves) versus those with Greenwood's (1972) $G_{disorder}$ function added to sillimanite (dashed lines). This diagram was computed by J.A.D. Connolly using Berman's (1988) data base and the VERTEX program of Connolly and Kerrick (1987).

Utilizing the thermodynamic treatment of Navrotsky and Kleppa (1967), Holdaway (1971) analyzed Al-Si disorder by expressing sillimanite as $Al(Al_{1-x}Si_x)(Si_{1-x}Al_x)O_5$. Disorder would be complete at x = 0.5; fully ordered sillimanite would have x = 0. The Gibbs free energy of disorder is

$$G_d = E_d + PV_d - TS_d \, . \qquad [6.4]$$

Using the superscript ° to refer to complete disorder, Holdaway (1971) defined:

$$E_d = xE^\circ \, ;$$

$$S_d = -2R[x \, ln \, x + (1-x) \, ln \, (1-x)] \, ; \qquad [6.5]$$

$$V_d = xV^\circ \, .$$

Differentiating equation [6.4] and assuming that equilibrium disorder occurs where $(dG_d/dx) = 0$, Holdaway derived:

$$x/(1-x) = \exp[-(E^\circ + PV^\circ/2RT)] \, . \qquad [6.6]$$

He used available molar volume and vibrational entropy data to compute the "third law" Clapeyron slope $dP/dT = \Delta S_{vib}/\Delta V_r$. Assuming that sillimanite formed in hydrothermal experiments has equilibrium Al-Si disorder, S_d was derived from the difference between the "third law" Clapeyron slope and the dP/dT slope fit to experimental data:

$$dP/dT = (\Delta S_{vib} + S_d^{Sil})/\Delta V_r \, . \qquad [6.7]$$

Computing the amount of disorder at the equilibrium value of x using equation [6.5], and assuming PV° is negligible, Holdaway used equation [6.6] to compute E° = 61.71 KJ. He noted that sillimanite from the Asama volcano in Japan (Aramaki, 1961) has an "... anomalously high cell volume..." (Holdaway, 1971, p. 121). Assuming this to be equivalent to the volume change due to Al-Si disorder, Holdaway then had the requisite data to compute V_d, S_d, and H_d as a function of T and x (Table 6.1). He concluded (p. 97) that: "Inconsistencies between calculated and experimental phase boundaries can be accounted for by sillimanite Al-Si disorder."

Table 6.1. Holdaway's (1971) computed volume (V_d in cm^3/mol), entropy (S_d in cal/deg·mol) and enthalpy (H_d in cal/mol) for Al/Si disorder in sillimanite at 1 atm pressure. The disorder parameter x corresponds to the fractional occupancy of Al on tetrahedral sites that would contain only Si in the ordered structure: $Al(Al_{1-x}Si_x)(Si_{1-x}Al_x)O_5$. The percent disorder is given by 2x. (From Holdaway, 1971, Table 8).

T °C	x	V_d	S_d	H_d
200	0.000	0.002	0.01	6
400	0.004	0.016	0.10	60
600	0.014	0.056	0.29	205
800	0.031	0.122	0.54	443
1000	0.051	0.205	0.80	746
1200	0.074	0.296	1.05	1077
1400	0.097	0.388	1.27	1411

In spite of Holdaway's admirable attempt to quantify the thermodynamics of Al-Si disorder in sillimanite, there are uncertainties with his approach. The S_d values for sillimanite were computed with equation [6.7] using the 1 atm experimental "bracket" of Weill (1966) coupled with Holdaway's (1971) brackets at elevated pressures. As discussed in Chapter 3, the interpretation of these experimental data is in question. Thus, the accuracy of Holdaway's computed S_d values is suspect. Furthermore, Holdaway's $\Delta V_{disorder}$ is based on Aramaki's (1961) determination of the unit cell volume of sillimanite from the Asama volcano, Japan. More recent X-ray determinations (Cameron and Ashworth, 1972; Beger, 1979) show that the molar volume of the Asama sillimanite is smaller than that determined by Aramaki (1961).

Saxena (1974) analyzed the effect of Al-Si disorder in sillimanite on the kyanite-sillimanite equilibrium. He represented disorder by the *exchange* equilibrium, Al(T1) + Si(T2) = Al(T2) + Si(T1), where T1 and T2 represent the respective Al and Si sites in fully ordered sillimanite. Assuming ideal mixing, the Gibbs free energy of mixing is

$$G_m^{Sil} = RT\,ln\,X_{Al\text{-}T1}\,X_{Al\text{-}T2}\ .$$

Assuming a small amount of disorder ($X_{Al\text{-}T2}$ = 0.01), he computed the kyanite-sillimanite equilibrium to be at a pressure exceeding 70 kbar at 700°C! Because this pressure is far higher than that determined by experiments at this temperature (P_{exp} = 7-8 kbar), Saxena (1974) concluded that an ideal mixing model is very inadequate. Thus, he chose to represent the free energy of disordered sillimanite by a regular, symmetric solution model:

$$G_m^{Sil} = -\ W[(X_{Al\text{-}T1})^2 + (X_{Al\text{-}T2})^2] + RT\,ln\,X_{Al\text{-}T1}\,X_{Al\text{-}T2}\ . \qquad [6.8]$$

Assuming that the sillimanite in the experiments of Richardson et al. (1968) was disordered, G_m^{Sil} = -217 cal/mol was computed by the difference in the 700°C location of the equilibrium between that derived by Anderson and Kleppa (1969) and Richardson et al. (1968). He suggested that, for small amounts of disorder, a reasonable range of W is 5000 and 9000 cal/mol. Using equation [6.8] with G_m^{Sil} = -217 cal/mol, he considered two "disordering schemes": (1) X_{Al-T2} = 0.05 and W = 6282 cal/mol, and (2) X_{Al-T2} = 0.10 and W = 5423 cal/mol. Saxena (1974) used this equation to show that small amounts of disorder in sillimanite yield much smaller P-T displacement of the kyanite-sillimanite equilibrium compared to that derived with sillimanite disorder obtained with the model of ideal mixing. Nevertheless, he computed that there was an appreciable (2 kbar) displacement of the kyanite-sillimanite equilibrium with a small amount (1%) of Al-Si disorder. Because this equilibrium is strongly affected by small amounts of disorder, Saxena (1974) concluded that variations in the Al-Si disorder of sillimanite in both starting material and run products of various experimental studies on this equilibrium could readily explain the different experimental locations of this equilibrium (Fig. 6.4).

We reexamine Saxena's approach by starting with a fundamental expression representing equilibrium for a reaction:

$$(\Delta G_r)_P^T = 0 = \Delta H^\circ_r - T\Delta S^\circ_r + P\Delta V_r + RT\, ln\, K \ .$$

In the case of ordered kyanite in equilibrium with partially disordered sillimanite we have:

$$0 = \Delta H^\circ_r - T\Delta S^\circ_r + P\Delta V_r + RT\, ln\, a^{Sil}_{Al_2SiO_5} \ . \qquad [6.9]$$

If we envision exchanging Si and Al between the T1 and T2 sites by disorder, and if we assume ideal mixing on each site, the activity of the Al_2SiO_5 component in sillimanite is

$$a^{Sil}_{Al_2SiO_5} = X_{Al-T1}\, X_{Si-T2} \ . \qquad [6.10]$$

Saxena (1974) considered the case where X_{Al-T2} = 0.01; thus, X_{Si-T2} = 1 - X_{Al-T2} = 0.99. Maintenance of stoichiometry requires that X_{Al-T1} = 0.99. Using $a^{Sil}_{Al_2SiO_5}$ computed from equation [6.10], coupled with the 1 bar values for ΔH_r°, ΔS_r° and ΔV_r given by Saxena (1974), equation [6.9] yields a pressure of 6.10 kbar at 700°C, whereas the computation with fully ordered sillimanite ($a^{Sil}_{Al_2SiO_5}$ = 1.0) yields 5.82 kbar. This 280 bar

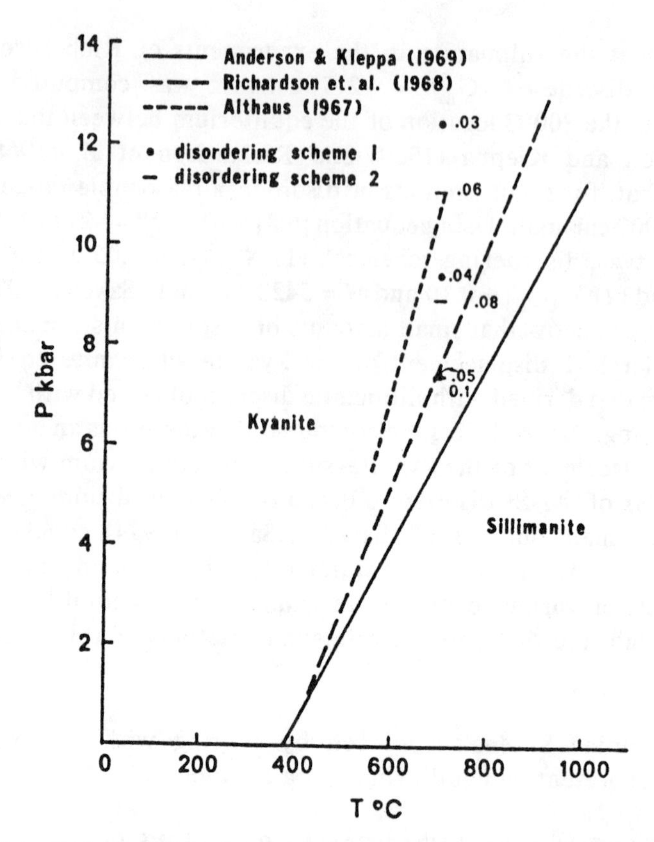

Figure 6.4. Saxena's (1974) analysis of Al-Si disorder in sillimanite and the kyanite-sillimanite equilibrium. The "disordering schemes" correspond to two different values of the amount of disorder ($X_{Al\text{-}T2}$): 0.05 for scheme 1 and 0.10 for scheme 2. $X_{Al\text{-}T2}$ is zero for complete order and 0.5 for complete disorder. The fractional numbers correspond to the amount of Al/Si disorder using a regular, symmetric solution model. (From Saxena, 1974, Fig. 1).

pressure difference is *far* less than the >70 kbar value computed by Saxena with the assumption of ideal mixing. He erroneously used the free energy of mixing for the *exchange reaction* rather than the difference in chemical potentials of the Al_2SiO_5 component between ordered and partially disordered sillimanite (as in equation [6.9]). I conclude that Saxena (1974) did not correctly demonstrate the inadequacy of the ideal mixing model for small amounts of Al-Si disorder in sillimanite.

Bowman (1975) estimated the configurational entropy of disorder in sillimanite (S_d) using arguments based on Clapeyron slopes of the andalusite-sillimanite equilibrium. As shown in Figure 3.25, he noted that the -13.6 bars/°C slope of the equilibrium determined by Holdaway (1971) significantly differs from the -7.4 bars/°C slope determined by his experimental study. Using equations [6.5] and [6.7], Bowman (1975) suggested that, in comparison with the sillimanite involved in his experimental study, that of Holdaway (1971) may have an additional 6-8% of Al-Si disorder. However, in light of the questions regarding Holdaway's (1971) and Bowman's (1975) experimental brackets (see Chapter 3), Bowman's computed S_d values for sillimanite are questionable.

For the andalusite-sillimanite equilibrium, Anderson et al. (1977) noted that the dP/dT slope constrained by the experimental data of Richardson et al. (1969) is inconsistent with that computed from their calorimetric data. In contrast, the dP/dT slope they computed is in good agreement with the experimental brackets of Holdaway (1971). They suggested that the experiments of Richardson et al. were deleteriously affected by the use of finely ground sillimanite as starting material: "The ultrafine or strained portion of a charge could have inverted to andalusite at a temperature well above the stability limit of well ordered sillimanite, simply because it is probably impossible to grow anything approaching a well-ordered sillimanite in quantities large enough to produce an increase in X-ray peak heights in feasible run-times" (Anderson et al., 1977, p. 591). Anderson et al. favored Holdaway's (1971) experimental results because Holdaway's computed ΔH for Al-Si disorder in sillimanite is in good agreement with the calorimetric results of Navrotsky et al. (1973). However, as discussed in the next section of this chapter, there are questions shrouding the interpretation of the experimental results of Navrotsky et al. (1973).

Robie and Hemingway (1984) concluded that the dP/dT slope of the andalusite-sillimanite equilibrium computed with third-law entropies is in good agreement with the experimental brackets of Holdaway (1971). This agreement suggests "... that Al and Si are ordered in sillimanite at least to 1100 K" (Robie and Hemingway, 1984, p. 298). As reviewed in Chapter 3, this conclusion is weakened by questions concerning Holdaway's (1971) experimental brackets on the andalusite-sillimanite equilibrium. In addition, Robie and Hemingway (1984) neglected the non-trivial configurational entropy introduced by solid solution of Fe^{3+} in the Fe-rich sillimanite used by Holdaway (1971). Furthermore, the conclusion of Robie and Hemingway (1984) assumes that both the sillimanite used as starting material and that grown during the experiments were fully ordered. There is no verification of these conclusions.

In spite of uncertainties in extracting S_d of sillimanite with the andalusite-sillimanite equilibrium (cf. Holdaway, 1971; Robie and Hemingway, 1984), comparison of computed Clapeyron slopes with experimental data for the kyanite-sillimanite equilibrium offers a way to extract configurational entropies of sillimanite. Robie and Hemingway (1984) showed that, for the kyanite-sillimanite equilibrium, the dP/dT slope computed with third law entropies is in good agreement with the slope constrained by experimental brackets (see Fig. 3.37). Accordingly, they suggested that Al-Si disorder has a negligible effect on the kyanite-sillimanite equilibrium at temperatures below 1100K (827°C). However, this conclusion is questionable in light of the large uncertainty in the extremal dP/dT slopes (15.1 bars/°C to 23.4 bars/°C) allowed by the experimental data available to Robie and Hemingway. As shown in Figure 3.27, the experimental data of A.L. Montana and S.R. Bohlen reveal significant curvature of the kyanite-sillimanite equilibrium above about 800°C. *Assuming* that this curvature is attributed to Al-Si configurational entropy of sillimanite, we may estimate the amount of disorder in this phase. A tangent to the curve was determined graphically near the upper limits of the experimentally-determined P-T conditions (Fig. 3.27). From the slope of this tangent [$(dP/dT)_{exp}$ = 31.6 bars/K], the vibrational entropies computed from heat capacities, and molar volumes corrected for compressibility and thermal expansion, a value of 4.8 J/mol·K is computed for the entropy of disorder of sillimanite (S_d) with equation [6.7]. Using equation [6.5] this S_d value yields x = 0.08, which corresponds to 16 percent disorder. It is important to emphasize that this value for disorder was derived from the experimental brackets of A.L. Montana and S.R. Bohlen at 11-15 kbar and 900-1000°C. As discussed in Chapter 3, the kyanite-sillimanite equi-librium may be unaffected by Al-Si disorder in sillimanite below 800°C. Assuming that this is true, Al-Si disorder in sillimanite would only be of significance to the Al_2SiO_5 phase equilibria at P-T conditions corresponding to the highest grades of metamorphism (i.e., the upper granulite facies and the sanidinite facies).

EXPERIMENTAL HEAT TREATMENT

Navrotsky et al. (1973) studied the enthalpy of Al-Si disordering in Brandywine Springs fibrolite. They subjected the fibrolite to hydrothermal treatment at P = 16-23 kbar and T = 1400-1500°C. As shown in Figure 6.5 there is no difference in the enthalpies of solution of the treated versus untreated material up to about 1300°C; however, enthalpy differences occur at T > 1300°C. This increase in ΔH_{obs} is accompanied by an increase in the unit cell volume (Fig. 6.6). Navrotsky et al. considered the increase in ΔH_{obs} above 1300°C (Fig. 6.5) to reflect tetrahedral Al-Si disorder. They used the disordering theory of Navrotsky and Kleppa (1967) to model the "plateau" at 1400-1550°C (Fig. 6.5). An

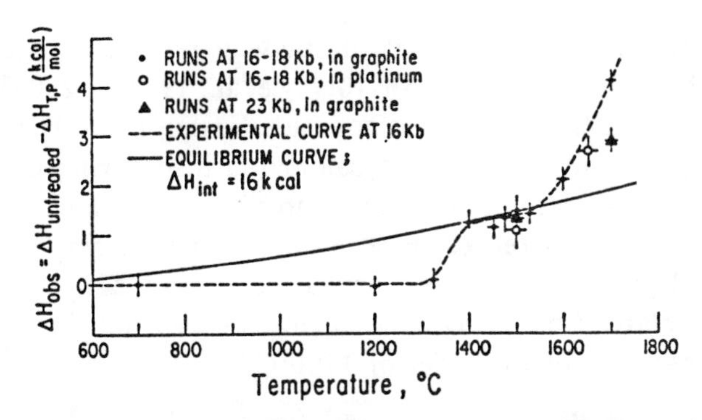

Figure 6.5. Molten oxide heat-of-solution data for samples of Brandywine Springs fibrolite that were heat treated at pressures of 16-18 kbar. Dashed line: experimental data; solid line: calculated for an Al/Si interchange enthalpy of 16 kcal/mol. (From Navrotsky et al., 1973, Fig. 1).

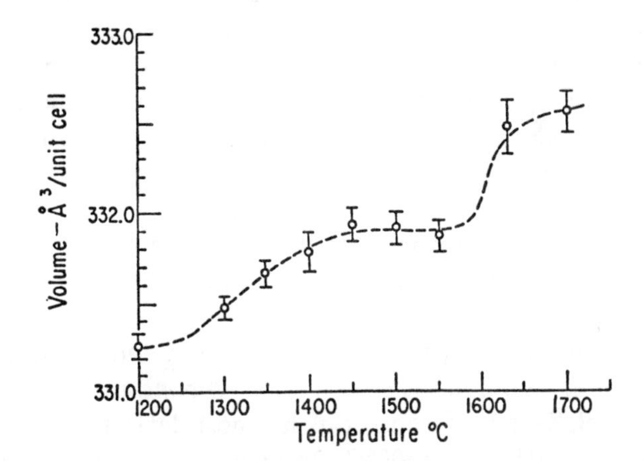

Figure 6.6. Unit cell volume of fibrolite that was subjected to hydrothermal treatment at P = 16-18 kbar. The cell volume at 1200°C is equivalent to that of untreated fibrolite. (From Navrotsky et al., 1973, Fig. 2).

"interchange enthalpy" (ΔH_{int}) of approximately 67 kJ/mol was computed for the disordering reaction:

$$Al_{Al} + Si_{Si} = Al_{Si} + Si_{Al} ,$$

where the left side of the equation refers to site occupancy in fully ordered sillimanite (Al_{Al} = aluminum on an aluminum site of the ordered sillimanite structure) whereas the right side refers to disordered site occupancy (Al_{Si} = aluminum on a silicon site of the ordered sillimanite structure). They considered that the good agreement of the calculated curve (solid line in Fig. 6.5) with the experimental data (dashed line) supports their contention "... that the plateau represents equilibrium or near-equilibrium disordering" (Navrotsky et al., 1973, p. 2504). However, it should be noted that ΔH_{int} was computed using measured ΔH_{obs} data. Thus, the agreement between the calculated disordering curve and the experimental data does not necessarily provide *independent* confirmation that the plateau region represents equilibrium disordering. Navrotsky et al. concluded (p. 2507) that the marked increase in ΔH_{obs} above 1550°C "...may be explained by a single, co-operative disordering process with a critical temperature of the order of 1700°C."

Holland and Carpenter (1986) subjected sillimanite to hydrothermal treatment at the same P-T conditions as that of Navrotsky et al. (1973). TEM examination of run products revealed the presence of abundant dislocations and antiphase boundaries (APB's) as well as small inclusions which they considered to be silica-rich glass. Thus, they questioned the interpretation of the calorimetric data of Navrotsky et al. (1973). In particular, they noted that the "enthalpy of vitrification" is typically larger than the enthalpy of ordering; thus, the enthalpy of solution measurements of Navrotsky et al. may in part reflect the presence of a small proportion of glass in the charge.

Using sillimanite from the same locality of that used in the X-ray study of Burnham (1963a), Beger (1979) studied the thermal transformations at 1000-1500° and 1 bar-15 kbar. Disordering of sillimanite is manifest by diminution in the intensity of the *l*-odd reflections, and increase in the *b* unit cell dimension. From mean Si-O and Al-O bond distances obtained from single crystal X-ray refinement of his most definitively disordered run product (run no. HS-4B; 1500°C, 24 kbar, 24 h), he made a "rough estimate" of 30% disorder. This estimate is compatible with the residuals of least-squares refinements assuming 20%, 30% and 40% disorder. However, Beger's analysis of errors in site occupancy suggests an uncertainty of ±11% in the degree of disorder. Because of positional disorder of the bridging oxygens (O_c), the Si^{IV}-O_c-Al^{IV} bond angle shows a relatively large change upon disordering. Beger's analysis of the size and orientation of the vibration ellipsoid of O_c *suggests* that the Si^{IV}-O_c-Al^{IV} linkage is preserved, thus supporting Greenwood's (1972) disorder model B. This has important implications regarding the thermodynamic properties of disordered sillimanite. In addition to experiments at 1500°C, Beger (1979, p. 109) concluded that some disorder is detectable in sillimanite heated at 1250°C and that small amounts of disorder may be induced as low as 1000°C.

Using single crystal X-ray diffraction, Guse et al.(1979) studied the transformation of sillimanite at 1 atm and 1600°C. After 5 h the *l*-odd lines of sillimanite became weaker. After 10 h sillimanite transformed into mullite (the *l*-odd reflections disappeared and the space group changed from *Pbnm* to *Pbam*). Samples heated 10-30 h yielded "mullite 1" with $a = 7.75$Å, whereas "mullite 2" with $a = 7.541$Å formed after 35 h. In light of Beger's (1979) analysis, diminution of the *l*-odd reflections, and increased *a* and *b* unit cell dimensions, suggest that sillimanite became disordered in the initial stages (the first 10 h) of heating. The observations of Guse et al. (1979) are commensurate with Beger's (1979) conclusion that the mullitization of sillimanite is preceded by Al^{IV}-Si^{IV} disorder.

NATURAL SILLIMANITE AND FIBROLITE

X-ray diffraction

"Since the differences in [X-ray] scattering power between Al and Si atoms is small, the refinement of tetrahedral site occupancies in disordered sillimanite is analogous to the recognition of dancers in a fog."

R.M. Beger (1979, p. 148)

X-ray structure refinements of sillimanite are incapable of detecting small amounts (< 5 %) of Al-Si disorder (Greenwood, 1972).

Cameron and Ashworth (1972) used X-ray diffraction and infrared (IR) spectroscopy to evaluate Al-Si disorder in fibrolite. They postulated that fully ordered fibrolite would belong to the space group *Pbnm* whereas full Al-Si disorder would correspond to space group *Pbam*. They concluded that the intensity of the *l*-odd reflections is inversely correlative with disorder such that the *l*-odd reflections would be zero for complete disorder. They state (p. 135) that: "The average intensity of the *l*-odd relative to *l*-even reflections is therefore a measure of the degree of disorder." Their X-ray data showed that fibrolite belongs to space group *Pbnm*, thereby substantiating that fibrolite is ordered. In a critique of Cameron and Ashworth's (1972) paper, Sahl and Seifert (1973) argued that complete disorder would preserve the *l*-odd maxima. For a number of (*hkl*) and (*hk0*) reflections they computed structure factors for selected amounts of disorder ranging from full Al-Si order to complete disorder. Contrary to Cameron and Ashworth, Sahl and Seifert (1973) found no systematic decrease in the relative intensities of the *l*-odd versus *l*-even reflections with increasing disorder. However, the studies of Beger (1979) and Guse et al. (1979) (discussed in the previous section) suggest that the relative intensities of the *l*-odd maxima are in fact a monitor of Al-Si disorder. Nevertheless, it remains to be demonstrated whether small amounts of disorder (< 5%) can be detected by measurements of the intensities of *l*-odd reflections.

Winter and Ghose (1979) carried out an X-ray refinement of Brandywine Springs fibrolite over the range 25 to 1000°C. To evaluate Al-Si disorder they computed the isotropic equivalent temperature factor (*B*) at various temperatures. At the temperature of absolute zero a criterion for complete Al-Si order is $B = 0$, whereas disorder would correspond to $B \neq 0$. As an example of a completely ordered phase, Winter et al. (1977) show that for low albite, *B* for tetrahedral Si and Al is a linear function of temperature, and extrapolation to zero kelvin yields $B = 0$. For zero kelvin Winter and Ghose (1979) derived *B* values for andalusite and kyanite (0.159Å^2 and 0.156Å^2, respectively) that are within 3σ of zero. However, the *B* value for fibrolite (0.246Å^2) is outside of the 3σ range of zero, suggesting some degree of Al-Si disorder.

Using X-ray analysis, Beger (1979) examined disorder in several samples of sillimanite and fibrolite. His study included samples from high-temperature volcanic environments. Sillimanite from the Asama volcano, Japan (the same locality as that studied by Aramaki, 1961 and Cameron and Ashworth, 1972) has cell dimensions larger than that of ordered sillimanite. Thus, Beger suggested that the Asama sillimanite has some disorder. However, the cell dimensions of the other sillimanite and fibrolite samples studied by Beger match those of ordered sillimanite. Thus, he concluded that the "collective evidence" argues that sillimanite and fibrolite are fully ordered.

<u>Neutron diffraction</u>

Finger and Prince (1972) carried out a neutron refinement of the structure of the Brandywine Springs fibrolite. In comparing the mean tetrahedral Si-O and Al-O distances with those of Burnham (1963a) for sillimanite, they noted that the Brandywine Springs fibrolite has larger Si tetrahedra and smaller Al tetrahedra, and concluded that this could be attributed to Al-Si disorder. However, Peterson and McMullan's (1986) single crystal neutron site refinement of sillimanite gave *no* evidence for disorder.

No Al^{IV}-Si^{IV} disorder was detected in a single crystal neutron refinement of the Sri Lanka sillimanite (Kerrick et al., ms.). However, analysis of error of the multiplicity parameter for tetrahedral site occupancies, coupled with the magnitudes of the scattering abilities of Al and Si, suggests that no less than 4-5% disorder can be detected. Nevertheless, this analysis is commensurate with MAS-NMR analysis (discussed in the next section) suggesting no detectable tetrahedral disorder in this sillimanite.

Spectroscopic studies

Cameron and Ashworth (1972) suggested that sillimanite and fibrolite are fully ordered because of the "sharpness" of IR spectral peaks. However, this conclusion is qualitative because a small but significant amount of Al-Si disorder (e.g., 5%) would produce negligible broadening of the IR absorption peaks (W.B. White, pers. comm.).

The author has been involved in a study of Al-Si disorder in sillimanite and fibrolite using magic angle spinning nuclear magnetic resonance (MAS-NMR) spectroscopy. As summarized in the reviews of Smith et al. (1983) and Kirkpatrick et al. (1985), MAS-NMR spectroscopy offers a powerful tool for examining short-range tetrahedral Al-Si disorder. MAS-NMR of the nuclides ^{29}Si and ^{27}Al can be used to examine the site occupancies of tetrahedrally-coordinated Si and Al. The spectra of ^{29}Si are particularly useful because they are uncomplicated by the broadening and splitting of resonances found with quadrupolar nuclides (such as ^{27}Al). We have obtained ^{29}Si MAS-NMR spectra of fibrolite and sillimanite. These samples include fibrolite (from Lewiston, Idaho), and gem-quality sillimanite (from Sri Lanka), which are advantageous because they have been studied by calorimetry (Topor et al., 1989; Hemingway et al., ms.) and neutron diffraction (Kerrick et al., ms.). In all samples, the location of the major peak at -87 ppm is in good agreement with the sillimanite MAS-NMR spectra of Smith et al. (1983) and Janes and Oldfield (1985).

To address the question of how Al^{IV}-Si^{IV} disorder in sillimanite would be manifest in ^{29}Si MAS-NMR spectra, we contrast the cation-oxygen-cation linkages for order versus disorder. As shown in Figure 6.7, all ^{29}Si nuclides in ordered sillimanite would be linked to three nearest neighbor (NN) Al tetrahedra. Disorder leads to three alternative NN cations (Fig. 6.7). Analysis of tetrahedral Al^{IV}-Si^{IV} disorder in tectosilicates and phyllosilicates (Kirkpatrick et al., 1985) suggest that, because of progressive electron deshielding, the disorder linkages should be located at more negative chemical shifts compared to the ordered linkage and that these peaks should be separated by *approximately* 5 ppm. Because

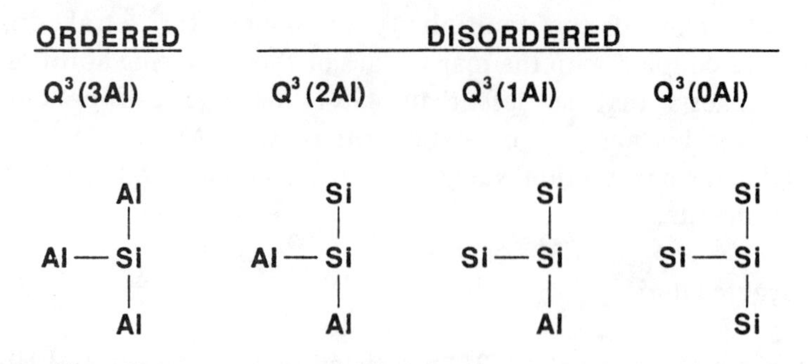

Figure 6.7 Schematic of nearest neighbor tetrahedral cation linkages for ordered and disordered configurations of the sillimanite structure.

sillimanite is expected to be largely ordered, the main peak at -87 ppm is assigned to the Q^3(3Al) linkage. Figure 6.8 shows a *hypothetical* MAS-NMR spectrum predicted for disordered sillimanite assuming that adjacent peaks are separated by 5 ppm. For the sillimanite structure, the peak separation between Q^3(3Al) and Q^3(2Al) has been refined by B.L. Sherriff using the method outlined by Sherriff and Grundy (in press). Two separate Q^3(2Al) configurations were considered: (a) disorder involving tetrahedra linking the double tetrahedral chains across bridging oxygens (Al^{IV}-O_c-Al^{IV} and Si^{IV}-O_c-Si^{IV} linkages are created), and (b) disorder preserving the Si^{IV}-O_c-Al^{IV} linkages (this is equivalent to disorder model B of Greenwood, 1972). The computed difference between the main peak position and the Q^3(2Al) peak is 8 ppm for (a) and 2 ppm for (b). The relatively large chemical shift for (a) arises from relatively large changes in electron shielding (upon disordering) resulting from the short Si^{IV}-O_c and Al^{IV}-O_c bond distances.

Figure 6.9 shows MAS-NMR spectra of sillimanite samples. Only the Sri Lanka sillimanite shows an extraneous peak (at ≈ -97 ppm). Because of the large chemical shift of this peak in relation to the main peak, Kerrick et al. (1990) suggest that the -97 ppm peak is not indicative of disorder. The Kilbourne Hole sillimanite is from aluminous xenoliths in basic volcanic rocks. In spite of the high temperature environment, as evidenced by partial melting (yielding glass) and complete disorder of the K-feldspars (Grew, 1979), the MAS-NMR spectrum of this sillimanite shows no evidence of disorder. The "left-skewness" of the main peak of this sample is opposed to that predicted for disorder, and may be attributed to the quadrupole moment of ^{27}Al. None of the peaks of the other sillimanite samples display peak asymmetry indicative of disorder.

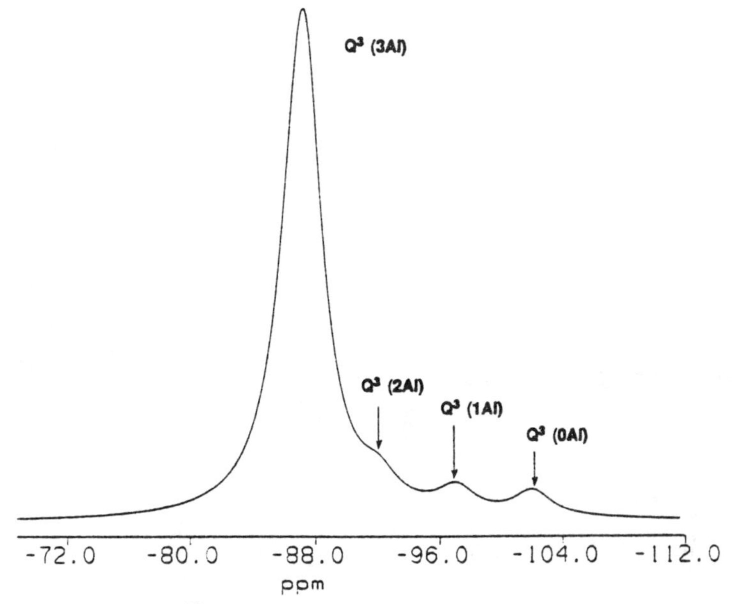

Figure 6.8. Hypothetical ^{29}Si MAS-NMR spectrum for sillimanite showing four peaks corresponding to the nearest neighbor tetrahedral cation-cation linkages shown in Figure 6.7. Adjacent peaks are separated by 5 ppm.

The MAS-NMR spectra of fibrolite show small peaks that are assignable to impurities. The peak at -108 ppm for the Guffey (Colorado) fibrolite (Fig. 6.10) corresponds to the position of the main MAS-NMR peak of quartz (Smith et al., 1983). This assignment is strengthened by the fact that fibrolite aggregates commonly contain intergrown quartz. The small peak at -95 ppm for the Brandywine Springs fibrolite corresponds to the main peak of pyrophyllite. Although I am unaware of any literature reports of pyrophyllite intergrown with the Brandywine Springs fibrolite, pyrophyllite is a very likely alteration product of the aluminum silicates. As with sillimanite, the main peaks of the fibrolite samples show no asymmetry. Thus, the MAS-NMR spectra of the sillimanite and fibrolite samples reveal no evidence of disorder. Our analysis of sensitivity suggests that the sillimanite and fibrolite samples have no more than 2-3% disorder. Improved signal/noise ratios provided by a long (24 h) run on the Sri Lanka sillimanite suggests a maximum of 1-2% disorder. The lack of disorder in the Sri Lanka sillimanite as implied by MAS-NMR spectroscopy is commensurate with the neutron refinement of a single crystal of Sri Lanka sillimanite (Kerrick et al., ms.). Because our MAS-NMR analysis includes sillimanite from granulite facies (Sri Lanka) and sanidinite facies (Kilbourne Hole) parageneses, it is concluded that Al^{IV}-Si^{IV} disorder is not a significant factor in the P-T stability of sillimanite formed in rocks subjected to crustal metamorphism.

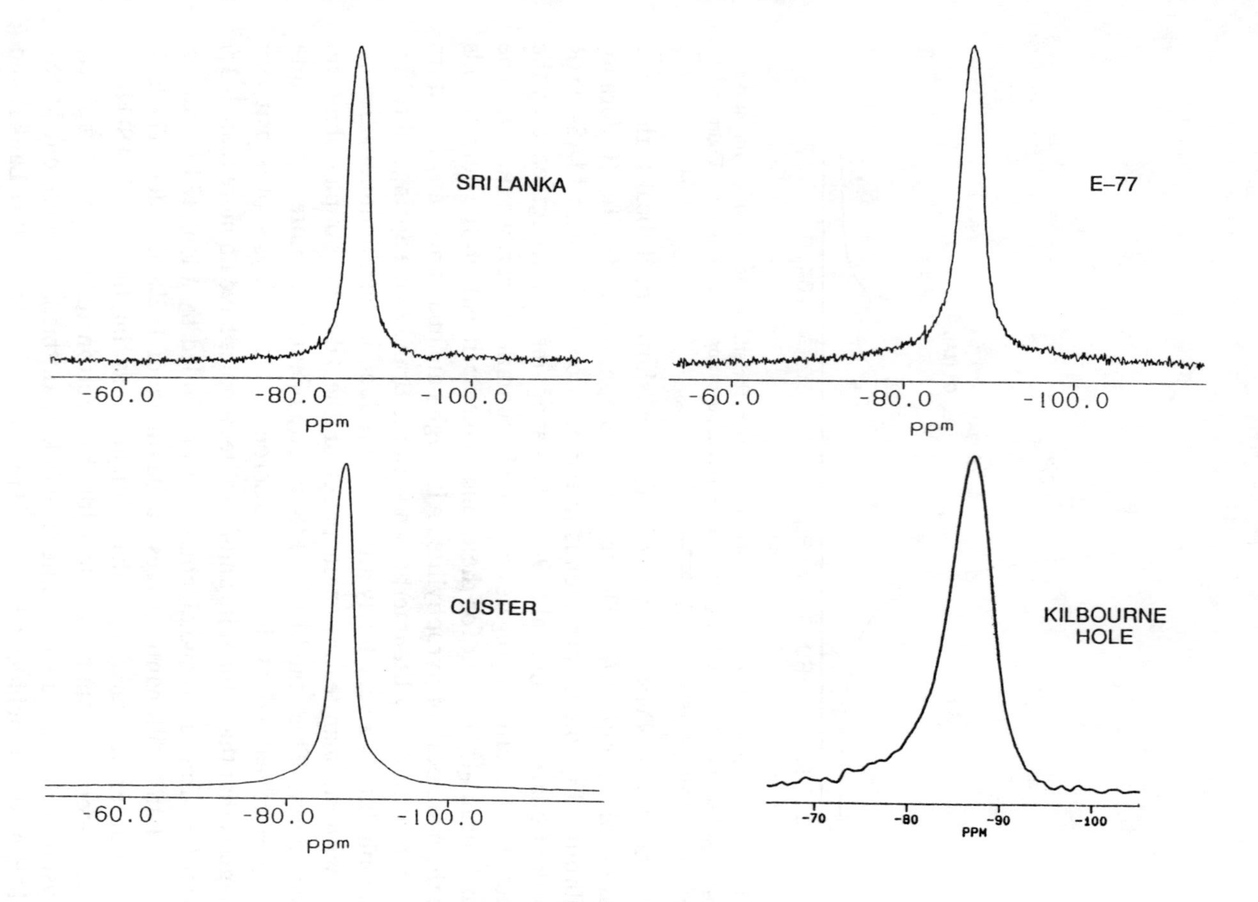

Figure 6.9. MAS-NMR spectra of sillimanite.

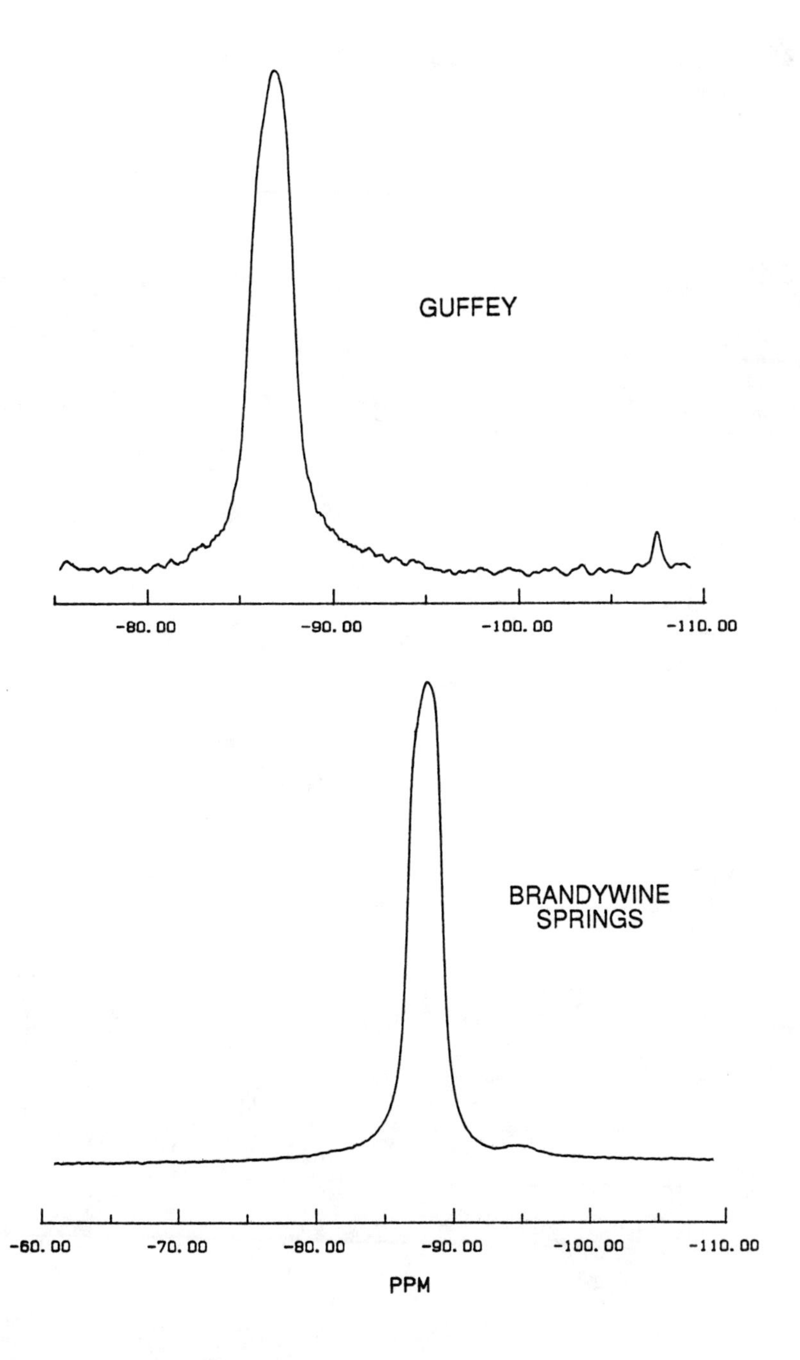

Figure 6.10. MAS-NMR spectra of fibrolite.

CHAPTER 7

THE FIBROLITE PROBLEM

INTRODUCTION

The term *fibrolite* refers to the fine-grained, acicular (fibrous) phase that is common in medium- and high-grade metapelitic rocks. There is a continuum of grain sizes between fibrolite with sub-micron diameters to coarse sillimanite with cm-scale dimensions. Consequently, no clear-cut distinction can be made between fibrolite and sillimanite on the basis of grain size. In spite of this continuum, there is a distinct bimodal distribution of grain sizes of fibrolite versus sillimanite in metapelites containing both "phases". Kerrick and Speer (1988) *arbitrarily* assumed that crystals with minimum dimensions less that 10 μm are fibrolite whereas sillimanite refers to larger crystals.

Many petrologists have implicitly assumed that fibrolite and sillimanite have identical P-T stability fields; thus, the term sillimanite has been used for fibrolite in many papers. Nevertheless, this assumption has been challenged by several workers. For example, Holdaway (1971, p. 103) concluded that "...fibrolite is probably a protosillimanite which crystallizes rapidly from an irreversible mineral reaction. It may well contain excess silica and water and may have Al-Si disorder." If Holdaway is correct, fibrolite and sillimanite would not have the same P-T stability field. The petrologic repercussions would be profound in that the thermobarometry of fibrolite-bearing metapelites has hinged on the assumption that fibrolite and sillimanite have identical P-T stabilities.

Because of the importance of fibrolite as a key phase in metamorphic thermobarometry, this chapter focuses on the *comparative* thermodynamic properties and phase equilibrium relations of fibrolite and sillimanite. Because the topical outline of this chapter follows the preceding four chapters, there is some duplication of subject material. This is justified because of the need for a *unified*, comparative analysis of fibrolite and sillimanite. The nature and interpretation of fibrolite in metamorphic rocks is discussed in the relevant topical sections of succeeding chapters.

EXPERIMENTAL HYDROTHERMAL STUDIES

Experimental studies on the kyanite-sillimanite equilibrium have utilized sillimanite or fibrolite-sillimanite mixtures as starting materials. For their experiments, Richardson et al. (1968) used a sample of intergrown fibrolite + sillimanite from Brandywine Springs, Delaware. Bell and Nord's (1974) optical examination revealed that this material contains about 30% (by volume) fibrolite. In contrast, Newton (1966b, 1969) used fibrolite-free sillimanite from Litchfield, Connecticut (R.C. Newton, pers. comm.). The relatively good agreement between the experimental results of Richardson et al. (1968) versus Newton (1966b, 1969) suggests that if there is a difference in chemical potentials of fibrolite and sillimanite, it is insufficient to have an appreciable effect on the kyanite-sillimanite equilibrium. The error band encompassing the experimental brackets of Richardson et al. (1968) and Newton (1966b, 1969) is about 1 kbar in width (Fig. 3.19). The *maximum* difference in molar Gibbs free energy between the Litchfield sillimanite and Brandywine Springs fibrolite-sillimanite can be computed by assuming a maximum pressure difference of 1 kbar between the kyanite-sillimanite equilibrium with these contrasting starting materials. Defining this 1 kbar uncertainty as ΔP, and taking the molar volumes of kyanite and sillimanite from Winter and Ghose (1979), the equation $\Delta G = \Delta P \ \Delta V_r$ yields a maximum ΔG value of 550 J/mol. This relatively small free energy difference is compatible with the fact that the experimental brackets of Richardson et al. (1968), which were obtained with fibrolite, are in good agreement with that of Bohlen et al. (ms.) in which sillimanite was used as starting material (see Fig. 3.27).

In the interpretation of experimental brackets with "seeded" starting material, we must consider that phases formed during hydrothermal experiments do not necessarily mimic the seed crystals. Thus, the experiments of Richardson et al. (1968) on the kyanite → sillimanite reaction did not necessarily form fibrolite (in spite of the presence of fibrolite seed crystals). Without characterization of the product phases, there remains ambiguity in the interpretation of these experimental results.

As discussed in Chapter 3, Holdaway (1971) experimentally investigated the andalusite-sillimanite equilibrium with two different starting materials: (1) andalusite + sillimanite, and (2) andalusite + fibrolite + sillimanite. As shown in Figure 3.16, Holdaway (1971) obtained anomalous results (i.e., initial weight gains followed by weight losses) in the experiments with fibrolite + sillimanite. Holdaway's contention of early reaction of fibrolite to andalusite is supported by microscopic examination which revealed no fibrolite in run products after 25 days (Fig. 3.16). He attributed the initial weight gains to the disequilibrium reaction: fibrolite → andalusite. Holdaway attributed the proposed high reactivity of fibrolite to a number of possible causes (interfacial energy, strain energy of bent fibers, Al-Si disorder, lattice defects, and "unstable composition").

CALORIMETRIC STUDIES

Using molten oxide calorimetry, Anderson and Kleppa (1969) measured the heat-of-solution of fibrolite (from Custer, South Dakota) and sillimanite (from Benson Mines, New York). Ten separate solution experiments were carried out for each mineral. Their mean value of the heat-of-solution of fibrolite is 29.00 ± 0.31 kJ/mol. Following correction of the sillimanite heat-of-solution data for 1.55 wt % Fe, the mean $\Delta H_{solution}$ is 28.58 ± 0.28 kJ/mol. Coupling this $\Delta H_{solution}$ value with that of kyanite, and with available entropy, heat content and molar volume data, they computed P-T equilibria for the kyanite-sillimanite and kyanite-fibrolite equilibria (Fig. 3.32). The validity of the 30°C temperature difference between the Sill I and Sill II curves (Fig. 3.32) is considerably lessened in light of precision uncertainties in the enthalpy of solution data. However, it should be noted that with precision uncertainties included there is no overlap in the heats-of-solution of fibrolite and sillimanite. Thus, Anderson and Kleppa's (1969) data *suggest* that there is a measurable difference in the enthalpies of solution of their fibrolite and sillimanite samples.

Salje's (1986) measurements suggest significant differences in the heat capacities of sillimanite and fibrolite (Fig. 7.1). He concluded that the enhanced C_p of fibrolite over sillimanite is attributable to lattice defects (stacking faults) or to fibrolite grain boundaries.

Recently, the author has been involved in cooperative calorimetric and X-ray studies of fibrolite and sillimanite with B.S. Hemingway, R.A. Robie, and H.J. Evans, Jr. (U.S. Geological Survey, Reston, Virginia) and L. Topor, O.J. Kleppa, and R.C. Newton (University of Chicago). The samples investigated are particularly advantageous in that they have also been characterized by petrography, chemical analysis, nuclear magnetic resonance spectroscopy, and transmission electron microscopy.

Using the low-temperature heat capacity calorimeter and the differential scanning calorimeter (DSC) at the U.S. Geological Survey, B.S. Hemingway measured C_p of the Sri Lanka sillimanite and Lewiston fibrolite from 7 to 1000 K (Hemingway et al., ms.). Unlike the large difference in C_p of fibrolite and sillimanite reported by Salje (1986), the heat capacities of the sillimanite and fibrolite reported by Hemingway et al. (ms.) are virtually identical at 298 K, and differ by less than 1 percent

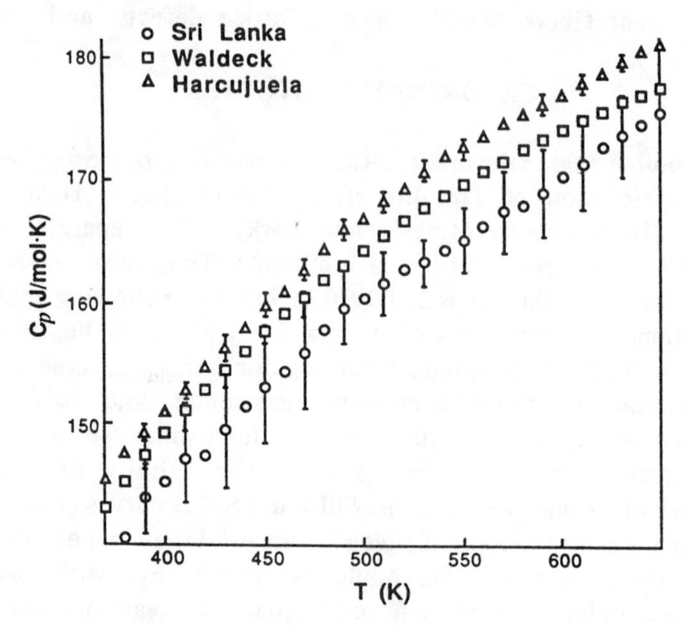

Figure 7.1. Heat capacities of fibrolite (Harcujuela, Spain), and sillimanite (Sri Lanka and Waldeck, Germany). Note that the largest difference in C_p is between the Sri Lanka sillimanite and Harcujela fibrolite. (Redrafted from Salje, 1986, Fig. 1).

at 1000 K. In evaluating Salje's (1986) method, they noted that there was a large difference between the mass of Sri Lanka sillimanite (3.53 mg) and the sapphire standard (40-60 mg). Hemingway et al. (ms.) contended that differences in the masses of sample versus standard introduced a significant error in Salje's (1986) C_p measurement of the Sri Lanka sillimanite. In contrast, their DSC measurements were carried out with similar masses of sample (29-42 mg) and sapphire standard (30.66 mg). In addition to the relatively close agreement in C_p between the Sri Lanka sillimanite and Lewiston fibrolite, their measured C_p for the Brandywine Springs fibrolite is also close to that of the Sri Lanka sillimanite. The data of Hemingway et al. (ms.) suggest that the vibrational entropies of Sri Lanka sillimanite and Lewiston fibrolite differ by only 0.03 J/mol·K at 298 K and 0.20 J/mol·K at 1000 K.

In contrast, Salje (1986) concluded that the 298 K vibrational entropies of fibrolite and sillimanite differ by 3.44 J/mol·K. Following the methods outlined by Salje and Werneke (1982a,b), Salje's (1986) 298 K entropies were computed by fitting the phonon densities of states with his high-temperature C_p data. The low-temperature heat capacities were obtained using Debye temperatures computed from the elastic constant data of

Vaughan and Weidner (1978). As noted by Hemingway et al. (ms.), derivation of 298 K entropies for silicates by low-temperature adiabatic calorimetry yields considerably more accurate low-temperature C_p (and, thus, S_{vib}) data compared to the method of Salje (1986). Because the data of Hemingway et al. (ms.) suggest that the heat capacities of fibrolite and sillimanite are virtually identical, I question Salje's (1986) assumption that stacking faults and/or grain boundary energies enhance the heat capacities of silicates. To test the latter assumption, Hemingway et al. (ms.) measured the C_p of novaculite (having a large molar grain boundary surface area) and compared the results to single crystals of (Brazilian) quartz with nil molar grain boundary surface area. Their measurements reveal that there are virtually no differences in the heat capacities of novaculite and Brazilian quartz. Thus, it is concluded that grain boundaries of fine-grained, monomineralic aggregates (such as novaculite and fibrolite) do not yield enhanced heat capacities compared to the coarse-grained equivalents.

Using the Calvet-type molten oxide calorimeter in the laboratory of O.J. Kleppa at the University of Chicago, L. Topor measured the heat-of-solution of the Sri Lanka sillimanite and Lewiston fibrolite. The results of this study are summarized in an abstract (Topor et al., 1989). Following correction of the heat-of-solution of fibrolite for a small amount of muscovite impurity, and considering uncertainties of ±2 standard deviations, our $\Delta H_{solution}$ value for this phase (28.95 ± 0.72 kJ/mol) is not significantly different from that of Sri Lanka sillimanite (29.60 ± 0.70 kJ/mol). Thus, precision errors preclude a *definitive* conclusion regarding differences in the enthalpies of formation of fibrolite versus sillimanite. The *mean* values of $\Delta H_{solution}$ suggest that fibrolite has a slightly less negative heat of formation than sillimanite. If this difference in $\Delta H_{solution}$ is real it could reflect the molar grain boundary enthalpy of fibrolite (Topor et al., 1989).

NON-STOICHIOMETRY

Because fibrolite is commonly intergrown with other phases (especially quartz), there is very limited *unambiguous* wet chemical analytical data on the Al-Si stoichiometry of fibrolite. In contrast, there are numerous electron probe analyses of fibrolite. The relative abundance of electron probe data reflects the fact that because of the small volume of irradiation, the electron beam can be placed on small areas of fibrolite that are virtually monomineralic (phase-pure).

Table 7.1 Chemical analyses (wt %) of fibrolite and sillimanite.

Author	Kwak (1971a)*		Kerrick (1987)**	Kerrick & Speer (1988)**	Kerrick & Woodsworth (1989)***		Cameron & Ashworth (1972)	Hemingway et al. (ms.)
Method****	EP		EP	EP	EP		EP	ES
Phase*****	F	S	F	S	F	S	F	F
No. of Analyses	13	10	9	8	9	8	1	1
Al_2O_3	62.47	62.64	61.62	62.32	62.15	62.95	62.4	61.5
SiO_2	36.99	36.83	37.50	36.75	36.62	36.59	36.8	37.9
TiO_2	0.06	0.10	0.03	0.01	----	----	0.02	< 0.01
MgO	ND	ND	0.08	0.01	0.14	0.01	< 0.02	0.03
Fe_2O_3	0.23	0.22	0.30	0.22	0.38	0.23	0.30	0.14
Cr_2O_3	ND	ND	0.06	0.09	0.05	0.02	0.03	ND
V_2O_5	ND	ND	ND	ND	0.04	0.03	ND	ND
MnO	ND	ND	0.02	0.03	0.02	0.02	ND	0.12
CaO	ND	ND	ND	ND	ND	---	ND	ND
Na_2O	ND	ND	ND	ND	ND		ND	0.11
K_2O	ND	ND	ND	ND	0.02	---	ND	0.47
TOTAL	99.75	99.79	99.88	99.43	99.42	99.85	99.6	100.27

* Samples from various localities.
** Samples from the Ardara aureole, Ireland.
*** Samples from the Mt. Raleigh pendant, British Columbia.
**** EP = electron probe
ES = emission spectroscopy
*****F = fibrolite, S = sillimanite

Analytical data for fibrolite and sillimanite are given in Table 7.1. Considering the *relative* analytical error of ±1 wt %, there are no *significant* differences in Al_2O_3 and SiO_2 concentrations of fibrolite and sillimanite. The Lewiston fibrolite, which has been the subject of recent calorimetric, X-ray, and MAS-NMR studies, may have somewhat higher Si/Al ratios that stoichiometric Al_2SiO_5 (Table 7.1). However, this fibrolite contains about 6.4 wt % muscovite. Because the SiO_2:Al_2O_3 wt % ratio in muscovite is larger than that of Al_2SiO_5, departure of the analyses of the Lewiston fibrolite from Al_2SiO_5 stoichiometry can be accounted for by the presence of muscovite.

As shown in Table 7.1, the analytical data of Kwak (1971a) on coexisting fibrolite and sillimanite from various localities, and that of Kerrick (1987) and Kerrick and Speer (1988) on samples from the Ardara aureole, suggest no significant differences in the minor element contents of these phases. The higher MgO and Fe_2O_3 contents of fibrolite in the Mt. Raleigh pendant (Table 7.1) *may* be attributed to intergrowths of fine-grained biotite. From ion microprobe analyses of B in samples from several localities, Kerrick and Houser (unpub.) have shown that fibrolite contains negligible quantities of boron (B_2O_3 < 500 ppm) and that there is no fractionation of boron between coexisting fibrolite and sillimanite.

There has been no definitive work on the interpretation of H_2O contents of fibrolite. Deer et al. (1982) report significant structural water contents (H_2O^+ = 0.33 and 0.43 wt %) for two fibrolite samples. Holdaway's (1971) contention that fibrolite has elevated "structural" water was countered by Deer et al. (1982, p. 723) who concluded that: "Small amounts of adsorbed and entrapped water are found frequently in ... fibrolite". Petrographic examination shows that fibrolite is commonly intergrown with "white mica"; thus, the H_2O^+ contents of fibrolite could be attributed to intergrowths of hydrous silicates. Chemical analysis of the Lewiston, Idaho, fibrolite yields H_2O^+ that is significantly in excess of that computed with mass balance assuming that the alkali and alkaline earth cations are assigned to likely hydrous silicates (e.g., K_2O is assigned to muscovite). Consequently, there remains uncertainty in the interpretation of the excess H_2O^+ in the Lewiston fibrolite. Definitive insight into this problem would require a detailed study involving determination of the presence and abundance of hydrous silicates (if any), and an IR study on the hydrous component of this fibrolite.

In summary, there is at present no analytical data suggesting significant departures of fibrolite from the Al_2SiO_5 stoichiometry. Nevertheless, variations in Al/Si ratios within the precision and accuracy limits of the electron probe (1-2 wt % relative) could have significant effects on the andalusite-sillimanite equilibrium. Further evaluation of this problem awaits more accurate chemical analyses of the Al_2O_3 and SiO_2 contents of phase-pure (monomineralic) fibrolite.

LATTICE DEFECTS

As discussed in Chapter 5, strain energy resulting from dislocations could significantly affect the chemical potentials of the Al_2SiO_5 polymorphs. Significant differences in dislocation densities of fibrolite and sillimanite could result in correspondingly large differences in the P-T stability fields of these phases.

Wenk (1983) described fibrolite with a dislocation density of $10^{10}/cm^2$. According to non-core elastic continuum calculations, the resultant elastic strain energy of about 100 J/mol (Fig. 5.4) could significantly affect the P-T stability field of this phase (Fig. 5.5). However, TEM examination of fibrolite from other localities (Doukhan et al., 1985; Kerrick, 1986) reveals much lower dislocation densities. Doukhan et al. (1985) reported dislocation densities of $1 \times 10^8/cm^2$ to $5 \times 10^8/cm^2$. As shown in Figures 5.4 and 5.5, such dislocation densities should have a negligible effect on the P-T stability relations. Thus, with the exception of fibrolite from the Bergell area of Switzerland (Wenk, 1983), there is currently no reason to suspect that dislocations will be an important factor in yielding differences in the stability relations of fibrolite and sillimanite.

Al-Si DISORDER

As discussed in Chapter 6, differing amounts of Al-Si disorder may be manifest by variations in unit cell volumes. Thus, the suggestion that fibrolite may be more disordered than sillimanite (Chinner et al., 1969; Holdaway, 1971; Greenwood, 1972) may be evaluated by comparing unit cell volumes of these phases.

From Gunier X-ray patterns, Moore and Best (1969) concluded that there were no significant differences in unit cell dimensions between the fibrolite and sillimanite samples they examined (Fig. 7.2). Subsequent X-ray studies by Cameron and Ashworth (Table 7.2), Thomas (Table 7.2) and Beger (1979) also revealed that the unit cell parameters of fibrolite and sillimanite are virtually identical. Recent X-ray refinements by H.T. Evans, Jr. (described in Hemingway et al., ms.) show that at 25°C the cell volumes of Sri Lanka sillimanite and Lewiston fibrolite differ by only 0.1 $Å^3$; at 1100°C the cell volumes differ by 1.10 $Å^3$. Because these

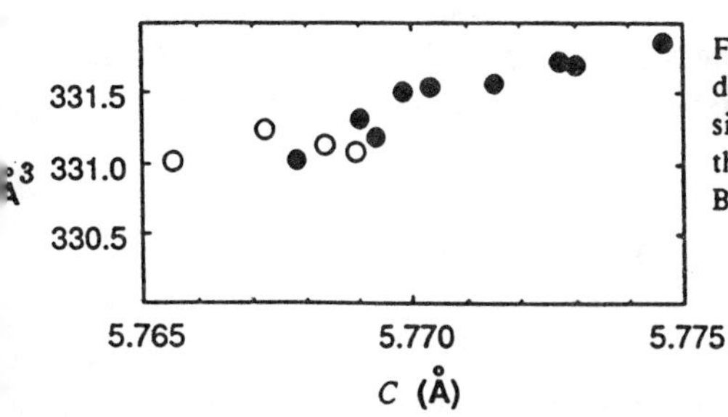

Figure 7.2. Unit cell volume versus c cell dimension of fibrolite (open circles) and sillimanite (filled circles). Note the overlap in the parameters. (Adapted from Moore and Best, 1969, Fig. 2).

Table 7.2 Unit cell dimensions of fibrolite and sillimanite.

Author(s)	Sample Locality	a(Å)	b(Å)	c(Å)	Volume (Å³)
		FIBROLITE			
Cameron & Ashworth (1972)	Dalradian, Scotland	7.4844	7.6722	5.7701	331.33
Thomas (1984)*	Essex, Connecticut	7.4843	7.6673	5.7681	330.998
Hemingway et al. (ms.)	Lewiston, Idaho	7.4830	7.6754	5.7685	331.32
		SILLIMANITE			
Author(s)	Sample Locality	a(Å)	b(Å)	c(Å)	Volume (Å³)
Burnham (1963a)	LaBelle Co., Quebec	7.4856	7.6738	5.7698	331.435
Holdaway (1971)	Montville Quadrangle, Connecticut	7.487	7.676	5.7735	331.8
Hemingway et al. (ms.)	Sri Lanka	7.4845	7.6713	5.7686	331.21

*Average of five samples.

differences are within experimental error, it is concluded that there are no significant differences in the cell volumes of fibrolite and sillimanite. Because unit cell volume increases with increasing disorder (Beger, 1979), this virtual equivalence of cell volumes suggests that there are no significant differences in the amount of Al-Si disorder between fibrolite and sillimanite. However, it is possible that small but significant amounts of disorder (e.g., 1-5%) could be masked by the precision error in unit cell determination.

Our MAS-NMR study (Kerrick et al., 1990) reveals no evidence of Al-Si disorder in fibrolite, and Bish and Burnham (unpub.) found no disorder in their neutron diffraction study of a fibrolite sample. Thus, currently-available data do not support the suggestions of Zen (1969), Holdaway (1971) and Greenwood (1972) that fibrolite may be more disordered than sillimanite.

GRAIN BOUNDARY ENERGY

Because of fine grain size, it is possible that the large molar grain boundary area of fibrolite results in significant molar grain boundary energy.

Grain boundary energy of fibrolite was considered by Holdaway (1971), who computed the molar surface energy of fibrolite using the range of surface energies that are typical for silicates (i.e., 200 to 1600 $ergs/cm^2$), assuming individual fibrolite crystals to have a square cross section (in sections cut perpendicular to the *c* axis), and considering the smallest grain diameters (0.3 μm) that he measured by optical examination of fibrolite samples. Considering fibrolite crystals as prisms with an infinite aspect ratio, he obtained a surface energy of 351 ± 71 J/mol. Using the equation: $d\Delta G/dT = -\Delta S_r$, coupled with available entropy data, Holdaway (1971) computed that a surface energy of 351 ± 71 J/mol for fibrolite would shift the andalusite-sillimanite equilibrium +120 ± 25°C. The recent high-temperature entropy data of Hemingway et al. (ms.) yields a similar shift of the andalusite-sillimanite equilibrium. It is important to note that Holdaway's (1971) computations were based on the smallest grain diameters (0.3 μm) observed in his samples. Because most of the fibrolite crystals in his samples had larger diameters, the molar surface area (and, hence, molar grain boundary energies) of his samples will be less than his computed values. We have measured grain diameters of fibrolite in several samples with a petrographic microscope and with TEM. We found many fibrolite crystals with diameters around 1 μm. Modeling fibrolite crystals as cylinders with infinite aspect ratios, and considering crystals of 1 μm diameter, the molar surface area is 2×10^6 cm^2. Using the range of surface energies considered by Holdaway (i.e., 200 to 1600 $ergs/cm^2$), the displacement of the andalusite-sillimanite equilibrium due to fibrolite surface energy would range from about 15°C to 110°C.

In evaluating the validity of such computations, we must consider that most of the surface energy values for silicates were obtained by measurements in vacuum (Rhee, 1975; Parks, 1984). In view of the current model that grain boundaries in rocks undergoing metamorphism do not have a continuous liquid film (White and White, 1981; Watson and

Brenan, 1987; Hay and Evans, 1988), we can consider grain boundaries of fibrolite aggregates as "dry" interfaces separating adjacent fibrolite crystals. Thus, the surfaces of fibrolite crystals should be treated as *grain boundaries* rather than free interfaces (as implicitly assumed by surface energy calculations) or solid-fluid interfaces.

In expressing the Gibbs free energy of a grain boundary as $G_{GB} = H_{GB} - T(S_{GB})$, it is important to note that grain boundary enthalpy is dominant compared to the entropy term (Adamson, 1982); thus, $G_{GB} \sim H_{GB}$. Accordingly, the author was involved in a molten oxide calorimetric study designed to measure the grain boundary enthalpy of fibrolite by comparing the heat-of-solution of fibrolite (Lewiston, Idaho) versus sillimanite (Sri Lanka). Based on the *mean* values of $\Delta H_{solution}$, the ΔH_f of the Lewiston fibrolite is approximately 1 kJ more positive than the Sri Lanka sillimanite (Topor et al., 1989). Unfortunately, the relatively large precision error eclipses the significance of this enthalpy difference. A *definitive* experimental measurement of the grain boundary enthalpy of fibrolite would be highly desirable

Barring direct experimental determination, the grain boundary energy of fibrolite can be evaluated theoretically. In so doing, the orientation of the crystal structures of adjacent grains is an important factor (e.g., Cooper and Kohlstedt, 1982). Fibrolite aggregates consist of swarms of acicular crystals with their *c* axes virtually parallel. In samples consisting of fibrolite aggregates, preliminary electron diffraction analysis by H.R. Wenk and M. Medrano confirms that the *c* axes are virtually coincident and that there is $\leq 15°$ of angular mismatch between the structures of adjacent fibrolite crystals. Thus, we may model the grain boundary energetics with "*symmetrical dislocation tilt wall*" theory. As shown in Figure 7.3, symmetrical tilt walls are modeled by a series of parallel edge dislocations. In this model the lattice planes of the two juxtaposed crystals are symmetrically disposed about the grain boundary, and the angle Θ is a measure of the misorientation of the structures of the two crystals. As shown in Figure 7.3, the distance between adjacent dislocations is approximated by (Nicolas and Poirier, 1976): $\mathbf{d} = \mathbf{b}/2(\sin \Theta/2)$. For small values of Θ (<15°), $\sin\Theta \approx \Theta$; thus, this equation becomes: $\mathbf{d} = \mathbf{b}/\Theta$. The tilt wall grain boundary energy (E_{TW}) is approximated by computing the "non-core" elastic strain energy in a cylindrical region coaxial with a dislocation line

$$E_d = [(\mu b^2/4\pi(1 - \nu)] \, [ln(R/r_o)] \, [1/R] \, .$$

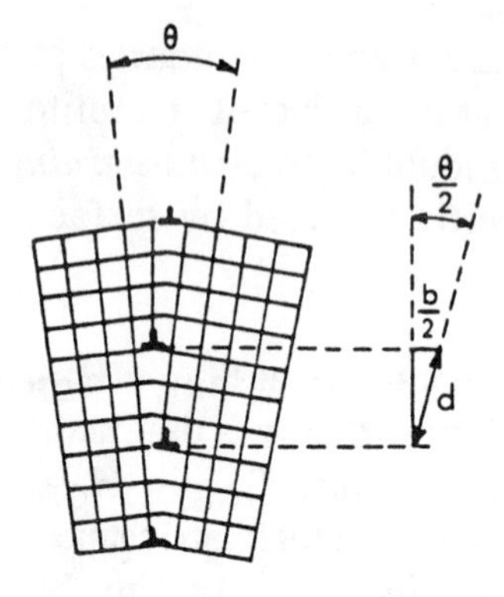

Figure 7.3. Symmetrical dislocation tilt wall model of a grain boundary. The heavy, inverted "T" symbols refer to parallel, straight edge dislocations, Θ = angular mismatch between lattice planes of adjacent grains, b = Burgers vector, and d = spacing between adjacent dislocations. (From Nicolas and Poirier, 1976, Fig. 3.40; copyright John Wiley & Sons).

Assuming that the outer radius is equivalent to the spacing of the dislocations (i.e., R = **d**; Nicolas and Poirier, 1976), the tilt wall energy (E_{TW}) is computed from:

$$E_{TW} = [\mu b/4\pi(1 - \nu)]\, \Theta\, [ln(b/r_o) - ln\, \Theta]\,. \qquad [7.1]$$

Taking the shear modulus (μ) and poissons' ratio (ν) for sillimanite from Birch (1966), and Vaughan and Weidner (1978), respectively, and assuming that r_o = 2**b** (Kerrick, 1986), the energy of the tilt wall can be computed as a function of Θ. As shown in Figure 7.4, the computed grain boundary energy attains a maximum value at about 10°. This maximum is a consequence of the functional form of equation [7.1]; i.e., at low Θ values the increase in E_{TW} with increasing Θ reflects the dominance of the term [Θ *ln* Θ] whereas at higher angles the term [Θ*ln*(b/r_o)] becomes dominant. Because the energies of dislocation cores are excluded, these calculations provide *minimum* estimates of grain boundary energies. For metals, the core energy amounts to 10-15% of the total dislocation strain energy (Kerrick, 1986). Because of covalent bonding, it is probable that the core energies of dislocations in silicates are larger than those of metals. Accordingly, grain boundary energies of about 1000 ergs/cm^2 may be conservative.

Considering that fibrolite crystals are cylinders with infinite aspect ratios, Figure 7.5 shows the molar grain boundary energy as a function of grain diameter. Calculations were made for three values of the specific grain boundary energy that are reasonable in light of the available data for surface energies of silicates. For grain diameters of 1 μm, which are common in fibrolite, a specific grain boundary energy of 1000 ergs/cm^2, as derived by symmetric dislocation tilt wall theory, yields a molar grain boundary energy of 200 J. If we arbitrarily choose a 1 atm reference

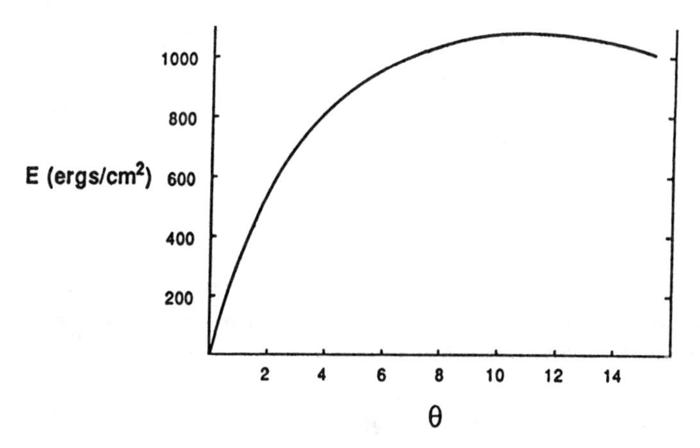

Figure 7.4 Energy of dislocation tilt wall grain boundary as a function of angular mismatch between lattice planes of adjacent grains (see Fig. 7.3).

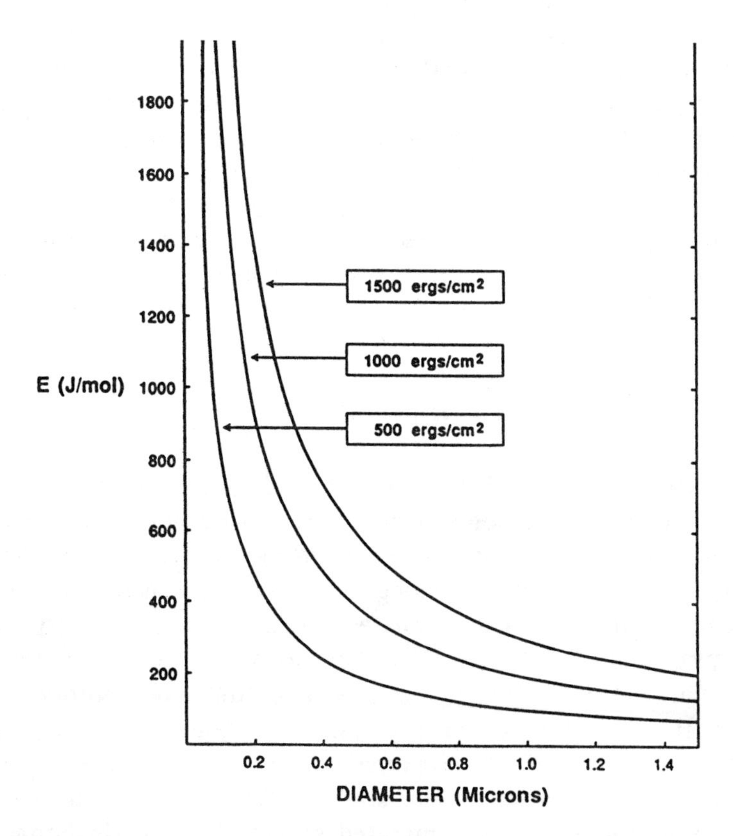

Figure 7.5. Molar grain boundary energy of fibrolite as a function of fiber diameter for three selected values of the specific grain boundary energy. Computations were made by modeling fibrolite crystals as cylinders with infinite aspect ratios.

temperature of 800°C for the andalusite-sillimanite equilibrium, 200 J/mol of grain boundary energy results in a 1 atm equilibrium temperature of 870°C for the andalusite-fibrolite equilibrium. Note that this equilibrium would be *metastable* in relation to the andalusite-sillimanite equilibrium. The implications of this analysis to metamorphic systems are explored in Chapters 8 and 9.

Although the symmetrical tilt wall model undoubtedly represents an unrealistic approximation of the actual grain boundaries between fibrolite crystals, this model may provide estimates of minimum grain boundary energies. Because many fibrolite crystals have grain diameters in the micron- to sub-micron range, the molar grain boundary energy of fibrolite *may* result in significant differences in the P-T stability of fibrolite versus sillimanite.

R.P. Wintsch (pers. comm.) suggested that grain boundary energy may explain why fibrolite crystals with grain diameters less than 0.5 μm appear to be rare. The large grain boundary energy of fibrolite with diameters less than 0.5 μm yields a significant driving force for recrystallization to form larger crystals. Because of the asymptotic decrease in grain boundary energy with increasing crystal diameters above about 0.5 μm (Fig. 7.5), the driving force for recrystallization diminishes exponentially. For crystal sizes above 1 μm there would be little driving force for recrystallization. Indeed, fibrolite with grain diameters on the order of 1 μm are quite common (Holdaway, 1971; Bell and Nord, 1974; Wenk, 1983; Doukhan et al., 1985; Kerrick, 1986).

CONCLUSIONS

There is currently no evidence that there are *consistent* differences in the crystal structure, stoichiometry and nature and abundance of lattice defects of fibrolite and sillimanite. I consider grain boundary energy to be the only variable that could produce a *significant* systematic difference in the chemical potentials of these phases. The recent heat capacity study of Hemingway et al. (ms.) suggests that there are no significant differences in the molar grain boundary *vibrational* entropy of fibrolite and sillimanite. However, the question regarding grain boundary enthalpy and configurational entropy of fibrolite remains. This problem becomes more significant with decreasing grain size of fibrolite. Even with a conservative estimate of a few hundred ergs/cm^2 for grain boundary energy, fibrolite with grain diameters less than 0.5 μm could have molar grain boundary energies that would significantly affect the P-T stability field of this phase. From the limited literature data (Holdaway, 1971; Bell and Nord, 1974; Doukhan et al., 1985), and from measurements made by the author, it would appear that most fibrolite crystals have grain diameters exceeding about 0.25 μm. The *average* grain diameter of fibrolite appears to be considerably larger than this minimum diameter. The general lack of fibrolite with crystal diameters less than about 0.25 μm may be a consequence of the exponential increase in molar grain boundary energy, and corresponding destabilization of fibrolite, as grain boundary diameter is decreased below 0.25 μm.

Without definitive knowledge of the P-T stability relations of fibrolite and sillimanite, insight into this problem may be gleaned from the field relations of these phases. Of particular relevance is the spatial relationship between the first appearance of fibrolite versus sillimanite in areas displaying prograde metamorphism. *Provided* that there are areas where petrographic and petrologic arguments can be made for the contemporaneous formation of these phases by the same reaction mechanism, the fibrolite and sillimanite isograds should be coincident if there are no significant differences in the chemical potentials of these phases. This question is addressed in Chapter 8.

CHAPTER 8

METAMORPHIC REACTIONS

INTRODUCTION

There are many reactions involving the aluminum silicates in prograde and retrograde metamorphism of aluminous lithologies. Some of them involve ferromagnesian phases and are conveniently depicted with Thompson's (1957) AFM projection. For metapelites there have been numerous attempts to calibrate a petrogenetic grid for these equilibria and to develop geobarometers and geothermometers for equilibria involving solid solution (Essene, 1989). Because this book is a review of the aluminum silicates, it is inappropriate herein to detail the many complex equilibria involving the Al_2SiO_5 polymorphs. Rather, this chapter focuses on Al_2SiO_5 polymorphic transformations, and retrograde reactions that cause replacement of the aluminum silicates by other phases. Particular attention is devoted to the *mechanisms* of reactions involving the aluminum silicates. Following the traditional petrologic assumption, fibrolite and sillimanite are treated as synonymous phases in the sections discussing each of the polymorphic transformations. The last section of this chapter focuses on fibrolite-forming reactions.

THE KYANITE → SILLIMANITE REACTION

The prograde reaction kyanite → sillimanite is a well known "reference" isograd reaction in pelitic rocks subjected to metamorphism at pressures above the Al_2SiO_5 triple point. Miyashiro (1961) used this reaction to define his medium pressure facies series (Fig. 1.2), which he formally designated as the "kyanite-sillimanite" facies series. In honor of the pioneering work of George Barrow, who first described this type of metamorphism in the Scottish Highlands, this is commonly referred to as the *Barrovian* facies series. The kyanite→sillimanite transformation is referred to as the *first sillimanite isograd* whereas the *second sillimanite isograd* is correlative with the reaction

$$\text{Muscovite} + \text{Quartz} = \text{Sillimanite} + \text{K-feldspar} + H_2O\ .$$

In favoring a model of progressive metamorphism, Carmichael (1969) derived a mechanism for the kyanite → sillimanite isograd reaction. As shown in Figure 8.1, he considered different reactions occurring in nearby subdomains. The textural evidence in one subdomain (Fig. 8.1a) suggests

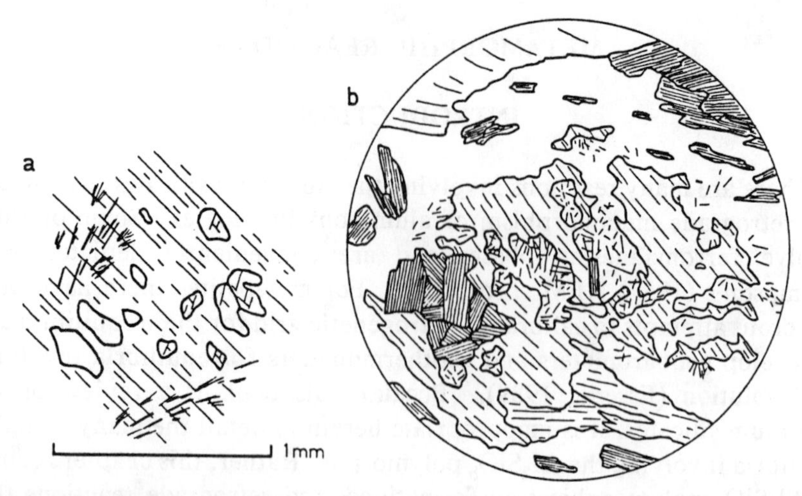

Figure 8.1. Textural features used by Carmichael (1969) to support his "ionic" reaction model for the kyanite → sillimanite transformation in metapelites. (a) Anhedral inclusions of kyanite (ruled pattern represents cleavages), quartz (clear), and "sprays" of fibrolite (black lines) enclosed by a single crystal of muscovite. Carmichael (1969) contended that the kyanite and quartz were replaced by muscovite + fibrolite. (b) Embayed muscovite porphyroblast surrounded by fibrolite (black lines) and quartz (clear). Carmichael (1969) concluded that the muscovite porphyroblast was replaced by fibrolite + quartz. (From Carmichael, 1969, Fig. 4).

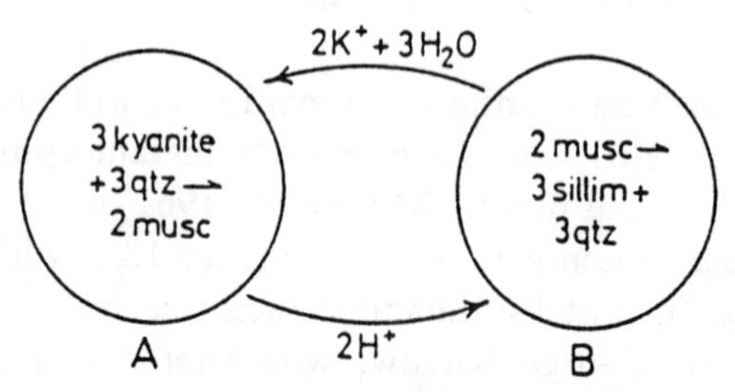

Figure 8.2. Carmichael's (1969) suggested reaction mechanism linking "subdomain" reactions A and B. Reaction A corresponds to Figure 8.1a, whereas B corresponds to Figure 8.1b. The exchange of K^+, H^+ and H_2O is assumed to occur via the fluid phase. The driving force for the overall reaction is ΔG for the kyanite → sillimanite reaction. (From Carmichael, 1969, Fig. 5).

growth of muscovite + fibrolite at the expense of kyanite. In contrast, the texture in the other subdomain (Fig. 8.1b) implies resorption of muscovite by fibrolite + quartz. Carmichael concluded that the texture shown in Figure 8.1a is represented by the reaction:

$$3 \text{ Kyanite} + 3 \text{ Quartz} + 2 K^+ + 3 H_2O = 2 \text{ Muscovite} + 2 H^+ ,$$

whereas that in Figure 8.1b is:

$$2 \text{ Muscovite} + 2H^+ = 3 \text{ Sillimanite} + 3 \text{ Quartz} + 2 K^+ + 3 H_2O .$$

As shown in Figure 8.2, he considered that the two subdomains were linked by exchange of components via an aqueous fluid phase. By adding the two subdomain reactions, the driving force for the overall system shown in Figure 8.2 is the Gibbs free energy change for the reaction kyanite = sillimanite. Carmichael's (1969) hypothesis has been attractive to many petrologists because it supports the concept of progressive metamorphism, for which there are convincing petrologic arguments (Yardley, 1989), and it provides an explanation for the lack of direct replacement of reactants by products. It is important to note that Carmichael (1969) gave no evidence that both subdomains shown in Figure 8.1 actually occur within any *single* thin section - Figure 8.1a shows a sample from Glen Clova, Scotland, whereas Figure 8.1b is a sample from Donegal, Ireland! The author has encountered no papers where the textures in Figures 8.1a (involving kyanite as a reactant) and 8.1b occur within subdomains of a single thin section. Thus, to my knowledge Carmichael's mechanism for the kyanite → sillimanite reaction has not been conclusively demonstrated by petrographic examination of rocks at the first sillimanite isograd. It is important to note that his model is based on the assumption of aluminum immobility. In the parlance of metasomatism, the reactions within each of the subdomains shown in Figure 8.2 assume a *constant aluminum reference frame.* Although he thoughtfully defended this assumption in his paper, the validity of this hypothesis is questionable (see Chapter 10).

Jansen and Schuiling (1976) described the transformation from kyanite-zone to sillimanite-zone rocks in a well-developed Barrovian metamorphic sequence in Naxos, Greece. This transformation was studied in both mica schists and in Al_2SiO_5-bearing quartz segregations. Kyanite and fibrolite coexist within a zone about 600 m thick (as measured perpendicular to isograd surfaces). Although textures suggest that the paramorphic replacement of kyanite by fibrolite occurred in some specimens, Jansen and Schuiling (1976) concluded that in most cases there is no textural evidence of a reaction relationship between kyanite and fibrolite. They did not discuss the fibrolite-forming reaction.

Fletcher and Greenwood (1979) carried out a detailed petrographic study of Barrovian-type metamorphism in the Penfold Creek area of British Columbia. Although they noted that "...both the fibrous (fibrolite) and the prismatic forms of sillimanite are present in high grade rocks", they did not distinguish these two phases in the text discussion of reaction textures (the generalized term "sillimanite" is used). As shown in Figure 8.3, they mapped three sillimanite zones. In Zone 1 "sillimanite" occurs within large crystals of muscovite. Textural evidence suggests that kyanite "...is apparently stable, judging by its unaltered appearance and well-defined grain boundaries" (Fletcher and Greenwood, 1979, p. 750). J.A. Grambling (personal communication) cautioned that if "sillimanite" formed by dissolution of kyanite, the apparently small amount of fibrolite in Zone 1 would imply that there would be negligible "corrosion" of kyanite grain boundaries. Specimens from Zone 2 are characterized by either the absence of kyanite or the partial replacement of kyanite by "sillimanite". Kyanite is absent in Zone 3. The *apparent* stability of kyanite in Zone 1 suggests that the incipient "sillimanite"-forming reaction does not involve kyanite as a reactant. These textural data imply that Carmichael's (1969) reaction mechanism is inapplicable to the first sillimanite isograd reaction in the Penfold Creek area.

Pigage (1982) studied fibrolite-forming reactions in the vicinity of the first "sillimanite" isograd in the Azure Lake area in southeastern British Columbia. Textural evidence implies that the isograd reaction involved production of the assemblage: fibrolite + biotite + ilmenite + muscovite from the reactant assemblage: staurolite + garnet + kyanite. He suggested that fibrolite initially formed by reactions involving the decomposition of staurolite and/or garnet; rutile was considered to be a reactant phase. Late-stage fibrolite formed by a garnet-producing reaction subsequent to the complete consumption of rutile. Pigage concluded that cation exchange reaction mechanisms such as those proposed by Carmichael (1969) could be applicable to the sillimanite isograd in the Azure Lake area. This is supported by the replacement of kyanite and staurolite by muscovite, similar to that in Figure 8.1a. Alternatively, Pigage (1982) suggested that muscovite may have formed later than fibrolite. The late-stage growth of coarse, poikiloblastic muscovite has been suggested in many studies of metapelites. With a cation-exchange mechanism, his detailed mass balance calculations show that the net sillimanite isograd reaction in the Azure Lake area is not represented by the simple polymorphic reaction kyanite → sillimanite.

Extremely abrupt kyanite-sillimanite isograds occur in the Truchas Peaks and Rio Mora Ranges in north-central New Mexico (Grambling, 1981; Grambling and Williams, 1985). In many samples, textural evidence suggests that kyanite was replaced by fibrolite. Consequently, the isograd reaction is represented by the paramorphic transformation of kyanite to fibrolite. However, Grambling (1981) observed that less commonly

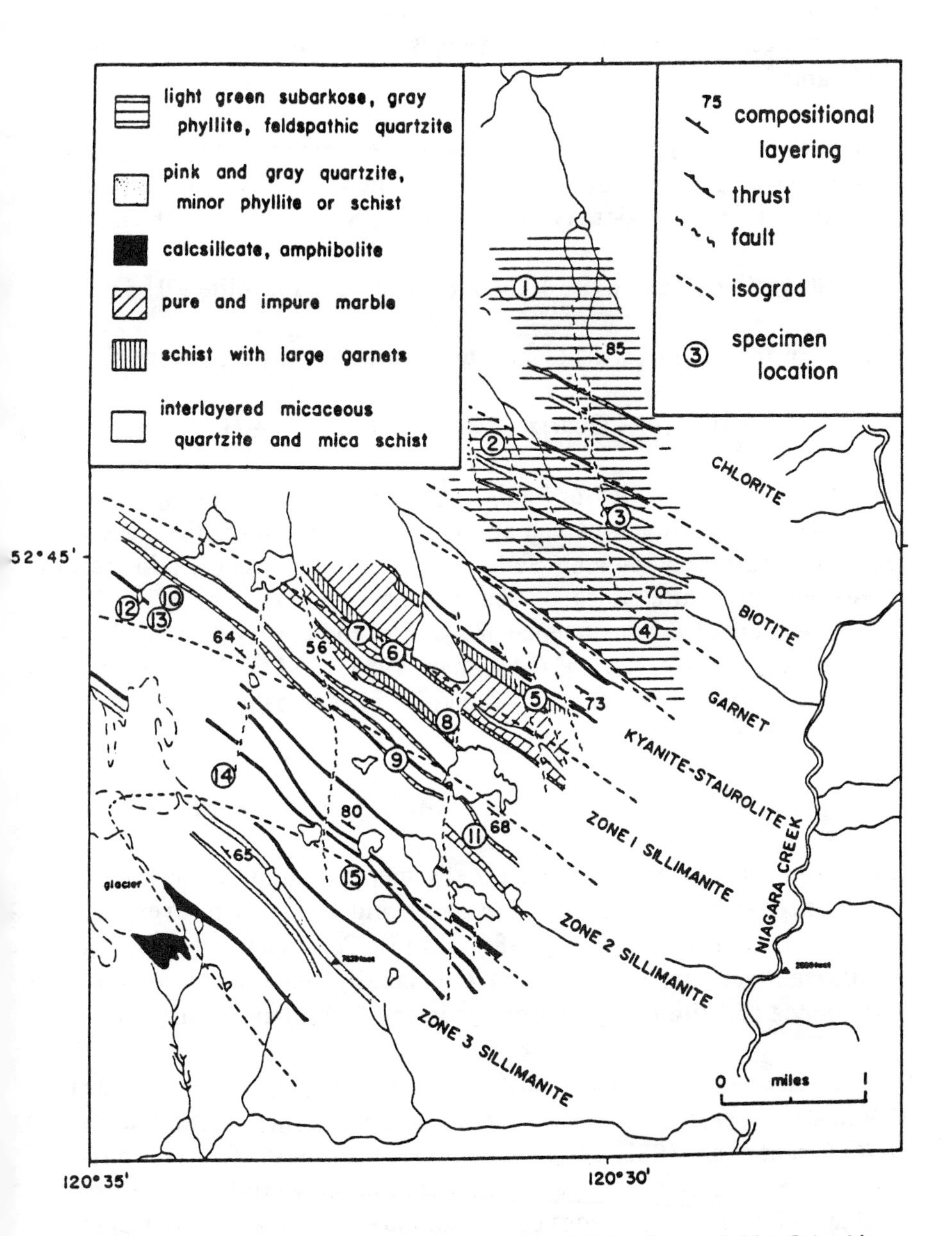

Figure 8.3. Metamorphic zones and isograds in the Penfold Creek area, British Columbia. "Zone 1 Sillimanite" is characterized by coexisting kyanite + sillimanite (with kyanite in apparent textural equilibrium), whereas in "Zone 2 Sillimanite" kyanite is either absent or partially transformed to sillimanite. (From Fletcher and Greenwood, 1979, Fig. 3).

muscovite partially pseudomorphs kyanite, and sillimanite is developed in the immediate vicinity of these paramorphs. In such cases, an ionic reaction mechanism similar to that suggested by Carmichael (1969) may be applicable.

McLellan (1985) investigated sillimanite isograd reactions in the Central Eastern Grampian region of Scotland where Barrow carried out his pioneering work on regional metamorphism. She appealed to the reaction,

$$\text{Staurolite} + \text{Muscovite} + \text{Quartz} = \text{Biotite} + \text{Fibrolite} + H_2O\ ,$$

to explain the intergrowths of biotite + fibrolite, and the reaction,

$$\text{Staurolite} + \text{Quartz} = \text{Garnet} + \text{Fibrolite} + H_2O\ ,$$

for rocks with fibrolite occurring in late garnets. Thus, McLellan concluded "...that much of the sillimanite in the sillimanite zone came from staurolite breakdown." Some of the pelitic rocks contain fibrolite inclusions in white mica. She emphasized that this fibrolite is "...independent of the occurrence of kyanite, and clearly does not form by polymorphic transformation" (McLellan, 1985, p. 813). The replacement of garnet by biotite + fibrolite + ilmenite in metapelites of Glen Esk is attributed to the reaction

$$\text{Almandine} + \text{Muscovite} = \text{Annite} + \text{Fibrolite} + \text{Quartz}\ .$$

McLellan concluded that Carmichael's mechanism is applicable to the occurrences of fibrolite enclosed in coarse muscovite. However, because this texture is *rare* (McLellan, 1985, p. 815), Carmichael's (1969) kyanite → sillimanite reaction mechanism is generally inapplicable to the first appearance of "sillimanite" (fibrolite) in the Grampian Highlands.

E.D. Ghent and colleagues carried out considerable research on the first sillimanite isograd in the Mica Creek area of British Columbia (Ghent 1976; Ghent et al., 1980, 1982). Kerrick and Ghent (unpub.) investigated textures in samples of metapelites collected in the vicinity of this isograd. In this area fibrolite precedes coarse, prismatic sillimanite. Petrographic examination revealed that fibrolite replaced biotite. There is no evidence that kyanite was unstable in this zone (Fig. 8.4). Specifically, kyanite was not replaced by muscovite, and kyanite grain boundaries show no evidence of resorption. Thus, Carmichael's mechanism for the kyanite → sillimanite isograd reaction appears to be inapplicable to metapelites of the Mica

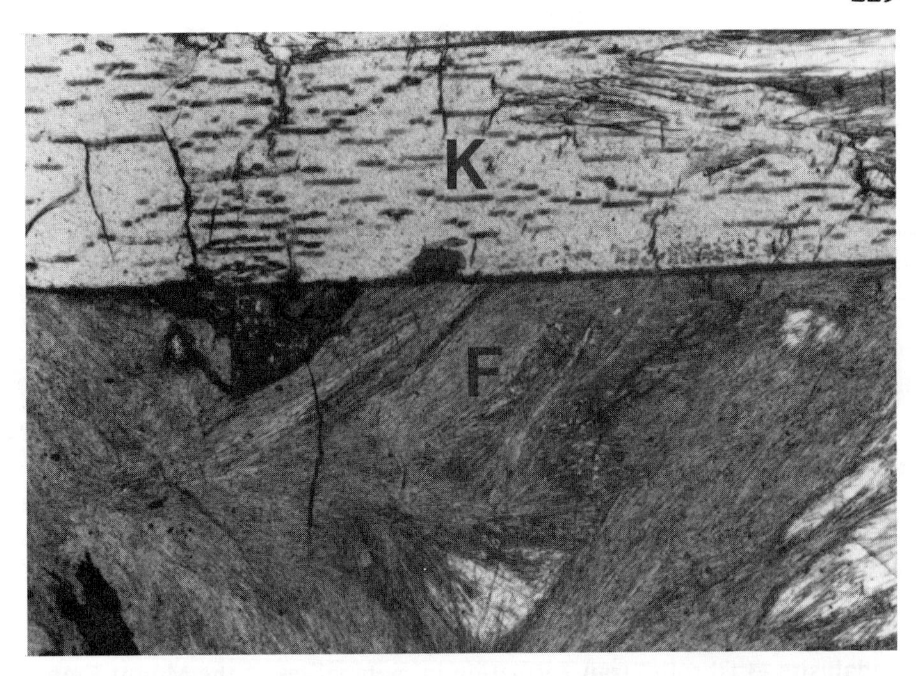

Figure 8.4. Photomicrograph of a sample of metapelite from the vicinity of the sillimanite isograd at Mica Creek, British Columbia. The straight grain boundary of the kyanite crystal (K) suggests no resorption by the adjacent fibrolite (F). (From Kerrick and Ghent, unpub.).

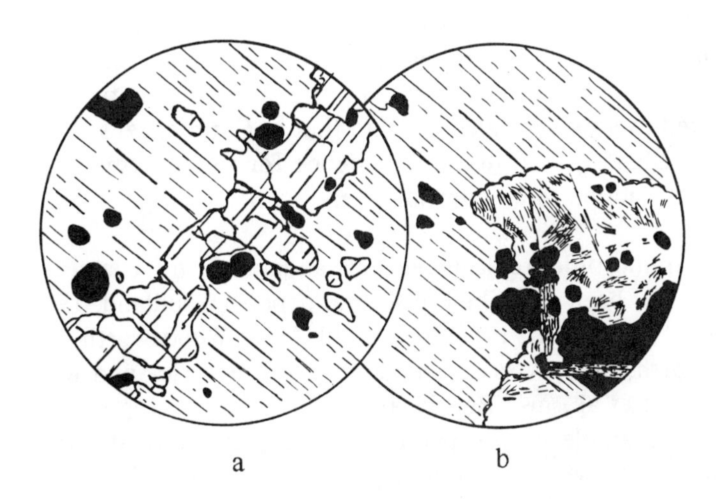

Figure 8.5. Replacement textures within a single thin section from Banffshire, Scotland. (a) Andalusite (widely spaced cleavage) replaced by a single crystal of muscovite (closely spaced cleavage). All andalusite crystals are in optical continuity. (b) Single crystal of muscovite (closely spaced cleavage) replaced by fibrolite (black "sprays") + quartz (clear). Note the scalloped shape of the muscovite-quartz grain boundary. (From Nockolds et al., 1978, Fig. 33.9; copyright Cambridge University Press, Cambridge, England).

Creek area.

THE ANDALUSITE → SILLIMANITE REACTION

Because of a negative dP/dT slope, the andalusite-sillimanite equilibrium provides a particularly useful thermobarometer. Therefore, the andalusite → sillimanite isograd is very important for delineating the P-T conditions of contact metamorphism and low-pressure regional metamorphism.

Could a reaction mechanism analogous to that of Carmichael (1969) be relevant to the andalusite → fibrolite reaction in muscovite-bearing rocks? I am aware of one reported example of an analogous texture (shown in Fig. 8.5). In this case, the andalusite-consuming reaction suggested by Figure 8.5a is analogous to that of Figure 8.1a, whereas that in Figure 8.5b is identical to the reaction suggested by Figure 8.1b.

Glen (1979) appealed to an "ionic" reaction mechanism for the andalusite → fibrolite transformation in metapelites of the Mount Franks area, New South Wales, Australia. Andalusite porphyroblasts show varying degrees of replacement by sericite ± fibrolite. Because textural evidence suggests that this transformation occurred at constant volume, Glen (1979) concluded that the pseudomorphic replacement reaction involved the release of Al (cf. Carmichael, 1969). Glen concluded that the replacement of andalusite by sericite was linked to adjacent subdomains where muscovite was replaced by fibrolite, and that the two subdomains were linked in a manner analogous to that in Carmichael's (1969) model. I consider Glen's (1979) textural evidence for the muscovite → fibrolite subdomain reaction to be questionable.

The paramorphic replacement of andalusite by coarse, prismatic sillimanite has been described in several papers (Woodland, 1963; Pitcher and Read, 1963; Ghent, 1968; Rosenfeld, 1969; Woodsworth, 1979; Tyler and Ashworth, 1983; Kerrick and Speer, 1988; Kerrick and Woodsworth, 1989; Vernon et al., ms.; Barnicoat and Prior, ms.). Figure 8.6 shows a superlative example of the paramorphic replacement of chiastolite by sillimanite. Because of the similarity in molar volumes of these phases, we are dealing with a close approximation to a "volume-for-volume" paramorphic replacement reaction.

Vernon (1987b) described the paramorphic replacement of andalusite by sillimanite in Precambrian metapelites of the northern portion of the

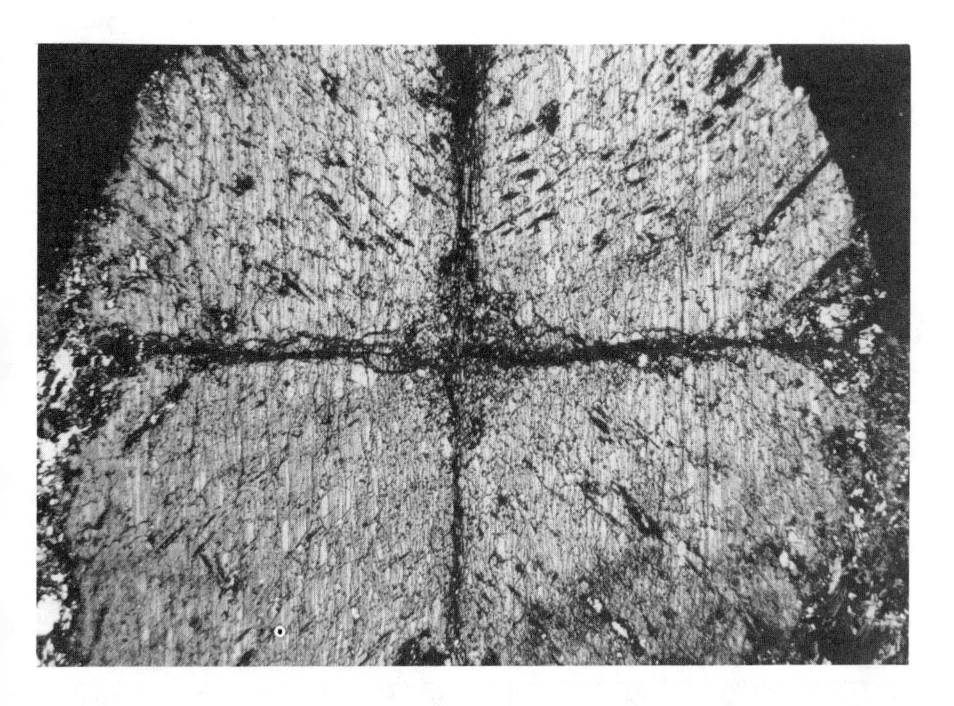

Figure 8.6. Chiastolite porphyroblast that was paramorphically replaced by sillimanite. Width of photo corresponds to about 10 mm. (From Rosenfeld, 1969, Plate 3A).

Sandia Mountains, New Mexico. Optical data suggest parallelism of the *c* crystallographic axes of the intergrown andalusite and sillimanite; however, the *a* and *b* axes are interchanged. Because the transformation of andalusite to sillimanite is confined to the highest grade rocks in this area, Vernon (1987b, p. 587) suggested "...that thermal energy may contribute to surmounting the energy barrier..." for sillimanite nucleation. Alternatively, he suggested that sillimanite nucleation may be "strain assisted" by preferentially occurring on stacking faults formed by dissociated dislocations in andalusite. With this nucleation mechanism, increase in the andalusite → sillimanite reaction progress with increasing metamorphic grade would reflect an increase in the abundance of stacking faults in the reactant andalusite with increasing metamorphic grade. This conclusion was defended on the basis of the study of Doukhan et al. (1985) in which it was concluded that the dislocation density of experimentally deformed andalusite increased with increasing temperature. However, without measurements of dislocation density of the andalusite with increasing metamorphic grade, Vernon's (1987b) suggested nucleation mechanism remains speculative.

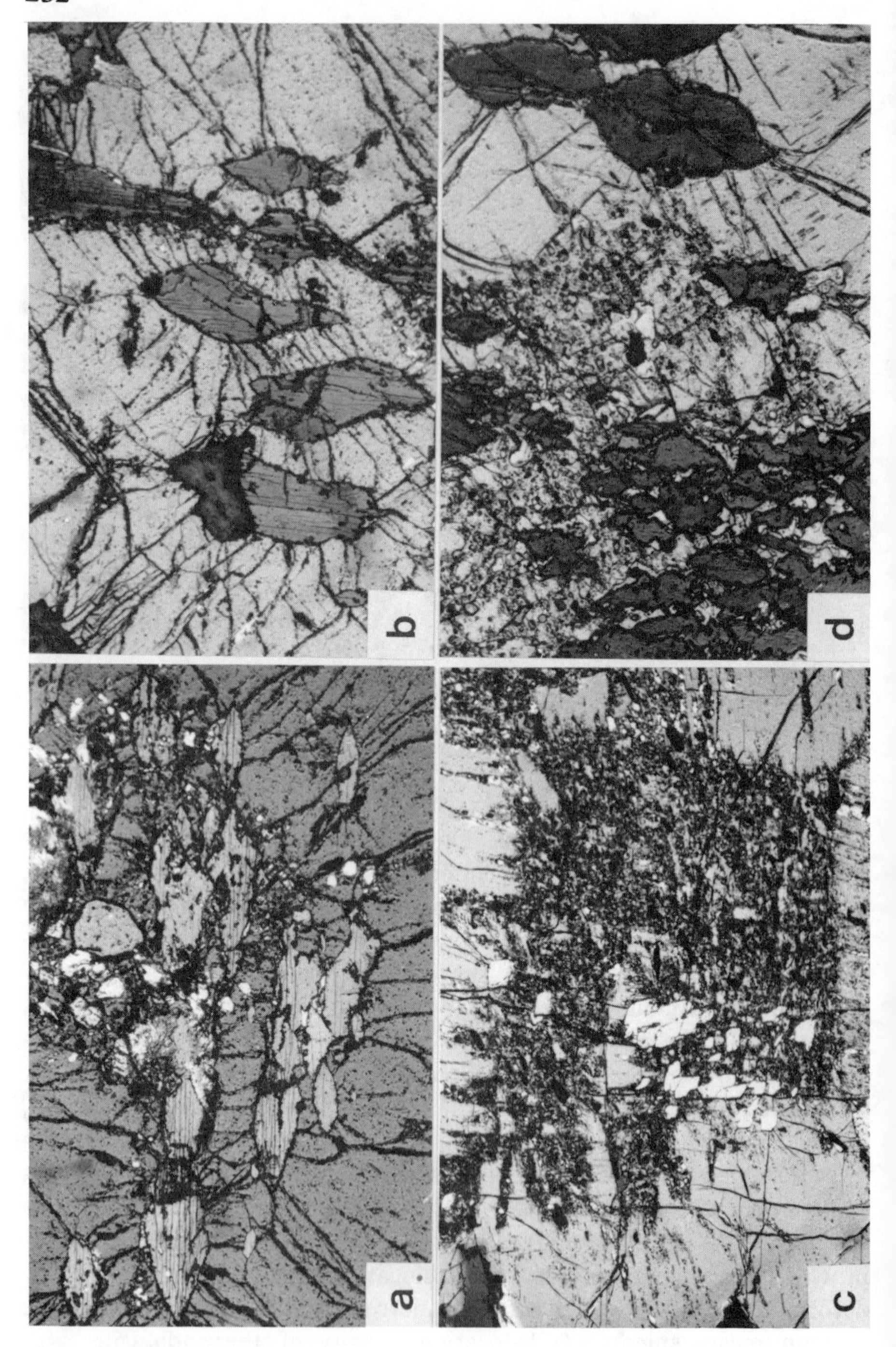
a
b
c
d

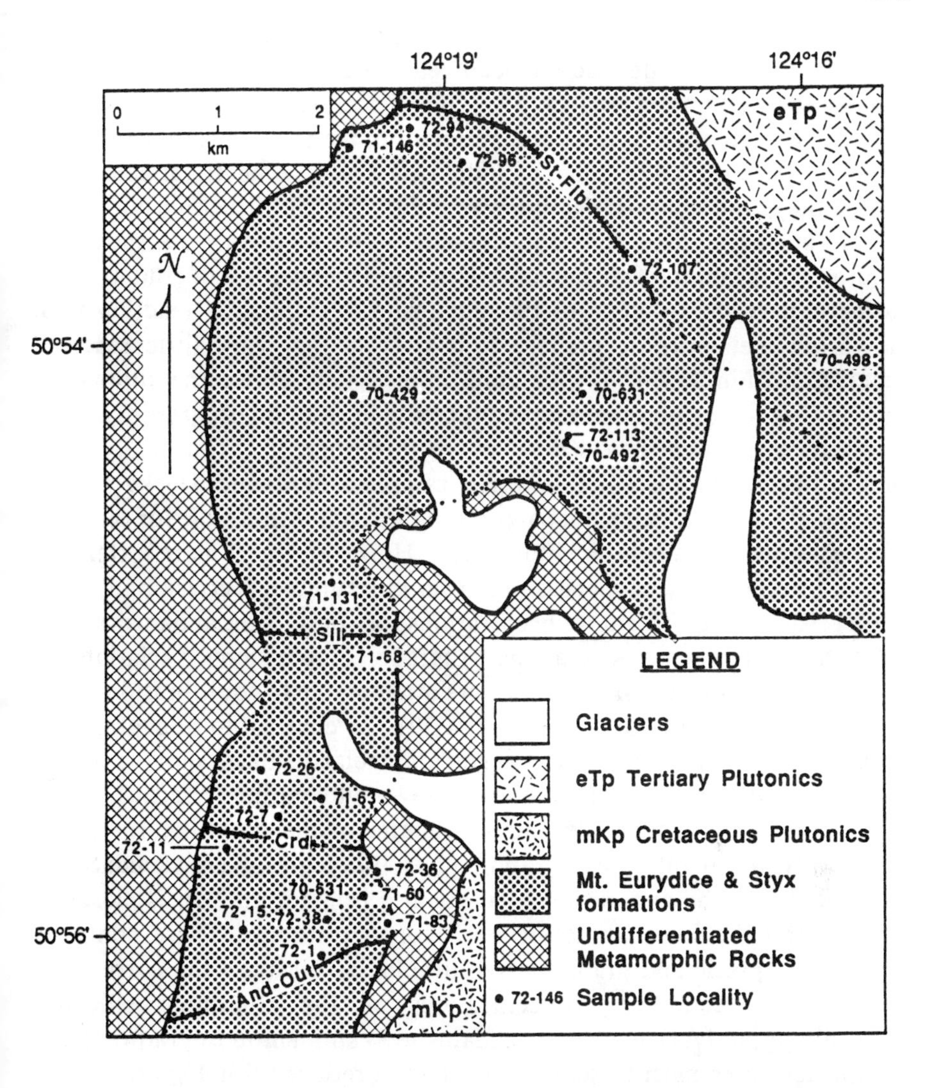

Figure 8.7. (*opposite page*). Photomicrographs showing the paramorphic replacement of andalusite by sillimanite in metapelites from the Mt. Raleigh area, British Columbia. (a) Crystals of sillimanite (light areas showing cleavage) enclosed in a single crystal of andalusite. (b) Sillimanite (dark areas) enclosed in a single crystal of andalusite. (c) Sillimanite crystals (clear areas in lower left quadrant) within an inclusion-rich core of chiastolite. Sillimanite is absent in the inclusion-poor portions of the chiastolite. (d) Sillimanite (dark) within a single chiastolite crystal. Note that the sillimanite crystals within the inclusion-rich core of the chiastolite (left half of photo) are considerably smaller than those in the inclusion-poor area (right half of photo). (From Kerrick and Woodsworth, 1989, Fig. 2).

Figure 8.8. (*above*) Geologic map of the Mt. Raleigh area. Abbreviations for the isograd index minerals are: And = andalusite, Crd = cordierite, Fib = fibrolite, Sil = sillimanite, St = staurolite. (From Kerrick and Woodsworth, 1989, Fig. 1).

In the contact aureole of the Kiglapait intrusion, Labrador, the andalusite → sillimanite reaction occurred by paramorphic replacement of andalusite by sillimanite (Speer, 1982; Kerrick and Speer, 1988). Based on the distribution of andalusite and sillimanite, the aureole is divisible into three zones (Fig. 4.37): zone I (andalusite only), zone II (andalusite + sillimanite), and zone III (sillimanite only). In the direction normal to isograd surfaces, zone II is approximately 300 m thick. As shown in Figure 4.38, graphite-free metaquartzites and metapelites of zone II show an increase in the progress of the andalusite → sillimanite reaction with increasing metamorphic grade. It is notable that very little reaction occurred within a 300 m-wide zone immediately above the sillimanite isograd (Fig. 4.38), and that marked increase in reaction progress occurred at distances greater than 400 m above this isograd. From analysis of andalusite-sillimanite pairs in three samples, Berg and Docka (1983) concluded that minor element partitioning was not responsible for the zone of coexisting andalusite + sillimanite. However, in a more extensive electron probe study of metapelites from this aureole, Kerrick and Speer (1988) found significant concentrations of Fe_2O_3 in andalusite and sillimanite and suggested that the prograde compositional variations in coexisting andalusite + sillimanite were attributable to equilibration along a T-X_{MAlSiO_5} partitioning loop (Fig. 4.39). In contrast to non-graphitic lithologies, interbedded graphite-sulfide metasiltstones show a much narrower zone of coexisting andalusite + sillimanite. Because Fe^{3+} concentrations of andalusite and sillimanite in the graphitic rocks are distinctly lower than those of graphite-free lithologies, Kerrick and Speer concluded that the andalusite → sillimanite isograd reaction is closer to univariant behavior with lower Fe^{3+} concentrations. These contrasts between graphitic and graphite-free lithologies illustrate the correlation between f_{O_2} and Fe^{3+} concentrations of the aluminum silicates. Accordingly, univariancy of the andalusite → sillimanite isograd reaction is expected to be most closely approached in reduced lithologies.

Kerrick and Woodsworth (1989) described aluminum silicates in low-pressure regionally metamorphosed pelites in the Mt. Raleigh pendant, British Columbia. Textural evidence clearly shows that andalusite was replaced by sillimanite (Figs. 8.7a,b). Andalusite and sillimanite coexist in a zone that is approximately 3 km wide (Fig. 8.8). Modal analyses show that there is an increase in andalusite → sillimanite reaction progress with increasing distance upgrade from the isograd marking the first appearance of sillimanite (Fig. 8.9). Electron probe analyses of coexisting andalusite + sillimanite reveal small concentrations of minor elements in both phases, with K_D (= $X^{Sil}_{Al_2SiO_5}$ / $X^{And}_{Al_2SiO_5}$) values close to unity. Thus, the zone of

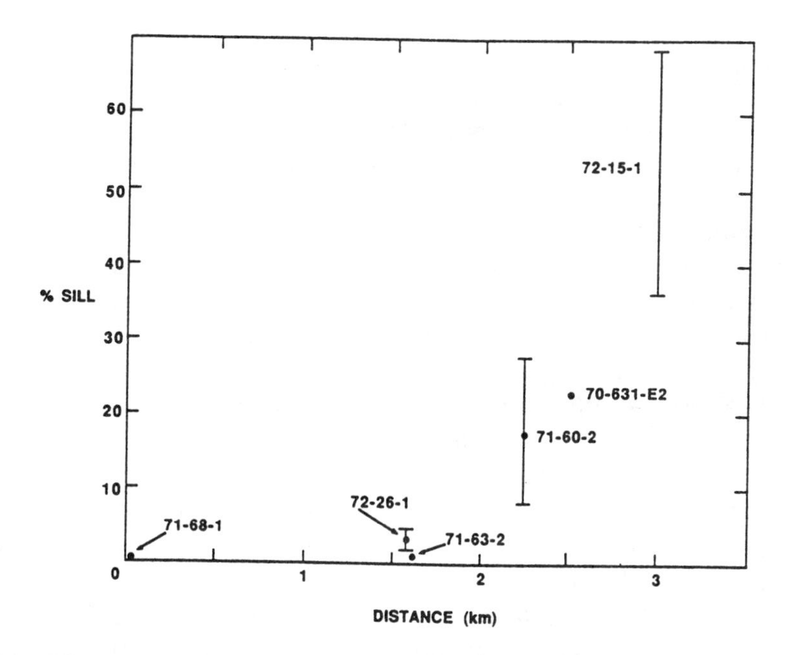

Figure 8.9. Modal percentage of sillimanite [= 100 x Sil/(Sil + And)] as a function of distance from the sillimanite isograd in the Mt. Raleigh area, British Columbia. For three samples, the bars encompass the range of modal variation between andalusite porphyroblasts. (From Kerrick and Woodsworth, 1989, Fig. 6).

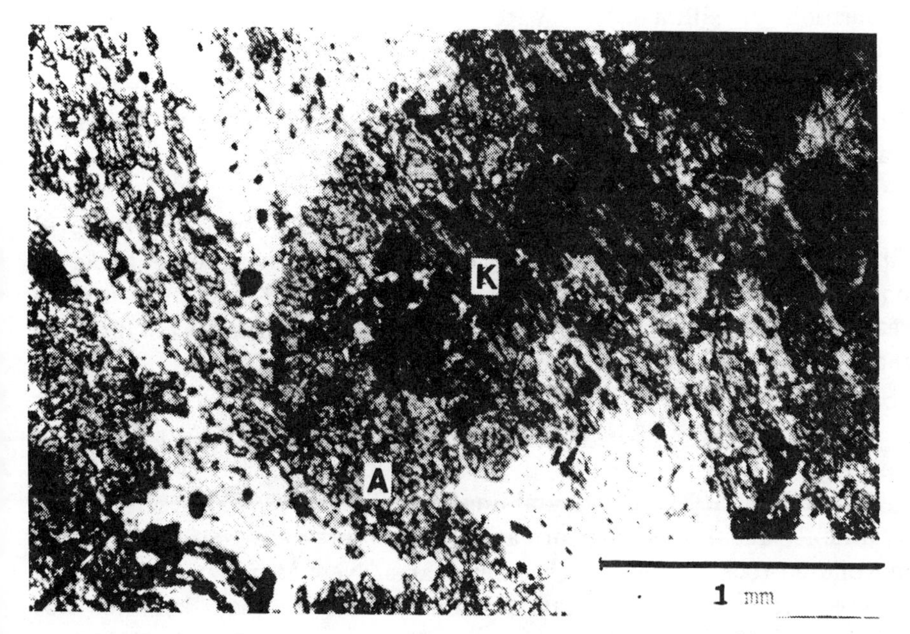

Figure 8.10. Textural evidence of for the arrested kyanite → andalusite transformation in a sample from the Truchas Peaks region, New Mexico. Remnant kyanite (K) is surrounded by poikiloblastic andalusite (A). (From Grambling, 1981, Fig. 3).

coexisting andalusite + sillimanite cannot be attributed to multivariancy due to minor element partitioning. We suggested that the first appearance of sillimanite is correlative with the univariant andalusite-sillimanite equilibrium and that the coexistence of andalusite and sillimanite at higher grades reflects metastable persistence of andalusite within the sillimanite P-T stability field. Garnet-biotite thermometry reveals a temperature gradient of approximately 20°C/km in the plane of exposure; therefore, the 3 km-thick andalusite + sillimanite zone represents metastable persistence of andalusite up to 60°C above the upper stability limit of andalusite (i.e., the andalusite-sillimanite equilibrium). We noted that little andalusite → sillimanite reaction occurred in a 1.5 km-thick zone immediately upgrade from the sillimanite isograd (Fig. 8.9). In comparison, reaction progress markedly increased at distances greater than 1.5 km above this isograd (Fig. 8.9). Textural data enabled us to provide an explanation for reaction progress as a function of metamorphic grade. The reactant andalusite is in the form of chiastolite porphyroblasts with inclusion-rich cores (Fig. 8.7c). At the sillimanite isograd, sillimanite is confined to the inclusion-rich areas of chiastolite (Fig. 8.7c). At higher metamorphic grade, sillimanite occurs in both the inclusion-poor and inclusion-rich areas (Fig. 8.7d). We concluded that the large surface area of andalusite in the inclusion-rich areas provided favorable sites for initial nucleation of sillimanite. With increasing temperature, sillimanite eventually nucleated within the inclusion-poor areas. The finer grain size of sillimanite in the poikiloblastic areas (Fig. 8.7d) was attributed to the phenomenon of *grain boundary pinning*, whereby growth of sillimanite was impeded by the presence of abundant inclusions.

Reinhardt and Rubenach (1989) described an abrupt andalusite → fibrolite isograd in low-pressure regionally metamorphosed rocks in northern Queensland, Australia. They noted (p. 157) that the isograd reaction is "remarkably sharp", in that andalusite+fibrolite coexist in a zone that is only a few meters thick.

THE KYANITE → ANDALUSITE REACTION

Grambling (1981) and Grambling and Williams (1985) described the kyanite → andalusite transformation in low-pressure *regionally* metamorphosed rocks of the Truchas Peaks area, New Mexico. As shown in Figure 8.10 there is clear textural evidence that kyanite was paramorphically replaced by andalusite. "Textures are interpreted to mean that kyanite grew first, andalusite grew later, as rocks followed a P-T path that crossed the kyanite-andalusite univariant boundary" (Grambling,

1981, p. 703). Grambling and Williams (1985) noted that kyanite + andalusite coexist in a zone that is > 500m thick. Assuming that a minimum geothermal gradient of 40°C/km is required to intersect the kyanite-andalusite equilibrium, this zone would correspond to a temperature interval of at least 20°C. Grambling and Williams (1985) attributed this zone to either metastable persistence of kyanite into the andalusite stability field or to multivariancy caused by solid solution. The latter seems likely in light of the high concentration of Fe^{3+} in andalusite (Fe_2O_3 = 1.63 to 1.87 wt %) and the large K_D (= $a^{And}_{Al_2SiO_5}/a^{Ky}_{Al_2SiO_5}$ = 2.20 to 2.53).

Several papers have described kyanite in contact aureoles (Tilley, 1935; Mackenzie, 1949; Woodland, 1963; Pitcher and Read, 1963; Lobjoit, 1964; Hollister, 1969; Naggar and Atherton, 1970; Loomis, 1972a; Barnicoat and Prior, ms.).

To the author's knowledge, textural evidence illustrating the prograde paramorphic replacement of kyanite by andalusite in contact metamorphism is only provided by contact metamorphism of a kyanite gneiss protolith within the aureole of the Ross of Mull granite, Scotland (MacKenzie, 1949; Barnicoat and Prior, ms.). This reaction occurred both in the gneiss and in Al_2SiO_5-bearing quartz segregations (MacKenzie, 1949). Barnicoat and Prior (ms.) attributed the fracturing and subgrain structure of the newly-formed andalusite to the negative ΔV of the kyanite → andalusite reaction.

In his description of the contact aureole of the Ronda ultramafic intrusion in southern Spain, Loomis (1972a) noted that, in progressing upgrade through the aureole, there is an increase in abundance of kyanite. Because of the andalusite → fibrolite prograde index mineral sequence in this aureole, the presence of kyanite in the higher grade rocks is incompatible with an isobaric, prograde path at pressures below that of the Al_2SiO_5 triple point. Therefore, Loomis (1972a) concluded that the gneissic, kyanite-bearing rocks were a mobile carapace that was transported ("dragged") upward during penetration of the intrusive to shallower crustal levels. He contended that kyanite stably crystallized at deeper structural levels and metastably persisted as the gneissic carapace rose to shallower depths and, thus, to P-T conditions outside of the kyanite stability field. Loomis (1972b) extended this model to other kyanite-bearing contact aureoles and thus questioned the traditional assumption of isobaric metamorphic conditions at the present level of exposure in kyanite-bearing contact aureoles.

Atherton et al. (1975) questioned Loomis' (1972a,b) model for the genesis of kyanite in contact aureoles. From structural arguments of the aureoles considered by Loomis (1972b), they concluded that the aureoles either underwent primarily *lateral* distention (i.e., a model of "ballooning" intrusives), or that kyanite crystallized in a static environment. In fact, they contended that in the Ronda aureole, which provided the impetus for Loomis' (1972a) model, there is no evidence that kyanite growth was pretectonic or syntectonic (as required by the model of Loomis). "Many occurrences of kyanite in aureoles may be explained in terms of simple equilibrium crystallization in the kyanite field beneath a relatively low pressure triple point and continued crystallization along an essentially isobaric path into the andalusite and sillimanite fields" (Atherton et al., 1975, p. 432). Nevertheless, Loomis' (1972a, 1972b) model could be valid for contact aureoles with anatectic migmatites adjacent to the intrusives. Flood and Vernon (1978) hypothesized that migmatites adjacent to the Cooma Granodiorite (New South Wales, Australia) were dragged upward during diapiric rise of the intrusive. However, they presented no cogent barometric arguments supporting their hypothesis. More recent studies of migmatitic aureoles (e.g., Evans and Speer, 1984; Pattison, 1989) suggest that there is no evidence for significant upward "drag" of the migmatitic envelopes during magmatic intrusion.

If the conclusions of Atherton et al. (1975) are valid, the presence of the assemblage kyanite + quartz in thermal aureoles is useful for constraining the P-T conditions of metamorphism. The upper pressure limit of the triple point is dictated by the presence of andalusite in contact aureoles. Assuming P-T conditions of 3.8 kbar and 500°C for the triple point (Holdaway, 1971), and a_{H_2O} = 1.0, kyanite + quartz within andalusite-bearing aureoles is confined to 380 -500°C and 2.25 - 3.8 kbar (Fig. 8.11). However, many pelites are graphitic, and therefore a_{H_2O} values less than unity occur as the result of reactions between graphite and fluid (Ohmoto and Kerrick, 1977). If the f_{O_2} conditions during metamorphism are between that of the QFM and Ni-NiO buffers, there is little difference in the temperature of the pyrophyllite dehydration equilibrium compared to that with a_{H_2O} = 1.0 (Fig. 8.12). However, for f_{O_2} conditions above Ni-NiO or below QFM, the pyrophyllite dehydration equilibrium lies at significantly lower temperatures compared to that with a_{H_2O} = 1.0 (Fig. 8.12). Using the approach of Ohmoto and Kerrick (1977), estimates of a_{H_2O} in graphitic pelites during contact metamorphism are in the range 0.7 to 0.9 (Tyler and Ashworth, 1982; Pattison, 1989). Ashworth and Evirgen (1985) concluded that, because of low a_{H_2O} (0.1-0.2), kyanite was stabilized in graphitic metapelites of the central Menderes Massif,

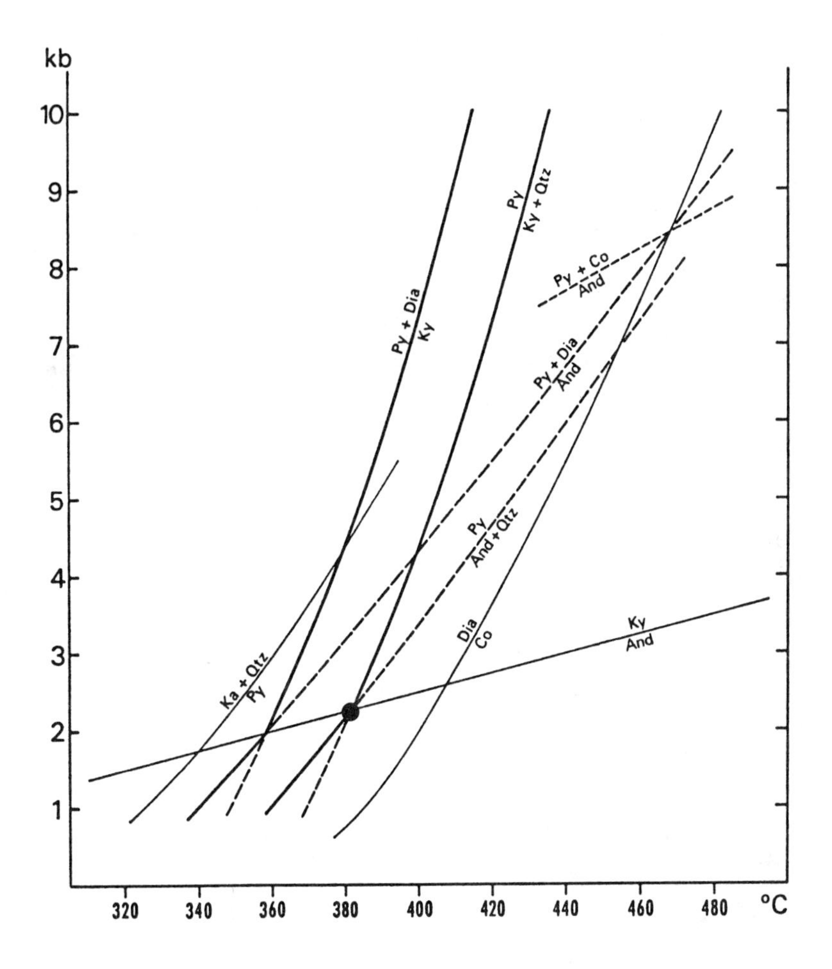

Figure 8.11. Phase equilibrium diagram for the system Al_2O_3-SiO_2-H_2O. The invariant point formed by the intersection of the pyrophyllite = Al_2SiO_5 + quartz + water equilibrium with the kyanite = andalusite equilibrium (filled circle) marks the lowest pressure for kyanite + quartz in a system with a_{H_2O} = 1.0. (From Haas and Holdaway, 1973, Fig. 3).

Turkey. However, because of an error in the equilibrium constant for the pyrite-pyrrhotite buffer, the computations of Ohmoto and Kerrick (1977) underestimated the fugacities of sulfur-bearing components in metamorphic fluids in equilibrium with graphite (Poulsen and Ohmoto, 1989). Poulsen and Ohmoto (1989) showed that in the typical pressure range of contact metamorphism (1-2 kbar), fluids in equilibrium with graphite + pyrite + pyrrhotite have X_{H_2O} values less than 0.7 at the higher grades of metamorphism.

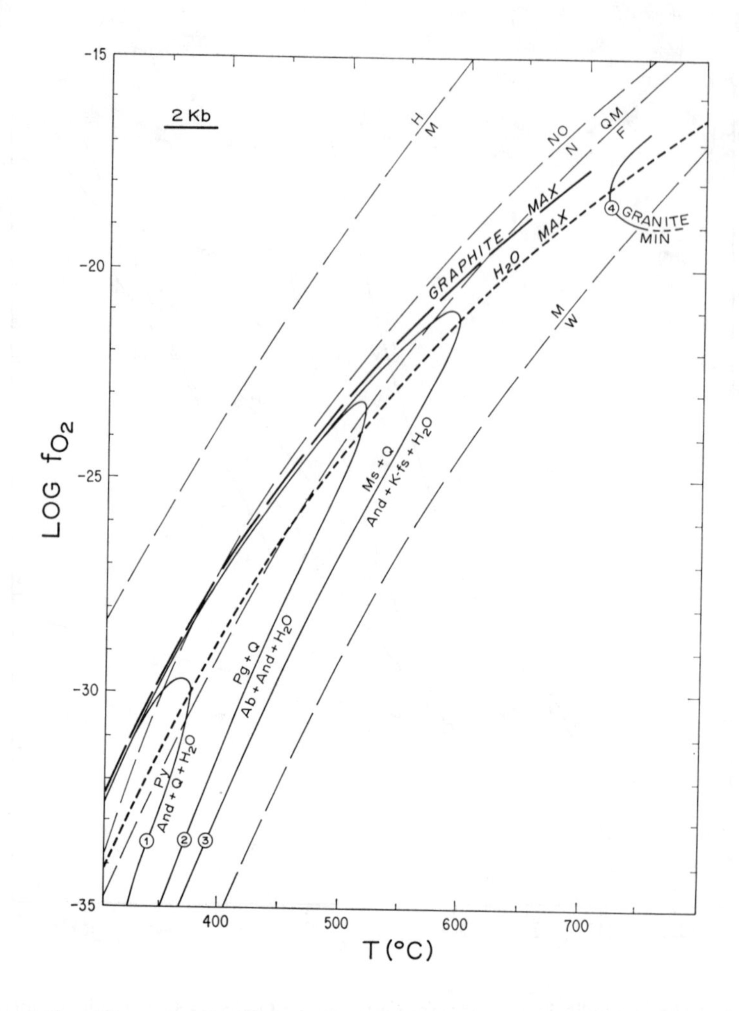

Figure 8.12. Ohmoto and Kerrick's (1977) depiction of equilibria relevant to graphitic metapelites at 2 kbar. Equilibrium (1) is the dehydration of pyrophyllite to andalusite + quartz + water. This diagram was derived from computations using X_{H_2O} values calculated for C-O-H-S fluids in equilibrium with graphite + pyrite + pyrrhotite. "GRAPHITE MAX" marks the upper f_{O_2} stability limit at each temperature, whereas "H_2O MAX" is the locus of maximal X_{H_2O} values. The oxygen buffers are: WM = wüstite-magnetite; QFM = quartz-fayalite-magnetite; NNO = nickel-nickel oxide; HM = hematite-magnetite. (From Ohmoto and Kerrick, 1977, Fig. 8).

The *retrograde* transformation of kyanite to andalusite has been suggested by some workers. Recently, the author described this transformation in spectacular Al_2SiO_5-bearing segregations in the Lepontine Alps, Switzerland (Kerrick, 1988). Discussion of this transformation is included in Chapter 10 in a comprehensive genetic model for these segregations.

THE ANDALUSITE → KYANITE REACTION

In contrast to the reactions discussed in the preceding sections of this chapter, the andalusite → kyanite reaction is unusual in that some workers have considered this reaction to have occurred upon prograde metamorphism whereas others contend that this reaction occurred during retrogression.

In a study of contact metamorphosed pelites in the Kwoiek area, British Columbia, Hollister (1969, p. 354) concluded that: "The most important feature...is the widespread textural relation of kyanite pseudomorphous after andalusite." As shown in Figure 8.13, Hollister (1969) considered two alternative prograde metamorphic paths to explain the aluminum silicate paragenetic relations that he deduced from textures. He considered the possibility of an equilibrium model with path C (Fig. 8.13). In this case, the andalusite → kyanite transformation would correspond to the crossing of the kyanite-andalusite equilibrium along the lower grade portion of path C. Hollister (1969) concluded that the implied pressure

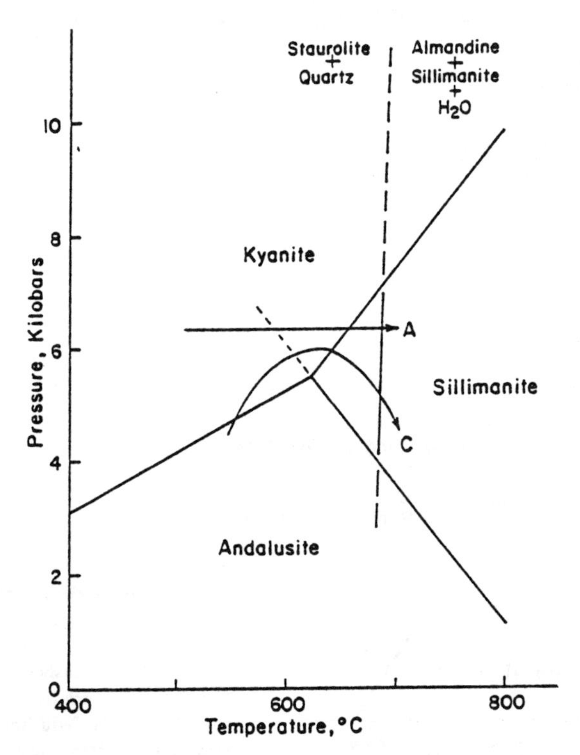

Figure 8.13. Two prograde P-T paths (A and C) considered by Hollister (1969) for contact metamorphism in the Kwoiek area, British Columbia. (From Hollister, 1969, Fig. 6).

increase along path C could have resulted from tectonic overpressure due to forcible emplacement of the adjacent intrusive(s) and/or an increase in P_{fluid} over $P_{lithostatic}$ due to rapid dehydration reactions in the metapelites. Hollister (1969) argued that the proposed mechanism for a pressure increase was unlikely. Consequently, he favored a model of isobaric metamorphism (path A in Fig. 8.13). He concluded that andalusite metastably formed in the kyanite P-T stability field along the lowest grade portion of path A, and that andalusite converted to kyanite within the kyanite stability field. As discussed in Chapter 9, the author suggests an alternative to Hollister's isobaric disequilibrium model.

In their study of the Wissahickon Schist in southeastern Pennsylvania, Crawford and Mark (1982) noted that in several specimens textural evidence suggests that andalusite was replaced by kyanite. Similar observations were made by Mitchell et al. (1988) in an occurrence of spectacularly large aluminum silicate crystals in the Wissahickon schist in south-central Virginia. Crawford and Mark contended that the andalusite + cordierite assemblage was produced by a low-pressure regional metamorphic event. Subsequently, andalusite was replaced by kyanite during a higher pressure Barrovian-type regional metamorphism. Crawford and Mark (1982) utilized a tectonic model to rationalize the thermobarometric relations of these two metamorphic events. As shown in Figure 8.14, they hypothesized that the increase in pressure yielding the

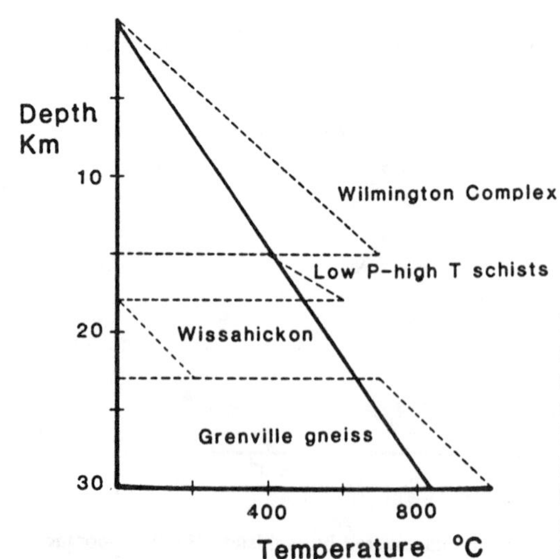

Figure 8.14. Crawford and Mark's (1982) hypothesized development of depth versus temperature gradients in tectonic units of the Pennsylvania Piedmont. The dashed lines depict the pre-thrusting gradients whereas the solid line is the "equilibrium" gradient established after thrusting. (From Crawford and Mark, 1982, Fig. 8).

Barrovian-type regional metamorphism of the Wissahickon resulted from tectonic thickening due to overthrusting of the Wilmington Complex.

Using a tectonic model similar to that of Crawford and Mark (1982), Baker (1987) and Beddoe-Stephens (in press) analyzed the andalusite → kyanite transformation in the Dalradian sequence of Scotland. In this area there is a kyanite-andalusite isograd (Fig. 8.15) which represents the approximate western limit of regional andalusite (a key index mineral for the low-pressure "Buchan" regional metamorphism of the Bannfshire region of Scotland). The Portsoy-Duchray Hill (P-DH) lineament, a major shear zone in the Banffshire region (see inset in Fig. 8.15), is the key tectonic feature in their models. Beddoe-Stephens (in press) concluded that andalusite formed in an early metamorphic event. Subsequently, rocks to the west of the Portsoy-Duchray Hill lineament underwent a pressure increase resulting in the transformation of andalusite to kyanite. Following Baker (1987), Beddoe-Stephens (in press) considered that this pressure increase resulted from tectonic thickening due to the northwestward thrusting of the Buchan zone rocks. The postulated pressure increase during overthrusting is supported by thermobarometric analyses of garnet zoning profiles using the method of Spear and Selverstone (1983).

Although some may object to appealing to tectonic thickening to explain prograde andalusite → kyanite transformations, estimates of the shapes and curvatures of normal "static" continental geotherms do not pass from the andalusite P-T stability field to that of kyanite. Thus, a model of tectonic thickening provides a reasonable explanation for the occurrences of the prograde transformation of andalusite to kyanite with an equilibrium model.

REACTIONS INVOLVING FIBROLITE

Fibrolite is considerably more common than sillimanite in metapelites. The predominance of fibrolite is reported in numerous studies of contact and regional metamorphism (Best and Weiss, 1964; Harte and Johnson, 1969; Grew and Day, 1972; Rumble, 1973; Frey et al., 1974; Grambling, 1981; Tyler and Ashworth, 1982; Pattison and Harte, 1985; Vernon et al., 1987; Stäubli, 1989; Carey et al., in press). For example, Rumble (1973) determined the distribution of fibrolite and sillimanite in a large area of regionally metamorphosed rocks of New England. Considering his sample localities containing fibrolite and/or sillimanite, fibrolite alone occurs in 43 localities, fibrolite + sillimanite is found in six samples, and sillimanite

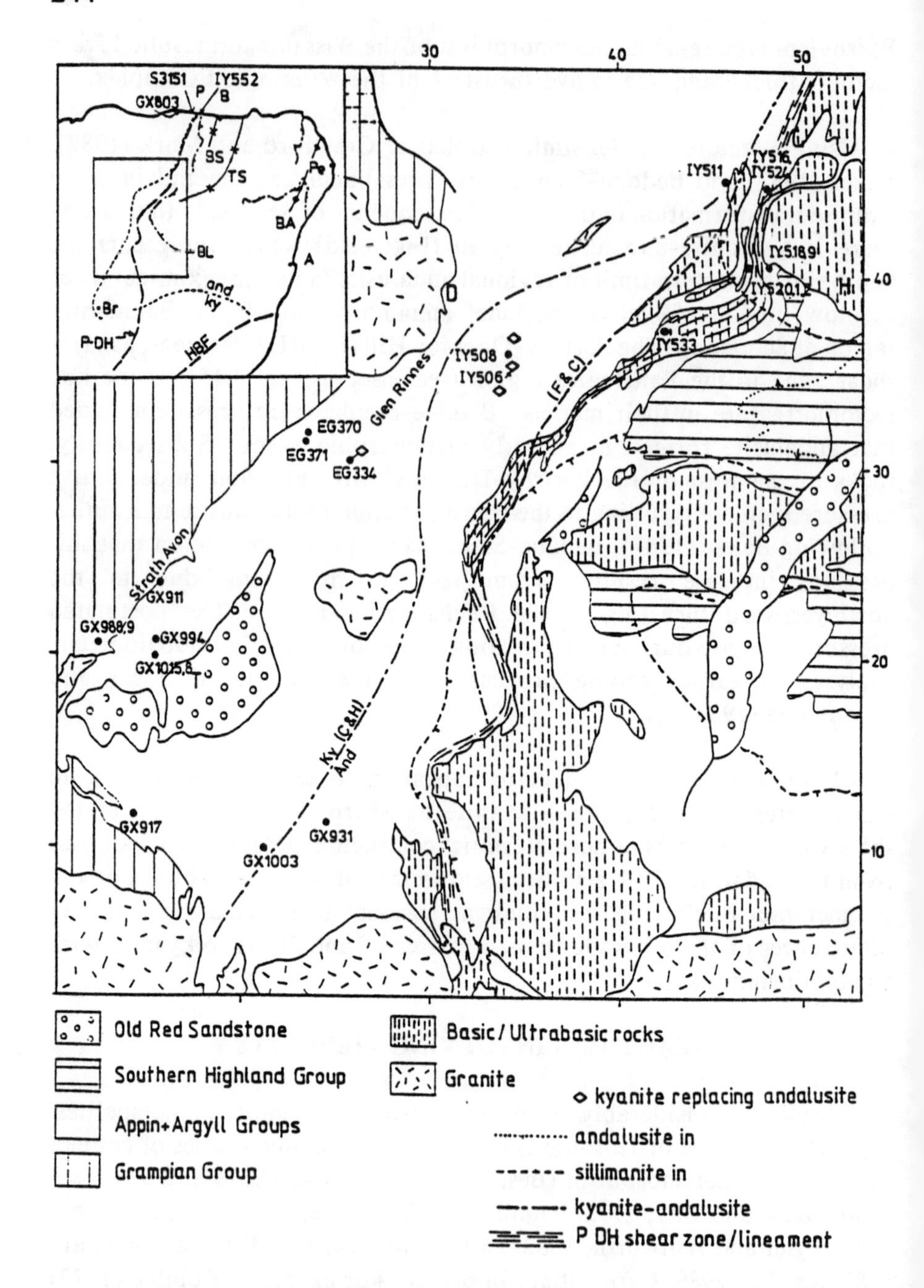

Figure 8.15. Geologic map showing the kyanite-andalusite isograd in a portion of the Grampian Highlands of Scotland. The inset shows the regional geologic setting of the main map. Note the Portsoy-Duchray Hill (P-DH) shear zone (lineament) and the localities of samples with textural evidence of the andalusite → kyanite reaction (diamond-shaped symbols). (From Beddoe-Stephens, ms., Fig. 1).

(alone) occurs at seven localities. As discussed in Chapter 7, questions remain regarding the thermodynamic properties of sillimanite versus fibrolite. Examination of the petrologic relations of fibrolite versus sillimanite offers insight into this question. For example, if fibrolite and sillimanite have identical chemical potentials, and if they form under equilibrium conditions, very similar distribution of these phases is expected in the field. However, if fibrolite and sillimanite have significantly different chemical potentials and/or if one or both of these phases form in disequilibrium, resultant complications are expected in the field distribution of these "phases".

It is unfortunate that fibrolite and sillimanite have not been distinguished in many petrologic studies. This undoubtedly reflects the implicit assumption that these "phases" have identical chemical potentials and, thus, identical P-T stability relations. The following discussion suggests that this assumption is not necessarily valid. Thus, the author appeals to investigators of future petrologic studies to consider fibrolite and sillimanite as separate phases and to carefully record the textural relations and the geographic distribution of fibrolite versus sillimanite. Grambling's (1981) study serves as an excellent model for such studies.

Much of the difficulty in the interpretation of fibrolite in metapelites arises from ambiguity in the interpretation of mineral textures. Vernon and Flood (1977) attempted to clarify fibrolite textures by defining two contrasting types of fibrolite. *Harmonious* fibrolite has textural features suggesting equilibrium and implying that the fibrolite was present prior to the final positioning of grain boundaries between other minerals in the rock. With harmonious fibrolite, impinging grain boundaries between other minerals apparently adjusted to that of the fibrolite. An example is a grain boundary between two feldspar crystals intersecting the prism faces of fibrolite at 90°. Vernon and Flood (1977, p. 227) concluded that harmonious fibrolite reflects "...clear microstructural equilibrium with the other minerals...". In contrast, *disharmonious* fibrolite has textural evidence suggesting growth after the grain boundaries between other phases were established. For example, feldspar-feldspar grain boundaries would meet the prism faces of disharmonious fibrolite at random angles. Vernon and Flood (1977) concluded that a number of common textures exhibited by fibrolite are ambiguous with regard to the timing of the growth of fibrolite in relation to the grain boundaries between other phases. Their examples include: (1) fibrolite in veinlets and knots ("faserkiesel"), (2) fibrolite needles enclosed within minerals and showing regular geometric arrangement that is apparently controlled by the atomic structure of the host crystal, (3) fibrolite concentrated in pressure shadows of porphyroblasts or augen, and (4) folded fibrolite aggregates contained within or adjacent to other minerals that display no optical evidence of deformation.

Some fibrolite-forming reaction mechanisms

Several unique reaction mechanisms for fibrolite formation have been suggested from analyses of textures coupled with petrologic and/or geochemical arguments. These mechanisms differ from the fibrolite-producing reactions discussed earlier in this chapter in that "special" conditions or processes (i.e., metasomatism, deformation or recrystallization) are called upon.

Base cation leaching. A number of studies have suggested that fibrolite formed from leaching of base cations from pre-existing silicates. Many studies have noted that fibrolite is commonly intergrown with biotite (Harker, 1950, p. 58; Tozer, 1955; Woodland, 1963; Best and Weiss, 1964; Chakraborty and Sen, 1967; Shelley, 1968; Scharbert, 1971; Loomis, 1972a; Kerrick, 1987; Kerrick and Woodsworth, 1989; Stäubli, 1989). Thus, fibrolitization of biotite has been a popular reaction mechanism for the formation of fibrolite by base cation leaching.

In a petrographic analysis of metapelites in the aureole of the Main Donegal Granite, Ireland, Tozer (1955) suggested that K, Fe, Mg and OH were expelled during the fibrolitization of biotite. The presence of small magnetite crystals in fibrolite aggregates, coupled with the fact that the pleochroism of biotite becomes less intense as the modal fibrolite/biotite ratio increases, suggest that fibrolitization may involve the fractional decomposition of biotite by reaction of the annite component to magnetite. He considered that muscovite + fibrolite intergrowths represent a second generation of fibrolite that formed by late-stage influx of K from the Main Donegal Granite.

Vernon (1979) hypothesized that fibrolite formed from base-cation leaching in rocks of the Cooma Complex, New South Wales, Australia. He presented textural evidence suggesting that fibrolite replaced several aluminous phases (biotite, muscovite, cordierite and feldspars). As shown in Table 8.1, Vernon considered that all of the replacement reactions involved the consumption of H^+. The reactions in Table 8.1 imply that fibrolite + quartz intergrowths are expected as products. However, Vernon (1979, p. 147) noted that quartz intergrowths are rare in fibrolite

Table 8.1. Equations for base-leaching reactions in hydrothermal alteration. (From Vernon, 1979, Table 1).

Reaction	Reactant	Product
$2NaAlSi_3O_8 + 2H^+ \rightleftharpoons Al_2SiO_5 + 2Na^+ + H_2O + 5SiO_2$	albite (in myrmekitic intergrowths)	sillimanite
$CaAl_2Si_2O_8 + 2H^+ \rightleftharpoons Al_2SiO_5 + Ca^{2+} + H_2O + SiO_2$	anorthite (in myrmekitic intergrowths)	sillimanite
$2KAlSi_3O_8 + 2H^+ \rightleftharpoons Al_2SiO_5 + 2K^+ + H_2O + 5SiO_2$	K-feldspar	sillimanite
$2K(Mg,Fe)_3AlSi_3O_{10}(OH)_2 + 14H^+ \rightleftharpoons Al_2SiO_5 + 2K^+ + 6(Mg,Fe)^{2+} + 9H_2O + 5SiO_2$	biotite	sillimanite
$2KAl_3Si_3O_{10}(OH)_2 + 2H^+ \rightleftharpoons 3Al_2SiO_5 + 2K^+ + 3H_2O + 3SiO_2$	muscovite	sillimanite
$(Mg,Fe)_2Al_4Si_5O_{18} + 4H^+ \rightleftharpoons 2Al_2SiO_5 + 2(Mg,Fe)^{2+} + 2H_2O + 3SiO_2$	cordierite	sillimanite

aggregates of the Cooma rocks; thus, the equations in Table 8.1 should be reformulated with aqueous $Si(OH)_4$ as a product. He postulated that the influx of H_2O, which may have originated in the crystallizing Cooma Granodiorite, was responsible for the widespread hydration of cordierite:

$$\text{Cordierite} + \text{red-brown Biotite} + H_2O = \text{Andalusite} + \text{green-brown Biotite} + \text{Quartz} + H^+.$$

Thus, a local source was postulated for the hydrogen ions that were involved in the fibrolitization reactions listed in Table 8.1. Although Vernon (1979, p. 149) referred to the widespread fibrolitization in the Cooma Complex as "...'regional' hydrogen metasomatism...", large-scale flux of H^+ from an external source is not required for fibrolitization.

Vernon et al. (1987) appealed to base cation leaching for the production of fibrolite in a contact metamorphosed leucogranite at Kentucky, New South Wales, Australia. They concluded that the concentration of fibrolite in veinlets suggests that this phase formed by the influx of hot, acidic volatiles generated during the emplacement of the adjacent intrusive. It is the author's opinion that Vernon et al. (1987) presented no irrefutable evidence supporting the process of base cation leaching. It is possible that the components to form the fibrolite were derived from the influx of solutes derived from an external source.

The author studied fibrolite in contact aureoles of Donegal, Ireland (Kerrick, 1987b). Textural evidence strongly suggests that most of the fibrolite in these aureoles formed by the decomposition of biotite. Other studies of contact aureoles have also suggested that biotite was pseudomorphed by fibrolite (Tozer, 1955; Woodland, 1963; Best and Weiss, 1964; Loomis, 1972a). In a few specimens, fibrolite occurs at grain boundaries of the other phases (Kerrick, 1987b, Fig. 2F). This "disharmonious" texture (Vernon and Flood, 1977) implies late-stage crystallization of fibrolite. Because aggregates of fibrolite and quartz appear to be the only solid product phases, the reaction,

$$2K(Mg_xFe_{1-x})_3AlSi_3O_{10}(OH)_2 + 14HCl = Al_2SiO_5 + 5SiO_2 + 2KCl + 6(Mg_xFe_{1-x})Cl_2 + 9H_2O \qquad [8.1]$$

was suggested for the fibrolitization of biotite. I concluded that Hcl was derived from the influx of acidic volatiles released during the later stages of crystallization of the granitic intrusives. As summarized by Kerrick (1987b, p. 251), recent studies on magmatic volatiles support my contention that acidic volatiles are evolved from crystallizing, water-saturated intrusives. In several specimens from the Ardara aureole, this reaction occurred in rocks lacking other Al_2SiO_5 phases; thus, fibrolitization cannot be considered as a polymorphic transformation. Local closed-system ionic equilibria linking two aluminum silicate phases, such as suggested by Carmichael (1969) for the kyanite → sillimanite transformation, are precluded for rocks with fibrolite as the sole aluminum silicate.

In briefly reviewing fibrolite in regionally metamorphosed pelites (Kerrick, 1987b), I concluded that late-stage replacement of biotite by fibrolite may have occurred by the influx of acidic volatiles. This contention is supported by Stäubli (1989) in his study of fibrolite in regionally metamorphosed pelites in the Himalayas of northwest India.

Foster (1990) provided an alternative interpretation of biotite + fibrolite intergrowths. He suggested that biotite may have reacted to fibrolite in some subdomains and simultaneously grown in others. Thus, biotite acted as a catalyst. With Foster's mechanism, biotite plays a role analogous to muscovite in Carmichael's (1969) reaction mechanism. Foster's analysis suggests that the correct interpretation of fibrolite-forming reaction mechanisms requires careful examination of the textures of all biotites within biotite+fibrolite assemblages.

· Deformation-induced fibrolitization. Vernon (1987a) appealed to strain to explain the origin of the common anastamosing fibrolite folia in high-grade metapelites. He noted that the fibrolite folia display evidence of strong deformation - the presence of tight crenulations is perhaps the most cogent evidence for high strain. In contrast, the interfolial zones lack evidence for strong deformation. Thus, Vernon (1987a) concluded that strain was concentrated in the fibrolite folia. In light of the evidence for low dislocation densities in fibrolite (Doukhan et al., 1985), he contended that the strain in the folia occurred primarily by intergranular slip (grain boundary sliding) and that deformation produced low dislocation densities within the fibrolite crystals. Vernon (1987a) suggested that other phases were present within the localized shear planes but that these phases strained by the production of dislocations rather than by grain boundary slip. High dislocation density increased the Helmholtz free energies such that these phases underwent dissolution in the fluid phase (see also Wintsch and Dunning, 1985). Accordingly, Vernon (1987a) concluded that pre-existing fibrolite became concentrated by dissolution of other phases in the fluid.

Following the model of Wintsch (1981), Wintsch and Andrews (1988) invoked a pressure solution model for the genesis of fibrolite in a granitic pegmatite within upper amphibolite facies gneisses in south-central Connecticut. Petrographic examination suggests that fibrolite replaced feldspars and biotite. From the well-developed lineation defined by the long axes of fibrolite needles, coupled with the folding of fibrolite aggregates, they concluded that fibrolite growth was syntectonic. Table 8.2 summarizes their postulated ionic reactions for the replacement of biotite and feldspar. Their paper detailed a model for the replacement of K-feldspar by sillimanite. Because of the apparent syntectonic crystallization of fibrolite, Wintsch and Andrews (1988) considered that localized, high deviatoric stress along microscale "shear zones" produced an increase in the chemical potential of the feldspar (Fig. 8.16). The resultant destabilization of feldspar drove reactions 1, 2 and 3 (Table 8.2) to the right, thereby forming sillimanite (actually fibrolite) and releasing alkalies and silica into the fluid phase. Wintsch and Andrews (1988) contended that dissolution of feldspars was enhanced by strain energy resulting from high dislocation densities at the sites of high normal stress. It should be noted that there are no *indisputable* arguments supporting their pressure solution model. In fact, Wintsch and Andrews (1988, p.

Table 8.2. Wintsch and Andrews' (1988) equations for ionic reactions producing sillimanite. (From Wintsch and Andrews, 1988, Table 3).

Reaction	ΔV(Solid) (cc/mole reactant)	ΔV(Solid) (%)	$\frac{V(Qz)}{V(Sill)}$
1. $2K_{0.8}Na_{0.2}AlSi_3O_8 + 2H^+ \; Al_2SiO_5 + 5SiO_2(aq)$	−83	−76	2.27
K-feldspar sillimanite			
$= + 1.6K^+ + 0.4Na^+ + H_2O$			
2. $5Na_{0.8}Ca_{0.2}Al_{1.2}Si_{2.8}O_8 + 6H^+ \; 3Al_2SiO_5$	−70	−70	1.67
plagioclase sillimanite			
$= + 11SiO_2(aq) + 4Na^+ + Ca^{++} + 3H_2O$			
3. $K_2(Fe_{2.6}Mg_{2.5}Al_{0.8})Si_{5.4}Al_{2.6}O_{20}(OH)_4 + 12.2H^+$	−223	−72	0.99
biotite			
$= 1.7Al_2SiO_5 + 3.7SiO_2(aq) + 2K^+ + 2.6Fe^{++}$	−223	−72	0.99
sillimanite			
$+ 2.5Mg^{++} + 8.1H_2O$			
4. $2K_2(Fe_{1.8}Mg_{3.0}Al_{1.0})Si_{5.4}Al_{2.6}O_{20}(OH_{3.2}F_{0.8}) + 6H^+$	−187	−62	0.91
biotite			
$= 1.8Al_2SiO_5 + 0.6Fe_3O_4 + 3.6SiO_2(aq) + 2K^+ +$			
sillimanite magnetite			
$3.0Mg^{++} + 0.8F^- + 1.2e^- + 4.6H_2O$			

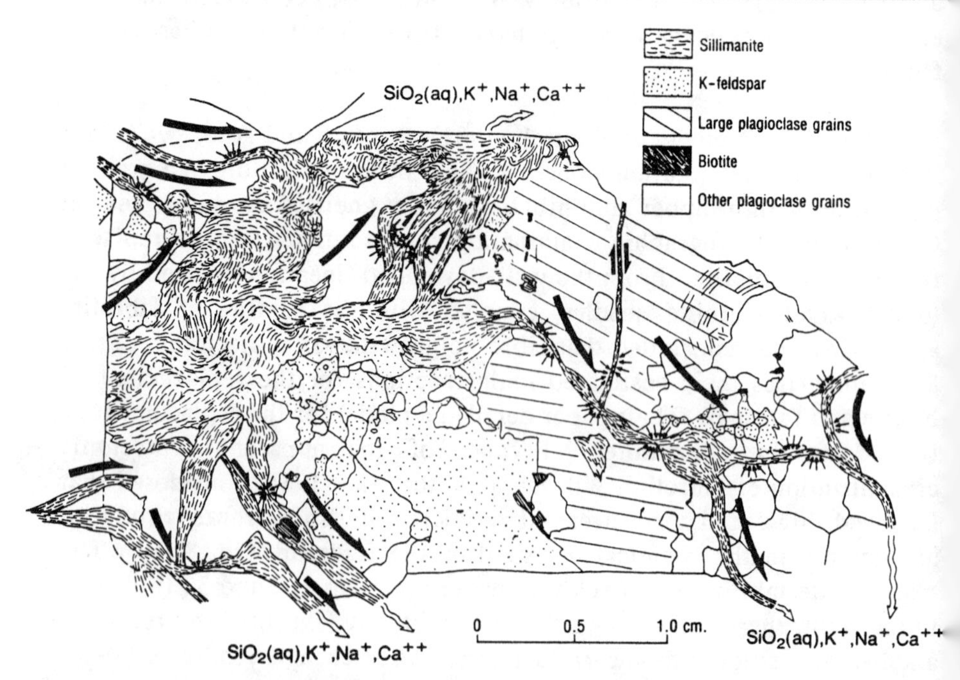

Figure 8.16. Fibrolite-bearing pegmatite from south-central Connecticut. Note (1) the truncation of feldspar grains by fibrolite strands and (2) the microfolds in the strands. The small convergent arrows depict possible sites of high stress and high strain energy due to abundant dislocations. The large arrows depict hypothesized displacement directions across fibrolite strands. (From Wintsch and Andrews, 1988, Fig. 3).

157) stated: "It is difficult to prove that non-hydrostatic stress existed in the pegmatite, and even harder to show that it operated during the production of sillimanite..." Further support for their model would be provided by the presence of locally high dislocation densities at grain boundaries of strained feldspars. However, even if TEM examination of the feldspars considered by Wintsch and Andrews (1988) did not reveal local regions of high dislocation density, it is nevertheless possible that their mechanism is applicable. During deformation, the regions of high dislocation density could have been entirely converted to fibrolite such that the remaining feldspar contains low dislocation density.

Flöttmann (in press) modeled the formation of fibrolite in shear zones within upper amphibolite facies gneisses of the Schwarzwald basement located in the southeastern part of the Federal Republic of Germany. The shear zones consist of anastamosing folia of biotite + fibrolite that enclose undeformed gneisses. The highly deformed shear zones have a mylonitic texture and consist of interlayered folia of fibrolite and "ribbon" quartz. In some cases, fibrolite aggregates define small-scale crenulations. Flöttmann provides cogent textural evidence that fibrolite replaced biotite. Delineation of the deformation path by finite strain analysis reveals minimum volume losses of 30% for the porphyroclasts and 60% for the matrix. Considering the amount of biotite in the gneiss protolith, Flöttmann (in press) concluded that the maximum modal amount of sillimanite formed by complete fibrolitization of biotite by the base cation leaching reaction is 19%. However, the modal amount of sillimanite in the shear zones is 7% higher than that computed by base cation leaching. Rather than appealing to the influx of externally-derived Al, Flöttmann considered that the excess modal fibrolite (i.e., the amount of fibrolite exceeding that computed from the fibrolitization of biotite) was produced by removal of quartz by dissolution into the fluid phase. As shown in Figure 8.17, there is an excellent correlation between the modal fibrolite/quartz ratio and the amount of strain recorded by quartz. Flöttmann interpreted Figure 8.17 to indicate that the breakdown of biotite was induced by strain and that low strain (Ra ~ 6) was required to initiate significant biotite decomposition. He concluded that the rate of the fibrolitization reaction was enhanced by factors such as increased surface area during mylonitization, deformation "surface defects", or the excess energy in dislocations, twins, and subgrain boundaries. He noted that mylonitization was accompanied by microfractures with characteristics suggesting that they propagated by stress corrosion by fluids at "abnormally" high pressure. Flöttmann envisioned the following model for the interplay between biotite dehydration and deformation.

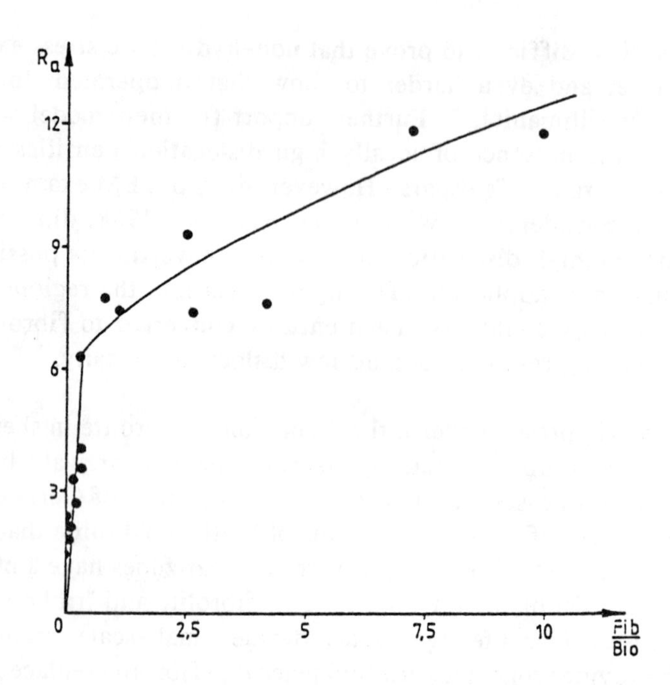

Figure 8.17. Aspect ratio of quartz grains (= R_a) as a func-tion of modal fibrolite/biotite (Fib/Bio) ratio. (From Flöttman, ms., Fig. 18).

Increasing strain within developing shear zones reached a threshold level for significant biotite dehydration. The evolved H_2O formed micro-fractures by stress corrosion. After formation of the microfractures, high fluid pressures yielded low effective stresses and thus microfractures opened by extension thereby producing a significant drop in fluid pressure within the microfractures. The resultant dilatancy resulted in "fluid pumping" whereby the shear zones became sinks for fluid infiltration.

In regionally metamorphosed pelites of the Kishtwar Window area of the Himalaya Range, Stäubli (1989, p. 83) noted that "...fibrolite is concentrated in zones of high, non-coaxial shear strain and occurs less abundantly in the strain-poor interfolial pods as random aggregates". Although not explicitly stated by Stäubli (1989), such features are similar to the fibrolite described by Wintsch and Andrews (1988) and by Flöttman (in press), thereby supporting a model of deformation-induced fibrolitization.

<u>Aluminum metasomatism</u>. Several papers have appealed to aluminum metasomatism for the formation of fibrolite. This topic is covered in Chapter 10.

<u>**The fibrolite ⇄ sillimanite reaction**</u>. In metapelites containing fibrolite + sillimanite, it is not uncommon to find these "phases" intergrown. In a few cases, sillimanite is fringed by fibrolite thereby suggesting that fibrolite replaced sillimanite (Ahmad and Wilson, 1982); however, in many papers describing this intergrowth, it was concluded that sillimanite replaced fibrolite (Chinner, 1961; Pitcher and Read, 1963, Fig. 4; Naggar and Atherton, 1970; Blümel and Schreyer, 1977; Grambling, 1981; Pigage, 1982; Vernon, 1987a). The truncation of fibrolite aggregates by the prism faces of sillimanite (Fig. 8.18) provides particularly clear evidence that sillimanite replaced fibrolite. On the basis of an increase in the modal sillimanite/fibrolite ratio with increasing metamorphic grade in the Ballachulish aureole, Scotland, Pattison (1989) concluded that fibrolite transformed to sillimanite. In the Mt. Raleigh pendant, British Columbia, Kerrick and Woodsworth (1989) found subhedral sillimanite within fibrolite aggregates. If Vernon's (1987a) textural interpretation of Figure 8.18 is correct, fibrolite cannot have formed during the waning stages of metamorphism (cf. Kerrick and Woodsworth, 1989). As noted by Vernon (1987a), it is difficult to establish *unequivocally* the relative timing of growth of sillimanite versus fibrolite from textural evidence alone. Because of the importance of this timing in the interpretation of the petrologic significance of fibrolite, further studies are necessary.

As discussed in Chapter 7, and illustrated in Figure 7.5, the molar grain boundary energy of fibrolite increases with decreasing fiber diameter. Because grain boundary energy is positive, the Gibbs free energy change for the fibrolite → sillimanite reaction is negative, and the magnitude of ΔG_r increases with decreasing fibrolite grain diameter. Accordingly, fibrolite is expected to recrystallize to coarser crystals. This recrystallization is in concert with most published interpretations of fibrolite-sillimanite intergrowths (see preceding paragraph).

RETROGRADE ALTERATION (REPLACEMENT) REACTIONS

Retrograde alteration of the aluminum silicates to Al-bearing phyllosilicates has long been recognized as common. The electron probe microanalyzer and the transmission electron microscope have been particularly useful for the identification of phases that formed by replacement of the aluminum silicates.

Muscovite and margarite are two very common phases formed from kyanite (Chinner, 1974; Gibson, 1979; Cooper, 1980; Baltatzis and Katagas, 1981; Yardley and Baltatzis, 1985; Janak et al., 1988) and

Figure 8.18. Euhedral porphyroblast of sillimanite (center) with sharp, planar grain boundaries. Note that the adjacent fibrolite appears to be truncated by grain boundaries of the sillimanite porphyroblast. (From Vernon, 1987a, Fig. 20).

andalusite (Velde, 1970; Guidotti and Cheney, 1976; Teale, 1979; Guidotti et al., 1979; Morand, 1988). Guidotti et al. (1979) and Baltatzis and Katagas (1981) contended that margarite is favored by the similarity in Al/Si ratios between margarite and the aluminum silicates. Even with the complete replacement of chiastolitic andalusite, the characteristic chiastolite cruciform structure of the precursor is commonly well preserved (e.g., Guidotti and Cheney, 1976). The alteration of kyanite and andalusite commonly occurs in veinlets, thereby supporting the theory that the K and Ca to form muscovite and margarite, respectively, and the OH component of these phases originated by the influx of aqueous fluids carrying solutes. However, in some cases the alteration products incorporated minor and trace elements from that of the reactant aluminum silicate. For example, Cooper (1980) described Cr-rich margarite that formed from Cr-rich kyanite, and Morland (1988) described altered chiastolite in which the elevated levels of Cr and V in margarite were inherited from the chiastolite.

Although colorless "white micas" are the most common replacement products of the aluminum silicates, other phases have been described. In altered kyanite from the Dalradian of Scotland, Chinner (1974) and Baltatzis and Katagas (1981) noted that plagioclase accompanies white mica as an alteration product. Ahn and Buseck (1988) and Ahn et al. (1988) described thin, submicroscopic lamellae of donbassite (aluminous chlorite); (001) of the donbassite is parallel to (110) of andalusite. They suggested that chloritization was favored by egress of fluid along the dominant (110) cleavage of andalusite. Furthermore, as with the replacement by margarite, they contended that the replacement reaction was favored by the similarity in Al/Si ratios of donbassite and andalusite. Because donbassite occurs as lamellae that are 0.1 to 0.3 μm thick, they emphasized that caution should be exercised in concluding that aluminum silicates are phase-pure based on the absence of included phases as deduced from routine petrographic examination. Wenk's (1983) description of sub-microscopic mullite inclusions within sillimanite epitomizes the importance of examining the aluminum silicates by electron microscopy.

Fleet and Arima (1985) described oriented hematite inclusions in sillimanite within a granulite from Labrador. Because the interface between the hematite and host sillimanite is parallel to a rational plane in both phases (i.e., (203) of hematite and (110) of sillimanite), they concluded that the hematite-sillimanite intergrowth was controlled by topotaxy. Fleet and Arima (1985) hypothesized that the hematite exsolved from sillimanite during one or more retrogressive events following the main granulite-facies metamorphism. It is notable that the sillimanite is Fe-rich (1.58 to 1.80 wt % Fe_2O_3). Thus, the sillimanite prior to exsolution apparently contained *substantial* Fe in solid solution. Accordingly, the P-T stability field of the original Fe-rich sillimanite would have been considerably expanded in relation to stoichiometric sillimanite (see Chapter 4).

In a limited number of localities (Macdonald and Merriam, 1938; Rose, 1957; Burt and Stump, 1983), inclusions of corundum have been found within andalusite. Rose (1957) described euhedral corundum crystals within andalusite in pegmatites at May Lake, Yosemite National Park, California. Also notable are crystals of diaspore within the andalusite. Because corundum and muscovite are intimately intergrown, he concluded that these phases crystallized contemporaneously. The formation of muscovite + corundum was believed to have occurred by "potash" metasomatism:

$$6 \text{ Andalusite} + 2\, H_2O + K_2O \rightarrow 2 \text{ Muscovite} + 3 \text{ Corundum}\ .$$

Diaspore was considered to have formed from the reaction of alumina and water. Burt and Stump (1983) described euhedral crystals of corundum within andalusite from a pegmatite in the Scott Glacier area, Antarctica. As with the May Lake locality, the corundum occurs intergrown with muscovite. Burt and Stump (1983) concurred with Rose's (1957) model for the formation of muscovite + corundum by potassium metasomatism.

CHAPTER 9

REACTION KINETICS AND CRYSTAL GROWTH MECHANISMS

INTRODUCTION

Disequilibrium and sluggish reaction kinetics represent one of the paramount problems with the aluminum silicates. This problem has frustrated experimentalists and has brought into question the validity of using the Al_2SiO_5 phase equilibria for interpreting the P-T conditions of metamorphism. This system epitomizes the "plague of the small ΔG's" (Fyfe et al., 1958, p. 22).

The dilemma introduced by sluggish kinetics is epitomized by Heitanen's (1956, p. 3) interpretation of aluminum silicates in rocks of the Boehls Butte area of Idaho: "In some thin sections all three modifications occur side by side, suggesting that they were crystallized close to the physical-chemical conditions in which all three may exist together." This conclusion was undoubtedly influenced by the general acceptance of an equilibrium model by most of the petrologic community. However, the coexistence of all three polymorphs instead could have resulted from a P-T path during metamorphism that traversed the stability fields of all three polymorphs, and the presence of coexisting polymorphs could be attributed to metastable persistence of the early-formed aluminum silicates. Because of sluggish reaction kinetics, the metamorphic P-T path of rocks of the Boehls Butte area *could* have been considerably removed from the P-T conditions of the triple point.

EXPERIMENTAL REACTION KINETICS

"... *placing numbers on this [Al_2SiO_5] diagram has given experimentalists great difficulty because reactions between the phases are slow and free energy differences very small.*"

W.S. Fyfe and W.S. MacKenzie (1969, p. 187)

For kinetic purposes, it is useful to separate metamorphic reactions into nucleation versus growth of solid products. Because of the prohibitively sluggish nucleation of product phases in the Al_2SiO_5 system, experimentalists have used starting material consisting of a mixture of the two solid phases involved in the particular reaction under investigation. Therefore, no experimental *nucleation* kinetic data exist for this system.

No single-crystal experimental data exist for the kyanite ⇄ sillimanite reaction. However, *qualitative* kinetic information can be extracted from the data of Bohlen et al. (ms.). As shown in Figure 3.27, there is a general decrease in the pressure uncertainty of the experimental brackets with increasing pressure and temperature. The bracket at the highest temperature (1000°C) is *remarkably* well constrained (±50 bars). The enhancement of reaction kinetics with increasing P and T is well known by experimentalists.

Limited data on *growth* kinetics are available from single crystal experiments on the kyanite ⇄ andalusite and andalusite ⇄ sillimanite reactions. Figures 3.15, 3.17, and 3.26 show that both the kyanite ⇄ andalusite and andalusite ⇄ sillimanite reactions are characterized by measurable forward reaction rates and little or no backward reaction rates. As discussed in Chapter 3, the negligible backward rates are responsible for the author's contention that reaction reversibility has not been convincingly demonstrated in single crystal experiments involving the aluminum silicates. A least-squares regression line fit to Heninger's (1984) data on the andalusite ⇄ sillimanite reaction (Fig. 3.26) suggests a temperature maximum of the reverse reaction. This maximum is similar to that determined in numerous single crystal experiments on reactions involving H_2O ± CO_2. The maxima of the backward reaction rates for the calcite + quartz ⇄ wollastonite + CO_2 and muscovite + quartz ⇄ andalusite + sanidine + H_2O reactions are clearly shown in Figure 9.1. As discussed by Schramke et al. (1987) the rate maxima of the reverse reactions can be rationalized by applying transition state theory to the net (overall) rate of reaction. Assuming that the temperature dependence of the rate constant is represented by the classical Arrhenius relation, the net rate of a heterogeneous reaction can be expressed as (Fisher and Lasaga, 1981):

$$R = A \exp(-E_A/RT)[1 - \exp(n\Delta G_r/RT)] \; . \qquad [9.1]$$

For the forward reaction, the term involving ΔG_r yields a progressive increase in R with increasing temperature above the equilibrium boundary. As temperature is decreased from the equilibrium boundary, R of the reverse reaction first increases because of the increase in ΔG_r; however, with further temperature decrease, the exponential term containing E_A becomes dominant and, thus, R approaches zero. These combined factors are responsible for the rate maximum of the reverse reaction.

An alternative explanation for the asymmetry of the forward versus backward reaction rates could be made on the basis of possible differences

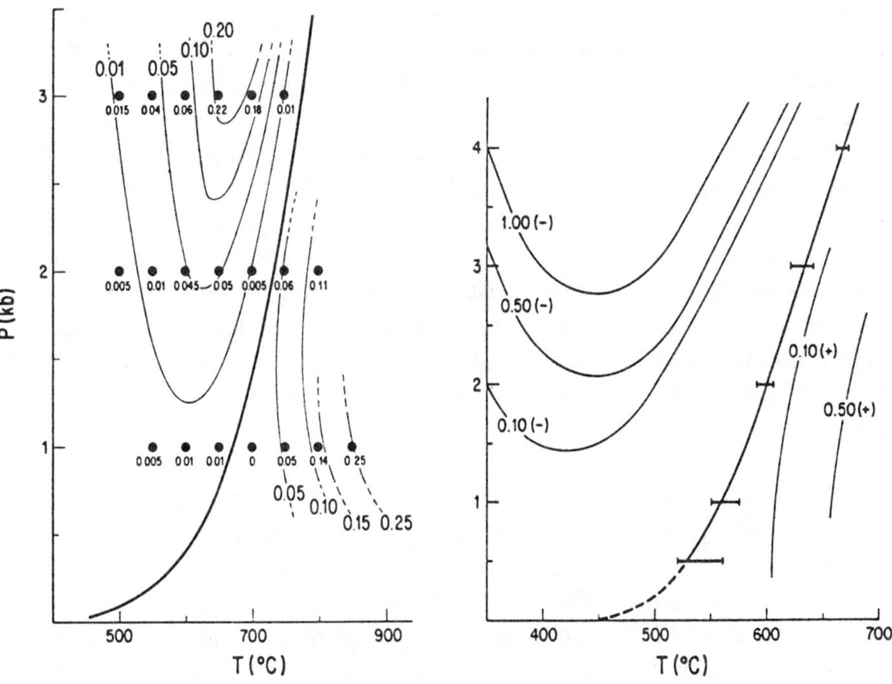

Figure 9.1. Rate maxima of reverse reactions. In both diagrams the isopleths illustrate the fractional extent of the reaction, and the heavy lines are the univariant equilibrium boundaries. (a) calcite + quartz ⇄ wollastonite + carbon dioxide. (From Tanner et al., 1985, Fig. 5). (b) muscovite + quartz ⇄ andalusite + K-feldspar + water. The horizontal bars are the experimental equilibrium brackets of Chatterjee and Johannes (1974). (From Schramke et al., 1987, Fig. 2).

in the rates of dissolution of andalusite (for the andalusite → sillimanite reaction) versus growth (for the sillimanite → andalusite reaction). It could be argued that the dissolution rate of andalusite is more rapid than the rate of growth. If so, this factor could explain the large differences in the forward versus backward reaction rates evident in Figures 3.17 and 3.26. However, the experimental data of Schramke et al. (1987) on the muscovite + quartz ⇄ andalusite + sanidine + H_2O reaction suggest that for andalusite there are no significant differences in the rates of growth (during dehydration) versus dissolution (during hydration). Therefore, it may be unlikely that differences in the rates of andalusite growth versus dissolution are a significant factor contributing to the differences in rates of the forward versus backward andalusite ⇄ sillimanite reaction.

Transition state theory arose from the theoretical analysis of homogeneous reactions in gases; thus, there are questions regarding the applicability of transition state theory to complex heterogeneous reactions. More realistic kinetic modeling of reactions involving solids, such as

Avrami's (1939, 1940) treatment, will undoubtedly improve the kinetic modeling of metamorphic reactions. Nevertheless, transition state theory rationalizes the limited kinetic data for the aluminum silicates. In addition to explaining the rate maxima of back reactions, the relative rates of the three reactions in the Al_2SiO_5 system are explicable using transition state theory. Equation [9.1] shows that the net reaction rate is an exponential function of ΔG_r. For any selected departure from the equilibrium temperature at 1 bar, the data of Hemingway et al. (ms.) show that $|\Delta G_r|$ for the andalusite $\rightleftarrows$ sillimanite reaction is 2.5 times lower than that of the kyanite $\rightleftarrows$ andalusite reaction and 3.5 times lower than the kyanite $\rightleftarrows$ sillimanite reaction. Thus, the rate of the andalusite → sillimanite reaction is expected to be significantly slower than that of the other two polymorphic reactions.

The effects of temperature versus pressure on reaction kinetics can be evaluated using Holdaway's (1971) single crystal data on the andalusite → sillimanite reaction. As shown in Figure 3.17, the rate of the andalusite → sillimanite reaction is faster at 3.6 kbar than at 1.8 kbar. For example, at a temperature 100°C above the intersection of the line fitted by least-squares regression with the line of zero weight change, the rate is about 1.8 times faster. This rate increase occurs in spite of a temperature decrease of about 110°C. Therefore, it is concluded that, for the andalusite → sillimanite reaction, pressure is a more significant kinetic factor than temperature.

A. Nitkiewicz (unpub.) carried out experiments on the effects of grain size on the kinetics of the andalusite → sillimanite transformation. As shown in Figure 9.2, runs with two contrasting grain sizes of fibrolite (30-44 μm and < 38 μm) were carried out at 721°C and 748°C. These data show a distinct increase in reaction rate with diminished grain size of fibrolite. In comparing a fixed reaction time (e.g., 1000 h), it is apparent that increasing temperature increases reaction rate and also amplifies the differences in reaction rates between the coarse-grained versus fine-grained fibrolite fractions. Nitkiewicz's experiments suggest that the surface area of powdered sillimanite has an important effect on reaction rate (cf. Wood and Walther, 1983). These results have significant implications for the interpretation of the andalusite $\rightleftarrows$ sillimanite reaction in the laboratory and in nature.

For prograde reactions, the growth rate of product phases will be inversely proportional to their size. Because the nuclei of sillimanite are exceedingly small, the initial growth rate of sillimanite is expected to be

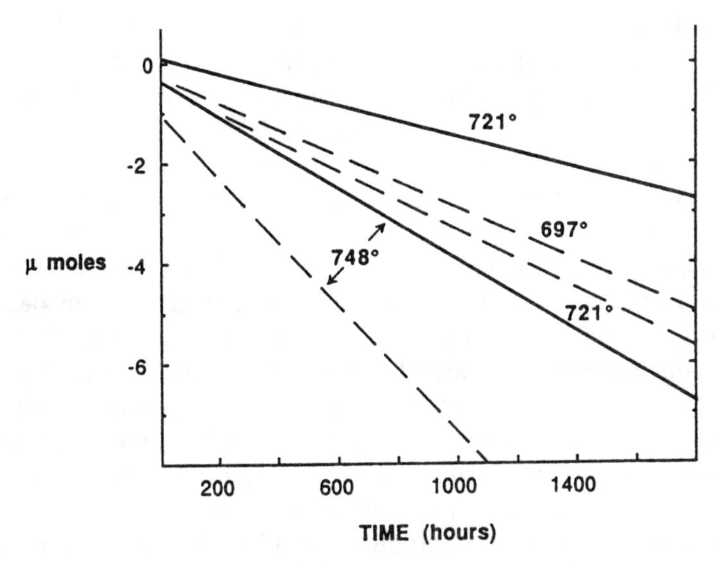

Figure 9.2. Extent of the andalusite → sillimanite reaction (measured by weight changes of andalusite single crystals) as a function of time. The experiments were carried out at 3 kbar using two different grain size fractions of "sillimanite": solid lines = 38-44μm, dashed lines << 38μm. (From Nitkiewicz, unpub.).

relatively rapid, and will diminish as grain size (and, thus, molar surface area) increases. I contend that this phenomenon provides an explanation for the fact that product phases at isograds, even in the short-lived development of small contact aureoles, are generally coarse grained. At the sillimanite isograd in the Mt. Raleigh pendant, the sillimanite crystals replacing andalusite average about 400 μm across (Fig. 8.6). I believe that following sillimanite nucleation, incipient growth of the sillimanite was rapid because of the high molar surface area of the first- formed sillimanite crystals.

FIELD EVIDENCE FOR REACTION KINETICS

"*In the present case [the Al_2SiO_5 system], there is much evidence from rocks pointing to disequilibrium and sluggish kinetics.*"

W.S. Fyfe (1969, p. 291)

This conclusion is based on the common coexistence of the aluminum silicates in metamorphic rocks and is supported by numerous field studies.

From a textural and spatial analysis of metamorphic minerals in the aureoles of the Donegal Granites, Ireland, Naggar and Atherton (1970) concluded that although the metamorphic minerals crystallized under equilibrium conditions, reaction overstepping was common. They concluded that the temporal sequence of the growth of the aluminum silicates was: kyanite → fibrolite → sillimanite. They believed that there were three successive reactions, each forming a unique mineral assemblage; i.e. the first reaction formed garnet, staurolite, kyanite and biotite, the second reaction formed andalusite and cordierite, and fibrolite and/or sillimanite formed from the third reaction. They concluded that "... the second reaction may be completely overstepped ..." as in pelitic xenoliths within the Thorr Granodiorite (Naggar and Atherton, 1970, p. 579). This conclusion was *apparently* based on the fact that andalusite (produced from the second reaction) is absent in these xenoliths. However, they described these xenoliths as "thoroughly recrystallized." Thus, it is possible that andalusite existed prior to the "recrystallization" that produced the highest grade (fibrolite-sillimanite) assemblages. Evidence to support this interpretation is found in the Ardara aureole, where the transformation of metapelites from the andalusite zone to the sillimanite zone was apparently accompanied by wholesale textural reconstitution (Kerrick, 1987b).

In the analysis of metamorphic reaction kinetics we must be aware of the limitations in extracting *nucleation* kinetic information from examination of metamorphic reactions. Fisher aptly summarized why nucleation kinetics is virtually intractable from petrographic analyses of metamorphic rocks: "The reason why the first andalusite nuclei formed at one point rather than another can no longer be determined; any hints to the answer have been obliterated" (Fisher, 1970, p. 100).

Although he argued for the equilibrium crystallization of fibrolite in an earlier paper on the Dalradian metamorphism of Scotland (Chinner, 1961), he subsequently concluded: "... the coexistence of sillimanite-andalusite or sillimanite-kyanite over wide tracts of country is unlikely to be a compositional effect and strongly suggests disequilibrium" (Chinner, 1966b, p. 167).

In a study of pelites that were thermally metamorphosed by intrusives of the Coast Batholith of British Columbia, Hollister (1969) concluded that disequilibrium crystallization was a major phenomenon in the formation of the aluminum silicates. As reviewed in the section on the andalusite → kyanite reaction in Chapter 8, he argued that prograde contact metamorphism was essentially isobaric. Based on textural evidence, it was concluded that there were two polymorphic reaction sequences, i.e., andalusite → kyanite → sillimanite, and andalusite → fibrolite. Hollister (1969) hypothesized that isobaric metamorphism occurred at a pressure above the Al_2SiO_5 triple point. He proposed that andalusite metastably crystallized in the P-T stability field of kyanite (point **a** in Fig. 9.3), andalusite transformed to kyanite when the prograde path intersected the metastable extension of the andalusite-sillimanite equilibrium (point **k** in Fig. 9.3), and sillimanite *stably* formed from kyanite upon intersection of the kyanite-sillimanite equilibrium (point **s**). Although the author finds no indisputable arguments against Hollister's (1969) model, there are reasonable alternatives. For example, it is possible that andalusite formed stably (within the andalusite P-T stability field) prior to contact metamorphism. Woodsworth (1979) proposed that during the Cretaceous, *significant* crustal thickening of the Coast Plutonic Complex could have been achieved by emplacement of the intrusives and possibly overthrusting. He proposed about 13 km of crustal thickening, which would translate to a lithostatic pressure increase of about 3.5 kbar. It is possible that in the Kwoiek pendant studied by Hollister (1969), this pressure increase postdated crystallization of andalusite. If so, the andalusite → kyanite transformation observed in rocks of the Kwoiek pendant could represent equilibrium whereby kyanite formed from andalusite upon crossing the kyanite-andalusite equilibrium (path A in Fig. 8.12). Hollister's conclusion that sillimanite formed from the equilibrium transformation of kyanite (point **s** in Fig. 9.3) was defended "... because a metastable polymorph would not be expected to nucleate in the presence of a more stable polymorph" (Hollister, 1969, p. 366). The validity of this conclusion is supported by textural data suggesting that kyanite was *replaced* by fibrolite (see Hollister, 1969, Fig. 2D).

Holdaway (1978) used kinetic arguments to explain the distribution of aluminum silicates in the Picuris Range, New Mexico. He derived the paragenetic sequence from textural details and concluded that prograde metamorphism occurred at pressures below the triple point, and that the prograde "path" went from the kyanite P-T stability field at lower grades to the sillimanite field at the highest grades. He noted that, although the transformation of kyanite to andalusite and/or sillimanite occurred in several specimens, kyanite is the only Al_2SiO_5 polymorph in many samples. Because of the lack of higher-grade aluminum silicates (i.e., andalusite or sillimanite) in such samples, Holdaway (1978, p. 1406) concluded that "... neither the kyanite-andalusite nor the kyanite-sillimanite phase boundary was surpassed very much during metamorphism." He stated that the proposed metastable persistence of kyanite could occur if none of the Al_2SiO_5 equilibria were overstepped by

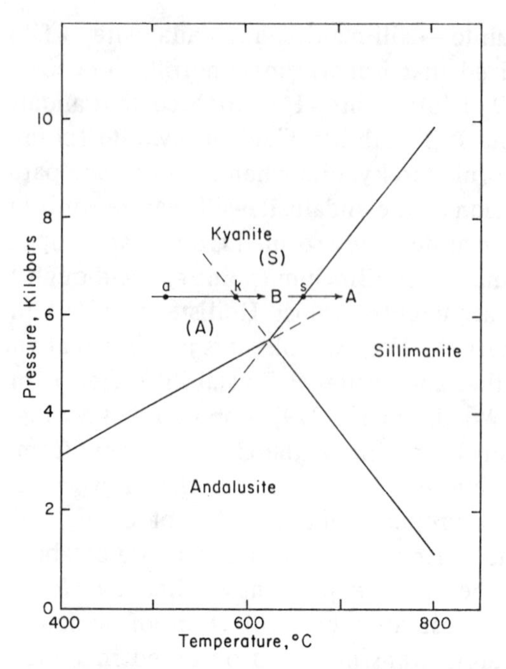

Figure 9.3. Hollister's (1969) inferred isobaric path for prograde metamorphism in the Kwoiek area, British Columbia. Points along this path refer to the first appearance of: a = andalusite, k = kyanite, s = sillimanite. (From Hollister, 1969, Fig. 3).

more than about 25°C. To explain the irregular spatial distribution of specimens containing kyanite transformed to andalusite and/or sillimanite, he alluded to "... local variations in access to water and other kinetic factors" (p. 1410). He provided no further elaboration on the factors controlling the "sporadic" transformation of kyanite to the other polymorphs. Holdaway (1978) noted that replacement of andalusite by sillimanite is the rarest of all the polymorphic replacement reactions in the Picuris Range. He argued that the rarity of this transformation is commensurate with the fact that the andalusite-sillimanite equilibrium has a lower ΔS_r than the other two polymorphic reactions; consequently, transition state theory predicts that, comparing all the Al_2SiO_5 polymorphic transitions, this reaction is most sluggish.

Grambling and Williams (1985) described an abrupt kyanite-sillimanite transition in the Rio Mora uplift, northern New Mexico (Fig. 9.4). Because coexisting kyanite + sillimanite have virtually identical compositions, the narrow (<180 m thick) zone of coexisting kyanite + sillimanite is attributed to the metastable persistence of kyanite into the sillimanite stability field. Assuming that the kyanite-sillimanite equi-

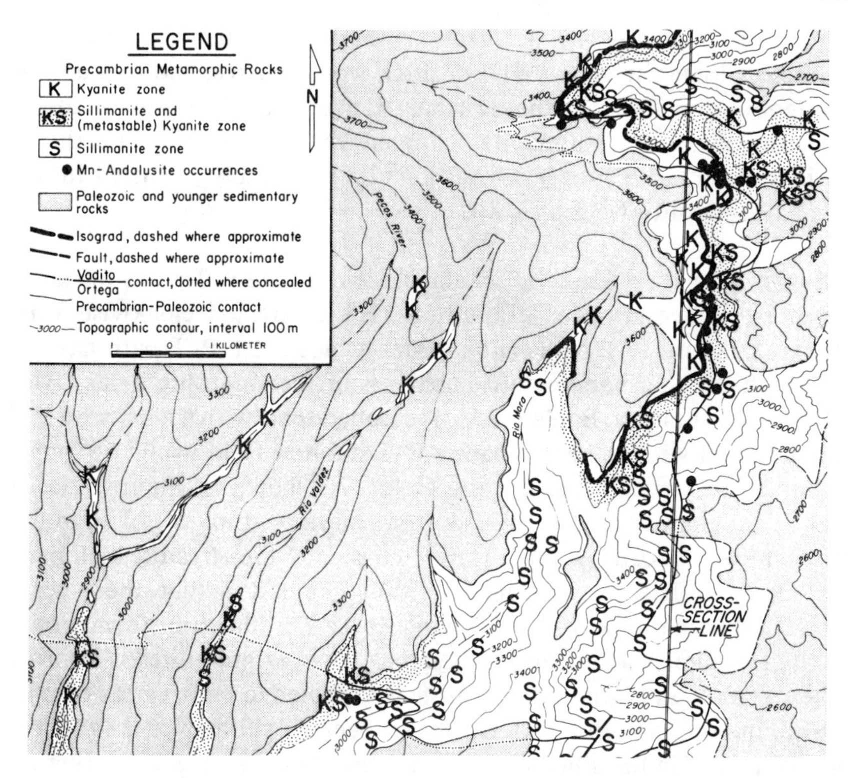

Figure 9.4. The distribution of kyanite, sillimanite and Mn-andalusite in the Rio Mora uplift, north-central New Mexico. Note that Mn-andalusite is confined to a manganiferous horizon adjacent to the Vadito-Ortega contact. (From Grambling and Williams, 1985, Fig. 2).

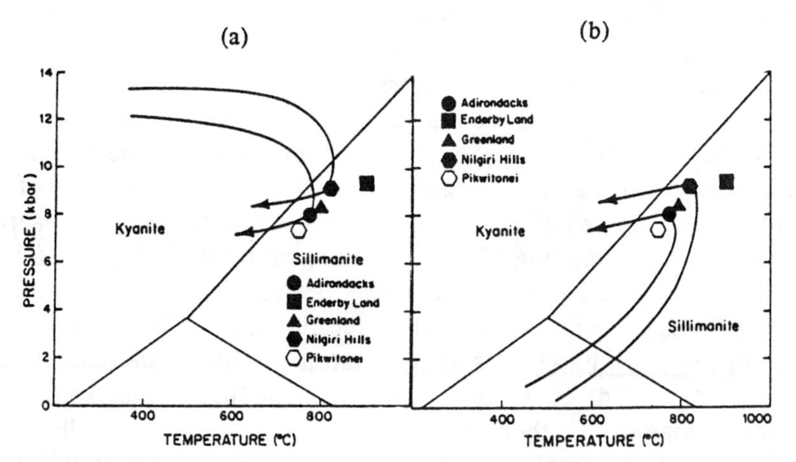

Figure 9.5. Bohlen's (1987) depiction of P-T conditions of granulites from several localities. (a) "Clockwise" P-T path resulting from nearly isobaric cooling from peak metamorphic conditions. (b) "Counterclockwise" P-T path consistent with magmatic heating of the lower crust before and during tectonic "loading", followed by nearly isobaric cooling from peak conditions. (From Bohlen, 1987, Fig. 7).

librium is intersected by a *linear* geothermal gradient ≤ 40°C/km, a kyanite + sillimanite zone ≤ 180 m thick would correspond to maximum overstepping of 7°C. In the Truchas Range, the 200 m-thick zone of coexisting kyanite + sillimanite, coupled with a maximum geothermal gradient of 40°C/km, suggests that kyanite metastably persisted into the sillimanite stability field by ≤ 8°C.

Bohlen (1987) used kinetic arguments to evaluate P-T paths during the metamorphic evolution of granulites. He contrasted clockwise versus counterclockwise P-T paths (Fig. 9.5). A clockwise P-T path has been popular in the thermobarometric analyses of metamorphic belts. Along with other arguments, Bohlen (1987) reasoned that, with a clockwise path, the time spent within the sillimanite P-T stability field would be short in relation to that within the kyanite field. As shown in Figure 9.5a, the "peak" P-T conditions determined from thermobarometry of granulites from various localities are fairly close to the kyanite-sillimanite equilibrium. Therefore, Bohlen (1987) concluded that there was a "vanishingly small" (< 100 J) driving force for the transformation of kyanite to sillimanite during the metamorphism of granulites. According to the P-T path in Figure 9.5a, kyanite is expected to be far more common than sillimanite in granulites. However, because sillimanite is in fact the dominant aluminum silicate in granulites, Bohlen (1987) preferred a counterclockwise P-T path, whereby a large portion of the sillimanite stability field was traversed. Much of this path (Fig. 9.5b) is well removed from the kyanite-sillimanite and andalusite-sillimanite equilibria, thereby explaining the prevalence of sillimanite in granulites. However, the data points in Figures 9.5a,b are about 75°C above the kyanite-sillimanite equilibrium. In the 8-10 kbar pressure range, ΔG_r of the kyanite-sillimanite reaction 75°C above the equilibrium is 700-800 J. Therefore, the free energy "driving force" for the kyanite → sillimanite reaction is considerably larger than that suggested by Bohlen (1987). Because of the virtually univariant behavior of the kyanite → sillimanite isograd reaction in several areas (Grambling 1981; Grambling and Williams, 1985; Ghent et al., 1980), Bohlen's (1987) conclusions regarding overstepping of the kyanite-sillimanite equilibrium are questionable. Bohlen's (1987) conclusions regarding the rates of the kyanite → sillimanite reaction can be evaluated in light of kinetic implications deduced from kyanite → sillimanite isograds. In their description of the kyanite → sillimanite isograd in the Mica Creek area of British Columbia, Ghent et al. (1980) suggested that kyanite and sillimanite coexist over a zone that is 10-20 meters thick. Their thermobarometric analysis suggests that, in the present plane of exposure, thermal gradients normal to the isograd were *approximately* 17°C/km. The zone of coexisting kyanite + sillimanite would thus correspond to 2-4°C overstepping. Accordingly, the sillimanite isograd at Mica Creek suggests that complete conversion of kyanite to sillimanite occurred within 5°C of the kyanite-sillimanite equilibrium boundary. Grambling and Williams (1985) concluded that the kyanite → sillimanite isograd reaction in the Truchas and Rio Mora uplifts in north-central New Mexico went to completion within a temperature interval less than 10°C. Thus, with the P-T path shown in Figure 9.5a, Bohlen's (1987) suggestion that the kyanite → sillimanite reaction progress would be sluggish in the "relatively short" period of a "few m.y." is unsubstantiated.

Kerrick and Woodsworth (1989) analyzed the kinetics of the andalusite → sillimanite transformation in the Mt. Raleigh pendant (British Columbia). As discussed in the section on the andalusite → sillimanite reaction in Chapter 8, this pendant displays a 1.5 km-wide zone above the sillimanite isograd in which little andalusite → sillimanite reaction occurred. As shown in Figure 8.8, reaction progress in this zone was less than 5%. Significant reaction occurred at distances above 2 km from the sillimanite isograd (Fig. 8.8). Kerrick and Woodsworth's thermobarometric analysis suggests that the 1.5 km zone of negligible andalusite → sillimanite reaction suggests that the andalusite-sillimanite equilibrium was overstepped by about 30°C before significant reaction occurred. This is compatible with the kinetic analysis of Wood and Walther (1983). In defining overstepping as the zone in which reactants and products coexist, they concluded that for the andalusite → sillimanite reaction, overstepping would be 19°C (for reaction control) and 40°C (for transport control). Because of the negligible andalusite → sillimanite reaction in a 1.5 km-wide zone immediately above the sillimanite isograd in the Mt. Raleigh pendant, Kerrick and Woodsworth cautioned that very careful petrographic analysis is necessary to delineate specimens with incipient andalusite → sillimanite reaction. In the case of the Mt. Raleigh pendant, failure to recognize the very small amount (≈ 1 modal %) of sillimanite at the sillimanite isograd would have led to significant mislocation (1.5 km!) of this isograd.

In the Trois Seigneurs Massif in southern France, Wickham (1987) described a prograde andalusite → sillimanite transformation with characteristics similar to those of the Mt. Raleigh pendant. Specifically, andalusite metastably persisted about 1 km above the sillimanite isograd. Marked increase in the andalusite → sillimanite reaction progress about 1 km above the sillimanite isograd results in an abrupt "andalusite-out" isograd. Considering his postulated *maximum* temperature gradient of 110°C/km, the 1 km-thick andalusite + sillimanite zone would correspond to a maximum temperature of about 110°C. Wickham (1987, p. 127) preferred lower thermal gradients of 80-100°C/km. Nevertheless, the 1 km zone of coexisting andalusite + sillimanite would imply *substantial* overstepping of the andalusite-sillimanite equilibrium.

In the Ballachulish aureole, Scotland, Pattison (1989) noted that the isograd marking the first appearance of sillimanite + fibrolite is coincident with the isograd representing the muscovite + quartz dehydration reaction. However, between the sillimanite-fibrolite isograd and the intrusive contact, andalusite persists and appears to be in textural equilibrium. Throughout this zone, andalusite forms more than 95 modal % of the aluminum silicates. Thus, Pattison (1989) implied that andalusite metastably persisted to temperatures significantly above the andalusite-sillimanite equilibrium. From a petrogenetic grid constructed for metapelites, Pattison (1989) concluded that metamorphism in the Ballachulish aureole occurred at a pressure of about 3 kbar (Fig. 9.6). With an equilibrium model, the isograd sequence suggests that the prograde P-T path crossed the invariant point formed by the intersection of the andalusite-sillimanite equilibrium with the muscovite + quartz = Al_2SiO_5 + K-feldspar + water equilibrium (filled square in Fig. 9.6). However, as shown in Figure 9.6, the andalusite-sillimanite equilibrium of Richardson et al. (1969) is incompatible with this invariant point for isobaric metamorphism at 3 kbar. Pattison (1989) addressed this dilemma by concluding that "... it seems clear that the Richardson et al. (1969) boundary is too high ... The Holdaway (1971) boundary remains a possibility if one argues for a kinetic delay in the growth of sillimanite ~100°C at 3 kbar. However, it has been decided to use ... [the dot-dash curve in Figure 9.6] ... as the best practical boundary between sillimanite-bearing and sillimanite-absent rocks, even though it is possible that this boundary may be particular to the sillimanite crystal size, degree of Al-Si disorder, minor element composition and rate of heating in the Ballachulish aureole." Pattison's placement of the andalusite-sillimanite equilibrium is questionable in light of the ±0.5 kbar uncertainty in the pressure estimate coupled with the fact that there is no textural evidence suggesting that the first appearance of sillimanite + fibrolite is correlative with the andalusite-sillimanite equilibrium. In fact, the presence of andalusite in textural equilibrium throughout the sillimanite-fibrolite zone suggests that the andalusite → sillimanite reaction did not occur in the Ballachulish aureole.

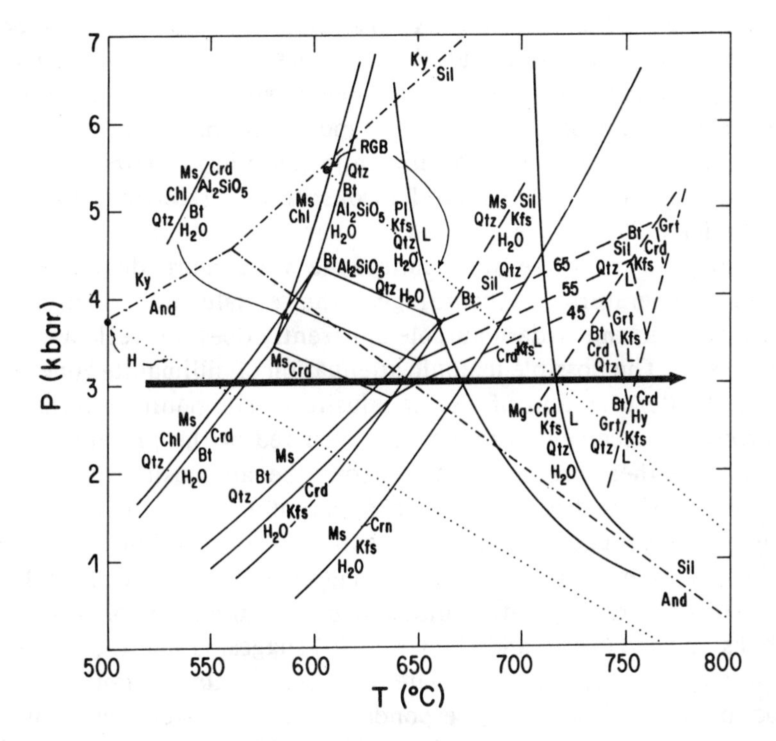

Figure 9.6. Pattison's (1989) P-T grid showing equilibria relevant to metamorphosed pelites in the Ballachulish aureole, Scotland. The numbers refer to the atomic [Mg/(Mg + Fe)] ratio in cordierite for equilibria involving this phase. The heavy line is Pattison's hypothesized isobaric prograde path in the Ballachulish aureole. The dotted line labeled "H" is Holdaway's (1971) andalusite-sillimanite equilibrium whereas the dotted line labeled "RGB" is the andalusite-sillimanite equilibrium of Richardson et al. (1969). Pattison's suggested location of the Al_2SiO_5 equilibria are shown with the dot-dash lines. (From Pattison, 1989, Fig. 5).

The measured variation in the degree of overstepping of the andalusite → sillimanite reaction between various areas offers a puzzling kinetic dilemma. Because the heating rate of contact metamorphism is expected to be several orders of magnitude faster than that of regional metamorphism, the andalusite → sillimanite reaction in low-pressure regional metamorphism should display considerably less overstepping than that in contact metamorphism. However, geologic evidence is not compatible with this conclusion. Kerrick and Speer's (1988) analysis suggests that in the Ardara aureole (Ireland), a zone of coexisting andalusite + sillimanite (if it exists) spans a *maximum* temperature interval of 50°C. In the Waterville-Vassalboro area of Maine, the zone of coexisting andalusite + sillimanite would correspond to a maximum temperature interval of only about 12°C (Kerrick and Speer, 1988). In the Ardara aureole, there are significant structural and textural differences between rocks of the andalusite and sillimanite zones. As discussed by Akaad (1956, p. 386), metamorphic rocks within about 200 m of the intrusion have structural features suggesting significant "mobilization"; however, metamorphic rocks outside of this zone lack evidence of mobilization. I noted that the transition from the andalusite to sillimanite zones in this aureole: "... was accompanied by wholesale textural reconstitution" (Kerrick, 1987b, p. 243). Holder (1979) suggested that, in comparison to aureole rocks at lower grade, rocks of the sillimanite zone were more ductile due to increased temperature. Accordingly, the andalusite → sillimanite transformation in the Ardara aureole apparently does not reflect a static thermal event. The possible tectonic attenuation of sillimanite zone rocks complicates interpretation of the andalusite → sillimanite isograd as a polymorphic replacement reaction that occurred at the present level of exposure. Furthermore, in passing from the andalusite zone to the sillimanite zone, marked increase in apparent mobility, coupled with textural evidence suggesting wholesale textural reconstitution, suggest that the transformation of andalusite to sillimanite could have been catalyzed by the infiltration of magmatic fluids and/or partial melting of sillimanite zone rocks. Akaad (1956) and Holder (1979) suggested that the boundary between the sillimanite and andalusite zones represents a tectonic "hiatus." It is possible that this hiatus corresponds to faults that were active during intrusion of the Ardara pluton (Holder, 1979, p. 119). Thus, interpretation of the andalusite → sillimanite isograd in the Ardara aureole is complicated by the possibility of tectonic disruption. In the Fanad aureole of Donegal, Ireland, the sillimanite zone is characterized by andalusite partially paramorphed by sillimanite. In contrast with the Ardara aureole, contact metamorphism in the Fanad aureole appears to have occurred in a static environment (Pitcher and Berger, 1972, p. 303). Thus, in comparison with the Ardara aureole, the *partial* transformation of andalusite to sillimanite in the Fanad aureole may better reflect *static* contact metamorphism, and thus the andalusite → sillimanite reaction in the Fanad aureole may be a more representative indicator of the sluggish andalusite → sillimanite reaction kinetics in contact metamorphism.

In the Mt. Raleigh pendant (British Columbia), the prograde thermal gradient deduced by Kerrick and Woodsworth (1989) suggests a temperature interval of 60°C over the 3 km-wide zone of coexisting andalusite + sillimanite. The large overstepping of the andalusite → sillimanite reaction in this pendant (British Columbia) is puzzling in light of the relatively extended thermal history implied by the interpretation of Woodsworth (1979) and Kerrick and Woodsworth (1989) that metamorphism was *regional*. If this interpretation is valid, the author contends that the regional heating may have been relatively rapid. Regional heating rates may have been enhanced by emplacement of intrusives during regional metamorphism. The large proportion of intrusive rocks in the Coast Plutonic Complex would be compatible with the hypothesis of increased regional heat flow due to magmatism. Indeed, enhanced heat flow by intrusives may have been an important factor in regional metamorphism in some metamorphic belts (De Yoreo et al., 1989).

Strain-assisted reactions

Although there have been suggestions that strain affects the rates of metamorphic transformations (Pitcher, 1965), there have been few field studies conclusively demonstrating that strain enhances the kinetics of Al_2SiO_5 polymorphic reactions.

Grambling's (1981) analysis of regional metamorphism in north-central New Mexico provides textural evidence of a strain-assisted mechanism for the kyanite → andalusite reaction. "Rarely strained" andalusite paramorphically replaced kyanite that is "commonly deformed" (Grambling, 1981, p. 703).

From a study of dislocations in natural and experimentally deformed aluminum silicates, Doukhan et al. (1985) proposed a strain assisted mechanism for the kyanite → sillimanite transformation. The experiments on which their model was based were carried out in a solid medium device. A sample of sillimanite was initially strained by 8% at P-T conditions well within the stability field of sillimanite ($P_{confining}$ = 5 kbar, T = 700°C) and subsequently strained by an additional 4% at P-T conditions ($P_{confining}$ = 9 kbar, 700°C) within the kyanite stability field but close to the kyanite-sillimanite equilibrium. They found that lamellae of kyanite nucleated on widely dissociated dislocations in sillimanite. Electron diffraction showed that the transformation was epitaxial; i.e., (100) of kyanite parallels (210) of sillimanite. As shown in Figure 9.7, they provided a "tentative" model for this transformation by considering that the stacking faults sever AlO_6 octahedral chains but not the SiO_4 and AlO_4 tetrahedral chains (cf. Figs. 9.7a,b). The formation of kyanite from the faulted structure (9.7b) is depicted in Figures 9.7c,d. The stress required for this process is that needed to overcome the Peierls forces (i.e., lattice friction forces) on the dissociated dislocations. In light of the low strain rates during the deformation of rocks, they suggest that relatively low flow stresses (less than 1 kbar) are sufficient to exceed the Peierls forces and thus induce the sillimanite → kyanite transformation. They

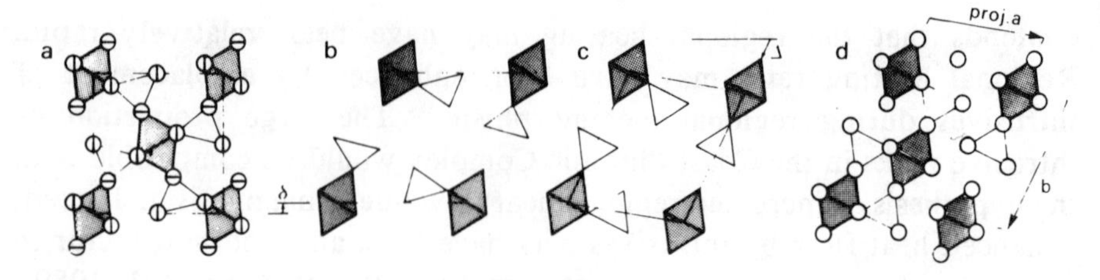

Figure 9.7. Possible model for the transformation of sillimanite to kyanite. (a) (001) projection of the sillimanite structure showing octahedra (shaded) and tetrahedra (unshaded). (b) (010) stacking fault within the central portion of the cell. Note that the octahedral chain present in (a) has been destroyed. (c) Two of the tetrahedra in (b) have been displaced. (d) Projection of the kyanite structure. (From Doukhan et al., 1985, Fig. 11).

noted that, in comparison to a "static" transformation mechanism involving diffusion, strain-induced transformations are expected to be most prevalent at low temperatures. Thus, they suggested that the low-temperature kyanite → andalusite reaction is the best candidate for a strain-induced Al_2SiO_5 polymorphic reaction in metamorphism. However, they caution (p. 94) "... our TEM investigations on rocks containing at least two Al_2SiO_5 polymorphs in contact have not provided any clear evidence that such strain-induced polymorphic transformations occur in nature."

Carey et al. (in press) appealed to this mechanism in their reexamination of the Al_2SiO_5 polymorphs of the Boehls Butte area of northern Idaho. They concluded that this area records three metamorphic events: M1, M2 and M3. Their analysis suggests that kyanite and sillimanite crystallized during the Barrovian-type M2 event. Thermobarometric analysis implies that the low-pressure andalusite-cordierite assemblage formed during virtually isothermal decompression subsequent to M2. Textures suggest that all andalusite was produced by the paramorphic replacement of kyanite. They contend "... that andalusite developed preferentially along cleavage planes, strained areas or other zones of weakness within kyanite ..." such that "... strain energy must have been critical to overcoming the kinetic barriers to [kyanite → andalusite] inversion." I question the textural evidence supporting their hypothesis that andalusite formed from *strained* kyanite. Carey et al. (in press) presented one photomicrograph supporting their postulated strain-induced reaction kinetics. Although it supports their conclusion that andalusite preferentially developed along cleavage planes, it provides no evidence that the reactant kyanite was (or is) strained.

In a study of the kyanite → andalusite transformation within Al_2SiO_5-bearing segregations of the Lepontine Alps of Switzerland (Kerrick, 1988), the author found no correlation between the amount of strain in kyanite and the extent of the kyanite → andalusite transformation. In particular, specimens were found with strongly bent kyanite that is unaltered to andalusite, whereas other samples showed significant replacement of *unstrained* kyanite by andalusite (see Chapter 10). As reviewed in Chapter 5, my analysis (Kerrick, 1986) suggests that there are no cogent arguments supporting the theory that dislocation strain energy affects the stability relations of the aluminum silicate polymorphs during metamorphism. It is my opinion that the study of Carey et al. (in press) does not conclusively demonstrate strain-induced polymorphic transformation. Nevertheless, their contribution will stimulate further research on this potentially important problem.

FIBROLITE METASTABILITY

There has been a long-standing controversy regarding the possibility that fibrolite crystallized and persisted as a metastable phase during metamorphism of pelitic lithologies. This problem is epitomized by the following selected quotes. Pitcher (1965, p. 333) concluded: "... it seems necessary to question whether equilibrium conditions are ever represented by fibrolite ... especially when the new phase [fibrolite] arises by a different method of growth, as a result of an independent reaction and after an interval of time. We are probably dealing here with the kinetics of polymorphic transformation." The lack of direct replacement of reactant minerals by product phases lead Atherton (1965) to question the concept of *progressive* metamorphism. He suggested that metamorphic rocks may have transformed into their present mineral assemblages *directly* from the protolith that existed prior to metamorphism. Chinner concluded: "... sillimanite [fibrolite] textural relations suggest that it may not be as reliable an indicator of physical conditions as andalusite and kyanite" (Chinner, 1966, p. 111). "In my opinion fibrolite is always a metastable mineral which probably forms from the reaction of a mineral or mineral assemblage which has been made very unstable by overstepping of the equilibrium boundary" (Holdaway, 1971, p. 127-128).

Contact metamorphism

In the Kiglapait aureole of Labrador, Speer (1982) noted that fibrolite is widespread, occurring considerably further from the intrusion than the isograd marking the first appearance of sillimanite. Thus, he intimated that fibrolite may not have formed under equilibrium conditions (Speer,

1982, p. 1903-1904). Berg and Docka (1983) also noted the encroachment of fibrolite into the outer portion of the Kiglapait aureole.

I examined the question of fibrolite metastability in two contact aureoles in Donegal, Ireland. As with the study of Speer (1982), I noted that the fibrolite isograd is considerably further from the intrusive than the sillimanite isograd (Figs. 9.8, 9.9). Garnet-biotite geothermometry in the Ardara aureole (Fig. 9.10) suggests that the maximum temperature at the fibrolite isograd was approximately 100°C below that of the sillimanite isograd. Providing that the sillimanite isograd represents a close approximation to the univariant andalusite-sillimanite equilibrium, Figure

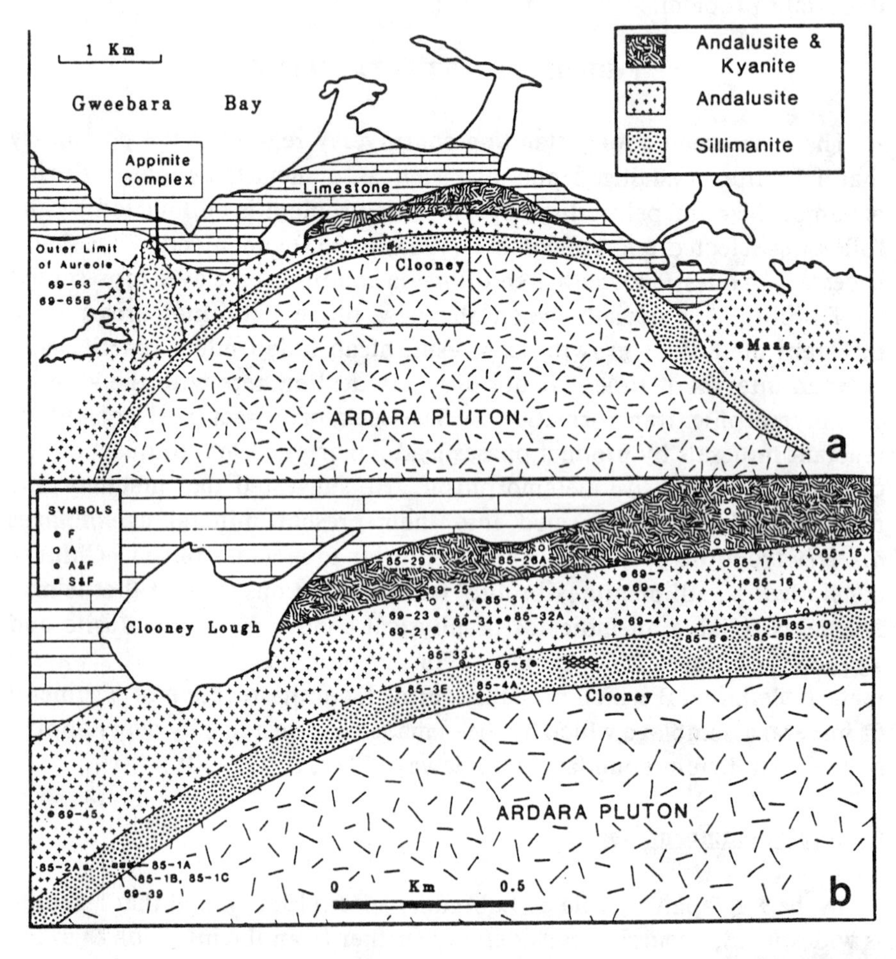

Figure 9.8. (a) Geology of the northern portion of the Ardara Pluton. The rectangular area encompassing Clooney is shown in the lower diagram. (b) Aluminum silicate zones in the Ardara aureole. (From Kerrick, 1987b, Fig. 1).

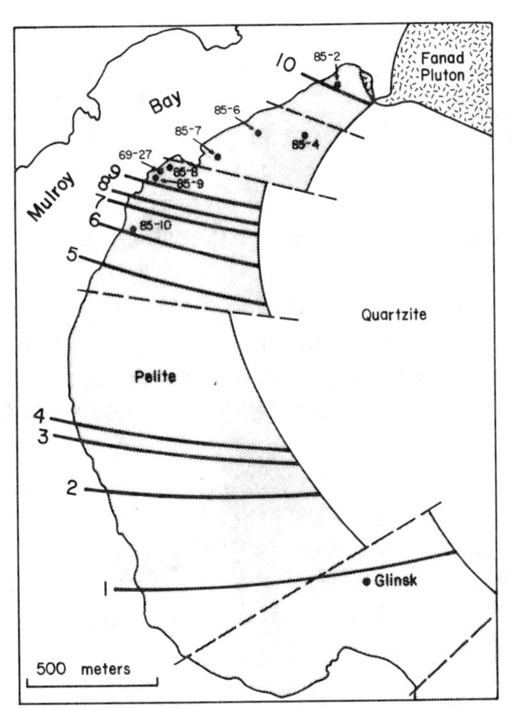

Figure 9.9. A portion of the contact aureole of the Fanad Pluton, Ireland. The isograds in the pelitic horizon are: (1) outer limit of new biotite growth, (2) south of this line new biotite is restricted to knots of chlorite, (3) new garnet appears, (4) andalusite appears and chlorite disappears, (5) cordierite appears, (6) fibrolite appears, (7) K-feldspar appears, (8) muscovite disappears, (9) andalusite becomes pleochroic, (10) sillimanite appears. (From Kerrick, 1987b, Fig. 3).

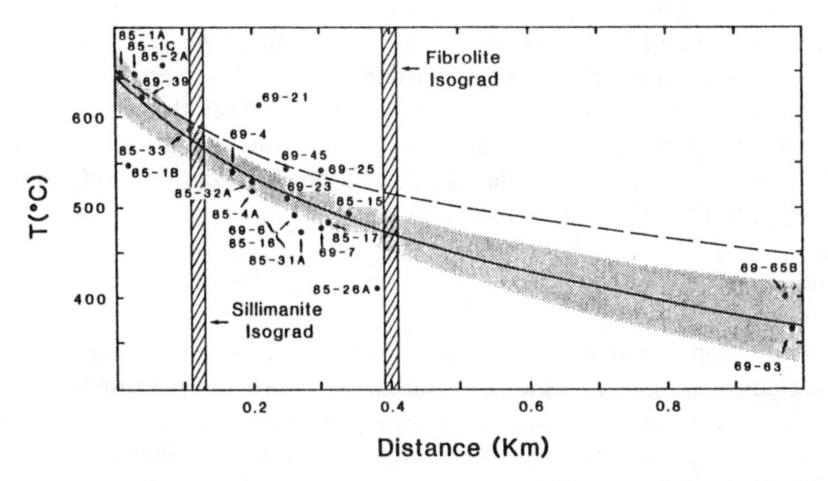

Figure 9.10. Temperature as a function of distance from the contact of the Ardara pluton, Ireland, showing the locations of the fibrolite and sillimanite isograds. Temperatures for the samples were determined using Ferry and Spear's (1978) garnet-biotite geothermometer. The solid line was obtained by least squares regression of the data points. (From Kerrick, 1987b, Fig. 4).

9.10 implies that fibrolite formed at temperatures well within the andalusite stability field. In elucidating the fibrolite-forming mechanism, I noted that in some basal sections of biotite, fibrolite is arranged in a regular triangular pattern (Fig. 9.11a). As shown in Figure 9.11b, this texture has been noted by other workers. To rationalize why fibrolite formed rather than the stable polymorph (andalusite), I followed Chinner's (1961) argument and concluded that fibrolite inherited the trigonal arrangement of the chains of Si and Al in the tetrahedral layers of the biotite structure. Andalusite nucleation was precluded because of the energy barrier necessary to "recoordinate" tetrahedral Al inherited from the host biotite into the hexahedral coordination of Al in andalusite.

In the Ross of Mull aureole, Scotland, Barnicoat and Prior (ms.) investigated the relationship between the kyanite → andalusite and andalusite → sillimanite transformations versus prograde dehydration reactions yielding andalusite and sillimanite. Kyanite + andalusite coexist over a zone about 300 m thick (Fig. 9.12). Particularly notable is the fact that the isograd marking the reaction:

$$\text{Muscovite} + \text{Quartz} = \text{Sillimanite} + \text{K-feldspar} + H_2O \ ,$$

occurs about 5 km farther from the intrusive than that of the andalusite-sillimanite equilibrium (Fig. 9.12). From the above relationships Barnicoat and Prior (ms.) concluded that disequilibrium played a dominant role in the development of the aluminum silicates in the Ross of Mull aureole. As shown in Figure 9.13, they estimated the temperature overstepping of the relevant metamorphic reactions by considering a 1 kJ free energy barrier for nucleation of products. As predicted by the fact that ΔG of the polymorphic reactions are smaller than devolatilization reactions, Figure 9.13 shows that the polymorphic transformations have considerably more overstepping than the dehydration reaction.

Regional metamorphism

There is considerable divergence of opinion regarding the question of fibrolite metastability in regional metamorphism. It is important to stress that fibrolite is considerably more common than sillimanite in most regionally metamorphosed pelites (Harte and Johnson, 1969; Grew and Day, 1972; Frey et al., 1974; Stäubli, 1989). Thus, we need to critically evaluate the possibility that fibrolite formed metastably during regional metamorphism.

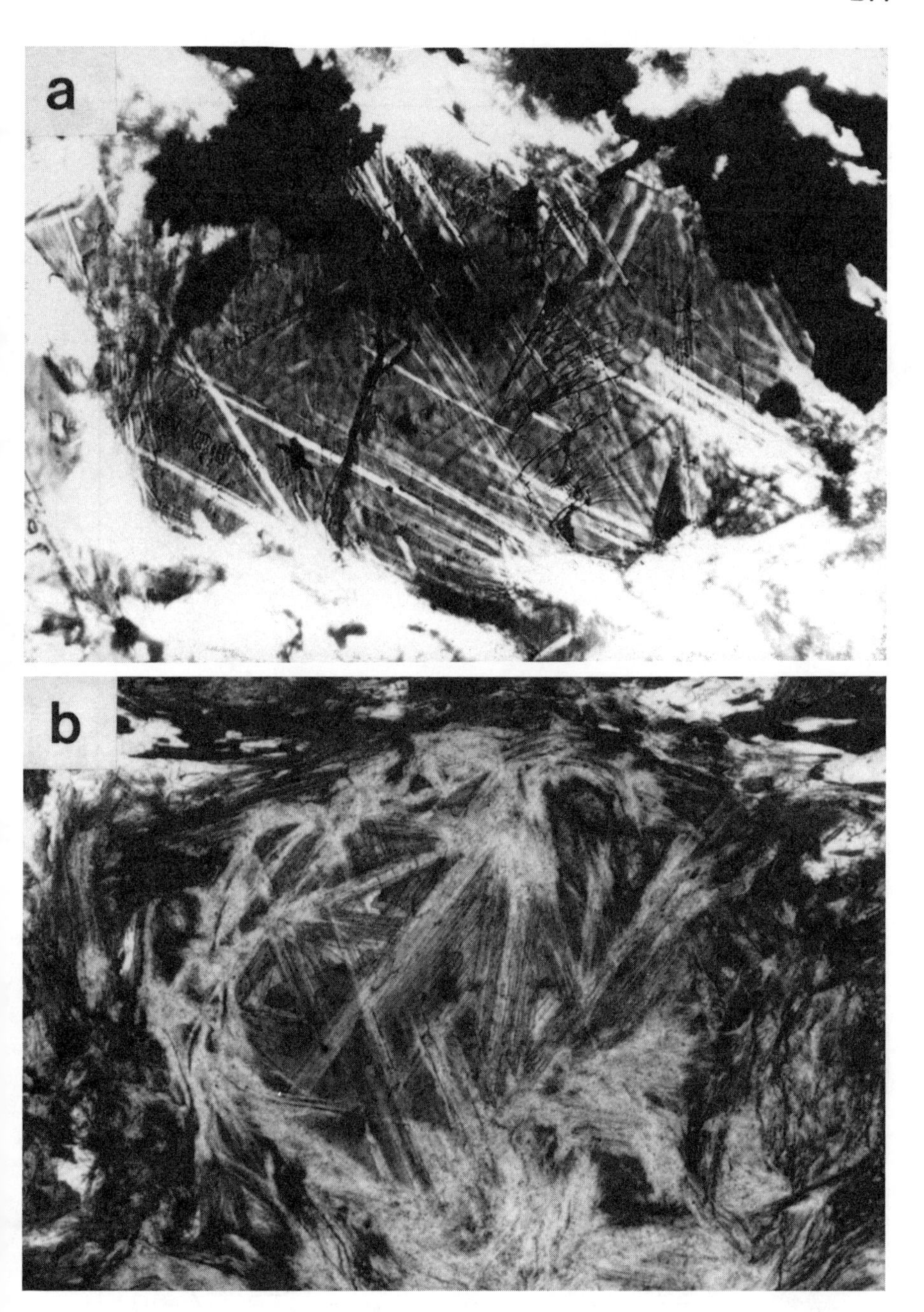

Figure 9.11. (a) Triangular arrangement of fibrolite within the basal plane of biotite in a metapelite from the Ardara aureole, Ireland. Width of the photo corresponds to 0.055 mm. (From Kerrick, 1987b, Fig. 2I). (b) Triangular arrangement of fibrolite within the basal plane of biotite in a sample of metapelite from Connemara, Ireland. Width of photo corresponds to approximately 2.0 mm. (From Yardley, 1977, Fig. 1C).

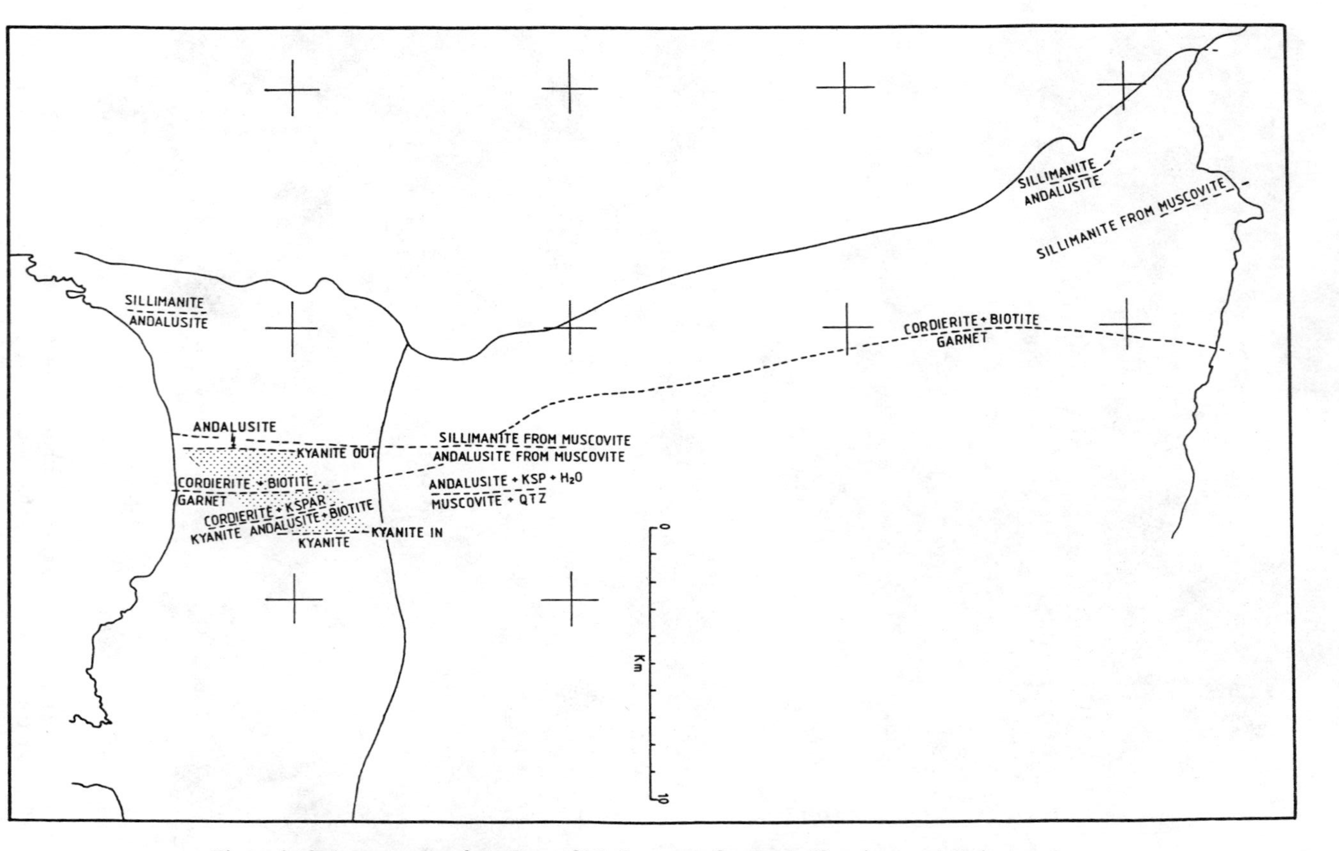

Figure 9.12. Isograd map of the Ross of Mull aureole, Scotland. Kyanite + andalusite coexist in the shaded area. (From Barnicoat and Prior, ms., Fig. 2b)

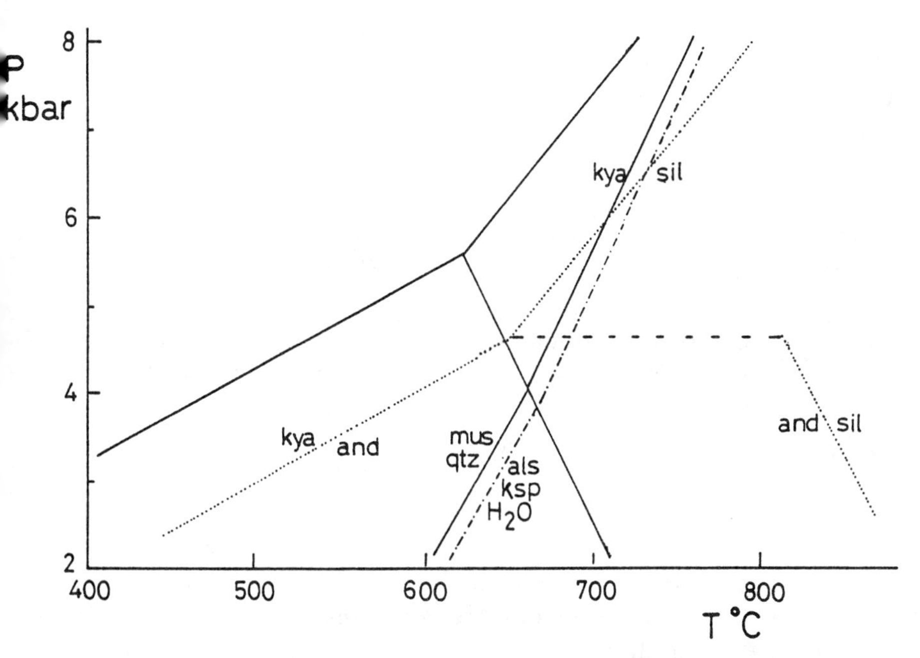

Figure 9.13. Displacement of equilibria resulting from 1 kJ of free energy required for nucleation of products [i.e., the high-temperature phase(s)]. Solid lines are equilibria, dotted lines are the "overstepped" Al_2SiO_5 polymorphic reactions, and the dot-dash line is the overstepped muscovite dehydration reaction. (From Barnicoat and Prior, ms., Fig. 9).

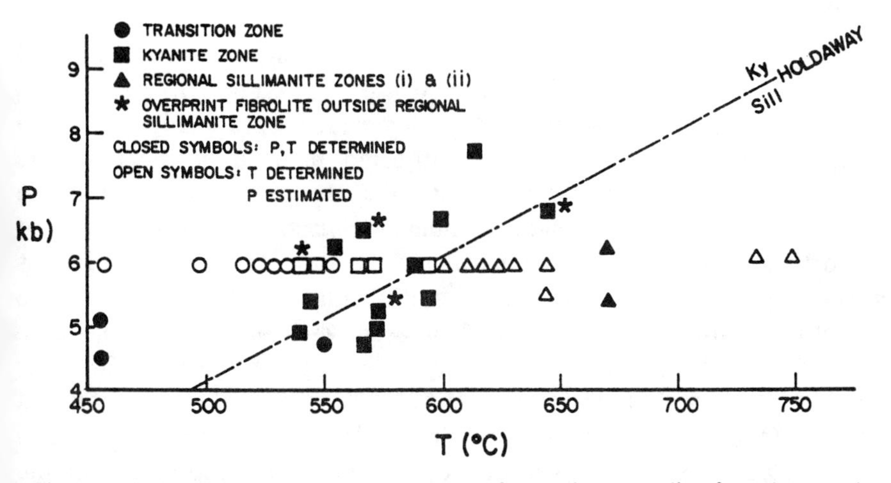

Figure 9.14. McLellan's (1985) thermobarometry of Dalradian metapelites from the central eastern Grampians, Scotland. Temperatures were determined with the garnet-biotite geothermometer, whereas pressures were determined using the garnet-plagioclase-aluminum silicate-quartz geobarometer. The kyanite-sillimanite equilibrium is from Holdaway (1971). Note that two of the fibrolite-bearing samples (asterisk) plot in the kyanite stability field. (From McLellan, 1985, Fig. 11).

In a study of sillimanite and fibrolite in Glen Clova, Scotland, Chinner (1961, p. 317) argued that metastable formation of fibrolite is unlikely because this phase occurs "... in the Scottish Dalradian in a zone of regional extent as the high-grade culmination of a well-displayed sequence of progressive metamorphism." However, in a subsequent paper, Chinner (1966) contended that fibrolite formed as a "late overprint" subsequent to the main regional metamorphism of the Dalradian. Although he concluded that the kyanite + fibrolite and andalusite + fibrolite assemblages represent disequilibrium (Chinner, 1966), he did not explicitly state that fibrolite formed outside of the sillimanite P-T stability field. Chinner's (1966) contention of a late sillimanite overprint has been upheld in most subsequent analyses of the Scottish Dalradian metamorphism (Bradbury, 1979; Wells, 1979; Dempster, 1985).

McLellan (1985) utilized the garnet-biotite geothermometer and the garnet-quartz-plagioclase-Al_2SiO_5 geobarometer to analyze the P-T conditions of the kyanite and sillimanite zones of the Dalradian. As shown in Figure 9.14, samples from the sillimanite zone plot within the sillimanite P-T stability field. Two samples of fibrolite-bearing rocks in the kyanite zone plot well within the kyanite stability field (Fig. 9.14). Because of the trace amounts of fibrolite in these samples, she intimated that fibrolite formed in the sillimanite stability field but that "... the rocks spent too little time in the sillimanite stability field for extensive re-equilibration ..." to kyanite (McLellan, 1985, p. 816).

Rumble (1973) studied aluminum silicates in a large area ($\approx$ 2500 km^2) of regionally metamorphosed rocks in New England (Fig. 9.15). However, much of his paper focused on the Mt. Moosilauke septum (Fig. 9.16). He considered that these rocks were subjected to a single metamorphic episode and that the aluminum silicates display textural features of the prograde and retrograde stages of metamorphism. With little apparent supporting evidence, he concluded that "The aluminum silicate minerals crystallized within their respective equilibrium stability fields. Once crystallized, however, they persisted metastably" (Rumble, 1973, p. 2428). Based on the spatial distribution of kyanite, andalusite, and sillimanite, Rumble (1973) and Hodges and Spear (1982) confirmed Thompson and Norton's (1968) placement of the "triple point isobar" [corresponding to invariant point (3) in Fig. 1.4] in the Mt. Moosilauke septum. Rumble noted that fibrolite is typically intergrown with biotite, muscovite or quartz. With the exception of one sample with a texture suggesting replacement of fibrolite by andalusite, no other samples have textures implying a direct reaction relationship between fibrolite and/or kyanite.

It is notable that fibrolite and sillimanite show very different regional distribution. If fibrolite is excluded, there is a reasonably consistent regional distribution of kyanite, andalusite and sillimanite. As shown in Figure 9.17a, kyanite is confined to the western portion of the map area, andalusite is restricted to the eastern and northeastern portion, and sillimanite is confined to the central portion of the area. With the exception of the area in the immediate vicinity of Mt. Moosilauke and Hurricane Mtn., sampling density is inadequate to accurately locate the kyanite-sillimanite and andalusite-sillimanite isograds. Nevertheless, the distribution of the aluminum silicates shown in Figure 9.16 does not indicate a "triple point isobar" passes through the Mt. Moosilauke septum. Rather, as concluded by Kerrick and Speer (1988), the andalusite-sillimanite isograd passes through this septum. Chamberlain and Lyons (1983) concluded that there were three major stages of metamorphism in south-central New Hampshire. The first metamorphic event (M1) was characterized by the formation of andalusite and the replacement of andalusite by sillimanite in the vicinity of the Devonian Kinsman plutons. Because of the replacement of andalusite by sillimanite, M1 occurred at pressures below the Al_2SiO_5 triple point. The M2 metamorphic event was marked by the regional development of sillimanite. From thermobarometric analysis, Chamberlain and Lyons (1983) estimated a pressure of 4.5 ± 1.5 kbar for M2. Because this mean pressure (4.5 kbar) lies above Holdaway's (1971) triple point, the kyanite-sillimanite isograd may be correlative with M2. M3 is marked by late-stage, localized retrogression that formed hydrous minerals (chlorite and muscovite). Thus, the regional distribution of the aluminum silicates in Figure 9.15 cannot be interpreted as having developed during a single metamorphic event. As shown in Figure 9.17, fibrolite is considerably more widespread than sillimanite. Because fibrolite is widely developed throughout the andalusite zone (Fig. 9.17b), there are significant differences in the distribution of sillimanite versus fibrolite (cf. Figs. 9.17a and 9.17b). Because fibrolite is commonly intergrown with micas, it is tempting to suggest that, as contended by Chinner (1961) for the Scottish Dalradian, fibrolite formed as a late overprint to the main regional metamorphism (M2). The presence of fibrolite and the lack of sillimanite in the andalusite zone brings into question the assumption that sillimanite and fibrolite crystallized under identical P-T conditions. Fibrolite within the andalusite zone could be attributed to progradation into the sillimanite zone, and metastable persistence of andalusite. If so, the lack of replacement of andalusite by sillimanite suggests that there are significant kinetic barriers for the andalusite → sillimanite reaction. Regardless of the various possible interpretations of the aluminum silicates in New Hampshire, the

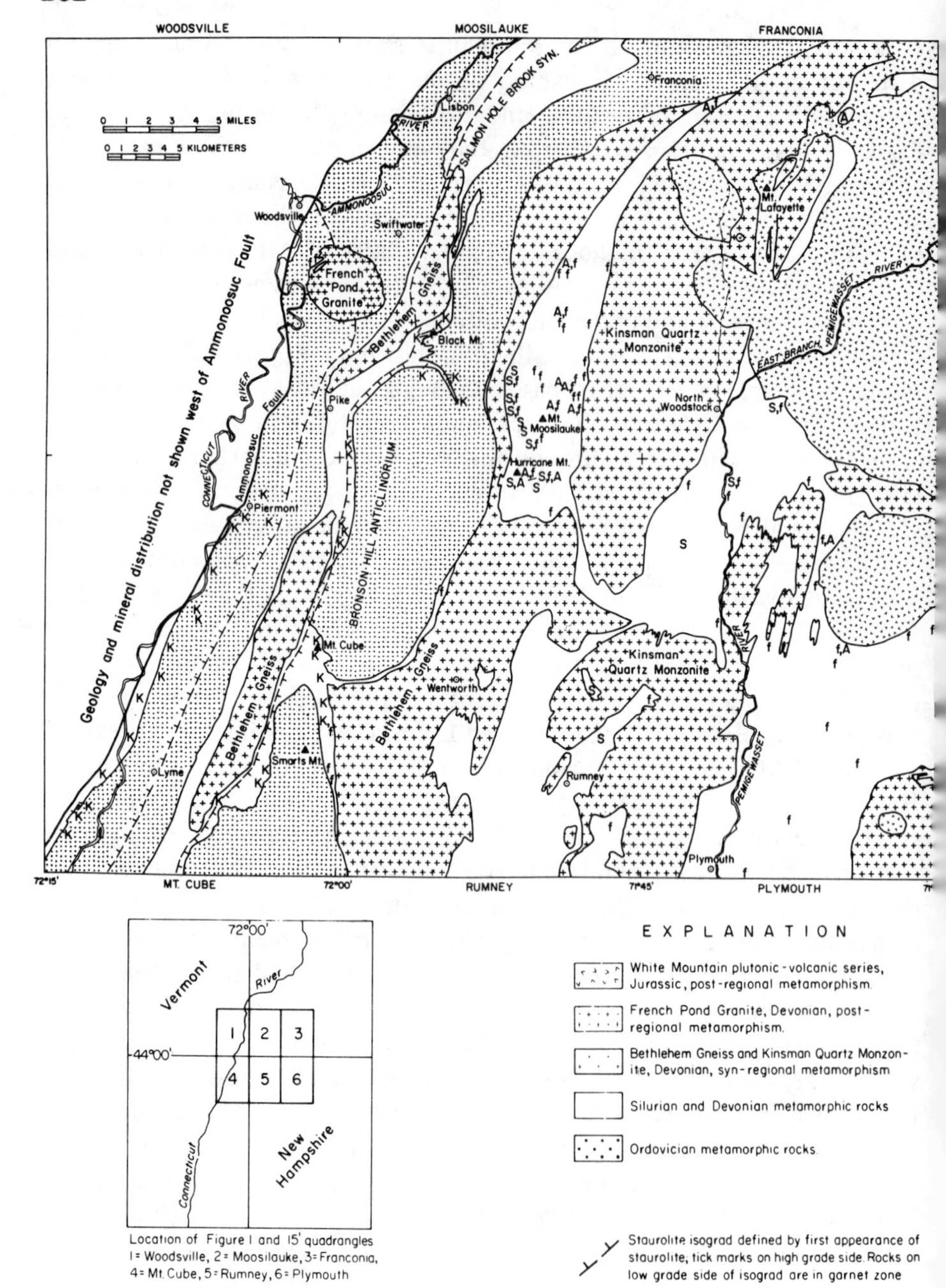

A Andalusite occurrence

f Fibrolitic sillimanite occurrence

K Kyanite occurrence

S Prismatic sillimanite occurrence

S,f,A Assemblage of prismatic sillimanite, fibrolitic sillimanite and andalusite occurs in one thin section.

[Figure 9.15 legend: see opposite page]

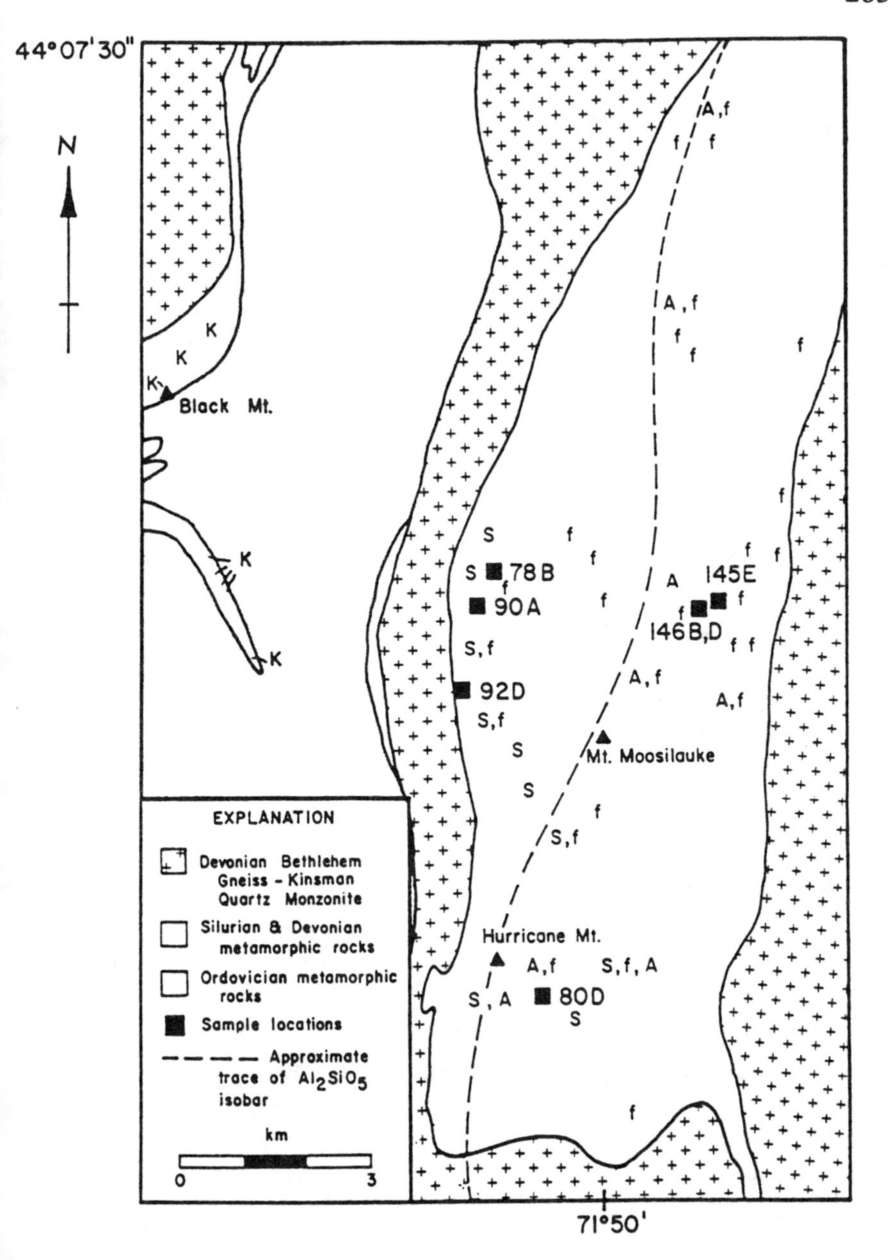

Figure 9.15. (*left*). Geologic map of an area in western Hew Hampshire, showing the regional distribution of the aluminum silicates. The Mt. Moosilauke septum, which is detailed in Figure 9.16, is shown in the central portion of the map. (From Rumble, 1973, Fig. 1).

Figure 9.16. (*above*). Geologic map of the Mt. Moosilauke septum, New Hampshire, showing the distribution of andalusite (A), sillimanite (S) and fibrolite (f), and the approximate location of the univariant andalusite-sillimanite equilibrium. (From Hodges and Spear, 1982, Fig. 2).

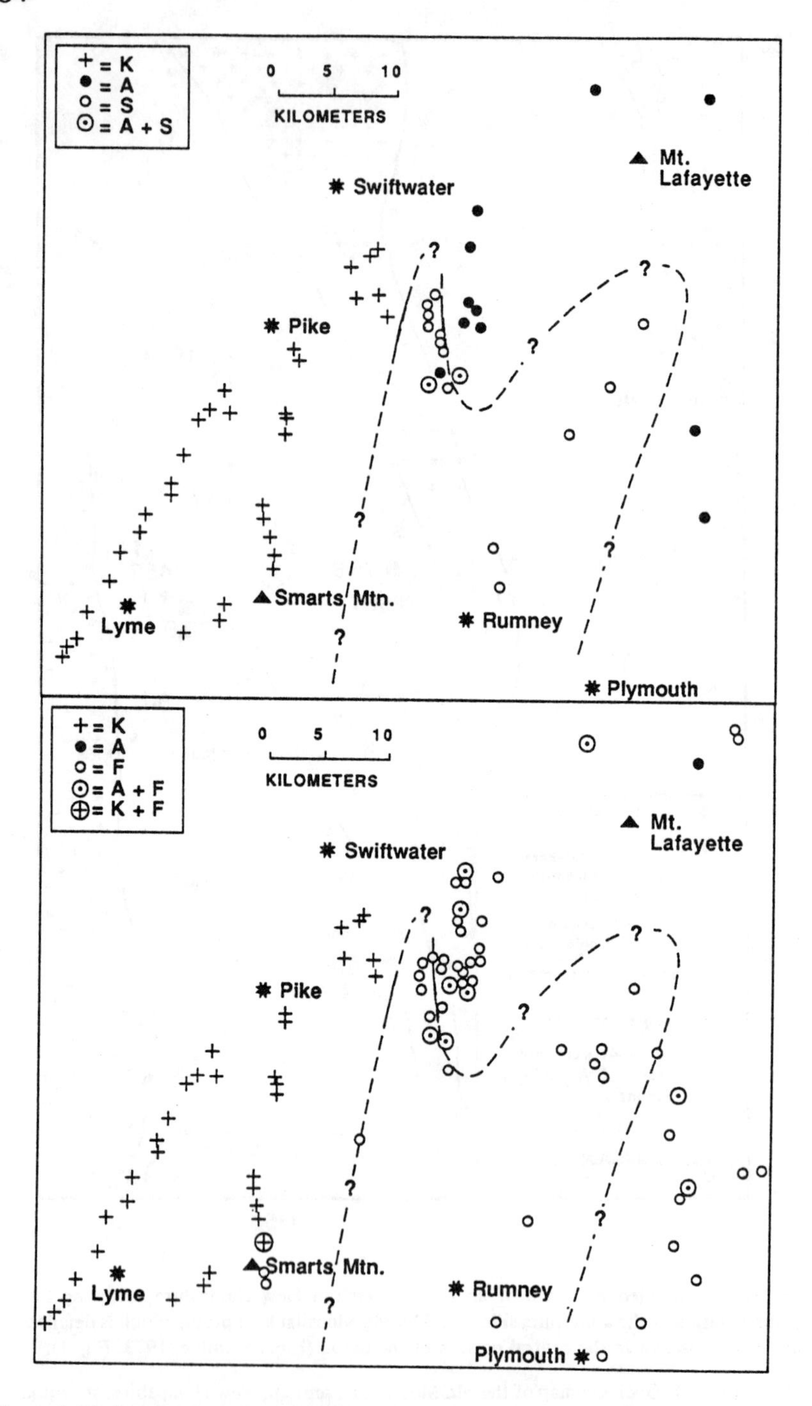

Figure 9.17. Distribution of kyanite (K), andalusite (A), sillimanite (S), and fibrolite (F) in northwestern New Hampshire. *Top*: fibrolite excluded. *Bottom*: sillimanite excluded. (Modified from Rumble, 1973, Fig. 1).

significant differences in the regional distribution of sillimanite versus fibrolite in the area studied by Rumble (1973) question the commonly-invoked assumptions that fibrolite and sillimanite formed under identical P-T conditions and that, for the purposes of thermobarometry, these phases may be considered as synonymous.

The Central Alps of Switzerland contain a classic sequence of Barrovian-type regional metamorphism. The isograds of this Tertiary *Main Alpine Metamorphism* (also referred to as the Ticino Thermal High) form an east-west elongation with metamorphic grade increasing southward (Fig. 9.18). The Insubric Line forms the southern boundary of the Lepontine metamorphic belt. In the "sillimanite" zone, fibrolite is far more common than sillimanite (Frey et al., 1974). Wenk et al. (1974) presented a detailed summary of the aluminum silicates in the Bergell Alps (Fig. 9.19), located at the eastern margin of the metamorphic belt. As shown in Figure 9.19, fibrolite occurs over a considerably larger area than sillimanite. It is particularly important to note that in the portion of the kyanite zone studied by Wenk et al. (1974) fibrolite is common and sillimanite is absent. In fact, fibrolite first appears at the kyanite-staurolite isograd; thus, fibrolite occurs throughout the entire kyanite zone. Within the kyanite zone they find no textural evidence for the direct replacement of kyanite by fibrolite. Furthermore, they stated (Wenk et al., 1974, p. 515, 517) "Kyanite is normally associated with biotite and seems to be forming as an alteration product of it ... Therefore, it is likely that the major metamorphic recrystallization of these specimens took place under pressure and temperature conditions where kyanite is the stable Al_2SiO_5 polymorph ... The intricate problems of the fibrolite-infested kyanite zone ..." are such that "... this rock series cannot be regarded as belonging to a 'sillimanite zone'." The zone which contains abundant sillimanite and which trends northeast from Valle della Mera (Fig. 9.19) coincides with the Gruf-Complex, which consists of migmatitic, isoclinally-folded gneisses. Thermobarometric analysis of the Gruf-Complex by Bucher-Nurminen and Droop (1983) suggests an early stage of metamorphism at about 10 kbar and 800°C, followed by marked decompression to yield the final metamorphic imprint at 3-4 kbar and 600-650°C (Fig. 9.20). They suggested that the sillimanite formed from kyanite during decompression. Thus, the sillimanite isograd shown in Fig 9.19 cannot be interpreted as an isograd reaction which occurred within a structurally *coherent* section. This example illustrates the pitfalls that can arise from the interpretation of isograds without a detailed tectonic analysis.

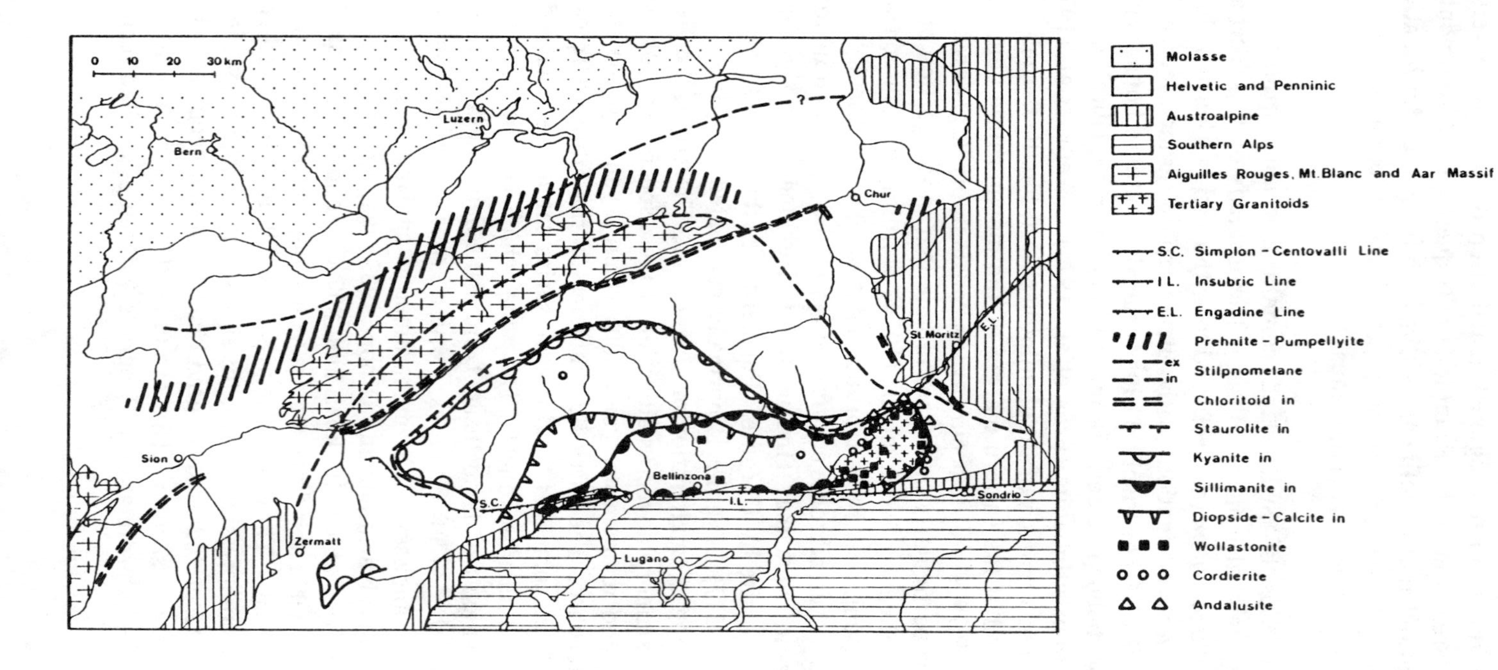

Figure 9.18. Isograds in regionally metamorphosed rocks of the Lepontine region, south-central Switzerland. (From Trommsdorff, 1980, Fig. 39).

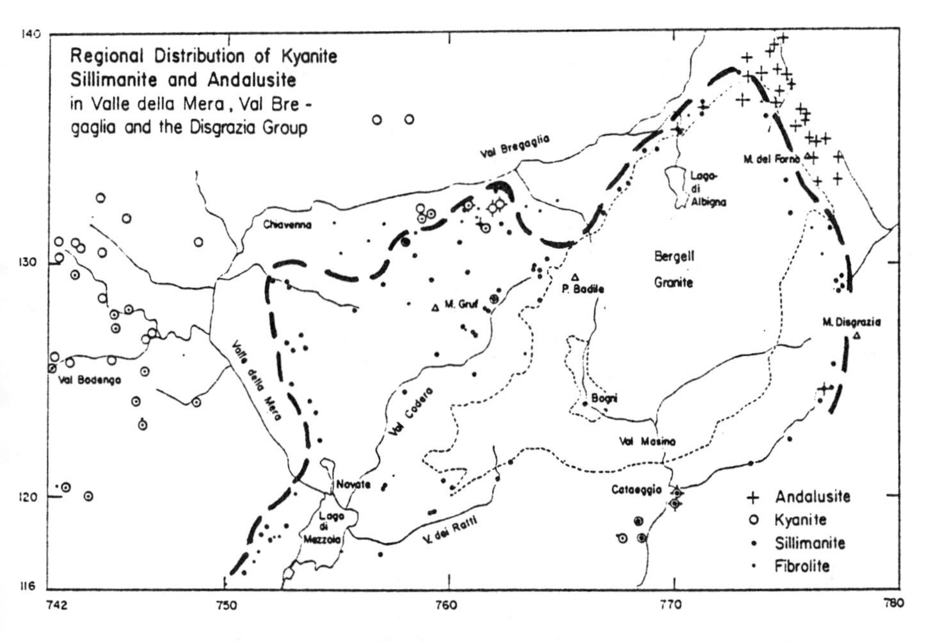

Figure 9.19. Distribution of aluminum silicates in the Bergell Alps of Switzerland. The heavy dashed line is the sillimanite isograd. (Modified from Wenk et al., 1974, Fig. 1).

In an unpublished study, I examined the distribution of fibrolite in metapelites in the region immediately west of that studied by Wenk et al. (1974). As in the study of Wenk et al. (1974) I found fibrolite to be widespread in the kyanite zone; furthermore, the first appearance of fibrolite is generally coincident with the staurolite-kyanite isograd. The fibrolite in this area is typically intergrown with biotite. I found no textural evidence for the replacement of kyanite by fibrolite. In addition, there was no suggestion that kyanite was resorbed or rimmed by muscovite. Thus, textural evidence suggests no instability of kyanite during the fibrolite-forming event. In the gneisses of the Bellinzona Zone, representing the southernmost and highest grade portion of the Lepontine metamorphic belt, Bühl (1981) interpreted biotite + fibrolite intergrowths as evidence of contemporaneous growth. However, it is also possible that these intergrowths resulted from the replacement of biotite by fibrolite. As contended by Chinner (1966) for the Scottish Dalradian, it is suggested herein that much of the fibrolite formed as a late "overprint" subsequent to the Main Alpine metamorphism. As shown in Figure 9.21, thermobarometric analysis of kyanite-zone rocks in the Neufenen Pass area suggests marked decompression following the Main Alpine metamorphism. The implied retrograde trajectory (enclosed by the shaded band in Fig. 9.21) passes through a significant portion of the

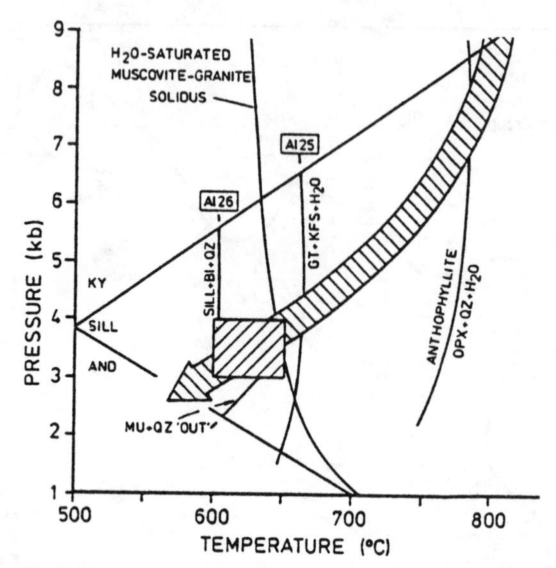

Figure 9.20. Thermobarometric evolution of sillimanite-bearing granulites of the Gruf-Complex, eastern Pennine Alps, Switzerland. The hachured rectangle represents the P-T conditions derived from mineral geothermometry and geobarometry. The hachured curve with arrow represents the P-T path during cooling and uplift of the Gruf-Complex. (From Bucher-Nurminen and Droop, 1983, Fig. 13).

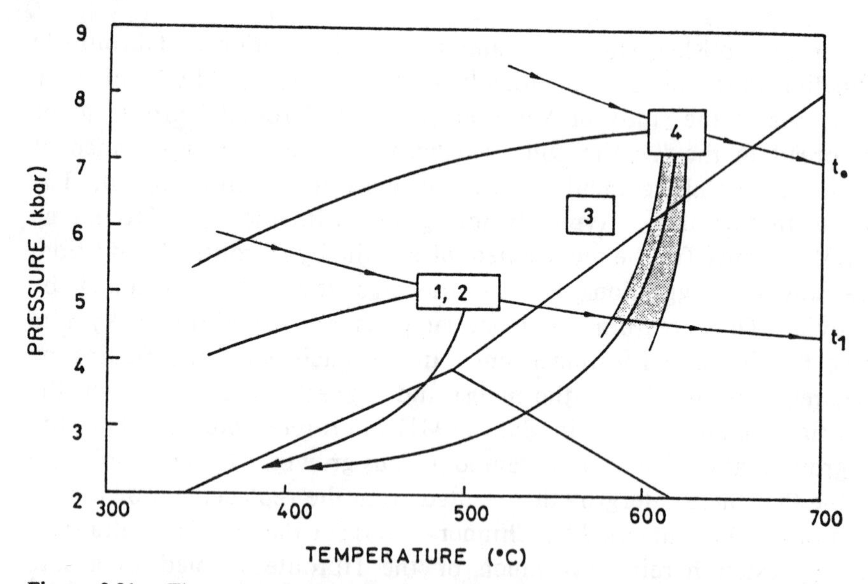

Figure 9.21. Thermobarometric analysis of metapelites from the Nufenen Pass area, Lepontine Alps, Switzerland. The numbers refer to the four metamorphic zones in this area; the index minerals and mineral assemblages for these zones are: (1) chlorite, (2) staurolite + chlorite, (3) staurolite + biotite, (4) kyanite. The open "boxes" are P-T estimates of the four zones derived from thermobarometry. Decompression trajectories are shown for metamorphic zones 1, 2 and 4. The shaded band enveloping the zone 4 trajectory represents the "well constrained" portion of the P-T path. (From Klaper and Bucher-Nurminen, 1987, Fig. 9).

sillimanite stability field. Because late fibrolite could have formed during decompression, there would be no reason to appeal to the *metastable* crystallization of fibrolite outside of the sillimanite P-T stability field.

Wells (1979) used numerical methods to model the P-T-time(t) evolution of the Caledonian regional metamorphism in the Central Highlands of Scotland. As shown in Figure 9.22, the computed P-T-t paths remain in the kyanite stability field. However, Wells (1979) concluded that with, small adjustments in the model (e.g., increasing the rate of erosion), the sillimanite P-T field could be intersected upon decompression. "This general trend towards lower temperatures would impose considerable kinetic limitations on sillimanite crystallization, with reaction probably occurring on the scale of only a few mineral grains. If this explanation for regional Caledonian sillimanite has any validity, the abundance of irregularly developed fibrolite rather than prismatic sillimanite may be a manifestation of the kinetic restrictions imposed by the general cooling of the rocks during their decompression through the sillimanite P-T stability field" (Wells, 1979, p. 669). He concluded that this kinetic model could explain the sporadic distribution of fibrolite in other areas of the Moine and Dalradian in the Scottish Highlands.

The preceding discussion has summarized fibrolite in three well known belts of Barrovian-type regional metamorphism (New England, south-central Switzerland, and the Scottish Highlands). In these areas fibrolitization is widespread and may have postdated the main regional metamorphism. The apparent overprint of fibrolite onto zones characterized by lower-grade Al_2SiO_5 index minerals (i.e., the kyanite zone in the Lepontine Alps and the andalusite zone in New Hampshire) brings into question the equilibrium versus disequilibrium crystallization of fibrolite. Even if fibrolite crystallized in the P-T conditions of sillimanite, the presence of fibrolite coexisting with other aluminum silicates, with typically no evidence of a reaction relation or resorption of one of the polymorphs, suggests that sluggish polymorphic reaction kinetics are a problem.

In one of the earliest attempts to evaluate the conditions of fibrolite formation by garnet-biotite geothermometry, Fleming (1973) addressed the question of fibrolite metastability in the andalusite-staurolite zone of the Mt. Lofty Ranges, South Australia. For garnet-biotite pairs, K_D [= $(Mg/Fe)^{Grt}/(Mg/Fe)^{Bt}$] values are indicative of metamorphic conditions below sillimanite grade. Thus, he concluded that fibrolite formed within the andalusite P-T stability field.

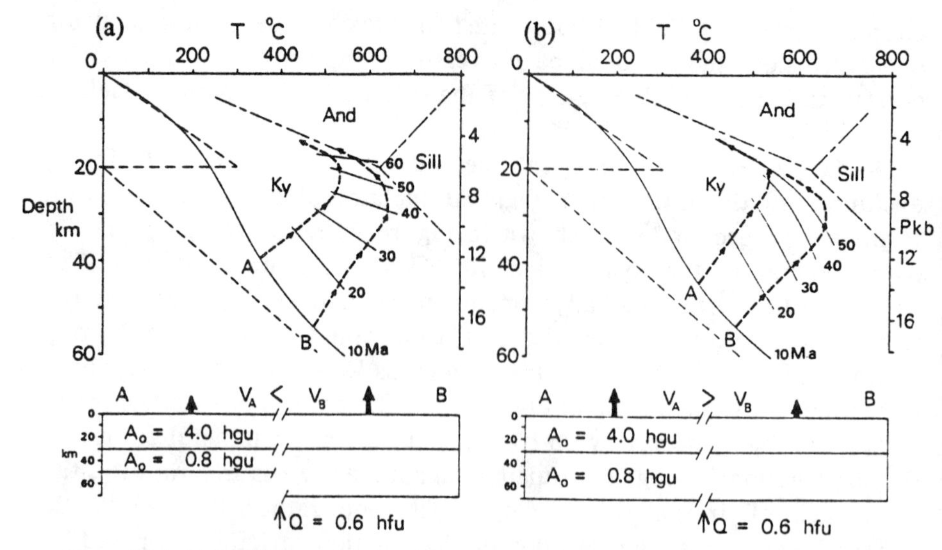

Figure 9.22. Wells' (1979) models for the post-metamorphic decompression of the Dalradian sequence of the Central Highlands of Scotland. (a) The effect of different crustal thicknesses and different syn-metamorphic erosion rates on the P-T paths. Erosion is assumed to begin 10 Ma after tectonic thickening. The bottom diagram illustrates the differences in crustal thicknesses and erosion rates for P-T paths A and B. The erosion rates of the two crustal blocks (V_A and V_B) are schematically represented by the heavy arrows (vectors). (b) The effect of differential erosion rates. (From Wells, 1979, Fig. 3).

In the Cooma Complex, Australia, Vernon (1978) concluded that cordierite was replaced by symplectic intergrowths of andalusite, biotite and quartz. Vernon (1978, 1979) believed that this reaction occurred during an extensive, late retrograde metamorphism. Textural evidence suggests that fibrolite formed later than andalusite (Vernon, 1979). Because there is no evidence of a late-stage thermal event, and provided that andalusite crystallized in equilibrium, Vernon (1979) argued that the late-stage fibrolite probably formed metastably.

In the Main Central Thrust (MCT) zone, eastern Napal, the prograde Barrovian-type aluminum silicate sequence is kyanite → fibrolite + kyanite → fibrolite + sillimanite (Hubbard, 1989). In relation to the MCT, fibrolite appears to be post-kinematic whereas kyanite is pre-or syn-kinematic. Thus, a late fibrolite overprint is suggested.

In gneiss domes of the Higher Himalayan Crystalline (HHC) sequence in India, Kundig (1989) noted that biotite + fibrolite replaced staurolite + kyanite. He concluded that, following Barrovian metamorphism producing kyanite and sillimanite, biotite + fibrolite formed during increasing temperature and/or decreasing pressure. In HHC paragneisses of southeastern Zanskar, India, kyanite and sillimanite are locally replaced by muscovite and fibrolite (Pognante and Lombardo, 1989). Thus, as with the Lepontine Alps and the Scottish Highlands, fibrolite formed in the late stages of Barrovian metamorphism in the Himalayan belt.

Young et al. (1989) concluded that fibrolite formed during retrograde metamorphism of migmatitic gneisses of the McCullough Range, southern Nevada.

In metamorphosed pelites of the Mt. Raleigh pendant of British Columbia, Kerrick and Woodsworth (1989) found that the fibrolite isograd occurs 5 km downgrade from the sillimanite isograd (Fig. 8.7). In many of the samples, fibrolite is intergrown with biotite. Textures provide cogent evidence that fibrolite replaced biotite. In most samples, fibrolite is not intergrown with coexisting chiastolite or sillimanite. As in the Donegal aureoles studied by Kerrick (1987b), it was concluded that fibrolite formed from the influx of late-stage, acidic volatiles. Because fibrolite is the only phase replacing biotite, Kerrick and Woodsworth concluded that reaction [8.1] is relevant to the fibrolitization of biotite in the Mt. Raleigh pendant. As shown in Figure 9.23, garnet-biotite thermometry at the estimated pressure of 2.5 kbar suggests temperatures at the fibrolite isograd that were *approximately* 100°C below those of the sillimanite isograd. Based on garnet zoning profiles, we argued that the temperatures derived for the samples near the fibrolite isograd (Fig. 9.23), and sample 71-68-1 collected at the sillimanite isograd, record maximum temperatures. They concluded that the sillimanite isograd is essentially coincident with the andalusite-sillimanite equilibrium, as discussed in THE ANDALUSITE → SILLIMANITE REACTION section of this chapter, and that fibrolite metastably formed at P-T conditions well within the andalusite stability field.

Digel et al. (1989) studied fibrolite in kyanite zone metapelites of the Mount Cheadle area, British Columbia, located about 40 km northwest of the Mica Creek area. Fibrolite occurs intergrown with biotite and (most commonly) in fibrolite-quartz nodules and layers. Fibrolite is sporadically distributed in the kyanite metapelites in this area. The thermobarometric estimate of 5.5 ± 0.6 kbar and 595 ± 12°C implies P-T conditions within the sillimanite stability field (Fig. 9.24). Digel et al. (1989) concluded that the sporadic distribution of fibrolite resulted from localized variations in factors influencing the kinetics of the fibrolite-forming reaction. "We suggest that nucleation and growth of fibrolite within the kyanite metapelites occurred only at sites where factors influencing reaction

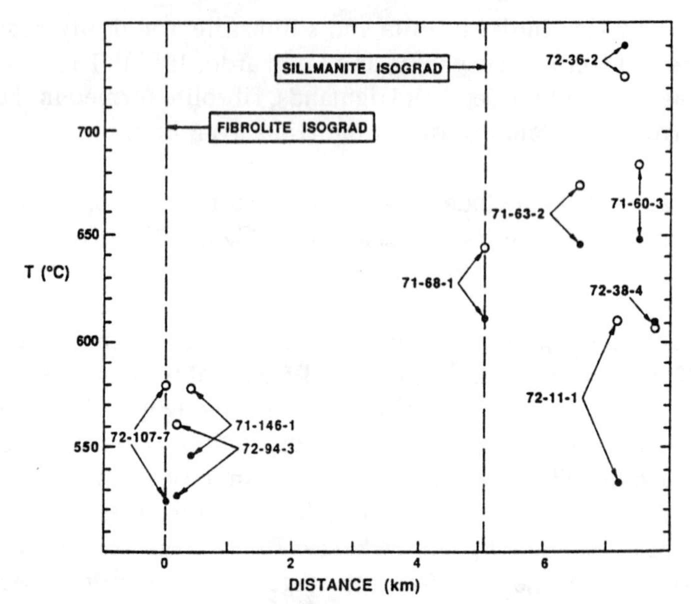

Figure 9.23. Garnet-biotite thermometric data (P = 2.5 kbar) as a function of distance upgrade from the sillimanite isograd in the Mt. Raleigh pendant, British Columbia. The filled versus closed circles were computed from two different equations for the garnet-biotite geothermometer. (From Kerrick and Woodsworth, 1989, Fig. 4).

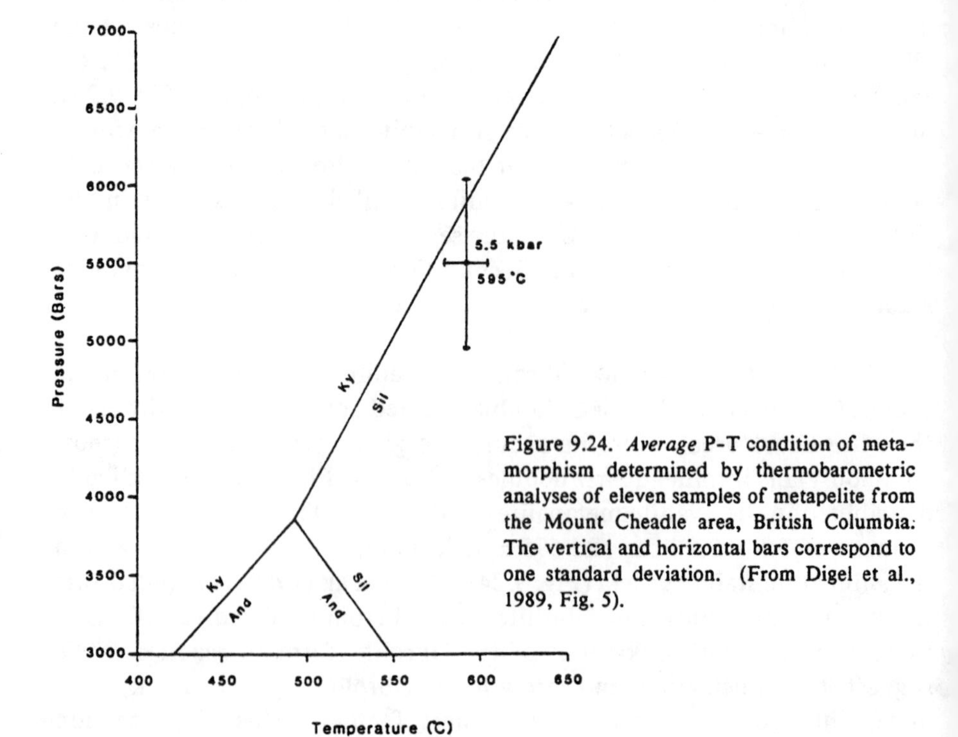

Figure 9.24. *Average* P-T condition of metamorphism determined by thermobarometric analyses of eleven samples of metapelite from the Mount Cheadle area, British Columbia. The vertical and horizontal bars correspond to one standard deviation. (From Digel et al., 1989, Fig. 5).

kinetics (i.e., localized deformation and the presence of fluid) were favorable" (Digel et al., 1989, p. 99). Channelized flow of hot, magmatic fluids was postulated as a potential major factor responsible for the variation in the kinetics of the fibrolite-forming reaction. The presence of kyanite in all fibrolite-bearing rocks suggests kinetic barriers for the kyanite → fibrolite reaction in this area. Assuming that their estimated P-T range for metamorphism is correct, the study of Digel et al. implies kinetic barriers for the kyanite → fibrolite reaction up to 50°C from the kyanite-sillimanite equilibrium; their *mean* P-T estimate (5.5 kbar and 595°C) suggests kinetic problems within 25°C from this equilibrium.

Widespread late-stage fibrolitization, which overprints lower grade rocks, does not occur in all areas of regional metamorphism. The following reviews examples of areas that lack such an overprint.

Grambling (1981) discussed the distribution of kyanite, andalusite, sillimanite and fibrolite in the Truchas Peaks region of New Mexico. As shown in Figure 9.25, there is an isograd separating rocks containing only kyanite to the south from those containing fibrolite and/or sillimanite to the north. Kyanite persists in a zone up to 3 km north of this isograd (Fig. 9.25). In this zone, kyanite metastability is marked by the partial replacement of kyanite by fibrolite, or by the alteration of kyanite to muscovite with sillimanite formed in the immediate vicinity. Grambling (1981) concluded that these textures were indicative of the respective kyanite → fibrolite or kyanite → sillimanite reactions. Although not explicitly stated, Grambling (1981) *presumably* assumed that the replacement of kyanite by muscovite reflected Carmichael's (1969) postulated mechanism for the kyanite → sillimanite isograd reaction. Grambling (1981) noted that this isograd marks the first appearance of *both* fibrolite and sillimanite; thus, he concluded "... that the 'fibrolite effect' which Holdaway encountered in experimental reactions is not a problem in nature" (Grambling, 1981, p. 705). This generalization is questionable in light of the apparent fibrolite "overprint" in many other areas of regional metamorphism.

Ghent et al. (1980, 1982) described a well-defined sillimanite isograd in the Mica Creek area, British Columbia. This isograd separates lower grade rocks containing kyanite as the only aluminum silicate from higher grade rocks containing sillimanite as the sole Al_2SiO_5 polymorph. There is an exceptionally narrow zone of coexisting kyanite and sillimanite; according to Ghent et al. (1980), these phases coexist over a zone that is 10-20 m thick (as measured normal to the isograd surface). More recently,

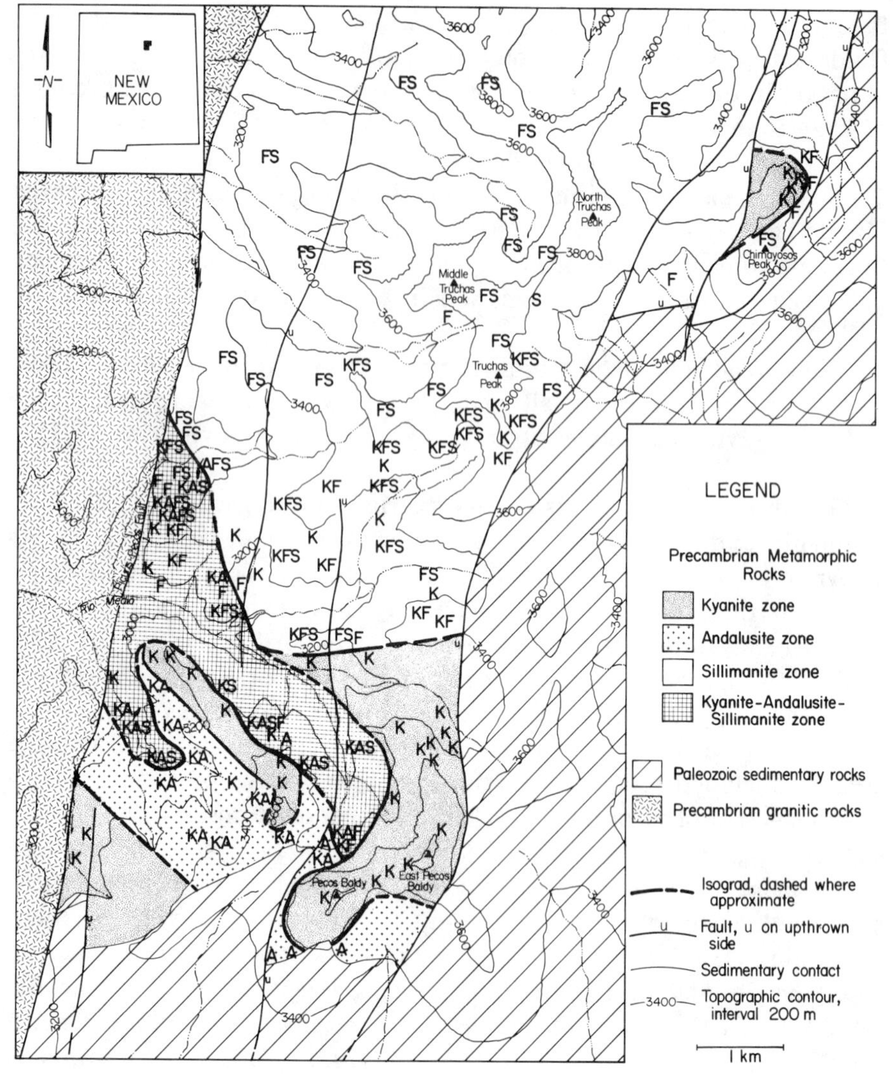

Figure 9.25. Geologic map of the Truchas Peaks region, New Mexico, showing the distribution of the aluminum silicates, and mineral zones and isograds defined by these phases. (From Grambling, 1981, Fig. 1).

Kerrick and Ghent (unpub.) carried out a detailed petrographic study of samples collected in the immediate vicinity of this isograd (Fig. 9.26). They confirmed that this isograd is marked by the appearance of fibrolite. In most samples examined there is irrefutable evidence that fibrolite replaced biotite. In specimens where the biotite → fibrolite reaction is well advanced, kyanite shows no evidence of resorption (Fig. 8.4) or

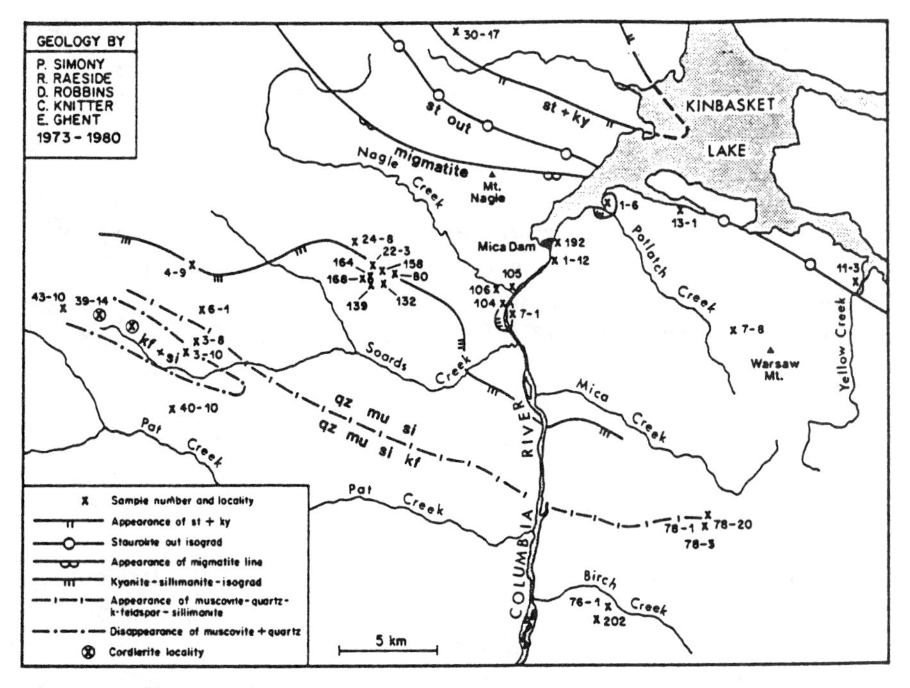

Figure 9.26. Isograds and sample localities in the Mica Creek area, British Columbia. (From Ghent et al., 1982, Fig. 1).

replacement by musçovite. Thus, samples collected at this isograd show no textural evidence for the kyanite → sillimanite reaction. This conclusion holds for sample 164 (Fig. 9.26), which was collected about 500 m upgrade from the isograd. However, sample 139, which was collected about 1 km above the isograd, contains remnants of kyanite surrounded by muscovite + fibrolite intergrowths. If the muscovite is secondary, as is commonly concluded for muscovite intergrown with fibrolite, this specimen provides textural evidence supporting the direct transformation of kyanite to fibrolite. Sample 168, which was also collected 1 km above the isograd, contains fibrolite as the only aluminum silicate. Thus, the transformation from metapelites containing only kyanite to those with only fibrolite occurs over a zone about 1 km thick. Because Ghent et al. (1980) have shown that the isograd is essentially vertical in this area, the 1 km-thick zone represents a measure of true thickness of the zone of coexisting kyanite + fibrolite. As measured perpendicular to the isograd, Figure 6 of Ghent et al. (1980) yields a temperature gradient of about 17°C/km. A 17°C temperature gradient across the 1 km-thick zone of coexisting kyanite + fibrolite is relatively insignificant. Thus, it is concluded that the sillimanite isograd in the Mica Creek area represents a reasonably close approximation to the univariant kyanite-sillimanite equilibrium in spite of the fact that textures do not support the isograd reaction as kyanite → sillimanite. In contrast to the Scottish Highlands and the Lepontine Alps, there is no widespread fibrolite "overprint" in kyanite zone rocks of the Mica Creek area.

In view of the contrast in the distribution of fibrolite between different areas of regional metamorphism (e.g., Mica Creek versus the Lepontine Alps), it is concluded that no generalization can be made regarding the kinetics of fibrolite formation in regional metamorphism. For areas displaying a widespread fibrolite overprint on lower grade rocks (e.g., the Lepontine Alps and the Mt. Raleigh pendant), the author recommends that, without detailed textural and thermobarometric analyses, fibrolite should not be used as an indicator of the P-T conditions of metamorphism.

Implications for an equilibrium model

In Chapter 7 it was concluded that because of grain boundary energy, fibrolite would be less stable than sillimanite. The implications for metamorphic systems are profound. With an equilibrium model, sillimanite (not fibrolite) should be the exclusive high-temperature polymorph. The widespread abundance of fibrolite in metaluminous rocks thus seriously questions a strict equilibrium model. If we allow for the metastable formation of fibrolite by the andalusite → fibrolite reaction, the isograd marking the andalusite → fibrolite reaction should be located at higher grades than the andalusite → sillimanite isograd. Because fibrolite typically *precedes* sillimanite in prograde sequences of andalusite-bearing metapelites, an equilibrium model is invalid and the above argument regarding the metastable andalusite → fibrolite transformation is inapplicable.

KINETIC MODELING OF POLYMORPHIC REACTIONS

Using arguments based on crystal structures, Winter and Ghose (1979) outlined mechanisms for polymorphic transformations amongst the Al_2SiO_5 polymorphs. Figure 2.20 illustrates their mechanism for the sillimanite → andalusite reaction. They hypothesized that, because of the presence of voids in the structure, the O_c atoms have a large amplitude of vibration normal to the $Si\text{-}O_c\text{-}Al_2$ plane. Thus, the $Al_2\text{-}O_C$ bond breaks and bridges across the adjacent "void" to bond with another O_C (see dotted arrow #1 in Fig. 2.20). A similar bond change mechanism occurs adjacent to other voids (e.g. Y in Fig. 2.20). This process forms the double chains of Si tetrahedra and Al hexahedra of the andalusite structure. This mechanism requires interchange of half of the Al_2 and Si atoms. Winter and Ghose (1979) concluded that because of low diffusivities of Al and Si, the sillimanite → andalusite transformation is kinetically sluggish. They concluded that Al/Si disorder in sillimanite would enhance the kinetics of this transformation. Winter and Ghose (1979) contended that because thermal expansion of the Al_2-O_C bond is considerably larger than that of other bonds within the Al hexahedra, instability of these polyhedra is a primary cause for the decomposition of andalusite to sillimanite. They suggested that disordered sillimanite may initially form but that sillimanite subsequently transforms into an ordered structure because of the instability of the Al-O-Al linkage (i.e., "aluminum avoidance").

Because of the lower volume of octahedrally-coordinated Al, Winter and Ghose (1979) argued that this polyhedron is favored with increasing pressure. They contended that the sillimanite → kyanite transformation occurs by shifting of Al_2 into nearby tetrahedral positions of kyanite. This requires breaking the Al_2-O_B bond to form Al_4-O linkages and rotation of the octahedral chains. However, in contrast to the sillimanite → andalusite reaction, no significant diffusion of Si and Al is required. They proposed a similar mechanism for the andalusite → kyanite reaction. Thus, they implied that the sillimanite → kyanite and andalusite → kyanite transformations should be faster than the sillimanite → andalusite reaction. Winter and Ghose (1979, p. 584) concluded "Among the few described coherent replacement textures, the andalusite → kyanite and sillimanite → kyanite transitions are much more common than the andalusite → sillimanite transition. This is to be expected, considering the Al-Si diffusion necessary for the latter transition." The author disagrees with the conclusion that the andalusite → sillimanite replacement reaction is the rarest of the Al_2SiO_5 polymorphic reactions. In fact, my review of the literature (see Chapter 8) suggests that there are *numerous* examples of this replacement reaction. However, their suggestion that disordered sillimanite first forms from the decomposition of andalusite is not unreasonable in light of Wenk's (1983) suggestion that submicroscopic mullite inclusions within sillimanite in pelitic xenoliths within the Bergell tonalite, Switzerland, formed from exsolution from a disordered, non-stoichiometric precursor. Detailed electron microscopy examination of sillimanite that has replaced andalusite will be necessary to further evaluate the hypothesis of Winter and Ghose (1979) regarding disorder in sillimanite.

Using single crystal experimental data for a variety of reactions, Wood and Walther (1983) developed a "master" rate equation for metamorphic reactions assuming a zeroth order rate equation. They concluded that reaction rate was controlled by the surface area of the single crystal. From the data plotted in Figure 9.27, they derived a linear equation of log K versus 1/T, which they defended using transition state theory. For prograde metamorphic reactions, *reaction overstepping* was defined as the temperature interval over which reactants and products coexist; the product phase would be completely consumed when temperature reached the upper temperature limit of overstepping. Walther and Wood (1984) considered a simple model of contact metamorphism with a heating rate of 200°C/10,000 yrs. They considered that the attainment of the thermal maximum occurs at 10,000 years subsequent to the emplacement of the adjacent intrusive and that cooling postdates this time period. As shown in Figure 9.28, Walther and Wood (1984) defined the overstepping temperature as that necessary to completely consume the reactant phase. In considering that the Al_2SiO_5 polymorphic reactions have ΔS_r values of *approximately* 1 cal/mol·K, they suggested overstepping of about 20°C for interface (surface) controlled reactions (= "surface detachment" in Fig. 9.28), whereas reaction rate control by diffusion through the fluid phase yields minimum temperature overstepping of approximately 30°C. Walther and Wood (1984, p. 255) concluded: "These results are consistent with the common occurrence of two or more aluminosilicate polymorphs in high temperature contact metamorphosed rocks ... " In the Mt. Raleigh pendant (Kerrick and Woodsworth, 1989), overstepping of the andalusite → sillimanite reaction, as marked by the *apparent* metastable persistence of andalusite above the andalusite-sillimanite equilibrium, is compatible with Walther and Wood's (1984) estimates of overstepping. In spite of this agreement, the potential weaknesses of Walther and Wood's (1984) treatment should be noted. In developing their rate equation using experimental single crystal weight change data, they assumed that the surface area of the single crystal was the rate limiting factor. However, the experiments of Schramke et al. (1987) on the muscovite + quartz ⇄ andalusite + K-feldspar + water reaction, which utilized weight changes of single crystals of quartz as the reaction monitor, revealed that the rate controlling factor was the surface area of powdered andalusite, not the surface area of the quartz single crystal (as assumed by Walther and Wood, 1984). In addition, Walther and Wood (1984) disregarded the pressure effect on reaction rate. Holdaway's (1971) single crystal experiments on the andalusite ⇄ sillimanite reaction at 1.8 kbar versus 3.9 kbar (Fig. 3.17) show that pressure has an important kinetic effect. The kinetic experiments of Tanner et al. (1985) and Schramke et al. (1987) on well known metamorphic devolatilization/volatilization reactions also show that pressure is an important kinetic variable. Wood and Walther's (1983)

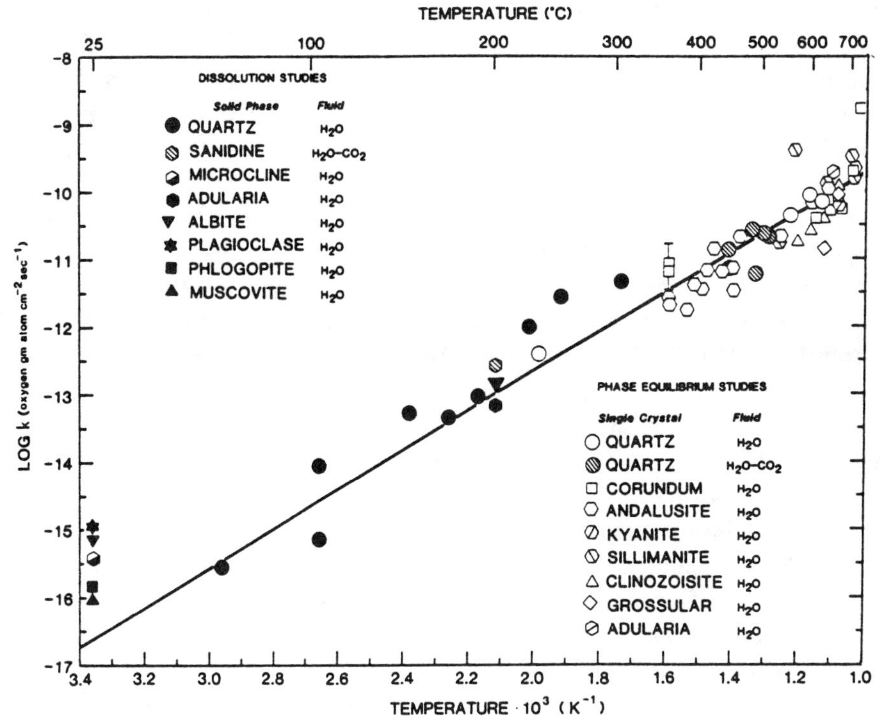

Figure 9.27. Logarithm of reaction rate constant versus reciprocal temperature. The numbered references for various experimental studies are given in Wood and Walther (1983). (From Wood and Walther, 1983, Fig. 2).

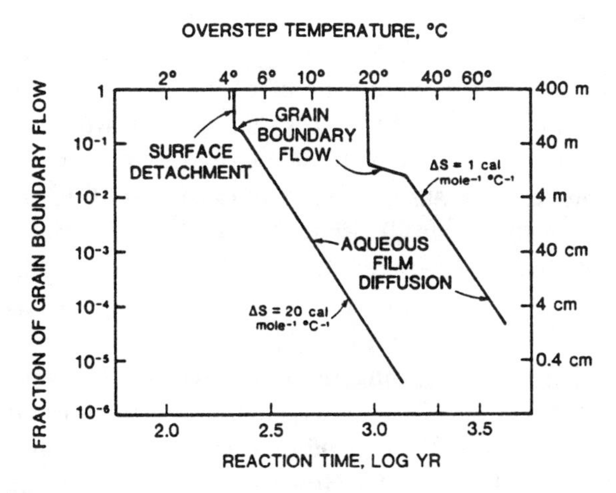

Figure 9.28. Walther and Wood's (1984) calculated temperature of overstepping of a dehydration reaction (ΔS = 20 cal/mol·°C) and solid-solid reaction (ΔS = 1 cal/mol·°C) as a function of the fraction of grain boundary flow and reaction time. The ordinate to the right is fracture spacing. (From Walther and Wood, 1984, Fig. 6).

kinetic equation is based on transition state theory. Derivation of their equation assumes a linear relationship between reaction rate and ΔG_r, a relationship that is only valid close to an equilibrium boundary (Fisher and Lasaga, 1981). Walther and Wood (1984) contended that, for a typical dehydration reaction, this linear relation holds to temperatures approximately 30°C above the equilibrium and that the temperature range of linearity is larger for solid-solid reactions. Thus, their treatment suggests that the linear assumption is valid for the 30°C temperature overstepping suggested by Kerrick and Woodsworth (1989) for the andalusite → sillimanite reaction in the Mt. Raleigh pendant. However, Schramke et al. (1987) had difficulty fitting experimental rate data on the muscovite + quartz hydration/dehydration reaction to a linear rate law. Thus, the validity of Wood and Walther's (1984) linear rate equation is questionable.

Lasaga (1986) utilized a refined non-linear rate law (derived from Monte Carlo calculations) and the experimental data of Schramke et al. (1987) to quantify reaction kinetics of the prograde "second sillimanite" isograd reaction (muscovite + quartz → sillimanite + K-feldspar + H_2O). As with Walther and Wood (1984) he defined temperature overstepping as the interval over which reactants metastably persist above the equilibrium temperature. Assuming that the kinetic results of the experiments of Schramke et al. (1987), which involved andalusite as the Al_2SiO_5 polymorph, are applicable to the analogous reaction involving sillimanite, and using two selected values for the surface area of andalusite, Lasaga (1986) computed a *minimum* temperature overstepping of 30°C for relatively small intrusions (up to 2 km across). He suggested that overstepping could account for the 300 meter-wide zone of coexisting muscovite + quartz + fibrolite + K-feldspar described by Tyler and Ashworth (1982) in the Strontian aureole in western Scotland. Ashworth and Tyler's (1983) estimates of the temperatures of the muscovite-out (645°C) and the cordierite-K-feldspar (690°C) isograds yield a thermal gradient of 35°C/km along a traverse perpendicular to the isograds and in the present plane of exposure. Accordingly, the 500 meter-thick muscovite + quartz + fibrolite + K-feldspar zone would correspond to about 23°C of overstepping. This overstepping interval is considerably less than that estimated by Lasaga (1986) for the overstepping of the muscovite + quartz dehydration reaction in the aureole of an intrusion the size of the Strontian granodiorite (about 8 km across). However, this overstepping is less than that modeled by Walther and Wood (1984) for a "typical" dehydration reaction in contact metamorphism. Thus, questions remain regarding the applicability of kinetic modeling of metamorphic reactions in contact aureoles. Nevertheless, it would appear that significant overstepping of the andalusite → sillimanite reaction has occurred in the relatively short lived metamorphic event of contact metamorphism as well as in regional metamorphism.

Because no experimental data exist on the kinetics of *nucleation* of polymorphic reactions involving the aluminum silicates, we presently have no way of evaluating the amount of overstepping of nucleation. For those attempting to correlate isograd reactions with univariant equilibria, it is important to emphasize that *all* prograde and retrograde reactions occur only with a finite overstepping of the equilibrium boundary. For example, let us consider the isobaric prograde metamorphic path of a kyanite-bearing pelite and examine the point at which the *exact* P-T conditions of the kyanite-sillimanite equilibrium are encountered. At this point there is no driving force for sillimanite to form (because $\Delta G_r = 0$). Thus, in order to form sillimanite, there must be a finite temperature overstepping of the equilibrium into the sillimanite stability field such that ΔG_r is negative. Aside from heating rate, the temperature of overstepping depends upon numerous other factors (e.g., availability of favorable nucleation sites, the magnitude of the kinetic barrier to form the activated complex, potential catalytic effects such as lattice defects and impurities in the reactant kyanite, etc.). About a univariant P-T equilibrium boundary there is a "zone of indifference" within which no spontaneous nucleation and growth of products will occur for a given heating rate. The experiments of Schramke et al. (1987) provide a clear experimental confirmation of the zone of indifference for the brucite ⇄ periclase + H_2O reaction. In prograde metamorphism, the "zone of indifference" may result in the displacement of an isograd to temperatures higher than that of the equilibrium temperature for the reaction under consideration. It is conceivable that the "zone of indifference" about the notoriously sluggish andalusite-sillimanite equilibrium is large, especially for the relatively rapid heating rates encountered in contact metamorphism. Although there are questions regarding the arguments supporting Pattison's (1989) contention that, in the Ballachulish aureole, the incipient andalusite → sillimanite reaction could have occurred 50°C above the andalusite-sillimanite equilibrium, his suggestion provides a provocative view on a potentially significant kinetic problem that requires attention by kineticists.

Rate maxima of the *reverse* andalusite ⇄ sillimanite and kyanite ⇄ andalusite reactions provide an explanation for the limited retrograde replacement reactions involving the aluminum silicates and thus for the survival of high-grade polymorphs during retrogression. For example, in the popular clockwise P-T path for regional metamorphism, rapid retrograde reactions would destroy the sillimanite. Experimental results suggesting negligible sillimanite → andalusite reaction rates are compatible with the limited observed replacement of sillimanite by andalusite.

CHIASTOLITE: CRYSTAL GROWTH MECHANISMS

Large chiastolite porphyroblasts are common in metapelites. The chiastolitic rocks examined by the author are characterized by fairly homogeneous distribution and size of the chiastolites. Considering the mesoscopic scale, these features suggest that the chiastolite nucleation sites were randomly distributed. The lack of a wide spectrum of crystal sizes suggests virtually simultaneous nucleation and growth of the andalusite crystals. It would appear that, following the formation of relatively widely separated nuclei, growth of the chiastolite crystals was rapid. Rapid growth of chiastolite would have suppressed nucleation of andalusite in the matrix rock. In essence, the facile growth of andalusite was such that the activation energy barrier for nucleation within the matrix rock was no longer attained by overstepping. Although these arguments are rudimentary and qualitative, they suggest that there are insignificant kinetic barriers to the nucleation and growth of chiastolite. Andalusite forms from dehydration reactions such as (Yardley, 1989, p. 80):

Chlorite + Muscovite + Quartz = Cordierite + Andalusite + Biotite + H_2O

Because of the relatively large ΔG_r for this dehydration reaction, rapid kinetics are expected. Numerous contact aureoles show that the isograd marking the first appearance of andalusite is very regularly distributed in relation to the contact between the intrusive and surrounding metamorphic rocks (Fig. 9.29). From this regularity, coupled with the qualitative kinetic arguments discussed above, I contend that the nucleation and growth of chiastolite is unimpeded by sluggish kinetics. Thus, in contrast to the conclusion of Hollister (1969), I suggest that formation of andalusite in medium grade pelitic rocks represents equilibrium crystallization within the andalusite P-T stability field.

In spite of the striking characteristic cruciform pattern of chiastolite, there have been relatively few attempts to explain the origin of the chiastolite habit. In addition to the "feathery", "x"-shaped pattern, some chiastolites display square, inclusion-filled cores (Figs. 2.7a and 2.8a).

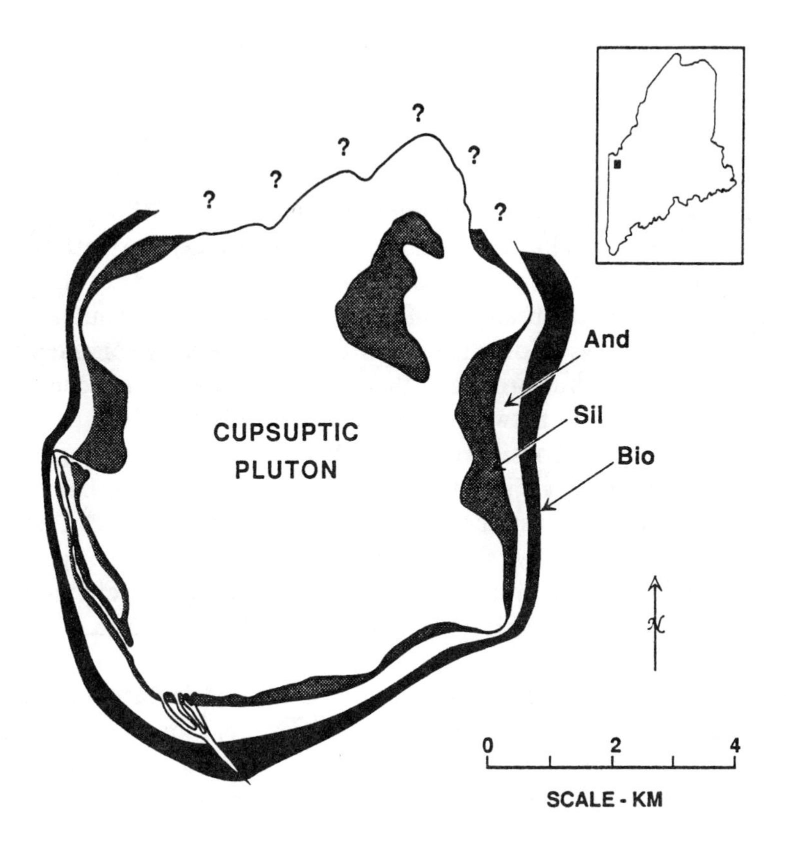

Figure 9.29. The Cupsuptic pluton and surrounding biotite, andalusite and sillimanite zones of the contact aureole. Inset map of Maine shows location of the area. Note conformity of the isograds. (From Bowers et al., 1990, Fig. 1).

Graphite, quartz and white mica are common constituents within the inclusion-rich areas. The four inclusion-poor sectors of chiastolite (Figs. 2.7 and 2.8) are in optical continuity. Many chiastolites contain elongated inclusions of quartz that are mutually parallel within each of the four sectors (Fig. 2.7c and 2.8). Within each sector, the direction of this elongation is parallel to the bounding {110} face.

Since it was proposed in 1939, Harker's theory was generally accepted for the following three decades (Spry, 1969, p. 175). Harker contended that the "feathery", "x"-shaped inclusion pattern of chiastolite arose from inclusions that were swept laterally from growing prism faces, and that the displaced inclusions accumulated at the edges of prism faces. Spry (1969, p. 176) questioned this mechanism because "... there is no obvious mechanism for propelling inclusions sideways along advancing crystal faces."

Because of the presence of parallel, elongated quartz within each of the four sectors of many chiastolites, Rast (1965) concluded that the direction of elongation represents the direction of fastest growth within each sector. "The orientation of inclusions ... in fact suggests that the growth in these directions was dendritic and the inclusions are relict remains of the matrix imprisoned between the branches of the dendrites. Thus the outer zone of the crystal first grows as dendrites positioned at the edges of the crystal. The skeletal crystal thus produced then grows through layers originating by slower growth but between the dendrites." (Rast, 1965, p. 85-86). Dendritic growth was considered to result from "supersaturation" within the matrix immediately adjacent to the chiastolites. Rast hypothesized that rapid dendritic growth resulted in decreased "supersaturation." Dendritic growth was followed by slower "layeritic growth" which produced elongation of the four crystal sectors thus forming the characteristic cruciform habit of some chiastolite porphyroblasts (Figs. 2.8a and 2.9).

Spry (1969, p. 177) concluded that the "x"-shaped inclusion pattern of chiastolite arose from preferential absorption of "carbonaceous particles" at the edges of growing prism crystal faces. "Poisoning" of these edge regions lead to retardation of growth of these areas. The consequent advance of the prism faces over the edges produced the cruciform crystal shape. Spry (1969) contended that the cruciform habit is better developed in smaller chiastolite crystals than larger crystals; thus, he hypothesized that "poisoning" of the growing edges by adsorbed impurities was more effective in the earlier stages of crystal growth.

Petreus (1974) believed that the feathery, "x"-shaped inclusion pattern of chiastolite arose from a secondary alteration process. He suggested that preferential encroachment of externally-derived fluids along the junctions separating the four sectors resulted in the alteration of andalusite to white mica. From these junctions, alteration spread along "linages" parallel to (110). Because the exterior portions of the crystal are the earliest to be attacked by the invading solutions, Petreus (1974) provided an explanation for the progressive broadening of the inclusion-rich bands from the center outward toward the edges of the prism faces (Fig. 9.30). The inclusion-rich bands of many chiastolites appear to be enriched in quartz and graphite. Thus, the author believes that Petreus' (1974) mechanism is inapplicable to such chiastolites. Figure 9.31 shows a chiastolite crystal with muscovite as the dominant constituent of the "x"-shaped bands and

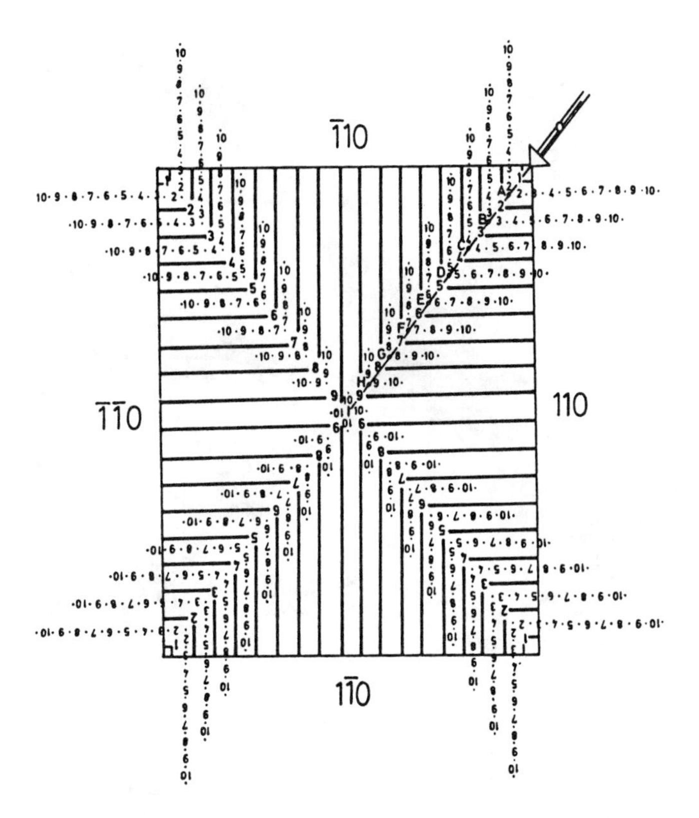

Figure 9.30. Petreus' (1974) model for the development of the "x"-shaped inclusion pattern of chiastolite. The numbers 1 through 10 refer to "time references" (the earliest is 1) following initial alteration of the chiastolite by the influx of fluid (arrow) along the junctions between the four sectors. (From Petreus, 1974, Fig. 5a).

a peripheral zone adjacent to the prism faces; consequently, Petreus' (1974) mechanism is seemingly relevant to the formation of muscovite associated with this crystal. However, there is no guarantee that the "x"-shaped pattern was not present prior to the late-stage alteration.

Nockolds et al. (1978) concluded that the inclusion-rich central areas of chiastolites were produced by rapid growth in the [001] direction. They contended that the inclusion-rich zones between the four sectors formed by dissolution and reprecipitation of matrix material upon growth of the {110} faces. The oriented spindlets of quartz within the four sectors were considered to be recrystallized matrix quartz. "Such spindlets are similar to the tubes of air trapped normal to the freezing interface in ice cubes, and their growth is energetically advantageous since the tubelet form involves a minimum diffusion distance and necessitates no further nucleation" (Nockolds et al., 1978, p. 358).

From examination of well-developed chiastolite porphyroblasts from the Mt. Raleigh pendant, British Columbia (Kerrick and Woodsworth, 1989), the author favors a mechanism similar to that proposed by Rast (1965) and Nockolds et al. (1978) for chiastolite. This analysis considers growth within the (001) plane. Within each of the four sectors, the most rapid growth occurs in a direction radially outward from the center of the porphyroblast. In some chiastolites this formed sectors that are elongated in the direction of growth (Figs. 2.8a and 2.9). Slower "lateral" growth of each sector occurred perpendicular to the [110] direction of each sector. Between each pair of adjacent sectors, this "lateral" skeletal growth formed inclusion-poor "islands" that are elongated parallel to one of the adjacent sectors. Growth perpendicular to [110] lead to coalescence of adjacent skeletal crystals. This coalescence resulted in included patches of the original matrix to the porphyroblasts. Because of impingement of adjacent sectors, the inclusion-rich patches are elongated parallel to the rapid growth direction [110] of the surrounding sector. Elongated inclusions of quartz (Figs. 2.7 and 2.8) formed by recrystallization of included quartz that was trapped between coalescing "islands" of inclusion-poor andalusite. The "x"-shaped, inclusion-rich zones within chiastolite are strikingly similar in appearance to the feathery, albite-rich zones within "trapiche" emeralds from Chivor, Columbia (Fig. 9.32). The mechanism proposed herein for the feathery, "x"-shaped, inclusion-rich arms of chiastolite is similar to that proposed by Nassau and Jackson (1970) for the albite-rich bands that are characteristic of the "trapiche" emeralds.

Rubenach and Bell (1988) investigated microstructural controls on the growth of chiastolite porphyroblasts in the aureole of the Tinaroo Batholith (northern Queensland, Australia). As shown in Figure 9.33 they presented superlative textural evidence that graphite accumulated in front of advancing crystal faces during growth of the andalusite porphyroblasts. They concluded that graphite of the host rock accumulated because of the dissolution of the advancing andalusite crystal faces. Their photo-

Figure 9.31. (*opposite, top*). Chiastolite porphyroblast in a sample from the Mt. Raleigh pendant, British Columbia. Muscovite is a primary constituent of the feathery cross and the area adjacent to the porphyroblast. The cuspate contact between the rim of the chiastolite and surrounding muscovite suggests that muscovite replaced chiastolite.

Figure 9.32. (*opposite, bottom*). Patterns characteristic of "trapiche" emeralds. The light regions are mixtures of corundum and albite. The dark, elongated sectors in the upper- and middle-right photos, and the "feathery" pattern of the light-colored areas in the lower right photo, are similar to chiastolite. (From Nassau and Jackson, 1970, Fig. 2).

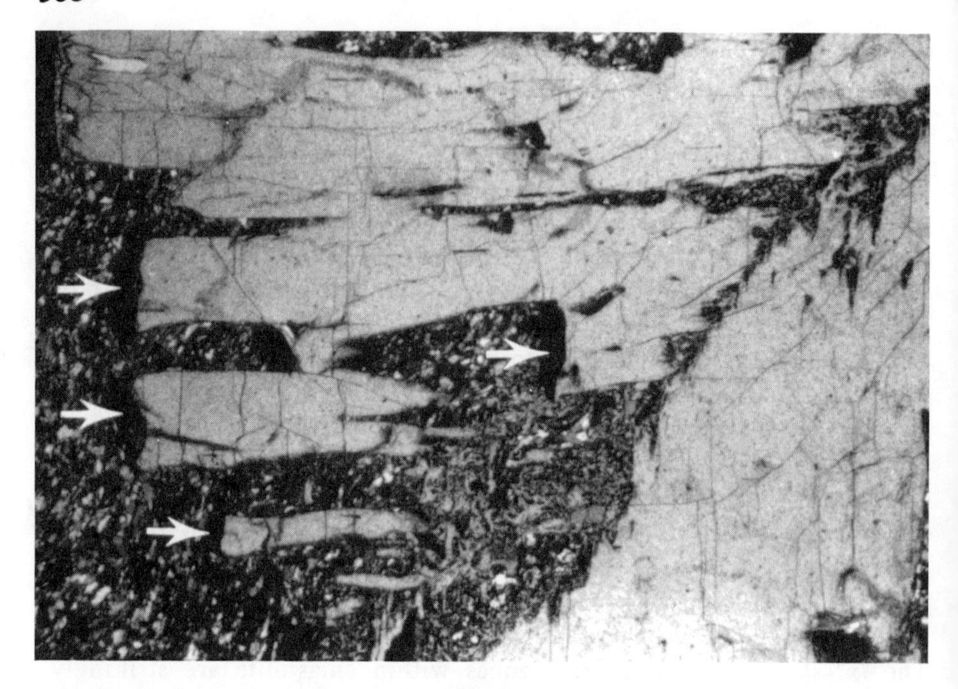

Figure 9.33. Skeletal chiastolite porphyroblast in a metapelite from the aureole of the Tinaroo Batholith, northern Queensland, Australia. Note the accumulations of graphite (black areas noted by arrows) at the ends of the skeletal arms of the chiastolite. (From Rubenach and Bell, 1988, Fig. 14).

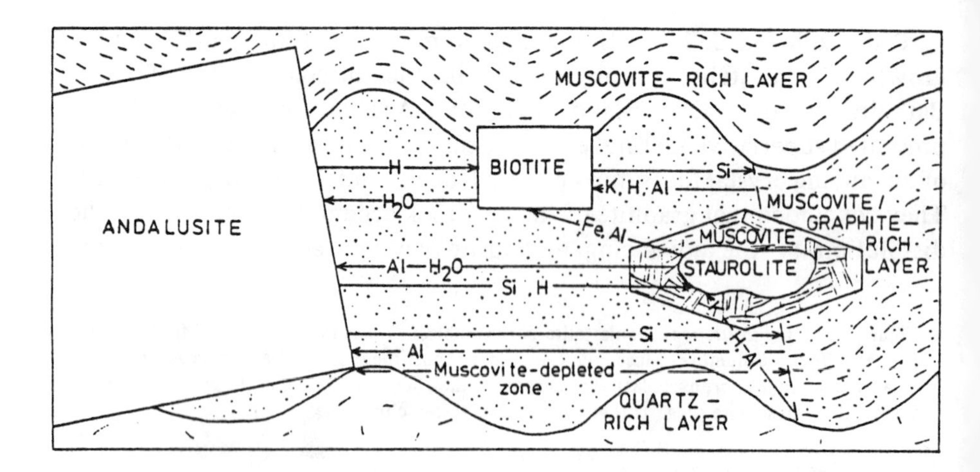

Figure 9.34. Rubenach and Bell's (1988) proposed reaction mechanism for the formation of andalusite in the aureole of the Tinaroo Batholith, northern Queensland, Australia. A constant volume reference frame was assumed at each growth and dissolution site. (From Rubenach and Bell, 1988, Fig. 17).

micrographs provide compelling evidence of the skeletal growth of chiastolite porphyroblasts (Fig. 9.33). It was concluded that chiastolite formed from the reaction:

$$12 \text{ Staurolite} + 9 \text{ Muscovite} + 7 \text{ Quartz} = 58 \text{ Andalusite} + 3 \text{ Biotite} + 12 \, H_2O \, .$$

Figure 9.34 summarizes Rubenach and Bell's (1988) postulated reaction mechanism. Primary constraints for their model are: (a) the existence of muscovite-depleted haloes adjacent to chiastolite porphyroblasts in graphitic layers (Fig. 9.34), and (b) the presence of strained muscovite (i.e. bent crystals with undulose extinction) in the graphitic layers. They hypothesized that the strained muscovite was replaced by quartz and that the Al released at this reaction site (dashed line in Fig. 9.34) diffused to the growing chiastolite porphyroblasts. Thus, their model stands as an example of the catalytic effect of strain on metamorphic reactions.

In contrast to andalusite, the myriads of fibrolite crystals within microscopic domains suggest that there are no significant barriers for the nucleation of fibrolite. The acicular habit of fibrolite is commensurate with the general observation that silicates with crystal structures containing parallel chains of tetrahedra commonly occur as crystals elongated in the direction of such chains (Givargizov, 1986). From crystal growth studies it is well known that acicular crystal forms are favored with supersaturation (Givargizov, 1986). Rast (1965, p. 86) suggested that such supersaturation was produced by temperature overstepping due to rapid heating. However, the abundance of fibrolite in the relatively slow heating of regional metamorphism suggests that rapid heating is not a factor in the formation of fibrolite.

A primary problem is why the same polymorph displays two contrasting habits, i.e., highly acicular crystals (fibrolite) versus larger prismatic crystals (sillimanite) with much smaller aspect ratios than fibrolite. Speer's (1983, p. 153) comment on factors controlling the morphology of carbonate crystals is relevant to this question: "If the crystal structure and chemistry were the only factors affecting morphology, crystal habits could be readily predicted, and, given the limited compositions of the orthorhombic carbonates, would be rather limited. But a crystal's morphology can be modified by varying the chemical and physical conditions prevailing during crystal growth. Such variations affect the growth rates of differing directions in the crystal and thus its morphology." It is possible that the different habits of fibrolite and sillimanite reflect differences in the compositions of metamorphic fluids. Variations in fluid composition could affect the growth rates of different crystal faces. Rapid growth of fibrolite along the *c* axis would be favored by conditions promoting rapid attachment of Al and Si at the ends of the octahedral and tetrahedral chains. The ease of attachment undoubtedly depends upon the nature of the speciation and adsorption at crystal faces. For example, variations in the activity of water my affect surface silanol groups. Variations in the concentrations of minor and/or trace elements in the fluid could affect crystal growth. Could elevated boron activities, as suggested by the common association with tourmaline, be a kinetic factor promoting rapid growth of fibrolite along the *c* crystallographic axis? Analysis of crystal growth with surface site energetics (e.g., Dowty, 1976a) could be helpful in addressing the cause of the acicular habit of fibrolite.

CHAPTER 10

ALUMINUM METASOMATISM

INTRODUCTION

Numerous workers have appealed to metasomatism for the genesis of some aluminum silicate parageneses. Two contrasting metasomatic models have been utilized. Some have considered Al to be either immobile or the least mobile element. In contrast, others considered Al to be one of the most mobile elements. This chapter assesses the arguments for and against each of these disparate hypotheses.

This chapter reviews parageneses showing evidence of the contrasting metasomatic processes of *diffusion* versus *infiltration.* As reviewed by Kerrick (1977), end member diffusion metasomatism may be envisioned by diffusive transport in rocks containing a *stagnant* fluid, whereas infiltration metasomatism occurs by transport due to flux of a flowing fluid. Al_2SiO_5-bearing veins and segregations with evidence suggesting that diffusion metasomatism was dominant are contrasted with those which are considered to have been dominated by infiltration metasomatism. The final section discusses the role of metasomatism in the genesis of fibrolite.

The following topical organization contrasts Al_2SiO_5-bearing segregations with evidence suggesting that the *bulk* of the segregations formed by either replacement of pre-existing rocks or by crystallization within fractures or cavities. In general, diffusion is considered to be important in many of the replacement segregations whereas infiltration was significant in the "cavity filling" segregations. It is important to stress that this topical subdivision is intended to imply the *dominant* metasomatic mechanism. In so doing, I do not imply that segregations form exclusively by one or the other of the contrasting metasomatic mechanisms (i.e., diffusion or infiltration). In fact, it is probable that both mechanisms were operative in the formation of most segregations.

Al_2SiO_5-BEARING VEINS AND SEGREGATIONS FORMED BY REPLACEMENT

Numerous studies have suggested that Al_2SiO_5-bearing veins and segregations formed by replacement of the host rock. In many cases, such segregations are in the form of ovoidal nodules having dimensions ranging from a few mm to about 0.25 m.

Losert (1968) carried out a detailed structural and petrologic study of sillimanite nodules in the Moldanubicum of the Bohemian Massif. The nodules dominantly consist of sillimanite + quartz. Poikiloblastic muscovite is a notable phase in many nodules. All nodules are characterized by the absence of feldspar. As shown in Figure 10.1, some nodules contain conspicuous biotite-rich selvages. Although many of the ellipsoidal nodules are oriented with long axes parallel to lithologic layering (S_1), some nodules are discordant to this layering. The features shown in Figures 10.2 and 10.3 *strongly* suggest that the nodules formed after the development of S_1 and after the tectonism that folded S_1 (Fig. 10.3). Chemical analyses show that the nodules have much lower alkali contents that the host rocks. Such chemical contrasts are compatible with the lack of feldspar in the nodules. Consequently, Losert (1968) concluded that the nodules formed by dealkalization due to the complete decomposition of feldspars. Losert (1968) contended that the inhomogeneous dealkalization was not controlled by cracks and joints; rather, he suggested that anisotropic fluid flow along S_1 (bedding) and S_2 (axial plane foliation) resulted in ellipsoidal nodules with long axes oriented parallel to the direction of maximum fluid transport. He believed that the fluid was generated during metamorphism, although he also suggested that magmatic fluid sources may have been important. Dealkalization was attributed to the decomposition of feldspar and biotite due to decreased activities of alkalies and/or increased a_{H^+} in the fluid phase. Because of low-pH fluids, Losert (1968) envisioned the nodules as forming by "acidic leaching".

Eugster (1970) analyzed equilibria between supercritical fluids and muscovite, K-feldspar and aluminum silicates. He emphasized the importance of using "ionic" equilibria to interpret petrologic processes. His application focused on the genesis of the sillimanite + quartz nodules described by Losert (1968). As shown in Figure 10.4, Eugster (1970) suggested two possible mechanisms to form the sillimanite from K-feldspar: (a) diminution in a_{K^+} and (b) decrease in f_{H_2O}. Localized partial

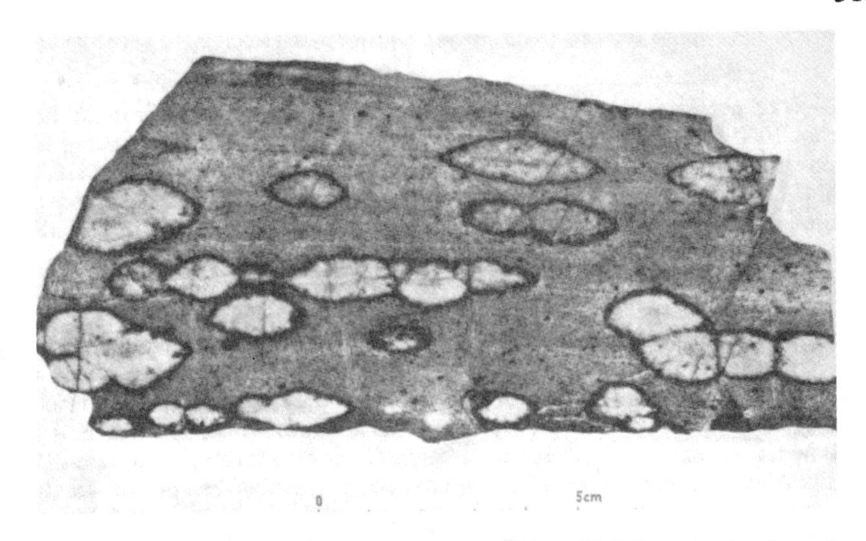

Figure 10.1. Nodules of fibrolite + quartz in a biotite-K-feldspar gneiss from the Moldanubicum of the Bohemian Massif. The dark rims surrounding the nodules are rich in biotite. (From Eugster, 1970, Fig. 10).

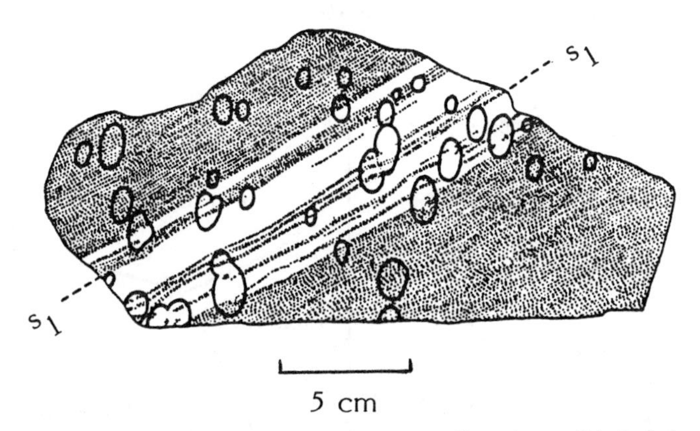

Figure 10.2. Fibrolite + quartz nodules (ovoids) that are discordant to lithologic layering (S_1) in metarhyolite from the Moldanubicum of the Bohemian Massif. The dashed line is parallel to S_1 (see text). (From Losert, 1968, Fig. 4).

melts poor in the K-feldspar component were considered to be sinks for K released by the replacement of K-feldspar by sillimanite in the adjacent melt-free leptynites (feldspathic granulites). Eugster (1970) concluded that the development of relatively few nodules (rather than the pervasive development of sillimanite throughout the leptynite) arose from the initial development of relatively sparse nuclei of sillimanite. Once these nuclei formed, the dealkalization of nearby K-feldspar readily occurred, thereby suppressing further nucleation of sillimanite. Although not explicitly

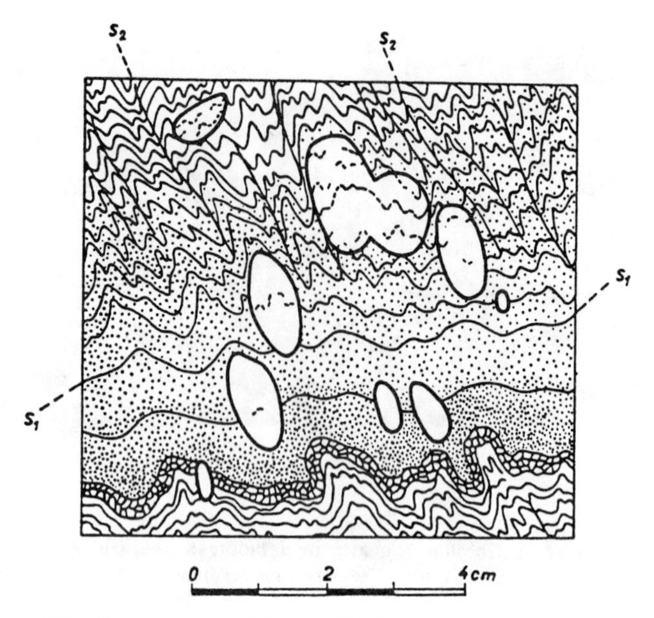

Figure 10.3. Fibrolite + quartz nodules (ovoids) in a metagreywacke. Continuity of the microfolded bedding pattern inside and outside of the nodules suggests that the nodules statically overgrew the matrix. (From Losert, 1968, Fig. 5).

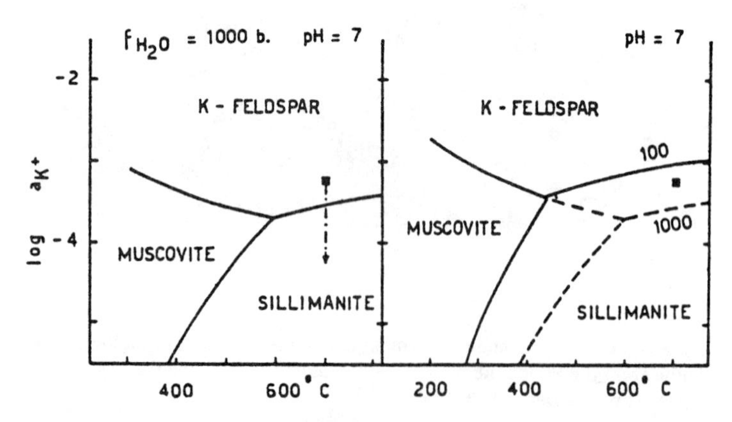

Figure 10.4. Eugster's (1970) mechanism for the transformation of K-feldspar into "sillimanite". *Left*: isothermal decrease in potassium ion activity. *Right*: decrease in f_{H_2O} from 1000 bars to 100 bars. (From Eugster, 1970, Fig. 11).

stated by Eugster (1970), suppression of sillimanite nuclei can be defended by the relevant ionic reaction

$$2 \text{ K-feldspar} + 2 H^+ = \text{Sillimanite} + 5 \text{ Quartz} + 2 K^+ + H_2O .$$

Initial nucleation of the product phases would be accompanied by a decrease in the ratio $(a_{K^+})^2 (a_{H_2O})/(a_{H^+})^2$, thereby suppressing subsequent nucleation of product phases. This hypothesis carries the implicit assumption that reaction and diffusion rates were fast enough to maintain $\mu_{Al_2SiO_5}$ at a level insufficient to nucleate additional sillimanite in the matrix.

Gimmel'farb et al. (1968) described sillimanite + quartz nodules in biotite gneiss of the northern Kodar Range, U.S.S.R. They concluded that these nodules represent metamorphosed aluminous cobbles derived from a conglomerate protolith. However, because these nodules are strikingly similar to those described by Losert (1968), I consider a metasomatic origin similar to that proposed by Losert (1968) and Eugster (1970) to be compelling.

Gresens (1971) studied kyanite-quartz segregations hosted by regionally metamorphosed Precambrian rocks of the Petaca Pegmatite district in northern New Mexico. His paper presented one of the earliest *comprehensive* genetic models for kyanite-quartz segregations. Gresens (1971) found that the segregations are confined to feldspathic schist (metarhyolite). As the kyanite-quartz segregations are approached, the schistosity of the metarhyolite becomes more pronounced and the schistosity becomes more contorted. Gresens (1971) concluded that kyanite growth was syntectonic. Consequently, the kyanite-quartz segregations were considered to have formed during shearing. He contended that the segregations formed by metasomatic interaction between metarhyolite host rocks and acidic fluids transported along the shear zones. The segregations are considered to represent the residue left from acid leaching of the metarhyolites. A primary reaction compatible with acid leaching of the feldspathic host rocks is:

$$\text{K-feldspar} + H^+ = \text{Kyanite} + \text{Quartz} + H_2O + K^+ .$$

Continued leaching of wall rocks would lead to an increase in the K^+/H^+ ratio of the fluid. Eventually, the K^+/H^+ ratio will be high enough to stabilize muscovite by the reaction:

$$\text{Kyanite} + \text{Quartz} + H_2O + K^+ = \text{Muscovite} + H^+ .$$

Accordingly, Gresens (1971) concluded that the sericite schist bordering the kyanite-quartz segregations was produced during the later stages of the hydrothermal activity. Using *apparently* qualitative arguments, Gresens

(1971) addressed the quantity of hydrothermal fluid required to form the mass of kyanite-quartz segregations observed in the Petaca Pegmatite District. He concluded (Gresens, 1971, p. 15) that "... the quantity depends on the acidity of the initial fluid. 'Reasonable' volumes of fluid are needed if the acidity is high ... much larger volumes are required as more 'reasonable' acidities are used ... "

Breaks and Shaw (1973) described sillimanite + quartz nodules and veins hosted by quartz monzonite of the Precambrian Silent Lake pluton in southern Ontario, Canada. The veins and nodules compose up to 30% of some outcrops. Structural features (Fig. 10.5) suggest that the nodules are boudinage formed by attenuation of sillimanite-quartz veins. Dark borders containing abundant biotite + muscovite are common features (Fig. 10.5). The veins dominantly consist of quartz and sillimanite; porphyroblasts of late muscovite are common. Tourmaline is a notable phase in the veins and nodules. At one locality they described a crenulated foliation that "... passes through nodules ..." (Breaks and Shaw, 1973, p. 117). This feature suggests that the nodules formed by replacement of quartz monzonite.

Although there are several similarities in the nodules described by Losert (1968) and Breaks and Shaw (1973), they have different structural aspects. The features shown in Figures 10.2 and 10.3 strongly suggest that the Bohemian Massif nodules formed in a *static* environment around relatively random "nucleation centers". Breaks and Shaw (1973) contended that the Silent Lake segregations were originally pegmatitic quartz-muscovite-K-feldspar veins that were fragmented and boudinaged during the emplacement of the pluton. During this deformation the veins were zones of structural weakness and thus were channelways for egress of aqueous fluids which produced sillimanite by alkali leaching. Breaks and Shaw believed that sillimanite was replaced by muscovite during the waning stages of this hydrothermal activity. However, parallelism of the veins, their tabular shape, and the apparent boudinage structures (Fig. 10.5), suggest instead that the Silent Lake segregations *may* have formed by channelized fluid flow along a set of parallel fractures. In contrast to the scenario of Breaks and Shaw (1973), this hypothesis is more compatible with their observation that at one locality the foliation continues through the nodules and, thus, the nodules formed subsequent to deformation.

Neither Losert (1968), Eugster (1970), or Breaks and Shaw (1972) provided an explanation for the biotite-rich selvages surrounding sillimanite nodules described in these studies. Losert (1968, p. 115) noted

Figure 10.5. Parallel fibrolite-quartz layers with dark, biotite-rich selvages (Silent Lake pluton, Ontario). Note the boudinage structure of the layers. (From Breaks and Shaw, 1973, Fig. 6; photo courtesy of F.W. Breaks).

that "... relics of incompletely replaced biotite ..." are common within the nodules. Thus, biotite and feldspar were unstable within the cores of the nodules. Decomposition of biotite and feldspar would be compatible with low a_{K^+}. Alkalies and ferromagnesian components diffusing away from the centers of the nodules could have reprecipitated biotite upon entering the higher a_{H^+} and/or higher f_{H_2O} environment of the surrounding rocks.

Fisher (1970) appealed to ionic equilibria in modeling the genesis of small, zoned andalusite-rich segregations (Fig. 10.6) enclosed in a sillimanite-bearing gneiss near Västervik, Sweden. He based his metasomatic model on the chemical data of Loberg (quoted in Fisher, 1970) suggesting that the segregations formed by migration of Fe, Mg, and Ca from the mantles to cores. Loberg's *in situ* differentiation model is supported by the fact that the bulk composition of the segregations (i.e., cores + mantles) is virtually identical to that of the surrounding sillimanite gneiss. A constant Al reference frame was defended by the fact that the $AlO_{3/2}$ content of the mantle is virtually identical to that of the entire segregation; furthermore, textures suggest that the biotite + andalusite intergrowths replaced microcline. Figure 10.7 depicts Fisher's (1970) representation of the mineral reactions forming the cores and mantles. Considering reactant ionic species, the mantles were local sinks for K^+,

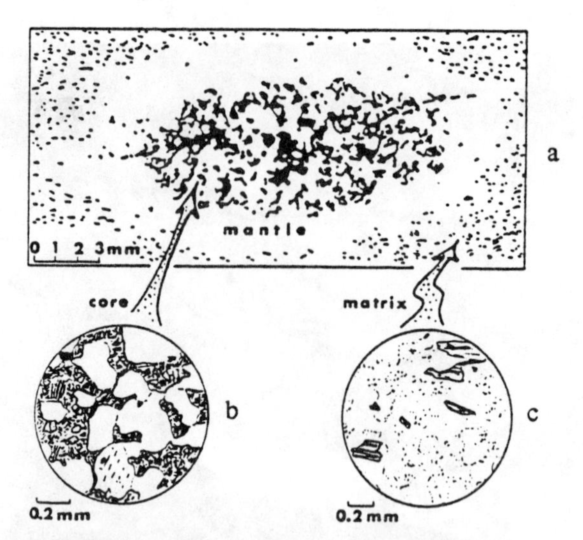

Figure 10.6. (a) Andalusite-rich segregation surrounded by a biotite-free mantle in a sample from Västervik, Sweden. (b) Sketch of a portion of a thin section from the core of the segregation, showing a fine-grained andalusite + biotite intergrowth (dark), quartz (clear) and plagioclase (streaks mark cleavage). (c) Sketch of a portion of a thin section of the host gneiss. Fisher (1970) contended that the texture of the andalusite-rich core (b) is similar to that of the gneiss (c). (From Fisher, 1970, Fig. 1).

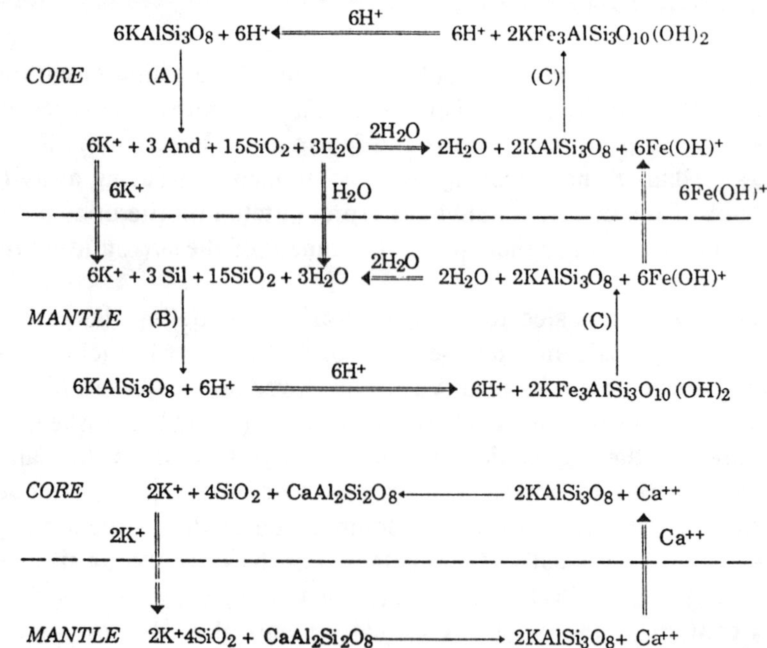

Figure 10.7. Fisher's (1970) hypothesized exchange cycle linking diffusive exchange of K and Fe (*top*), and K and Ca (*bottom*), between cores and mantles of segregations near Västervik, Sweden (see Fig. 10.6). The arrows with double lines indicate direction of migration of chemical species whereas the arrows with single lines represent chemical reactions. (Redrawn from Fisher, 1970, Figs. 5 and 6).

whereas the cores were sinks for $Fe(OH)^+$ and $Mg(OH)^+$. Textural data suggest that andalusite of the segregations postdated the formation of sillimanite in the surrounding gneisses. Thus, Fisher (1970) considered that the *overall* driving force for the segregation-forming process was ΔG for the sillimanite → andalusite reaction. He contended that the sparse development of andalusite nuclei was attributed to sluggish kinetics of the sillimanite → andalusite reaction. Fisher (1970) noted that the segregations formed with small gradients in the activities of the ionic components between the cores and mantles. The andalusite segregations at Västervik require diffusion over relatively small distances (ca. 0.5 cm). Application of Ficks Law showed that these segregations could have formed with realistic diffusivities in a geologically short time period (1000 years).

Using irreversible thermodynamics, Fisher (1973, 1977) quantified diffusion-controlled metamorphic differentiation. His computational method was based on a series of equations specifying diffusivities, rates of production and mass conservation of diffusing species, and rates of metamorphic reactions. To model the andalusite segregations at Västervik, Fisher (1977) again assumed a constant Al reference frame and determined the fluxes of species across the mantle/core boundary from comparisons of the bulk compositions of the mantles and cores. Using the basic tenant of the second law of thermodynamics, i.e., that the total entropy production of irreversible reactions is positive, Fisher (1977) solved square matrices of the kinetic ("phenomenological") coefficients of equations linking the fluxes and chemical potentials of major oxide components. As shown in Figure 10.8, Fisher (1977) evaluated growth of the segregations by comparing reaction rate control versus diffusion rate control. From this analysis, the Västervik segregations suggest that diffusion was the rate controlling mechanism. The maximum diffusion distance of these segregations, as deduced by the core radii of about 0.5 cm, coupled with the line for diffusion control in Figure 10.8, yields a time of approximately 66,000 yr. As concluded by Fisher (1977), this time period is very reasonable for a metamorphic episode.

Yardley (1977) invoked ionic equilibria and local diffusion to explain the development of small-scale fibrolite segregations in the Connemara Schists of western Ireland. With increasing metamorphic grade, fibrolite first appears as rims on garnet and as intergrowths with biotite. With increased reaction garnet becomes replaced by fibrolite + biotite. With a constant Al reference frame, Yardley (1977) noted that the replacement of garnet would be accompanied by a 50% decrease in volume. He contended that such a substantial volume decrease would result in deflection

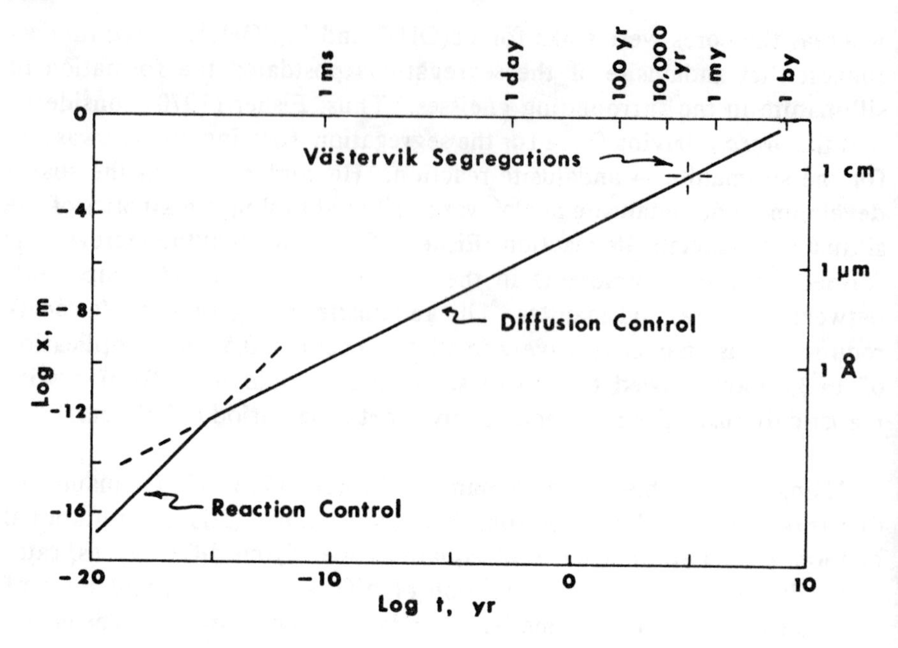

Figure 10.8. Fisher's (1977) analysis of the rate of growth of the radii (x) of andalusite-rich segregations near Västervik, Sweden. (From Fisher, 1977, Fig. 8).

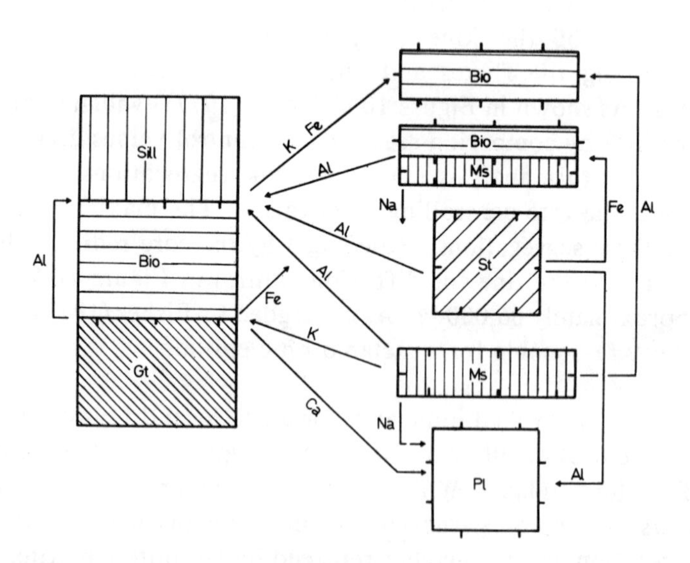

Figure 10.9. Yardley's (1977) schematic representation of ionic reaction cycle for the development of fibrolite during metamorphism of pelites in the Connemara region of western Ireland. Ticks represent the direction of movement of grain boundaries. (From Yardley, 1977, Fig. 2).

("bowing") of the schistosity toward the replaced garnet porphyroblasts. Because the schistosity is not deflected near the pseudomorphed garnets, he argued that volume was conserved during the replacement of garnet by biotite + fibrolite. Thus, he concluded that Al was mobile during metamorphism. Further evidence for Al mobility is provided by textures of andalusite porphyroblasts that occur in nearby areas of the pelitic unit that was the focus of Yardley's (1977) study. Inclusions within the andalusite porphyroblasts are oriented parallel to the surrounding schistosity. Therefore, growth of the andalusite porphyroblasts occurred with little or no volume change. These examples suggest that a constant volume reference frame is applicable to the pseudomorphism of garnet by fibrolite + biotite and the growth of andalusite porphyroblasts. Figure 10.9 depicts Yardley's (1977) overall ionic reaction cycle. The Al to form sillimanite is considered to have been derived from the breakdown of other aluminous phases. In analyzing the factors controlling the mobilities and directions of diffusion of various components, he concluded that there were significant differences in the metasomatic development of the andalusite-bearing versus fibrolite-bearing metapelites. In particular, andalusite replaced staurolite, a reaction that involved flux of Al *toward* staurolite rather than *away from* staurolite as implied by the fibrolite-bearing rocks. Thus, Yardley (1977) concluded that diffusion was controlled by the stoichiometry of metamorphic reactions and the spatial distribution of product phase nucleation.

In a series of papers, Foster (1977, 1981, 1982, 1983, 1986, 1990) used a local equilibrium, irreversible thermodynamic model to interpret small-scale segregations in fibrolite-bearing pelitic schists near Rangeley, Maine. Rocks in this area are part of a Paleozoic metasedimentary sequence that underwent low-pressure regional metamorphism and subsequent contact metamorphism (Foster, 1977). From a thorough petrographic analysis, Foster (1977) concluded that the fibrolite segregations formed by the replacement of biotite adjacent to garnets. Continued reaction formed segregations with fibrolite-rich cores surrounded by biotite-rich mantles. As shown in Figure 10.10, the development of the segregations was attributed to cation-exchange reactions linking the segregations and adjacent (growing) garnets and (unstable) staurolite. The reaction leading to the pseudomorphic replacement of staurolite provides a sink for alkalies released by the fibrolite-forming reaction occurring at the boundary between the cores and mantles of the segregations. In utilizing a mean molar reference frame, Foster (1977) concluded that the $AlO_{3/2}$ component had the largest mass transfer. His computational approach was similar to that of Fisher (1977), i.e., he outlined a series of equations

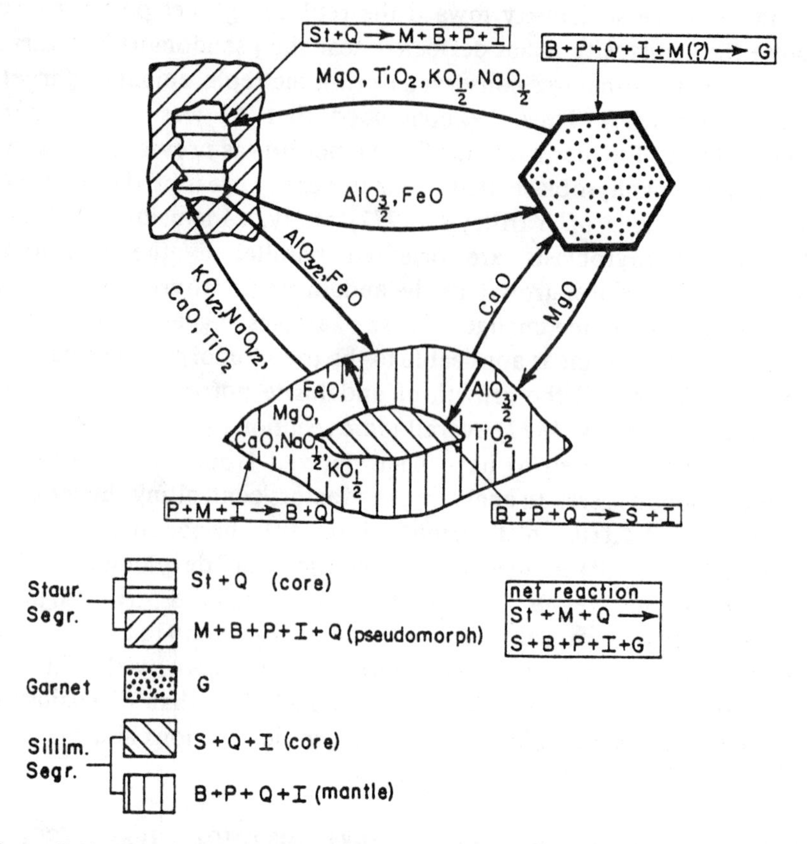

Figure 10.10. Foster's (1981) cation exchange mechanism for the formation of sillimanite segregations in metapelites near Rangeley, Maine. Abbreviations for the minerals are: St = staurolite, S = sillimanite, G = garnet, B = biotite, M = muscovite, P = plagioclase, I = ilmenite, Q = quartz. (From Foster, 1981, Fig. 1).

which included the Gibbs-Duhem equations in the rock matrix adjacent to the segregations, conservation equations relating the fluxes of components between the segregations and matrix, and conservation equations for SiO_2 and H_2O. In all of Foster's papers it was assumed that the overall system was closed to all components except H_2O. Support for the cation exchange reactions summarized in Figure 10.10 is given by the fact that the sum of the four ionic reactions (Fig. 10.10) closely matches that of the overall (net) reaction. His computations showed that $AlO_{3/2}$ has the *largest* local mass transfer of all of the ten major oxide components that he considered. Foster's 1982 paper illustrated the utility of the local-equilibrium, irreversible-thermodynamic model for predicting textures of the fibrolite segregations. The predicted textures were confirmed by petrographic examination. Foster (1986) refined his metasomatic model to explain why some garnets are resorbed whereas others have textures indicative of growth. He concluded that garnets were resorbed where fibrolite replaced *nearby* biotite; in contrast, garnets experienced growth if the biotite → fibrolite reaction was further removed.

Foster's (1990) most recent analysis shows that the formation of biotite-rich rims surrounding sillimanite + quartz nodules may involve simultaneous dissolution of biotite in a zone immediately adjacent to the core (Fig. 10.11). He concluded that biotite is nearly conserved in the *overall* reaction linking the nodules with the adjacent muscovite pseudomorphs after staurolite + quartz. Thus, Foster contended that biotite served as a catalyst for the fibrolite-forming reaction rather than a reactant. Foster's (1990) analysis emphasizes that the interpretation of fibrolite-forming reactions must be based on examination of textures within all subdomains of individual specimens.

Foster's analysis provides a possible explanation for the biotite-rich selvages around nodules hosted by feldspathic rocks described by Losert (1968) from the Bohemian Massif and by Breaks and Shaw (1972) in southern Ontario. That is, ferromagnesian components diffusing from decomposition of biotite within the cores of the nodules could react with feldspars to form biotite in the host rocks immediately adjacent to the biotite-rich selvages. Although Foster formulated the biotite-forming reaction with plagioclase as a reactant (Fig. 10.11), this reaction could be reformulated with K-feldspar. As suggested by Foster for the sillimanite segregations at Rangeley, Maine, the biotite-producing reaction could have involved consumption of muscovite and ilmenite in the host rocks.

Rubenach and Bell (1988) carried out a detailed study of the mechanisms of growth of chiastolite crystals in metapelites from the aureole of the Tinaroo Batholith, Queensland, Australia. Adjacent to the chiastolite porphyroblasts there is little or no deflection of the foliation. Because of the implied constancy of volume in the formation of the porphyroblasts, they concluded that "... it is obvious that Carmichael's reference frame of constant alumina is invalid for these rocks ..." (Rubenach and Bell, 1988, p. 664).

The analyses of Yardley (1977), Foster (1977) and Rubenach and Bell (1988) provide arguments against the popular use of a constant-Al reference frame in the metasomatic modeling of microscopic scale phenomena in metapelites.

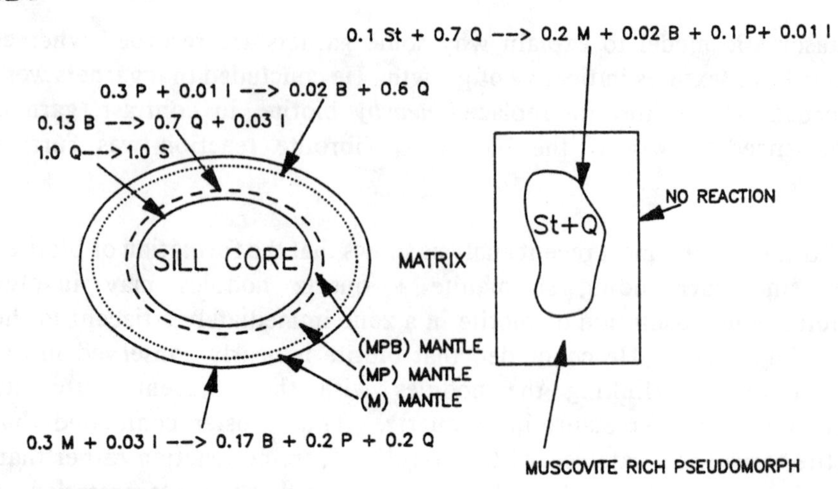

Figure 10.11. Foster's (1990) interpretation of local reactions involved in the formation of fibrolite + quartz nodules (*left*) and adjacent muscovite pseudomorphs after staurolite + quartz (*right*). As indicated at the bottom, the net reaction linking the nodules and pseudomorphs produced minor amounts of biotite. Within the nodule biotite is produced in the rim and simultaneously consumed in the region adjacent to the core. (From Foster, 1990; diagram courtesy of C.T. Foster).

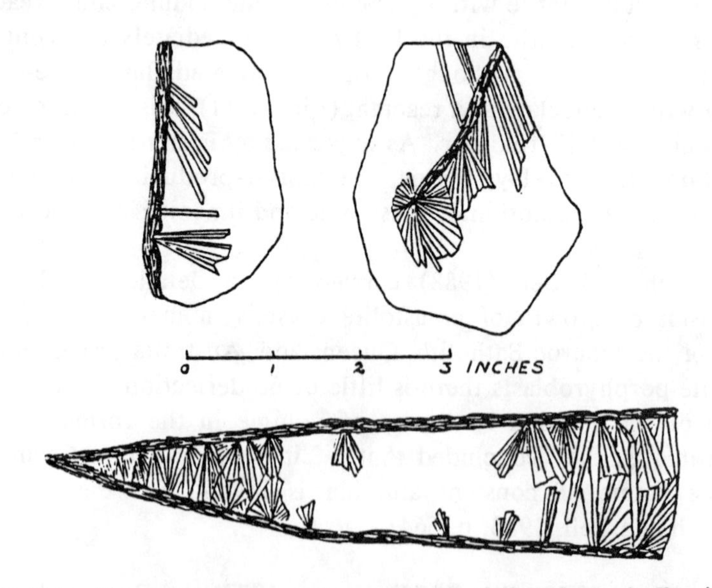

Figure 10.12. Kyanite-quartz segregations on the isle of Unst, Scotland. Kyanite forms rosettes enclosed in quartz (clear). The dark pattern is chlorite (note that the chlorite forms vein selvages in the upper left and bottom sketches). Note that kyanite (K) forms "sprays" that appear to have grown from the walls of the segregations (upper left and bottom sketches). (From Read, 1932, Fig. 3).

Al_2SiO_5-BEARING VEINS AND SEGREGATIONS FORMED BY CRYSTALLIZATION WITHIN FRACTURES AND CAVITIES

In comparison to the veins and segregations described in the previous section, a contrasting paragenesis is aluminum silicates in veins and segregations with features suggesting crystallization within fractures and cavities. The studies summarized below appeal to *infiltration* metasomatism as a dominant mechanism, and are thus treated separately from veins and segregations (discussed in the preceding section) where diffusion is believed to have been dominant.

Most veins and segregations described in this section have a *pegmatitic* texture. The following studies implicitly or explicitly appeal to crystallization from a supercritical aqueous fluid. However, a popular alternative mechanism is crystallization of volatile-rich silicate melts (Černý, 1982). This alternative genetic model is reviewed in Chapter 11.

Read's (1932) study of kyanite-quartz veins in the isle of Unst (Shetland Islands, Scotland) represents one of the earliest detailed accounts of this type of metasomatic deposit. The veins are hosted in kyanite-chloritoid schists; many veins are parallel to schistosity, although some are discordant. The veins are bordered by coarse-grained, kyanite-rich, quartz-poor selvages. Kyanite crystals apparently grew as "rosettes" from vein walls (Fig. 10.12 bottom and upper left) or what appear to be veinlets transecting quartz (Fig. 10.12 upper right). In reviewing various possible mechanisms for vein formation, Read (1932) concluded that they formed by segregation from the host rocks during metamorphism. This conclusion was supported by the *apparently* qualitative mass balance argument that the combined compositions of the veins + selvages are virtually identical to the composition of the host rocks. Accordingly, he concluded that prior to metasomatism the selvages consisted of unaltered pelitic host rock and that during metamorphism the components that form the veins were derived by "endogenous secretion" from the selvages. In comparing the mineralogy of the veins and host rocks, Read (1932, p. 326) concluded: "A short and incomplete survey of the accounts of quartz-veining in metamorphic rocks shows very clearly the correspondence between the mineral content of the quartz veins and that of the country-rocks ... " Although not explicitly stated by Read, his discussion of a genetic model *suggests* that he did not favor crystallization within cavities. However, I consider that the rosettes of kyanite radiating into the segregations from the selvages (Fig. 10.12) provide strong evidence for crystallization within cavities.

Veins similar to those described by Read (1932) were found by Miyashiro (1951) in staurolite-kyanite schists from the Fukushinzan District, Korea. Euhedral crystals of kyanite and quartz form druses lining open cavities. Miyashiro concluded that the druses represent cavities that were filled with pockets of fluid during the formation of the veins. The druses, coupled with the coarse, pegmatitic character of these and many other Al_2SiO_5-bearing veins and segregations, suggest that transport of components in significant quantities of aqueous fluid was an integral aspect of their genesis.

Fonteilles (1965) described segregations with coarse-grained, euhedral kyanite hosted by staurolite schists in the Baud area, Morbihan, France. He concluded that the veins formed within tension fractures, and that precipitation of the vein kyanite occurred because fluid pressure within the fractures was markedly less than that of the host rocks.

In the Truchas range, north-central New Mexico, quartz veins hosted by quartzite and schist contain fibrolite-rich selvages up to 5 cm thick (Grambling, 1981). Grambling concluded (p. 706) that the selvages "...may have formed from a type of contact metamorphism, induced by hot fluids migrating along channels ..." However, because of the presence of kyanite and andalusite in the adjacent host rocks, he proposed an alternative origin for the selvages: "... local concentrations of fluids may have acted as a simple catalyst in the polymorphic reactions." Because tourmaline is an accessory in the selvages, and because the selvages have a much higher modal amount of fibrolite compared to kyanite + andalusite in the host rocks, I favor a metasomatic origin.

Sauniac and Touret (1983) carried out a study of fluid inclusions in a pegmatitic kyanite-quartz segregation from the Main Central Thrust of the Himalayas. The segregation is virtually a bimineralic kyanite + quartz rock. In contrast to Read's (1932) generalized correlation between the mineralogy of segregations and host rocks, kyanite is not a widespread constituent of the host rocks. Rather, "... kyanite appears only in some biotite-rich layers ..." (Sauniac and Touret, 1983, p. 35). Because the segregation is conformable to the surrounding schistosity, they contended that the segregation formed syntectonically. As shown in Figure 10.13, thermobarometric analysis suggests that the segregations formed during the later stages (stage II in Fig. 10.13) of a retrograde decompression P-T

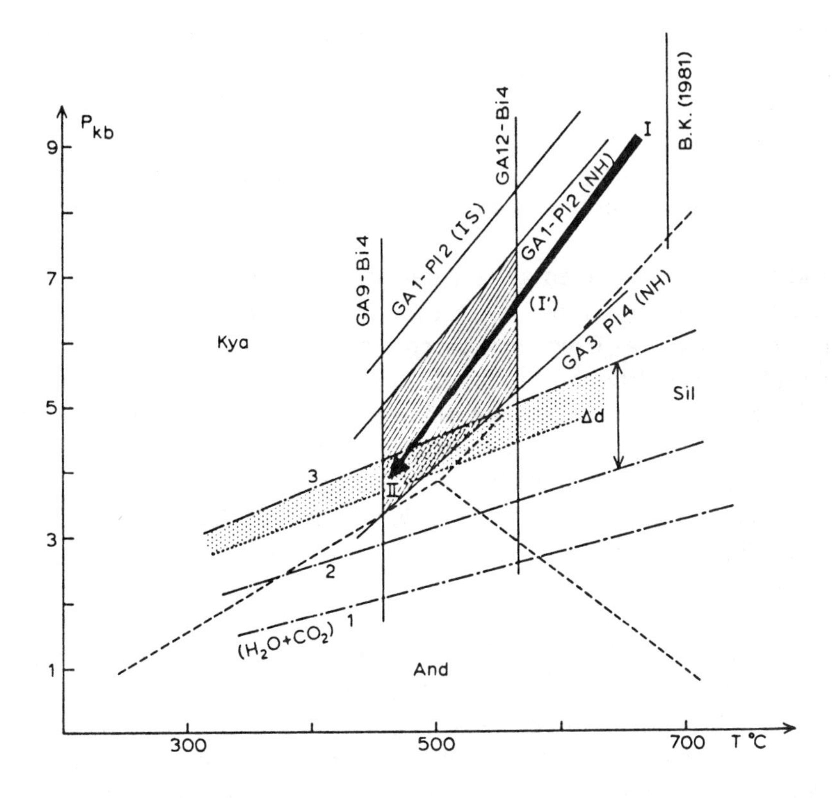

Figure 10.13. Thermobarometric analysis of a kyanite-quartz segregation from the Main Central Thrust, Himalayas. The dashed lines are univariant Al_2SiO_5 equilibria. The lines labeled "GA-BI" were determined from garnet-biotite geothermometry of the host metapelites, whereas those labeled "GA-PL" were derived from the garnet-plagioclase-kyanite-quartz geobarometer. The dot-dash lines are isochores for H_2O-CO_2 fluid inclusions within quartz crystals from the segregations. The stippled area is the most probable location of the isochore for the earliest fluids in the segregation. The heavy arrow represents the hypothesized decompression trajectory: I is the peak metamorphic conditions, I' is the maximum garnet-biotite temperature of the host rocks, and II is the final P-T condition recorded by thermobarometry of the host rocks and by the earliest fluid inclusions within the segregations. (From Sauniac and Touret, 1983, Fig. 9).

trajectory. Although they did not elaborate on the mechanism, it was concluded that the segregation was a fluid sink during retrogression.

Andalusite-bearing veins and segregations have been described from numerous localities. This paragenesis occurs in a variety of disparate host rocks: ultramafics (Murdock, 1936), metabasites (MacDonald and Merriam, 1938), metapelites (Akaad, 1956; Rose, 1957; Woodland, 1963; Yardley et al., 1980; Spear, 1982; Mitchell et al., 1988), and granitic rocks (Burt and Stump, 1983).

MacDonald and Merriam (1938) concluded that an andalusite-bearing pegmatite vein in the western Sierra Nevada, California, originated by the influx of Al-bearing magmatic fluids. Derivation of the vein components from the surrounding wall rocks was excluded on the basis of their relatively Al-poor compositions (the host rocks are amphibolites) and the apparent lack of metasomatic alteration of the nearby schists.

Akaad (1956) described andalusite-bearing segregations and veins in the northern aureole of the Ardara pluton, Ireland. He noted that the segregations occur as irregular-shaped masses that are elongated parallel to the schistosity of the host rocks. Some veins are concordant to the host rock foliation whereas others are discordant. Andalusite is concentrated in the margins of the veins and appears to have grown as clusters projecting from the vein walls. With relatively little discussion of the genetic mechanism, Akaad (1956) hypothesized that the veins and segregations formed by "metamorphic differentiation" of the host metapelites.

Stout et al. (1986) studied fluid inclusions in Al_2SiO_5-bearing segregations and metapelitic host rocks from the Mica Creek area, British Columbia. Mineral equilibria suggest peak metamorphic P-T conditions of 7-8 kbar and 600-700°C (Fig. 10.14). They concluded that sillimanite and kyanite formed in the segregations and in the pelitic host rocks during the peak of metamorphism. Uplift and cooling subsequent to metamorphism produced unmixing of the H_2O-CO_2-NaCl fluid. Intersection of isochores for samples with unmixed fluids (i.e., samples with coexisting H_2O-rich fluids and CO_2-rich fluids) suggest a *minimum* fluid inclusion exsolution temperature of 410°C [3 in Fig. 10.14]. In one locality, segregations containing andalusite + quartz are hosted by kyanite-bearing pelitic schists. Compared to samples from other localities in the Mica Creek area, fluid inclusions within andalusite-quartz segregations are unique in that they contain CH_4 and N_2. The low densities and the isochores of the CH_4- and N_2-bearing fluid inclusions suggest that the andalusite-quartz segregations formed at 1-2 kbar and approximately 400°C [4 in Fig. 10.14]. Thus, the P-T history recorded by the segregations at Mica Creek suggests an earlier Barrovian-type metamorphism in which crystallization of kyanite or sillimanite within the segregations was contemporaneous with formation of kyanite or sillimanite in the respective host rocks, followed by the late-stage formation of andalusite in the segregations upon retrograde decompression. Stout et al. (1986) did not address metasomatic mechanisms for the origin of these segregations. However, E.D. Ghent (pers. comm.) considers that they

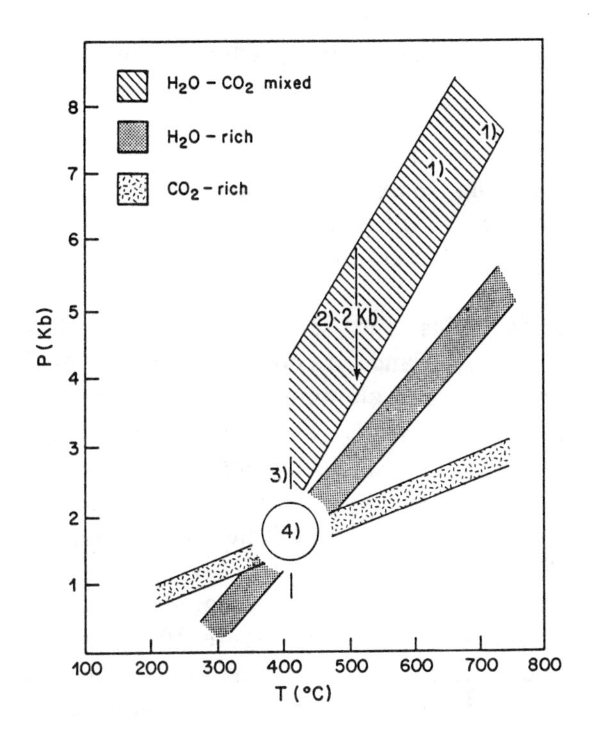

Figure 10.14. Thermobarometric analysis of Al_2SiO_5-bearing segregations and metapelitic host rocks from the Mica Creek area, British Columbia. The patterned areas are isochores of H_2O-rich, CO_2-rich and mixed H_2O-CO_2 fluids. The points along the hypothesized uplift path are: 1) peak metamorphic P-T conditions (determined from mineral equilibria), 2) 2 kbar envelope below the isochore of the densest fluid inclusion, 3) minimum temperature of the "empirical" solvus, and 4) final P-T conditions as derived from analysis of fluid inclusions. (From Stout et al., 1986, Fig. 5).

formed by crystallization within open cavities rather than by replacement of pre-existing pelitic rocks.

Mitchell et al. (1988) described spectacularly large (up to 30 x 17.5 cm) crystals of andalusite in quartz veins within the Wissahickon schist of south-central Virginia. Most of the andalusite crystals were paramorphically replaced by kyanite and (subordinate) sillimanite. Mitchell et al. (1988) did not discuss a genetic model for the veins. However, because of the general textural and mineralogical similarities with Al_2SiO_5-bearing quartz veins described in other localities, a common origin is likely.

Al_2SiO_5-bearing segregations in the Lepontine Alps, Switzerland: a case study

"In the interest of environmental protection, and to avoid unscrupulous mineral collectors and commercial mineral dealers, the locations of particularly worthwhile localities are not given"

Free translation of H.-H. Klein (1976, p. 440)

Al_2SiO_5-bearing veins and segregations are widespread in the Lepontine Alps of Switzerland. As epitomized by the above quote from Klein (1976), many of the aluminum silicate segregations consist of spectacular, gem-quality crystals. The segregations occur in masses up to 10 meters across (Keller, 1968). Specimens from this region are common in mineral museums; accordingly, Al_2SiO_5-bearing segregations from the Lepontine Alps are perhaps the most famous of this type of paragenesis.

The Lepontine Alps display a classic prograde sequence of Barrovian-type metamorphism. As shown in Figure 10.15, metamorphic grade increases from north to south; the highest grade rocks are immediately north of the Insubric Line - a major "suture" zone in the southern Alps. Fibrolite is the primary index mineral of the "sillimanite" zone.

As shown in Figure 10.15, Al_2SiO_5-bearing segregations are common in the region 10-30 km north of Locarno and Bellinzona. The segregations occur in the deepest nappes of the Pennine sequence (i.e., the Simano, Cima-Lunga and lower Adula nappes). As shown in Figure 10.15, kyanite is the first aluminum silicate to appear in the segregations in traversing upgrade. The first appearance of kyanite in the segregations generally coincides with the kyanite-staurolite isograd of the host metapelites. At the highest grades of metamorphism, the segregations are characterized by fibrolite as the sole Al_2SiO_5 phase (Fig. 10.15). Andalusite is confined to segregations within the central portion of the kyanite zone of Figure 10.15.

Keller (1968) carried out a detailed study of segregations in the Campo-Tencia-Pizzo Forno area, located about 15 km northwest of Biasca (Fig. 10.15). On the basis of mineralogy, he distinguished three types of segregations: (1) virtually pure quartz, (2) kyanite + andalusite + quartz + feldspar, and (3) kyanite + quartz + staurolite + muscovite ± paragonite. As shown in Figure 10.16, vein type (2) is much more widespread than type (3). Aside from the aluminum silicates, the kyanite-andalusite segregations [type (2)] are characterized by the presence of plagioclase

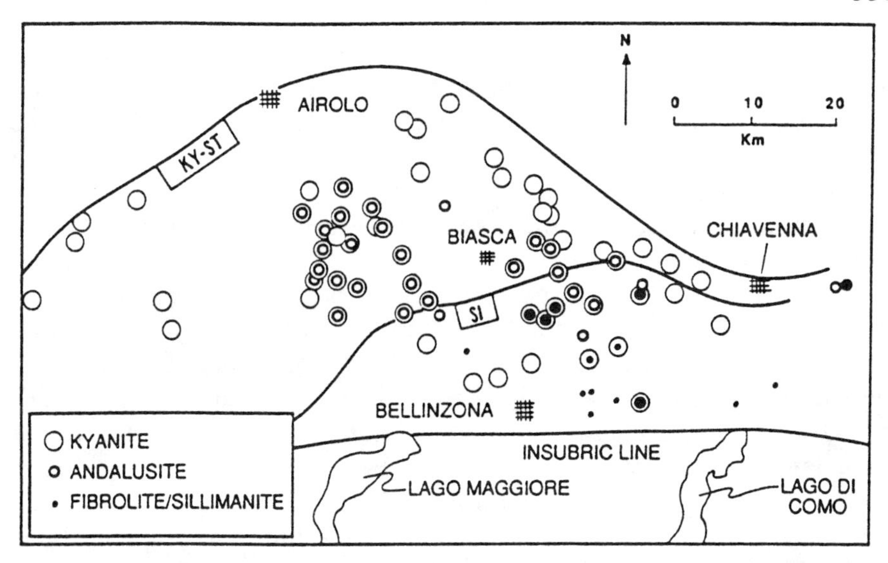

Figure 10.15. Distribution of aluminum silicates in segregations in the Lepontine Alps, Switzerland. The kyanite-staurolite (KY-ST) and sillimanite (SI) isograds are from Frey et al. (1980). (From Kerrick, 1988, Fig. 1).

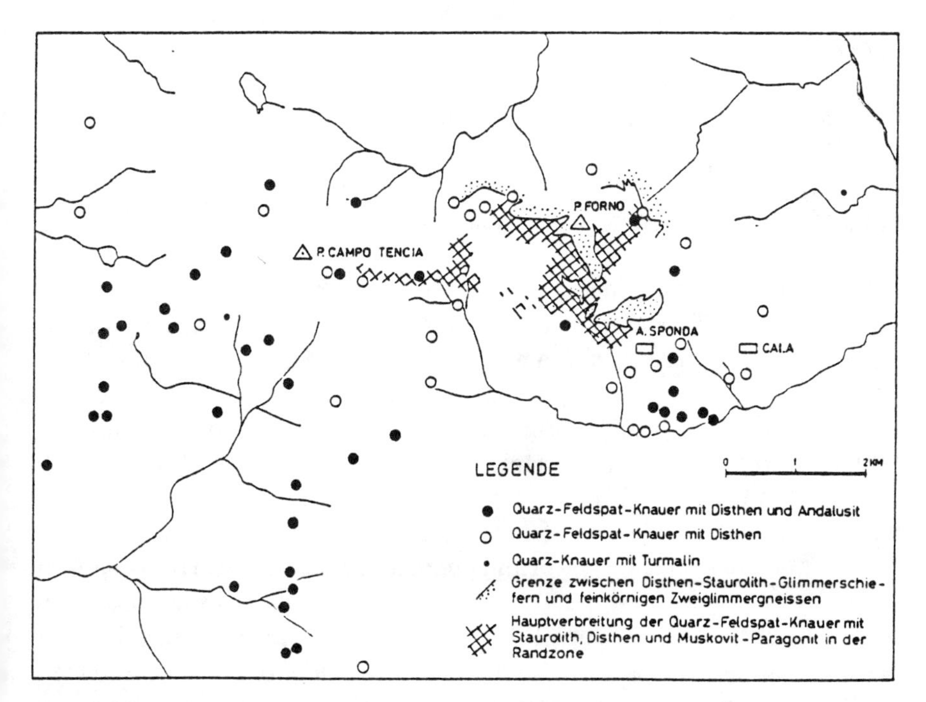

Figure 10.16. Distribution of andalusite ("Andalusit") and kyanite ("Disthen") in quartz ("Quarz") segregations in the Campo Tencia-Pizzo Forno area of the Lepontine Alps, Switzerland. (From Keller, 1968, Fig. 16).

(oligoclase). Tourmaline and late muscovite are notable accessories in types (2) and (3). The kyanite-andalusite segregations are further distinguished by the presence of dark rims of chloritized biotite and the common occurrence of quartz lining open cavities (druses). With the exception of andalusite, all of the major and minor phases of the segregations also occur in the host rocks (Fig. 10.17); furthermore, the composition of plagioclase in the segregations is virtually identical to that of the host rocks (Fig. 10.18). The kyanite-andalusite segregations are most commonly hosted by quartzo-feldspathic gneisses. It is notable that such segregations rarely occur in mica schist host rocks. The kyanite-andalusite segregations typically display mineralogical zoning (Fig. 10.17): the *cores* primarily consist of quartz + feldspar, kyanite ± muscovite is abundant in the *inner border* (zone II in Fig. 10.17), and the dark *outer border* (zone I in Fig. 10.17) consists of abundant chloritized biotite. In contrast to kyanite-andalusite segregations, staurolite-kyanite segregations [type (2)] are restricted to the area bounded by P. Forno, P. Campo Tencia, and A. Sponda (crosshatched area in Fig. 10.17). Type (2) segregations are characterized by the lack of andalusite and the abundance of staurolite. These segregations typically display mineralogical zonation. Keller (1968) distinguished three zones (Fig. 10.17): *cores*: virtually all quartz; *inner border* (zone II in Fig. 10.17): staurolite + kyanite + muscovite ± paragonite; *outer border* (zone I in Fig. 10.17): biotite + staurolite + kyanite. *All* minerals contained within this type of segregation are also present in the host rocks (Fig. 10.17). In contrast to the leucocratic host rocks of the kyanite-andalusite segregations, staurolite-kyanite segregations are hosted by mica-rich schists.

Keller (1968) contended that the veins and segregations formed by metamorphic differentiation of the host rocks. The formation of the Al_2SiO_5 masses by a segregation process is strongly supported by the clear-cut discordance between the margins of the segregations and the schistosity of the host rocks (Fig. 10.19). In light of structural arguments supporting the *syntectonic* formation of the segregations, Keller (1968) adopted Ramberg's (1956) tectonic mechanism for "dilation pegmatites"; that is, the dilatant opening of fissures provides "sinks" for metasomatic transport of the segregation-forming components in fluids from the host rocks. He contended that the biotite-rich selvages represent the residuum formed by metasomatic "depletion" of components from a protolith that was identical to the host rock. As shown in Figure 10.20, he contended that the paramorphic replacement of kyanite by andalusite occurred along a P-T path of progressively diminishing pressure and temperature.

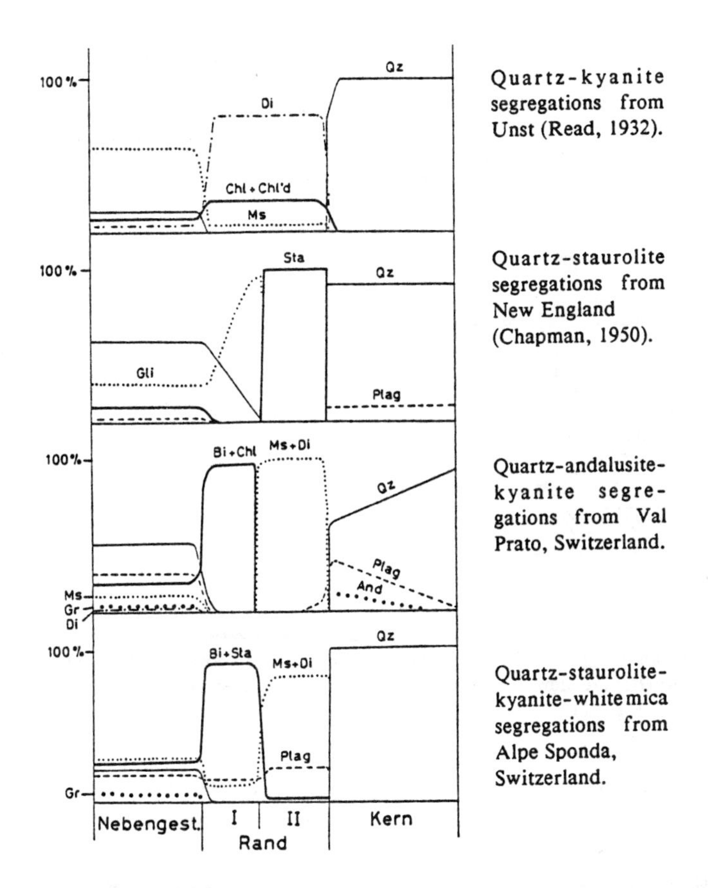

Figure 10.17. Schematic cross sectional depiction of the mineralogy and mineral abundances of segregations from the Val Prato (III) and Alpe Sponda (IV) areas of the Lepontine Alps, Switzerland. For comparison, the segregations described by Read (1932) from Unst, Scotland, and by Chapman (1950) from New Hampshire, are shown. And = andalusite, Bi = biotite, Chl = chlorite, Chl'd = chloritoid, Di = kyanite, Gli = white mica, Ms = muscovite, Plag = plagioclase, Qz = quartz, Sta = staurolite. Nebengest. = host rock; Rand = rim of segregations; Kern = core of segregations (From Keller, 1968, Fig. 19).

Klein (1976) carried out a comprehensive analysis of Al_2SiO_5-bearing segregations in the kyanite zone of the Lepontine Alps. He showed that there was a systematic correlation between metamorphic grade and the mineralogy of the segregations (Fig. 10.21). As shown in Figures 10.22 and 10.23, Klein (1976) considered that the segregations consist of three mineralogical zones. The "reaction" and "core" zones are characterized by coarse kyanite that is virtually devoid of inclusions. In contrast, the host rocks and "rim zones" (Fig. 10.22) consist of poikiloblastic kyanite of the same grain size (Fig. 10.24). Similarity of grain size between the "rim" zone and host rock, and the poikiloblastic habit of kyanite within these

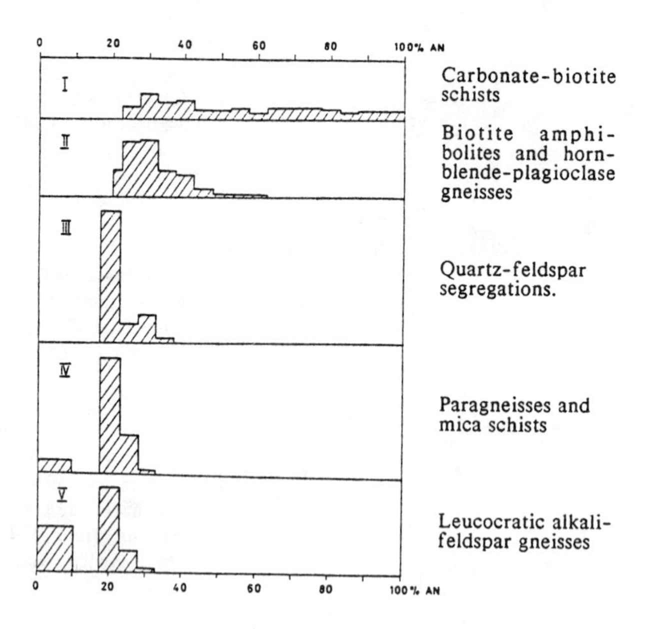

Figure 10.18. Histograms of An-content of plagioclase in various metamorphic rock types in the area shown in Figure 10.16. Note that the plagioclase compositions of the Al_2SiO_5-bearing segregations (III) are similar to those of the host paragneisses and mica schists (IV). (From Keller, 1968, Fig. 11).

zones, argues that the "rim" zone represents host rock that was metasomatically altered during the segregation-forming event.

Klein (1976) appealed to non-homogeneous stress to explain the origin of the segregations. He noted that aluminum silicates are rare in mica schist host rocks but common in quartz-rich, mica-poor lithologies. He argued that quartzo-feldspathic lithologies had a relatively competent rheology during the segregation-forming deformation event. Accordingly, local stress-enhanced dilatant zones were believed to be favored sites where segregations formed. Confinement of the kyanite → andalusite transformation to the segregations was attributed to the dilatant opening of the segregation sites coupled with locally elevated strain fields. He considered that this polymorphic transformation occurred during decompression which accompanied uplift and marked erosion of the Lepontine region. In contrast to quartzo-feldspathic host rocks, Klein (1976) hypothesized that the presence of abundant micas in the mica schists resulted in a more homogeneous distribution of stress. Because of the comparatively reduced magnitude of the "stress risers", the transformation of kyanite to andalusite did not occur in segregations hosted by mica schists.

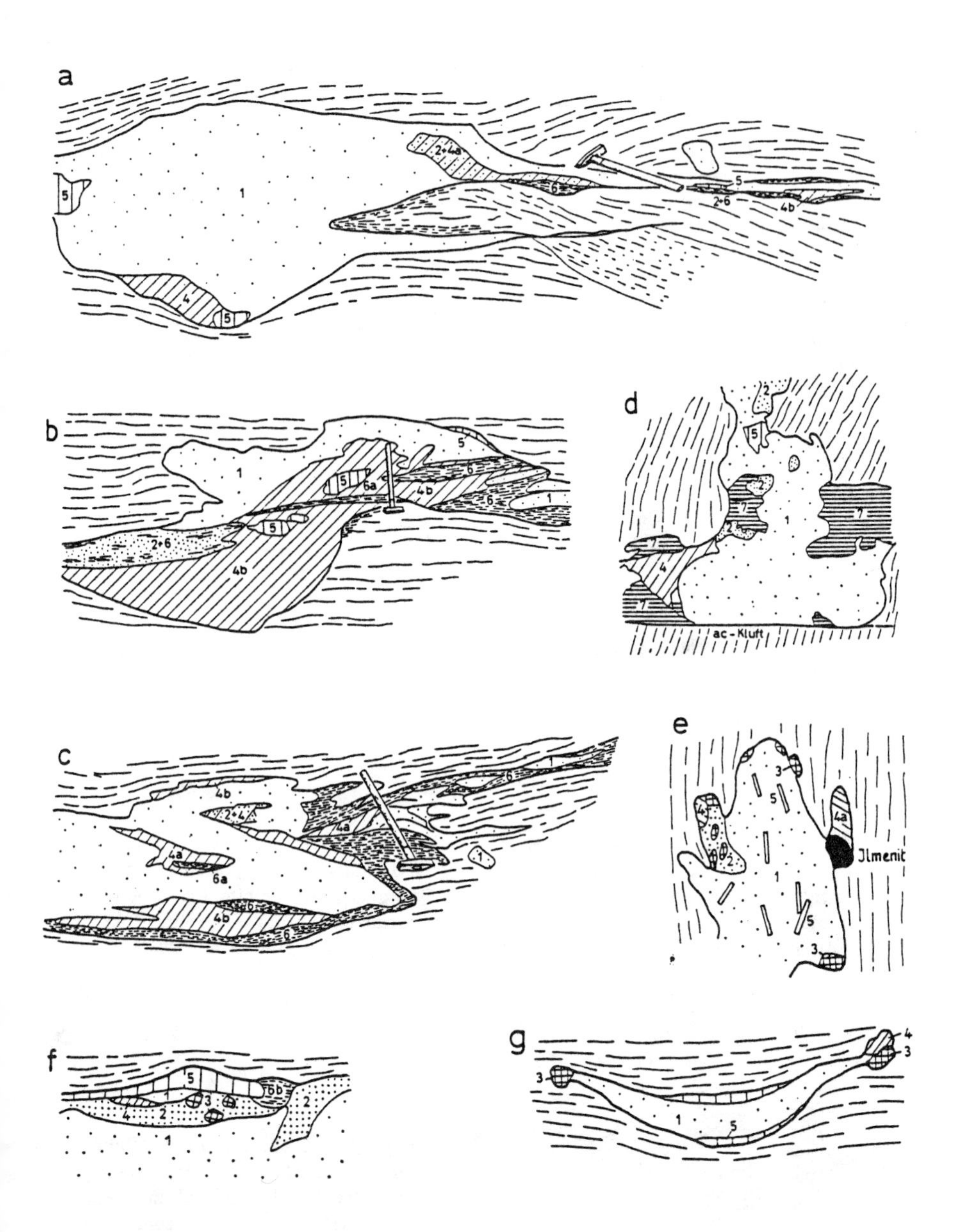

Figure 10.19. Keller's (1968) sketches of segregations from the Alpe Sponda (a, b and c), Alpe Toira (d) and Val Prato (e, f, and g) areas of the Lepontine Alps, Switzerland. The numbers correspond to the following minerals or mineral assemblages: 1 = quartz, 2 = plagioclase, 3 = andalusite, 4 = white mica, 4a = white mica + kyanite, 4b = white mica + kyanite + staurolite, 5 = kyanite, 6 = biotite, 6a = biotite + staurolite, 6b = biotite + tourmaline, 7 = chlorite. (From Keller, 1968, Fig. 15).

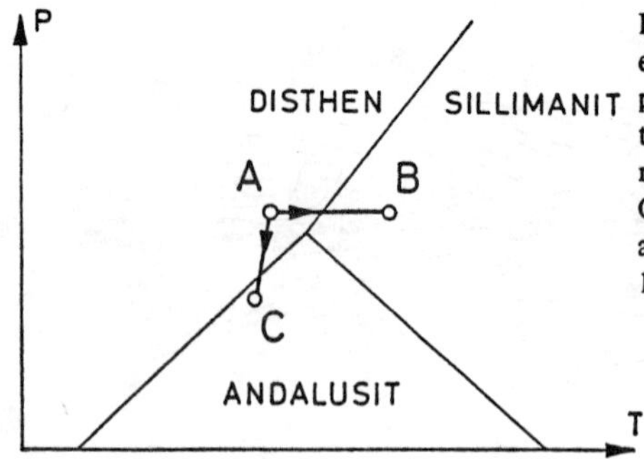

Figure 10.20. Schematic Al_2SiO_5 phase equilibrium diagram showing the "normal" prograde P-T gradient (A → B) for the Barrovian-type regional metamorphism in the Lepontine region of Switzerland, and the pressure drop (A → C) required for the transformation of kyanite to andalusite within segregations. (From Keller, 1968, Fig. 18).

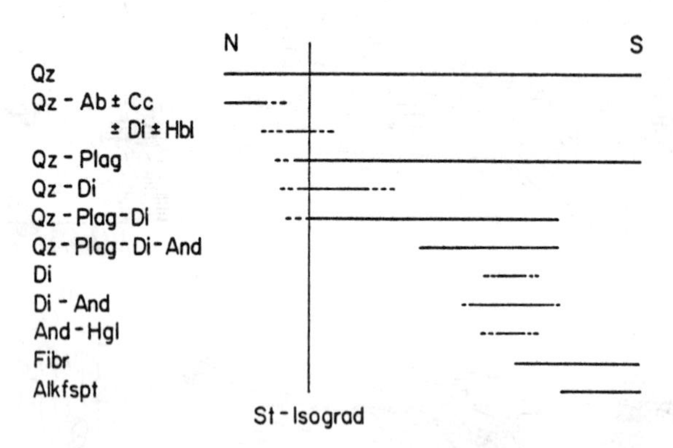

Figure 10.21. Schematic representation of the mineralogy of segregations as a function of metamorphic grade in the Lepontine Alps, Switzerland. Metamorphic grade increases from left to right. Mineral abbreviations are as follows: Qz = quartz, Ab = albite, Cc = calcite, Hbl = hornblende, Plag = plagioclase, Di = kyanite, And = andalusite, Fibr = fibrolite, Alkfspt = K-feldspar. The staurolite isograd marks the first appearance of staurolite and kyanite in the host metapelites. (From Klein, 1976, Fig. 1).

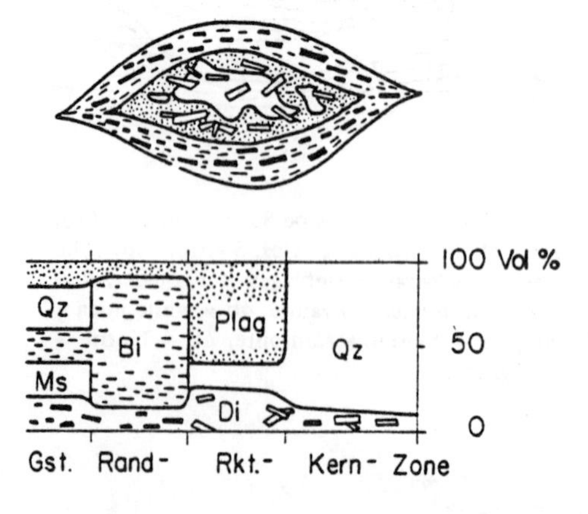

Figure 10.22. *Top*: Schematic representation of a typical segregation from the kyanite zone of the Lepontine Alps, Switzerland. *Bottom*: Schematic cross sectional sketch showing mineralogical zonation of the segregations. The abbreviations on the abscissa are: Gst = host rock, Rand = rim zone, Rkt. = intermediate "reaction" zone, Kern = core. Mineral abbreviations are: Ms = muscovite, Bi = biotite; abbreviations for the other minerals are explained in Figure 10.21. (From Klein, 1976, Fig. 2).

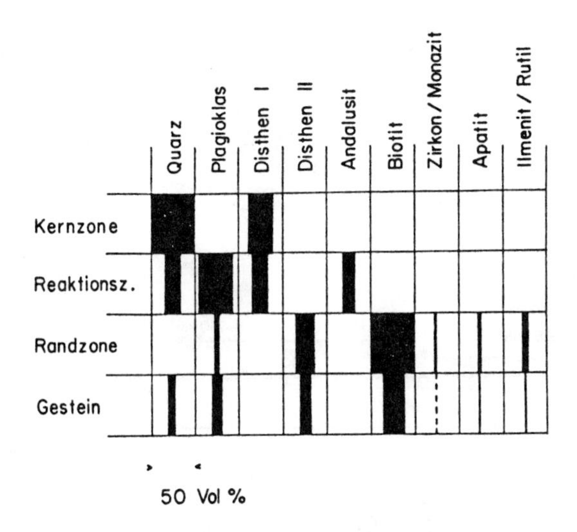

Figure 10.23. Distribution of minerals in kyanite-bearing segregations from the Lepontine Alps, Switzerland. "Disthen I" is inclusion-free kyanite whereas "Disthen II" is poikiloblastic kyanite. Note that the inclusion-free kyanite is restricted to the host rocks and rim zone, whereas poikiloblastic kyanite is confined to the intermediate "reaction" zone and the core. (From Klein, 1976, Fig. 3).

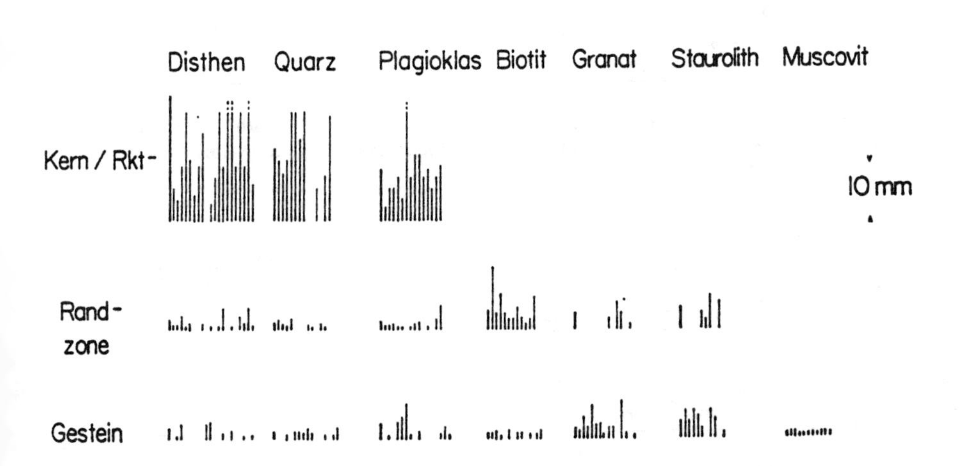

Figure 10.24. Grain size of minerals in Al_2SiO_5-bearing segregations and metapelite host rocks from the Lepontine Alps, Switzerland. (From Klein, 1976, Fig. 4).

Wenk (1970) concluded that andalusite is confined to rocks at higher elevations in the central Tessin region of the Lepontine Alps. As shown in Figures 10.25 and 10.26, this distribution has been confirmed by Klein (1976) and by Thompson (1976). However, whereas Klein (1976) considered andalusite to form during retrograde metamorphism, Thompson

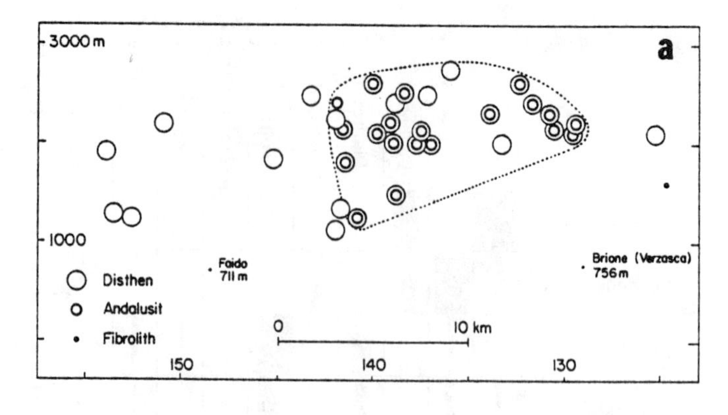

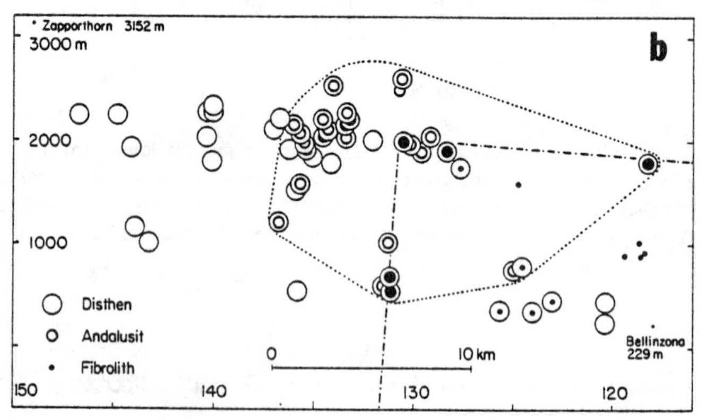

Figure 10.25. Al_2SiO_5 mineralogy of segregations as a function of depth. These diagrams were obtained by projection onto two north-south profiles in the Lepontine Alps, Switzerland. (From Klein, 1976, Fig. 7).

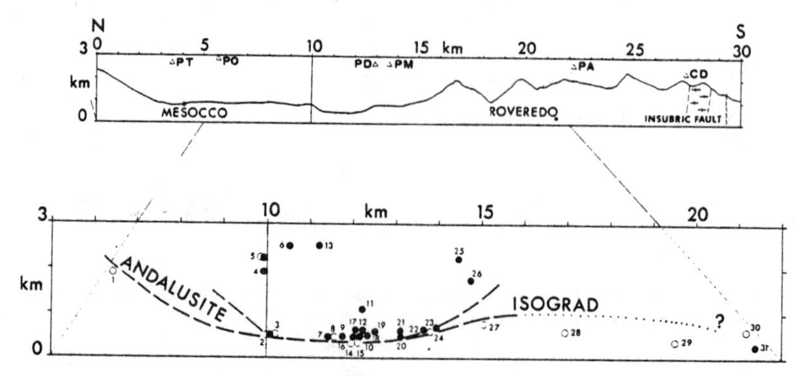

Figure 10.26. Cross section of topography (*top*) and depth distribution of the Al_2SiO_5 mineralogy (*bottom*) of segregations in the Val Mesolcina area of the Lepontine Alps, Switzerland. The portion of the bottom diagram to the right of the vertical line at 10 km was obtained by projection onto a north-south profile, whereas that to the left of the vertical line was projected onto a profile trending northeast-southwest. The filled circles represent samples containing andalusite whereas those shown with open circles lack andalusite. (From Thompson, 1976, Fig. 3).

(1976) concluded that andalusite formed during prograde metamorphism or at the peak of metamorphism.

The author recently carried out an analysis of the genesis of Al_2SiO_5-bearing segregations in the Lepontine Alps (Kerrick, 1988). The presence of segregations within discordant faults and fractures (Fig. 10.27e), the pegmatitic texture of the segregations (Fig. 10.27a), and the presence of druses and tourmaline, support the argument that the segregations formed by the transport of components in an aqueous fluid phase. As in the studies of Keller (1968) and Klein (1976), it was concluded that the distinctive biotite-rich selvages formed by metasomatic depletion of Si, alkalies, and Al from a protolith that was identical to that of the host rocks. In light of recent experimental studies, I concluded that Al was most likely present in the fluid as an alkali-aluminum complex with the same molar alkali/aluminum ratio as that of alkali feldspars. Al could have entered the fluid phase through the congruent dissolution of kyanite in the host rocks (Kerrick, 1988). However, because of Klein's (1976) hypothesis that the kyanite concentrated in the outer "rim zones" of segregations is probably a residue from a metasomatically-depleted kyanite schist host rock, I question my hypothesis that the dissolution of host rock kyanite provided a source of Al in solution. As shown in Figure 10.22, the reaction rim lacks muscovite and quartz, which are primary constituents of the adjacent host rocks (Fig. 10.22). Therefore, the "reaction zones" may have formed by the dissolution of muscovite and quartz into the fluid phase. In light of the occurrence of segregations in boudin necks (Fig. 10.27f), it was suggested that such segregations formed by the mechanism outlined in Figure 10.28. Dilatant opening of fractures would have provided a fluid pressure gradient to drive fluids into the dilatant openings. This is equivalent to the mechanism of "tectonic pumping" that was outlined by Fyfe et al. (1978, p. 299-301). Considering the mass of Al contained within selected segregations, coupled with experimental data on the concentration of aqueous Al at P-T conditions during formation of the segregations, it was concluded that large volumes of infiltrating fluid were necessary. The most reasonable fluid source was metamorphic dehydration reactions occurring in the host rocks. Breakdown of alkali silicates was suggested as a source for alkalies in the fluid. For example, K in the fluid phase could have been provided by:

$$2\ \text{Muscovite} = 3\ \text{Kyanite} + 3\ \text{Quartz} + 2\ KOH_{aq} + H_2O\ .$$

The apparent decomposition of muscovite in forming the rim zone (Fig. 10.22) would have provided a *local* source for K. In contrast to the above

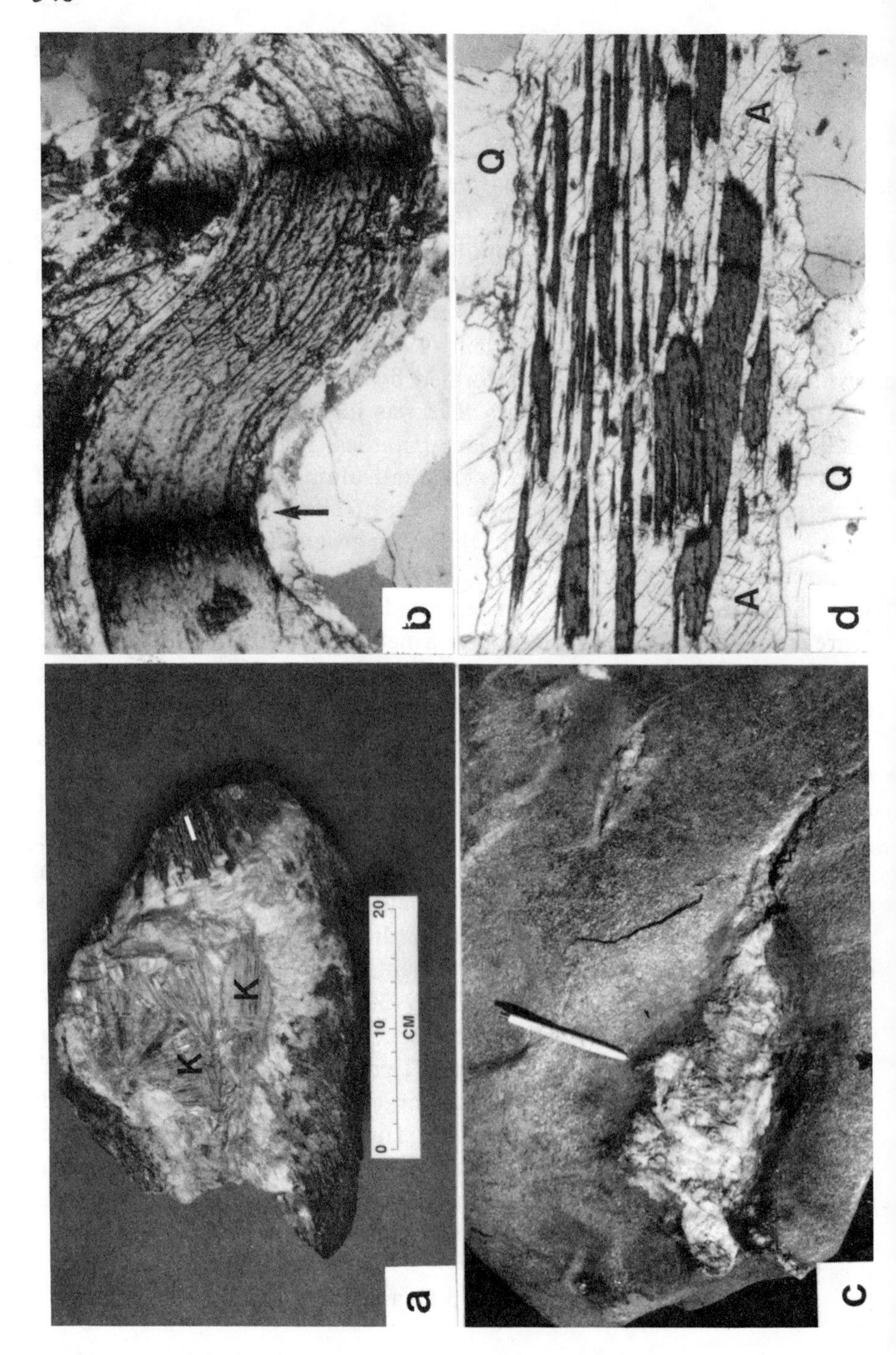
a
K
K
0 10 20
CM
b
c
d
Q
A
A
Q

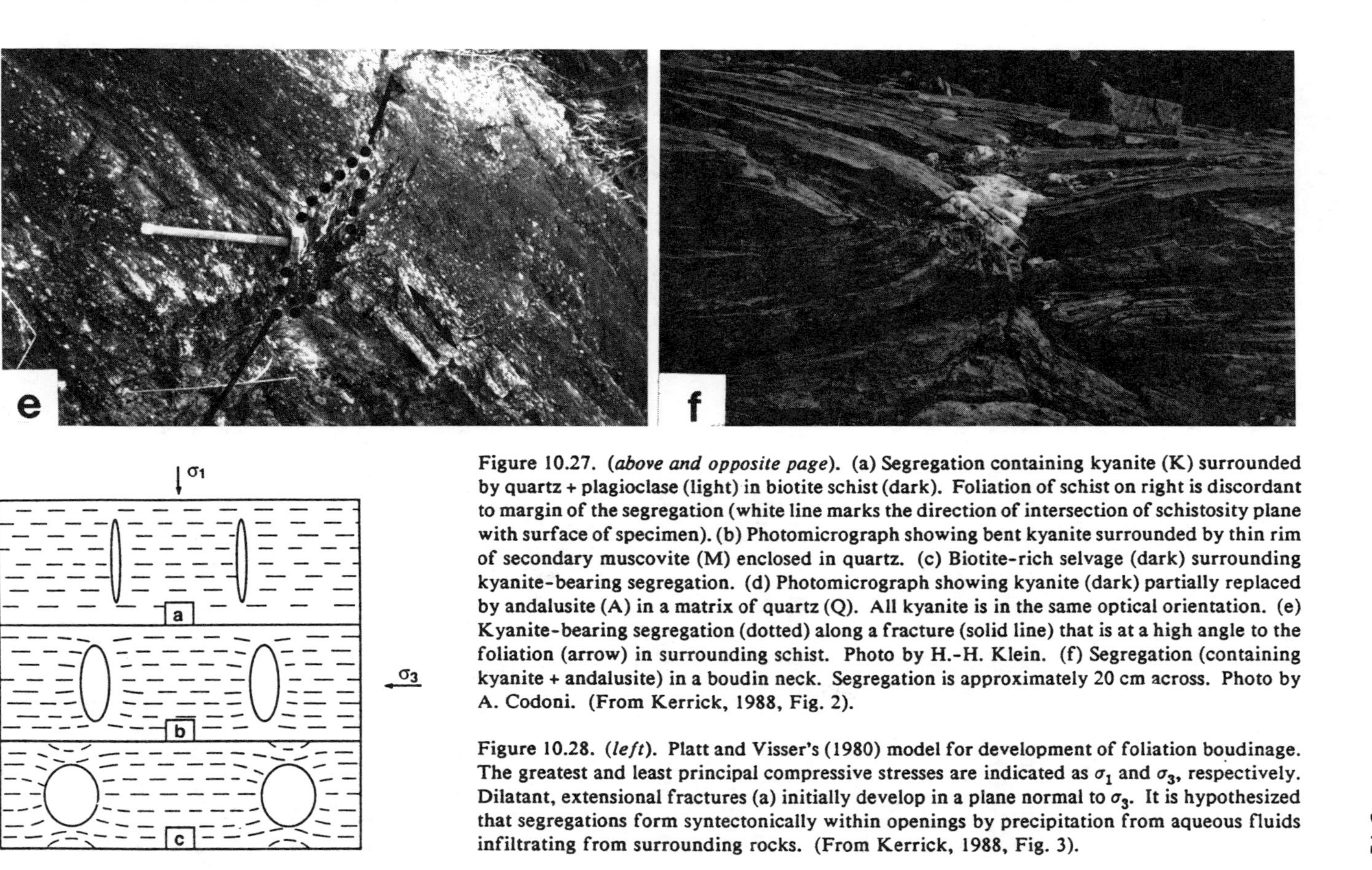

Figure 10.27. (*above and opposite page*). (a) Segregation containing kyanite (K) surrounded by quartz + plagioclase (light) in biotite schist (dark). Foliation of schist on right is discordant to margin of the segregation (white line marks the direction of intersection of schistosity plane with surface of specimen). (b) Photomicrograph showing bent kyanite surrounded by thin rim of secondary muscovite (M) enclosed in quartz. (c) Biotite-rich selvage (dark) surrounding kyanite-bearing segregation. (d) Photomicrograph showing kyanite (dark) partially replaced by andalusite (A) in a matrix of quartz (Q). All kyanite is in the same optical orientation. (e) Kyanite-bearing segregation (dotted) along a fracture (solid line) that is at a high angle to the foliation (arrow) in surrounding schist. Photo by H.-H. Klein. (f) Segregation (containing kyanite + andalusite) in a boudin neck. Segregation is approximately 20 cm across. Photo by A. Codoni. (From Kerrick, 1988, Fig. 2).

Figure 10.28. (*left*). Platt and Visser's (1980) model for development of foliation boudinage. The greatest and least principal compressive stresses are indicated as σ_1 and σ_3, respectively. Dilatant, extensional fractures (a) initially develop in a plane normal to σ_3. It is hypothesized that segregations form syntectonically within openings by precipitation from aqueous fluids infiltrating from surrounding rocks. (From Kerrick, 1988, Fig. 3).

reaction proposed in by 1988 paper, I consider the reaction

$$2\ \text{Muscovite} = (\text{KAl(OH)}_4 \cdot \text{SiO}_2)_{aq} + 2.5\ \text{SiO}_2 + 2.5\ \text{Al}_2\text{SiO}_5\ ,$$

to be a more reasonable source of K in the fluid. Precipitation of Al_2SiO_5 + quartz within the segregations was modeled by

$$(\text{KAl[OH]}_4 \cdot [0.5 + n]\text{SiO}_2)_{aq} = \text{KOH}_{aq} + 0.5\text{Al}_2\text{SiO}_5 + n\text{SiO}_2 + 1.5\text{H}_2\text{O}\ .$$

This reaction could be reformulated into separate reactions (i.e., one with $(KAl[OH]_4)_{aq.}$ and another with $(SiO_2)_{aq.}$) in light of Woodland and Walther's (1987) suggestion that the alkali-aluminum complexes may contain no Si. Precipitation of Al_2SiO_5 + quartz within the segregations (rather than muscovite) was attributed to reduced f_{H_2O} within the dilatant openings.

My genetic model for the Lepontine Al_2SiO_5-bearing pegmatites was influenced by the experimental data of Anderson and Burnham (1983) and Anderson et al. (1988) which suggests that Al is present in the fluid as alkali-aluminum complexes. However, based on the correlation between the dielectric constant of the solvent and solute speciation (Walther and Schott, 1988), it is instead possible that Al exists as $Al(OH)_n$ complexes (J.V. Walther, pers. comm.). In an experimental study, Morgan and London (1989) showed that the solubilities of Al and Si are enhanced by the presence of B in alkaline aqueous fluids. In addition to metal hydroxide and alkali complexes, they also suggested that the increased aluminosilicate solubilities could be produced by interaction with alkali borate oxyanions. The latter suggestion is compatible with the elevated B contents revealed by the common presence of tourmaline in Al_2SiO_5-bearing pegmatites. An improved model for aluminum metasomatism thus awaits further studies on the nature of Al complexes in supercritical aqueous solutions.

As shown in Figure 10.29, I argued that andalusite formed from kyanite during the later stages of decompression of the Pennine nappe sequence. Localized, transient dilatancy within the segregations yielded corresponding "excursions" into the andalusite P-T stability field. However, I concluded that the P-T trajectory of the host rocks remained in the kyanite stability field (curved solid line in Fig. 10.29); thus, the kyanite → andalusite transformation was precluded in the host rocks. The influx of large amounts of aqueous fluid during retrograde decompression is supported by the marked alteration of biotite to chlorite in biotite-rich

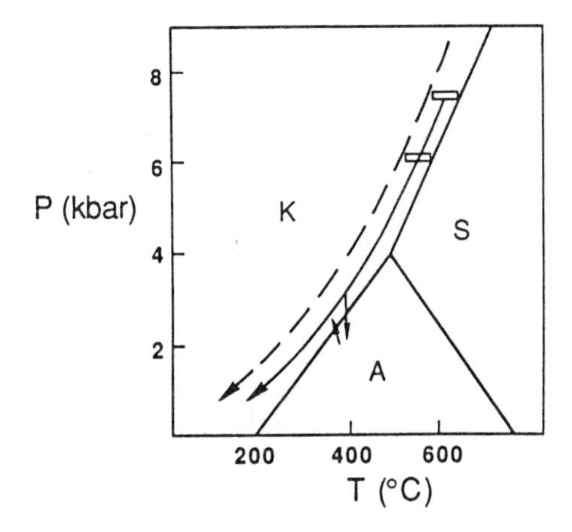

Figure 10.29. Postulated thermobarometric regime attending decompression following the Main Alpine metamorphism of the Lepontine Alps, Switzerland. The open rectangles are Klaper and Bucher-Nurminen's (1987) estimates of kyanite-grade metamorphism in the Nufenen Pass area. The rectangle at 7.3 kbar represents thermobarometry from garnet core compositions, whereas that at 6 kbar is from garnet rim compositions. The solid curve is the decompression path for andalusite-bearing segregations. The short arrows depict proposed P-T excursion into the andalusite stability field. The dashed line represents decompression path of segregations beneath the andalusite isograd of Thompson (1976). K = kyanite, A = andalusite, S = sillimanite. (From Kerrick, 1988, Fig. 4).

selvages surrounding the segregations. The gently-dipping andalusite isograd delineated by Thompson (1976) in the region northeast of Bellinzona (Fig. 10.26) was rationalized by comparison of the P-T trajectory of rocks shallower than the andalusite isograd (solid curve in Fig. 10.29) with the andalusite-free segregations at deeper levels (dashed line in Fig. 10.29). It was hypothesized that the P-T trajectory of rocks at greater depths was relatively removed from the kyanite-andalusite equilibrium boundary. Thus, pressure diminution resulting from localized dilatancy was inadequate to enter the andalusite P-T stability field. Thompson (1976) suggested that confinement of the kyanite → andalusite reaction to the segregations could reflect differences in the amount of strain energy of kyanite in the segregations versus host rocks. In essence, strained kyanite in the segregations would have transformed to andalusite within the segregations, whereas unstrained kyanite in the host rocks would have remained unaltered. However, I noted that this mechanism is inapplicable because numerous andalusite-bearing segregations contain undeformed kyanite (Fig. 10.27d) and because some strongly bent kyanite is unaltered to andalusite (Fig. 10.27b).

Although not explicitly stated by Kerrick (1988), I suggest that the quartzo-feldspathic central regions of the segregations (i.e., the "reaction" and "core" zones in Fig. 10.22) formed by crystallization within dilatant openings whereas the biotite-rich rims (Fig. 10.22) *may* be metasomatically-altered host rock. Accordingly, infiltration metasomatism was dominant in the formation of the quartzo-feldspathic portions. Diffusive transport *could* have been significant in the development of the rims. Additional metasomatic modeling of these segregations is clearly needed. However, regardless of the detailed metasomatic mechanism(s) involved, the pegmatitic Al_2SiO_5-bearing segregations of the Lepontine Alps obviously required a large influx of Al-bearing fluids. As such, they provide an excellent example of infiltration-dominated metasomatism, and they provide cogent evidence of significant transport of Al during metamorphism.

Kyanite-bearing veins in eclogites

Al_2SiO_5-bearing veins and segregations are hosted by rocks other than pelites and quartzo-feldspathic lithologies. Perhaps most notable are the occurrences of kyanite in veins and segregations within eclogites.

Kyanite-quartz-omphacite veins occur within eclogite-bearing rocks of the Zermatt-Saas ophiolite zone in the Swiss Alps (Barnicoat and Fry, 1986; Fry and Barnicoat, 1987; Barnicoat, 1988). Mineral assemblages of these veins suggest a_{H_2O} between 0.55 and 1.0. The elevated a_{H_2O} values, coupled with the presence of fluid inclusions in omphacite, suggest the presence of a H_2O-rich fluid phase during vein formation (Barnicoat and Fry, 1986). The presence of euhedral omphacite and garnet occur in the veins, coupled with the lack of alteration haloes in host rocks adjacent to the veins, suggest that the veins formed in equilibrium with the host rocks under the highest grade metamorphic conditions (Barnicoat, 1988). Although kyanite is relatively rare in the host rocks, Fry and Barnicoat (1987) contended that kyanite was an abundant early mineral in the eclogites but was consumed in a subsequent lawsonite-forming reaction. Thus, the minerals occurring in the veins were presumably widespread constituents of the host rocks. This supports Read's (1932) and Vidale's (1974) observations that the mineralogy of veins mirrors that of the host rocks.

Kyanite-quartz-omphacite veins are common in eclogites of the Adula nappe in the Lepontine Alps (Heinrich, 1986). Many veins are discordant to the foliation of the host eclogites. Because all minerals in the veins occur in the host eclogites, and because of the presence of euhedral kyanite crystals up to several centimeters long, needles of omphacite that apparently grew inward from vein walls, and lack of alteration of the host rocks adjacent to the veins, Heinrich (1986) suggested that the veins formed by crystallization from a fluid that was in equilibrium with the adjacent kyanite eclogites.

Kyanite-quartz-omphacite veins hosted by kyanite eclogites also occur in the Tauern Window of Austria (Holland, 1979a). Holland (1979a) found fluid inclusions in omphacite and zoisite in samples of veins and host eclogites. He concluded (Holland, 1979a, p. 20): "It is difficult to understand the growth of idioblastic omphacite and kyanite, with crystals of up to five or more centimeters in size, in veins, boudin necks and tension fractures without invoking the presence of...a fluid".

Kyanite-bearing veins have been described in some Norwegian eclogites (Krough, 1982; Griffin, 1987; Austrheim, 1987). The presence of kyanite-zoisite-omphacite-rutile lenticles, with crystals up to several centimeters across, and euhedral omphacite and rutile, suggest crystallization in the presence of an aqueous fluid phase (Krough, 1982).

In summary, petrologic evidence suggests that kyanite veins in eclogites formed by crystallization from solutes in aqueous fluids that were in equilibrium with the kyanite eclogite host rocks. Such veins argue against the classical assumption that eclogites represent fluid-absent ("dry") metamorphism.

The presence of kyanite and other aluminous minerals (omphacite, zoisite) in veins within eclogites shows that Al is highly mobile under eclogite-facies metamorphism. Experimental data on the aqueous solubilities of assemblages with alkali silicate minerals (Fig. 10.30) suggest that the solubility of Al increases with increasing pressure. Schneider and Eggler's (1986) experimental determination of the solute contents of fluids in equilibrium with peridotite minerals implies significant concentrations of aqueous Al at the P-T conditions of eclogite facies metamorphism. Accordingly, the presence of aluminous minerals in veins within eclogites is compatible with the elevated aqueous Al concentrations suggested by experimental solubility studies. The common occurrence of omphacite in such veins is compatible with the existence of alkali-aluminum complexes in aqueous fluids of eclogite facies metamorphism.

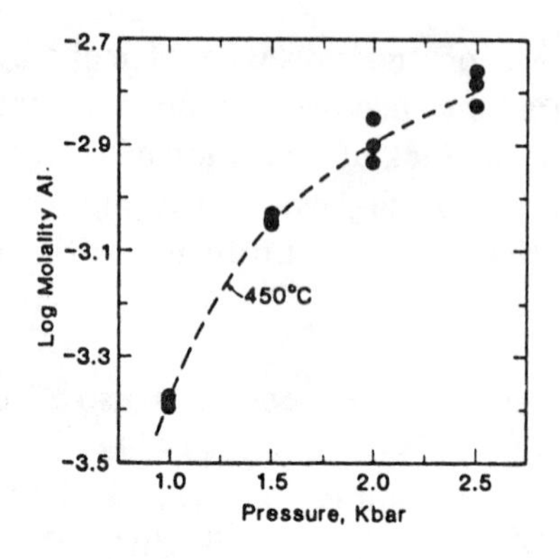

Figure 10.30. Concentration of aluminum in aqueous solutions in equilibrium with albite, paragonite and quartz, as a function of pressure at 450°C. (From Woodland and Walther, 1987, Fig. 2).

FIBROLITE AND ALUMINUM METASOMATISM

Several studies have concluded that significant transport of Al is suggested by fibrolite occurrences.

Best and Weiss (1964) described pegmatitic fibrolite-quartz-K-feldspar segregations in the Lake Isabella roof pendant, southern Sierra Nevada, California. "These segregations are seldom lenticular ... but more commonly appear as ramifying networks, several inches in diameter, in places with a thick central trunk lying transverse to a weak relict foliation" (Best and Weiss, 1964, p. 1249). The form of the segregations, and their pegmatitic character, suggest that they formed by crystallization within fluid-filled cavities formed by an anastamosing fracture network.

In an extensive study of the aluminum silicates in the aureoles of the Donegal Granites (Ireland), Pitcher and Read (1963) suggested that much of the fibrolite is of "metasomatic-pneumatolitic" origin. Petrographic evidence formed a primary basis for this conclusion. As shown in Figures 10.31 and 10.32, the occurrence of fibrolite in vein-like aggregates and at grain boundaries argues that fibrolite formed in domains that have traditionally been assumed to have been channelways for fluid transport. In the aureole of the Thorr Granodiorite, Pitcher and Read (1963, p. 271) noted: "Quartz-fibrolite and sericite-fibrolite veinlets ramify through ..." high grade hornfelses, and that these veinlets "... become more important near the boundary of the granitic rocks ...". Pitcher and Read (1963) noted a close correlation between fibrolite, tourmaline and late muscovite. Replacement of andalusite by muscovite + fibrolite is common (Fig. 10.33). As illustrated in Figure 10.34, overprinting of andalusite zone rocks by late-stage fibrolite is common in the Donegal aureoles. In

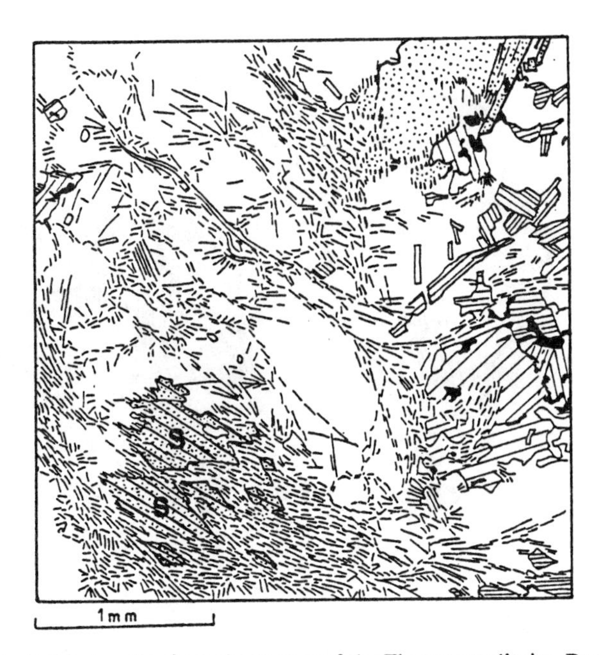

Figure 10.31. Pelitic hornfels from the aureole of the Thorr granodiorite, Donegal, Ireland. Pitcher and Read (1963) concluded that the sillimanite prisms (S) formed from fibrolite (illustrated with the dashed pattern). Note the vein-like network of fibrolite in the central and upper left portions of the sketch. (Modified from Pitcher and Read, 1963, Fig. 4).

Figure 10.32. Pelitic hornfels from the aureole of the Thorr granodiorite, Donegal, Ireland. Swarms of fibrolite needles transect muscovite (M). B= biotite. (Modified from Pitcher and Read, 1963, Fig. 8).

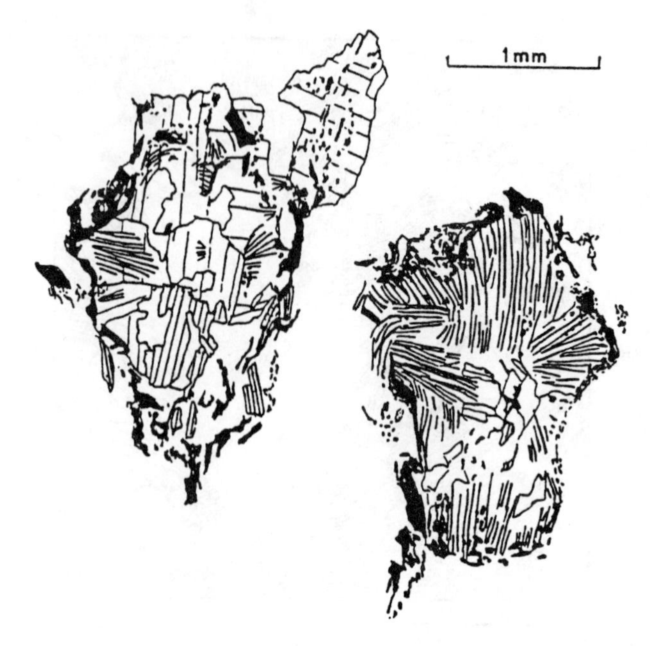

Figure 10.33. Andalusite porphyroblasts that were replaced by fibrolite (closely spaced lines) and large plates of muscovite (widely spaced cleavages). Fringes of graphite (black) outline the original peripheries of the andalusite porphyroblasts. (From Pitcher and Read, 1963, Fig. 6).

summarizing their views on fibrolite in these aureoles, Pitcher and Read (1963, p. 293) concluded: "The veinlike mode of occurrence, the common association with white mica and tourmaline, and the varying trespass of this assemblage upon pre-existing zones, together suggest a reconstitution under the influence of fluids moving outward from the igneous bodies."

In reviewing aluminum silicates in the Donegal aureoles, Pitcher and Berger (1972) concurred with Pitcher and Read's (1963) late-stage fibrolite-muscovite-tourmaline association. Pitcher and Berger (1972, p. 307) concluded: "This late-state encroachment of a metasomatic zone seems to be a general feature of the Donegal aureoles." They hypothesized that this metasomatism could have occurred by the influx of magmatic volatiles, although they suggested that metasomatic transport could have been enhanced by anatexis of high-grade rocks (Pitcher and Berger, 1972, p. 325).

In a reexamination of fibrolite in the Ardara and Fanad aureoles (Donegal, Ireland), the author observed abundant "anastomosing stringers" of fibrolite in rocks of the sillimanite zone (Kerrick, 1987b). In addition,

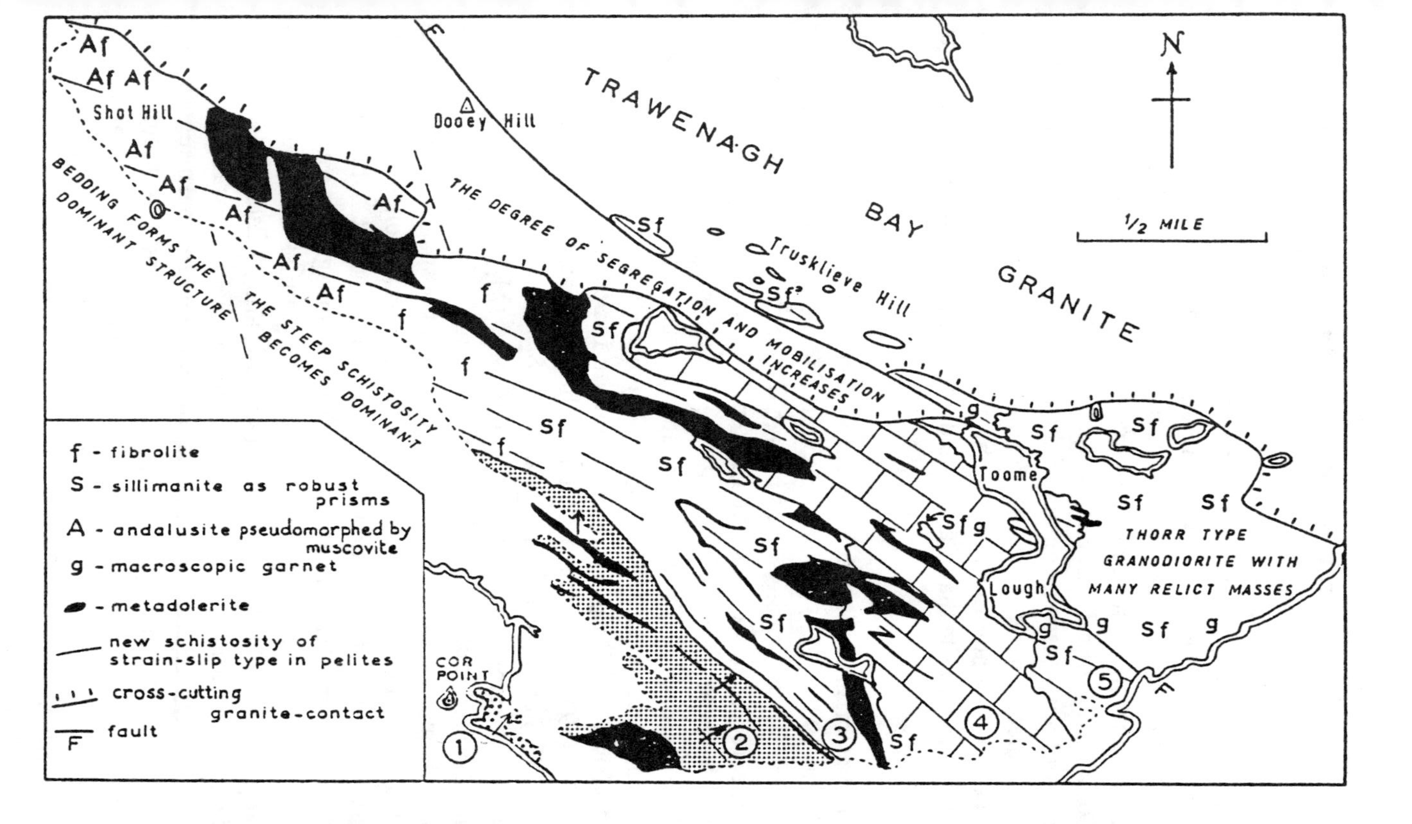

Figure 10.34. Geologic map of a contact metamorphosed pelitic horizon in the Lettermackaward area, Donegal, Ireland. Metamorphic grade increases from northwest to southeast. Note that fibrolite "overprints" the andalusite and sillimanite zones of this aureole. (From Pitcher and Read, 1963, Fig. 5).

fibrolite is concentrated at grain boundaries in some sillimanite-zone rocks. I considered such textures to represent metasomatic influx of Al into such rocks. However, it was noted that fibrolite within lower grade rocks of the andalusite zone is typically intergrown with biotite. Thus, in contrast to Pitcher and Read (1963) who suggested widespread metasomatism, I concluded that fibrolite formed by metasomatic influx of Al was is confined to the highest grade rocks. I concurred with Pitcher and Read's (1963) metasomatic fibrolite-tourmaline-muscovite assemblage. In the Ardara aureole, a magmatic source for B is supported by Akaad's (1956) description of a 3 - 25 cm-thick tourmaline ± quartz layer along the northern and northeastern contact of the Ardara pluton. I concluded that late-stage metasomatic fibrolite formed from the influx of acidic, magmatic fluids containing aluminum, boron and alkalies. The influx of K generated muscovite that is associated with the late fibrolite-tourmaline assemblage.

In studies of granulite facies rocks at Broken Hill, Australia, Ahmad and Wilson (1981, 1982) provided support for the metasomatic formation of fibrolite from B-bearing fluids in high-grade rocks. Two metamorphic events are recorded in rocks at Broken Hill: an early granulite facies metamorphism (M_1) and later amphibolite facies retrograde metamorphism (M_2). The presence of fibrolite aggregates crosscutting other minerals, and the concentration of fibrolite at grain boundaries, suggest that fibrolite formed after crystallization of the granulite facies assemblage. Ahmad and Wilson (1981) contended that fibrolite aggregates are parallel to a schistosity (S_2) that overprints M_1. Because S_2 is believed to have formed during the late amphibolite facies metamorphism (M_2) they contended that fibrolite formed during this retrograde event. Using particle track methods, Ahmad and Wilson (1981) determined the spatial distribution of B and U on the microscopic scale. They found that fibrolite has relatively elevated B contents and that U is concentrated at fibrolite grain boundaries. Ahmad and Wilson (1981) favored a magmatic source for the metasomatic fluids (i.e., post-metamorphic granites and pegmatites). However, in a subsequent paper (Ahmad and Wilson, 1982) they considered base-cation leaching as a feasible mechanism for the formation of fibrolite. Regardless of the source, the Broken Hill rocks suggest that fibrolite formed by metasomatic transport of Al, U and B in a fluid phase. The B enrichment is in concert with the metasomatic fibrolite-tourmaline assemblage in the Donegal aureoles and in the segregations of the Lepontine Alps.

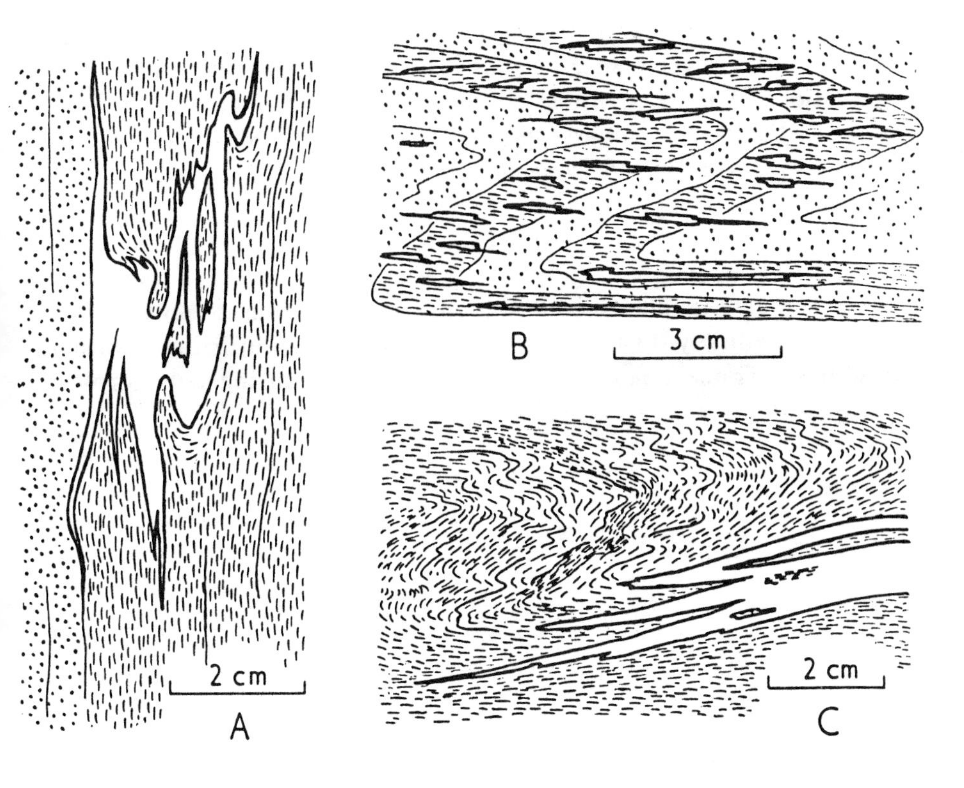

Figure 10.35. Fibrolite-quartz veins (clear) within gneissic host rocks of the Dalradian sequence, Glen Clova area, Scotland. (From Chinner, 1961, Fig. 1).

Veins and segregations dominantly consisting of fibrolite + quartz are common in high grade metapelitic rocks.

As shown in Figure 10.35, fibrolite-quartz veins are common in the high grade portion of Barrow's classic area in the Scottish Highlands. The clearly discordant nature of these veins strongly suggests significant metasomatic transport of Al.

Fibrolite + quartz segregations are abundant in high grade rocks of the Lepontine Alps. The segregations contain anastamosing folia of fibrolite within quartz. One sample contains veinlets consisting of fibrolite + quartz. There is no textural evidence that fibrolite replaced pre-existing kyanite and/or andalusite. Thus, I contend that fibrolite is a primary phase that crystallized from infiltrating fluids. The presence of fibrolite + quartz in veinlets suggests continued metasomatic activity subsequent to the formation of the segregations. These segregations have similar size, geometry, and structural settings, compared to the kyanite + quartz segregations at lower grades. Hence, I believe that the fibrolite + quartz segregations of the Lepontine "sillimanite" zone originated by the same process as the kyanite + quartz segregations at lower grades; i.e., the influx of aqueous fluids derived from surrounding host rocks.

EPILOGUE

The parageneses described in this chapter imply considerable mobility of Al. Accordingly, I suggest *considerable* caution in utilizing a constant Al reference frame to model metasomatic processes during metamorphism of pelites and quartzo-feldspathic lithologies.

CHAPTER 11

ANATECTIC MIGMATITES, MAGMATIC PEGMATITES AND PERALUMINOUS GRANITOIDS

INTRODUCTION

The Al_2SiO_5 polymorphs occur in anatectic migmatites, peraluminous granitoids, and magmatic pegmatites. Thus, in addition to their importance in "sub-solidus" metamorphism, this mineral group is of significance in igneous petrology.

ANATECTIC MIGMATITES

Aluminum silicates (especially fibrolite) are common in both the leucocratic (*leucosomes*) and melanocratic (*melanosomes*) portions of migmatites. Many petrologic studies of pelitic migmatites within the last decade have argued for an anatectic origin (e.g., Tracy, 1985). However, "sub-solidus" metamorphic differentiation is considered to have been important in other migmatites (referred to as *metamorphic migmatites*). Yardley (1978) and McLellan (1983) review the criteria for distinguishing anatectic migmatites from metamorphic migmatites.

Numerous sillimanite-consuming reactions have been proposed for anatectic migmatites, e.g., oligoclase + K-feldspar + sillimanite = quartz + H_2O + melt (Tyler and Ashworth, 1982), biotite + sillimanite + K-feldspar + H_2O = melt, biotite + sillimanite + H_2O = cordierite + melt (Wickham, 1987), and biotite + plagioclase + sillimanite + quartz = garnet + melt (LeBreton and Thompson, 1988; Vielzeuf and Holloway, 1988). The consumption of sillimanite in these reactions provides one mechanism for the production of *peraluminous* melts (i.e., melts with the molar ratio $Al_2O_3/(CaO+Na_2O+K_2O)$ exceeding 1.0). Alternatively, anatexis may proceed by sillimanite-producing reactions, such as those involving muscovite (Kerrick, 1972)

$$\text{Muscovite + Quartz + Plagioclase + } H_2O \text{ = Sillimanite + Melt}$$

and

$$\text{Muscovite + Quartz + Plagioclase = Sillimanite + K-feldspar + Melt .}$$

coexisting andalusite + fibrolite in the leucosomes of a cordierite-bearing migmatite. Textures suggest that andalusite was replaced by cordierite + fibrolite. Brown (1983) described migmatites from Brittany, France, that contain kyanite and sillimanite occurring as inclusions in plagioclase. He concluded that kyanite initially formed at about 10 kbar and 730 ± 50°C and that sillimanite formed during subsequent decompression (Fig. 11.1). Barber and Yardley (1985) described cordierite, fibrolite and andalusite in leucosomes of migmatites in the eastern Connemara area of Ireland. Textural evidence supports their contention that cordierite and andalusite are of magmatic origin. They concluded that fibrolite formed during an early metamorphic event (estimated to have occurred at 4.5-6 kbar and T ~ 750°C), whereas andalusite is believed to have formed at P < 3.5 kbar during subsequent decompression. In leucosomes of andalusite-bearing migmatites of Mount Stafford, central Australia, Vernon and Collins (1988) cited an abundance of crystal faces bounding andalusite (Fig. 11.2), K-feldspar and cordierite as evidence for crystallization of these phases from an anatectic melt. McLellan (1989) concluded that there are two groups of migmatites in the Grampian Highlands of eastern Scotland. She argued that the first-formed migmatites resulted from subsolidus differentiation during kyanite-grade metamorphism. The leucosomes of these migmatites lack aluminum silicates. Other migmatites have sillimanite in the leucosomes and are considered to have formed from anatexis during sillimanite-grade metamorphism.

PERALUMINOUS GRANITOIDS

There have been many reports of andalusite and fibrolite in peraluminous granitic intrusives (Brammall and Harwood, 1932; Exley and Stone, 1966; Haslam, 1971; Kennan, 1972; Harris, 1974; Clarke et al., 1976; Abbott and Clarke, 1979; Clarke, 1981; Currie and Pajari, 1981; LeFort, 1981; Price, 1983; Brandstätter and Zemann, 1984; Propach and Gillessen, 1984; Zaleski, 1985; Kerrick and Speer, 1988; Hubbard, 1989). Aside from containing one or more of the aluminum silicates (andalusite, fibrolite, and sillimanite), peraluminous intrusives typically contain one or more of the following characteristic minerals: garnet, cordierite, topaz, tourmaline, muscovite, and spinel (Clarke, 1981; Zen, 1988). In most petrologic studies of intrusives containing aluminum silicates, the Al_2SiO_5 phase equilibrium diagram has been used to deduce the P-T conditions of magmatic crystallization.

There has been a long-standing controversy regarding andalusite in granitoids. That is, andalusite could represent: (a) residual xenocrysts

Wickham (1987) described progressive anatexis of pelitic rocks in a continuous prograde sequence from peraluminous leucosomes in migmatites at an intermediate structural level to peraluminous leucogranite plutons at deeper structural levels. Fibrolite is a common phase in both the leucosomes and plutons.

Fibrolite is by far the most common aluminum silicate in pelitic migmatites (Ashworth, 1976; Tyler and Ashworth, 1982; Jamieson, 1984; Weber et al., 1985; Wickham, 1987; Perreault and Martignole, 1988). Few migmatites have been described with kyanite or andalusite. In a gneiss complex in Cape Breton Island, Nova Scotia, Jamieson (1984) described

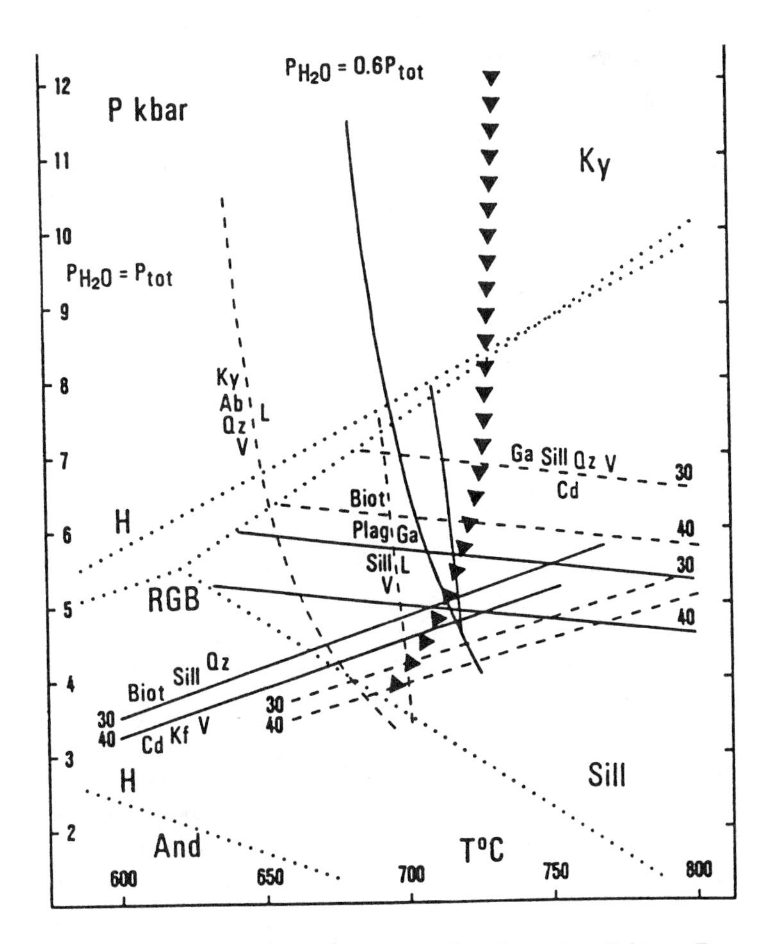

Figure 11.1. Phase equilibria applicable to migmatization in southern Brittany, France. The filled triangles depict the hypothesized decompression path. (From Brown, 1983, Fig. 16; copyright Shiva Publishing Limited).

Figure 11.2. Photomicrograph of a pegmatitic segregation from the Mt. Stafford area, central Australia, showing abundant, euhedral andalusite with interstitial quartz (light). Width of photo corresponds to 4.4 mm. (From Vernon and Collins, 1988, Fig. 5).

formed by assimilation of xenoliths derived by stoping of pelitic wall rocks, (b) restite formed from anatexis of pelitic rocks at depths greater than the present levels of exposure, or (c) the product of magmatic crystallization.

There are few published studies of peraluminous granitoids that present *cogent* arguments for a non-magmatic origin of the aluminum silicates. Lorenzoni et al. (1979) concluded that andalusite and fibrolite in peraluminous intrusives in southern Italy (Calabria) are xenocrysts derived from parent metamorphic rocks that underwent anatexis. The author considers their arguments supporting this contention to be equivocal. Their conclusion that the "swarms" of fibrolite represent "metamorphic structures" is questionable. They contended that the presence of unaltered fibrolite and andalusite inclusions in plagioclase suggests that the aluminum silicates were present at "...the first stage of consolidation of the magma..." (Lorenzoni et al., 1979, p. 425). However, as in the peraluminous intrusives of southern Italy, plagioclase in the Macusani Volcanics of Peru (Pichavant et al., 1988) also contains inclusions of andalusite and fibrolite. Pichavant et al. (1988a) considered that the aluminum silicates *coprecipitated* with plagioclase. This conclusion could also apply to the peraluminous granitoids described by Lorenzoni et al. (1979). Because textures of fibrolite in the adamellites of northwestern Maine are similar to those of the regionally metamorphosed fibrolite-bearing metapelitic country rocks, Guidotti (1978) concluded that the fibrolite is a residual phase that remained after anatexis of metapelites.

Aluminum silicates in granitoids could be produced by metamorphism. Zaleski (1985) concluded that andalusite an a peraluminous granitoid in the Moy Intrusive Complex, Scotland, was produced by contact metamorphism.

In several localities, peraluminous granitoids have been described with textures suggesting that andalusite was a product of magmatic crystallization (Haslam, 1971; Clarke et al., 1976; Kerrick and Speer, 1988). The existence of aluminum silicates as magmatic phases is bolstered by their occurrence in volcanic rocks. Zeck (1970) described fibrolite in dacite from Cerro del Hoyazo, Spain. Based on textural evidence, he concluded that the fibrolite crystals "...are probably precipitation products of the magma..." (Zeck, 1970, p. 229). Fibrolite and andalusite occur in the Macusani Volcanics in Peru (London et al., 1988; Pichavant et al., 1988a,b). Based on textural evidence, Pichavant et al. (1988a) concluded that fibrolite formed during the "early magmatic" stage whereas andalusite crystallized during the "late magmatic" stage. This sequence of crystallization is in concert with a P-T path which started in the sillimanite stability field and ended in the andalusite stability field. The euhedral to subhedral shapes of the andalusite provide convincing evidence that it was in equilibrium with the melt.

Evidence presented in the preceding paragraph supports the contention that there is a P-T stability field of andalusite + melt. With the solidus for metaluminous granodiorites (Fig. 11.3), only the Al_2SiO_5 phase equilibrium diagram of Richardson et al. (1969) yields a stability field of andalusite + melt. This has been used as an argument against Holdaway's (1971) phase diagram (Vernon et al., ms.). In addressing this argument, Holdaway (1971, p. 128) appealed to expansion of the P-T field of melt "...by volatiles such as fluorine". In favoring Holdaway's (1971) phase diagram over that of Richardson et al. (1969), Price (1983) concluded that the apparent equilibrium of andalusite with peraluminous melts is attributed to temperature lowering of the solidus due to elevated B and F in the melts and/or expansion of the andalusite P-T stability field by solid solution. To address the question of solid solution, Kerrick and Speer (1988)

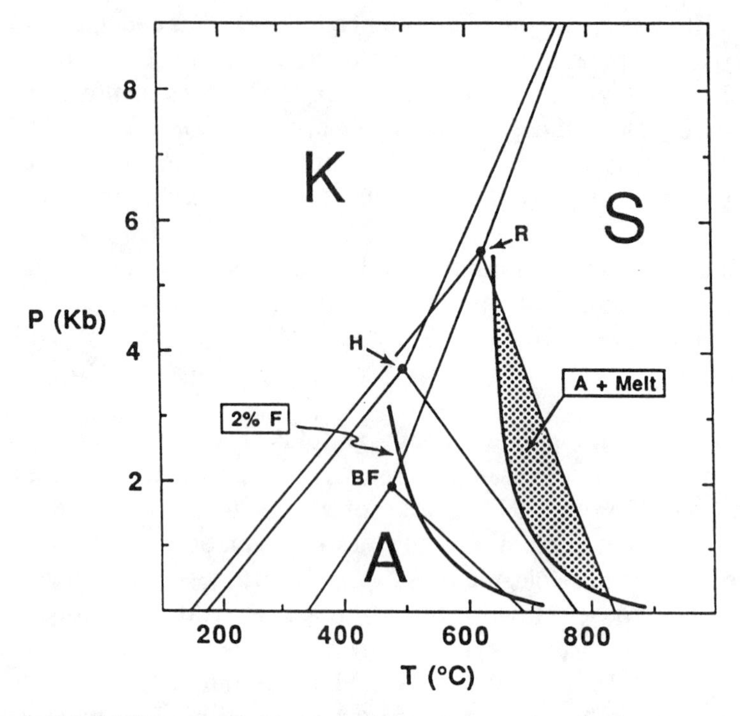

Figure 11.3. Al_2SiO_5 phase equilibrium diagrams of Brown and Fyfe (BF), Holdaway (H), and Richardson et al. (R). The granodiorite solidus with a_{H_2O} = 1.0 and with 2 wt % F are from Manning and Pichavant (1983). The shaded area is the stability field of andalusite + melt.

analyzed minor element contents of andalusite and fibrolite in selected samples of peraluminous granitoids. Their data suggest that because of the relatively low minor element contents of andalusite, and computed values of K_D (= $X^{Fib}_{Al_2SiO_5}$ / $X^{And}_{Al_2SiO_5}$) close to unity for andalusite-fibrolite pairs, minor element solid solution has an insignificant effect on the andalusite P-T stability field. Thus, Kerrick and Speer rationalized the occurrence of andalusite in peraluminous granitoids by accepting Holdaway's (1971) Al_2SiO_5 phase equilibrium diagram and appealing to expansion of the P-T stability field of melt by F and B. Kerrick and Speer's conclusions regarding minor element partitioning are supported by virtually end-member compositions of andalusite and sillimanite in the Macusani volcanics (Pichavant et al., 1988a). Enhanced concentrations of F and B are supported by the common occurrence of topaz and tourmaline in peraluminous granitoids. Zen (1988, p. 34) also appealed to the lowering of the solidus temperature by addition of F to explain the presence of primary magmatic andalusite in plutons of peraluminous granitoids "...that apparently consolidated at pressures significantly less than 3-4 kbar...".

However, he argued that elevated B_2O_3 contents (> 1 wt %) are unlikely in peraluminous melts.

MAGMATIC PEGMATITES

"The most important genetic controversy has involved igneous versus metamorphic origins for some of the deeper-seated pegmatites..."

G.E. Brown and R.C. Ewing (1986, p. 234)

Chapter 10 summarizes studies of Al_2SiO_5-bearing pegmatites in which the authors implicitly or explicitly appealed to crystallization from aqueous solutions within fractures and cavities. However, a popular alternative mechanism is the crystallization of pegmatites from hydrous silicate melts. A magmatic origin is particularly invoked for *granite* pegmatites, so-named because of the correspondence between their bulk compositions and the thermal minimum in the $NaAlSi_3O_8$-$KAlSi_3O_8$-SiO_2-H_2O system (Brown and Ewing, 1986). Cěrný (1982, p. 406) concluded that the formation "...of granitic pegmatites from supercritical fluids, hydrothermal solutions and/or aqueous vapor phase in the absence of melt has no support in the wealth of field and analytical data available today". Cěrný gave the following reasons for this conclusion: (1) there is no experimental or field evidence for leaching of the requisite solutes from metamorphic or igneous host rocks, (2) crystallization from aqueous solutions is precluded because the solutes would be purged by reaction with host rocks through which the fluids pass prior to influx of the fluids into fractures and cavities, (3) there is no field evidence supporting the *endogenous* secretion of pegmatite-forming fluids from host rocks into fractures and cavities, (4) granitic pegmatites have a narrow range of compositions closely corresponding to minimum melts, (5) pegmatite crystallization sequences correspond to that expected from the progressive crystallization from granitic melts, and (6) chemical, petrologic, and fluid inclusion data support initial crystallization from granitic melts followed by increasing involvement of hydrothermal solutions evolved from the granitic magma.

Following the correlation between pegmatite type and depth of formation proposed by Ginsberg et al. (1979), Cěrny and Hawthorne (1982) suggested that kyanite-bearing pegmatites within kyanite-grade Barrovian metamorphic regimes originated by anatexis or from pegmatite-generating granitoids. There are, however, arguments against a magmatic origin for such pegmatites.

Kyanite-bearing pegmatites have a wide variation in bulk composition. Some consist of kyanite + quartz (Read, 1932; Sauniac and Touret, 1983) and thus lack the feldspathic components required for granite minimum compositions. However, others contain muscovite + plagioclase + quartz and thus have a bulk composition of the "granite" system. In such cases, it could be argued that muscovite (rather than K-feldspar) crystallized from the hydrous granite pegmatite magmas. However, in kyanite pegmatites in the Lepontine Alps, there is a wide range of relative modal proportions of muscovite, plagioclase, quartz and Al_2SiO_5 (Keller, 1968; Klein, 1976; Codoni, 1981). Thus, there is no *apparent* clustering of the bulk compositions to granite minimum compositions.

Kyanite-bearing pegmatites in Barrovian-type regional metamorphism occur in discordant veins and segregations that are in sharp contact with host rocks. No cogent field evidence exists for anatectic origin of these pegmatites from the surrounding host rocks. Following Cĕrný's (1982, p. 413) model for rare-earth pegmatites, it could be argued that, because of low viscosities, the pegmatite magmas were generated at greater depths and subsequently underwent significant upward migration prior to crystallization. However, kyanite-bearing pegmatites typically form small, isolated pods and veins. It is difficult to envision a mechanical mechanism for significant upward migration of small ($< 1\ m^3$) packets of pegmatite magma. The occurrence of kyanite-bearing pegmatites in boudin necks (Fig. 10.27f) particularly supports a local derivation.

There are relatively few reports of andalusite in pegmatites occurring in low-pressure metamorphic regimes. Crystallization from hydrous granite pegmatite magmas is likely for some andalusite-bearing pegmatites.

The study of Vernon and Collins (1988) provides an example of the formation of andalusite-bearing pegmatitic leucosomes by low-pressure anatexis.

Rose (1957) studied andalusite-bearing pegmatite veins at May Lake, Yosemite National Park, California. The veins are zoned with andalusite-rich margins and cores dominantly consisting of quartz + K-feldspar + plagioclase. Tourmaline is a notable accessory in the pegmatites. The pegmatite veins are discordant; Rose (1957) contended that they are joint fillings. The host rocks are mainly andalusite-cordierite hornfelses.

Because of the discordant nature of the segregations, and the fact that most andalusite crystals are oriented with *c* crystallographic axes at high angles to the vein walls, these segregations are apparently fracture fillings. Rose (1957) hypothesized that the pegmatites formed by injection of granitic magma along fractures and that the aluminous composition of the pegmatites arose from reaction between the magma and pelitic wall rocks.

Andalusite-bearing pegmatite occurs within biotite granites in the Scott Glacier area, Antarctica (Burt and Stump, 1983). The geologic setting and major mineralogy of the pegmatite is compatible with a magmatic origin. Burt and Stump (1983) attributed the relative rarity of andalusite-bearing pegmatites to the restricted P-T stability field of andalusite + melt. However, because of the low solidus temperatures for peraluminous melts (Webster et al., 1987; London et al., 1988), coupled with the temperature lowering of the solidus due to B and F (Pichavant, 1987; London, 1987), the P-T stability field of andalusite + hydrous peraluminous melts is undoubtedly considerably larger than that implied by Burt and Stump (1983).

Cěrny and Hawthorne (1982, p. 170) concluded that pegmatitic sillimanite occurs in "...circumstances strongly suggesting a late origin, connected with deformation and local recrystallization of the primary pegmatite assemblages". This mechanism is supported by Wintsch and Andrew's (1988) genetic model for fibrolite in a granitic pegmatite from south-central Connecticut (discussed in Chapter 8). In the sillimanite zone of the Lepontine Alps, crosscutting veinlets of fibrolite + quartz occur within quartz-rich segregations, thereby suggesting a late origin for the fibrolite. I know of no examples of fibrolite- or sillimanite-bearing pegmatites with evidence for crystallization of these phases from a hydrous melt.

REFERENCES

[To maximize the utility of this volume, the references are comprehensive, listing many uncited papers as well as referenced ones. Readers are thereby provided with a single, extensive bibliographic source]

Abbott, R.N., and Clarke, D.B. (1979) Hypothetical liquidus relationships in the subsystem Al_2O_3-FeO-MgO projected from quartz, alkali feldspar and plagioclase for $a_{H_2O} \leq 1$. Canadian Mineral. 17, 549-560.

Abraham, K., and Schreyer, W. (1975) Minerals of the viridine hornfels from Darmstadt, Germany. Contrib. Mineral. Petrol. 49, 1-20.

Abs-Wurmbach, I., and Langer, K. (1975) Synthetic Mn^{3+}-kyanite and viridine $(Al_{2-x})Mn_x^{3+})SiO_5$ in the system Al_2O_3-MnO-MnO_2-SiO_2. Contrib. Mineral. Petrol. 49, 21-38.

______, Langer, K., and Tillmans, E. (1977) Structure and polarized absorption spectra of Mn^{3+}-substituted andalusites (viridines). Naturwissenschaften 64, 527-528.

______, Langer, K., Seifert, F., and Tillmanns, E. (1981) The crystal chemistry of (Mn^{3+}, Fe^{3+})-substituted andalusites (viridines and kanonaite), $(Al_{1-x-y}Mn_x^{3+}Fe_y^{3+})_2(O \mid SiO_4)$: crystal structure refinements, Mössbauer, and polarized optical absorption spectra. Zeits. Kristallogr. 155, 81-113.

______, Langer, K., and Schreyer, W. (1983) The influence of Mn^{3+} on the stability relations of the Al_2SiO_5 polymorphs with special emphasis on manganian andalusites (viridines), $(Al_{1-x}Mn_x)_2(O/SiO_4)$: an experimental investigation. J. Petrol. 24, 48-75.

Acharyya, K.S., Mukherjee, S., and Basu, A. (1990) Manganian andalusite from Manbazar, Purulia District, West Bengal, India. Mineral. Mag. 54, 75-80.

Ackermand, D., Herd, R.K., Reinhardt, M., and Windley, B.F. (1985) Genese der sapphirinführenden gesteine in Bereich der Caraiba Kupfermine, Bahia, Brasilien. [Abs.] Fort. Mineral. 63, 3.

Agrell, S.O., and Smith, J.V. (1960) Cell dimensions, solid solution, polymorphism, and identification of mullite and sillimanite. J. Amer. Ceram. Soc. 43, 69-76.

Ahmad, R., and Wilson, C.J.L. (1981) Uranium and boron distributions related to metamorphic microstructure-evidence for metamorphic fluid activity. Contrib. Mineral. Petrol. 76, 24-32.

______, and Wilson, C.J.L. (1982) Microstructural relationships of sillimanite and fibrolite at Broken Hill, Australia. Lithos 15, 49-58.

Ahn, J.H., and Buseck, P.R. (1988) Al-chlorite as a hydration reaction product of andalusite: a new occurrence. Min. Mag. 52, 396-399.

______, Burt, D.M., and Buseck, P.R. (1988) Alteration of andalusite to sheet silicates in a pegmatite. Amer. Mineral. 73, 559-567.

Akaad, M.K. (1956) The northern aureole of the Ardara Pluton of county Donegal. Geol. Mag. 93, 377-392.

Albee, A.L., and Chodos, A.A. (1969) Minor element content of coexisting Al_2SiO_5 polymorphs. Amer. J. Sci. 267, 310-316.

Alderman, A.R. (1942) Sillimanite, kyanite and clay deposits near Williamstown, South Australia. Trans. Roy. Soc. S. Australia 66, 3-14.

______ (1950) Clay derived from sillimanite by hydrothermal alteration. Mineral. Mag. 29, 271-279.

Althaus, E. (1966) Der stabilitätsbereich des pyrophyllits unter dem einfluβ von säuren II. Mitteilung: Folgerungen für die petrogenese, insbesondere von pyrophyllit-und andalusitlagerstatten. Contrib. Mineral. Petrol. 13, 97-107.

______(1967) The triple point andalusite-sillimanite-kyanite. Contrib. Mineral. Petrol. 16, 29-44.

______(1969) Experimental evidence that the reaction of kyanite to form sillimanite is at least bivariant. Amer. J. Sci. 267, 273-277.

Altherr, R., Okrusch, M. and Bank, H. (1982) Corundum- and kyanite-bearing anatexites from the Precambrian of Tanzania. Lithos 15, 191-197.

Andersen, T., Burke, E.A.J., Austrhein, M., ((1989) Nitrogen-bearing aqeous fluid inclusions in some eclogites from the Western Gneiss region of the Norwegian Caledonides. Contrib. Mineral. Petrol. 103, 153-165.

Anderson A.B. (1985) Point defects in crystals: a quantum chemical methodology and its applications. In: Schock, R.N. (ed.), Point Defects in Minerals. Amer. Geophys. Union, Geophysical Monograph 31, Mineral Physics 1, 18-25.

Anderson, G.M., and Burnham, C.W. (1983) Feldspar solubility and the transport of aluminum under metamorphic conditions. Amer. J. Sci. 283-A, 283-297.

______, Pascal, M.L., and Rao, J. (1988) Aluminum speciation in metamorphic fluids. In: Helgeson, H.C. (ed.), Chemical Transport in Metasomatic Processes. Reidel, Boston. 297-321.

Anderson, P.A.M., and Kleppa, O.J. (1969) The thermochemistry of the kyanite-sillimanite equilibrium. Amer. J. Sci. 267, 285-290.

______, Newton, R.C., and Kleppa, O.J. (1977) The enthalpy change of the andalusite-sillimanite reaction and the Al_2SiO_5 diagram. Amer. J. Sci., 277, 585-593.

Angel, R.J., and Prewitt, C.T. (1986) Crystal structure of mullite: A re-examination of the average structure. Amer. Mineral. 71, 1476-1482.

Anisimova, A.S., Nkrasova, L.P, Ploshko. V.V., and Shport, N.P. (1975) Fluorine-zoisite and fluorine-kyanite from the eclogites and hydrothermal veins of the Caucasus [in Russian]. Zap. Vses. Mineral. Obshch. 104, 97-99.

Aramaki, S. (1961) Sillimanite and cordierite from volcanic xenoliths. Amer. Mineral. 46, 1154-1165.

______, and Roy, R. (1959) Revised equilibrium diagram for the system Al_2O_3-SiO_2. Nature 184, 631-632.

______, and Roy, R. (1963) A new polymorph of Al_2SiO_5 and further studies in the system Al_2O_3-SiO_2-H_2O. Amer. Mineral. 48, 1322-1347.

Aranovich, L. Ya., and Podlesskii, K.K. (1983) Experimental study of the equilibrium garnet + sillimanite + quartz = cordierite. Dokl. Acad. Sci. USSR, Earth Sci. Sect. 259, 136-139.

______ (1989) Geothermobarometry of high-grade metapelites: simultanteously operating reactions. In: Daly, J.S., Cliff, R.A., and Yardley, B.W.D. (eds.), Evolution of Metamorphic Belts, Geol. Soc. London Special Publication 43, 45-61.

Ashworth, J.R. (1975) The sillimanite zones of the Huntly-Portsoy area in the north-east Dalradian, Scotland. Geol. Mag. 112, 113-224.

______ (1976) Petrogenesis of migmatites in the Huntly-Portsoy area, north-east Scotland. Mineral. Mag. 40, 661-682.

______ (1979) Textural and mineralogical evolution of migmatites. In: Harris, A.L., Holland, K.E., and Leake, B.E. (eds.) The Caledonides of the British Isles, reviewed. Scott. Acad. Press, Edinburgh, Scotland, 357-361.

______, and Chinner, G.A. (1978) Coexisting garnet and cordierite in migmatites from the Scottish Caledonides. Contrib. Mineral. Petrol. 65, 379-394.

______, and Tyler, I.M. (1983) The distribution of metamorphic temperatures around the Strontian Granodiorite. Geol. Mag. 120, 281-290.

______, and Evirgen M.M. (1985) Plagioclase relations in pelites, central Menderes Massif, Turkey. I. The peristerite gap with coexisting kyanite. J. Metamorphic Geol. 3, 207-218.

Atherton, M.P. (1965) The chemical significance of isograds. In: Pitcher, W.S, and Flinn, G.W. (eds.) Controls of Metamorphism, 169-202, John Wiley & Sons, New York.

______ (1977) The metamorphism of the Dalradian rocks of Scotland. Scottish J. Geol. 13, 331-370.

______, and Brotherton, M.S. (1972) The composition of some kyanite-bearing regionally metamorphosed rocks from the Dalradian. Scott. J. Geol. 8, 203-213.

______, Naggar, M.H., and Pitcher, W.S. (1975) Kyanite in some thermal aureoles. Amer. J. Sci. 275, 432-443.

Atkin, B.P. (1979) Oxide-sulphide-silicate mineral phase relations in some thermally metamorphosed graphitic pelites from Co. Donegal, Eire. In: Harris, A.L., Holland, K.E., and Leake, B.E. (eds.) The Caledonides of the British Isles, reviewed. Scott. Acad. Press, Edinburgh, Scotland, 375-384.

Atzori, P., Lo Giudice, A., and Pezzino, A. (1985) Peraluminous leucocratic rocks in the Calabria-Peloritani high-grade metamorphic complex: a review. Periodico Mineral. 54, 115-118.

Austrheim, H. (1987) Eclogitization of lower crustal granulites by fluid migration through shear zones. Earth Planet. Sci. Lett. 81, 221-232.

Autran, A., and Guitard, G. (1957) Sur la signification de la sillimanite dans les Pyrénées. Comptes Rendus Séanc. Soc. géol. Fr., 141.

Avrami, M. (1939) Kinetics of phase change I. J. Chem. Phys. 7, 1103-1112.

______ (1940) Kinetics of phase change II. J. Chem. Phys. 8, 212-224.

Backström, H. (1896) Manganandalusite from Vestana (Swed.). Geol. För. Förh., Stockholm. 18, 386-389.

Baker, A.J. (1985) Pressures and temperatures of metamorphism in the eastern Dalradian. J. Geol. Soc. London 142, 137-148.

______ (1987) Models for the tectonothermal evolution of the eastern Dalradian of Scotland. J. Metamorphic Geol. 5, 101-118.

______, and Droop, G.T.R. (1983) Grampian metamorphic conditions deduced from mafic granulites and sillimanite-K-feldspar gneisses in the Dalradian of Glen Muick, Scotland. J. Geol. Soc. London, 140, 489-497.

Baker, J., Powell, R., and Sandiford, M. (1987) Corona textures between kyanite, garnet and gedrite in gneisses from Errabiddy, Western Australia. J. Metamorphic Geol. 5, 357-370.

Balconi, N. (1941) Sintesi della sillimanite. Rend. Soc. Mineral. Ital. 1, 82-86.

Ballevre, M., Pinardon, J.-L., Kienast, J.-R., and Vuichard, J.-P. (1989) Reversal of Fe-Mg partitioning between garnet and staurolite in eclogite-facies metapelites from the Champtoceaux nappe (Brittany, France). J. Petrol. 30, 1321-1349.

Baltatzis, E., and Katagas, C. (1981) Margarite pseudomorphs after kyanite in Glen Esk, Scotland. Amer. Mineral. 66, 213-216.

Banerji, A.K. (1954) On a kyanite pegmatite in Ghadigih, Singhbhum district, Bihar. Sci. and Cult. 20, 241-242.

Barber, J.P., and Yardley, B.W.D. (1985) Conditions of high grade metamorphism in the Dalradian of Connemara, Ireland. J. Geol. Soc. London 142, 87-96.

Barker, A.J., and Anderson, M.W. (1989) The Caledonian structural-metamorphic evolution of south Troms, Norway. In: Daly, J.S., Cliff, R.A., and Yardley, B.W.D. (eds.), Evolution of Metamorphic Belts, Geol. Soc. London Special Publication 43, 385-390.

Barker, D.S. (1987) Rhyolites contaminated with metapelite and gabbro, Lipari, Aeolian Islands, Italy: products of lower crustal fusion or of assimilation plus fractional crystallization? Contrib. Mineral. Petrol. 97, 460-472.

Barnicoat, A.C. (1988) The mechanism of veining and retrograde alteration of Alpine eclogites. J. Metamorphic Geol. 6, 545-558.

______, and Fry, N. (1986) High-pressure metamorphism of the Zermatt-Saas ophiolite zone, Switzerland. J. Geol. Soc. London 143, 607-618.

______, Cartwright, I., and O'Hara, M.J. (1987) Kyanite in the mainland Lewisian complex. Scott. J. Geol. 23, 209-213.

______, and Prior, D.J. (ms.) Contact metamorphism around the Ross of Mull granite, Scotland.

Barrera, J.L., Bellido, F., and Klein, E. (1985) Contact metamorphism in synkinematic two-microgranites produced by younger granitic intrusions, Galicia, NW Spain. Geol. Mijnb. 64, 413-422.

Barton, M.D. (1986) Phase equilibria and thermodynamic properties of minerals in the BeO-Al_2O_3-SiO_2-H_2O (BASH) system, with petrologic applications. Amer. Mineral. 71, 277-300.

Bearth, P. (1932) Geologie un petrographie der Keschgruppe. Schweiz. miner. petrogr. Mitt. 12, 256.

Beddoe-Stephens, B. (in press) Pressures and temperatures of Dalradian metamorphism and the andalusite-kyanite transformation in the Northeast Grampians. Scott. J. Geol.

Beger, R.M. (1979) Aluminum-silicon ordering in sillimanite and mullite. Ph.D. Thesis, Harvard University, 312p.

______, Burnham, C.W., and Hays, J.F. (1970) Structural changes in sillimanite at high temperature. Abstr. Geol. Soc. Amer. 2, 490.

Bel'kov, I.V. (1963) Kyanite schists of the Keyvy suite. Izd. Akad. Nauk. USSR.

Bell, P.M. (1963) Aluminum silicate system: experimental determination of the triple point. Science 139, 1055-1056.

______, and Williams, D.W. (1971) Pressure calibration in piston-cylinder apparatus at high temperature. In: Ulmer, G.C. (ed.), Research Techniques for High Pressure and High Temperature. Springer-Verlag, New York. p. 195-215.

______, and Nord, G. (1974) Microscopic and electron diffraction study of fibrolitic sillimanite. Carnegie Inst. Wash. Ann. Rept. Dir. Geophys. Lab. 1973-74, 444-448.

Benard, F., Moutou, P., and Pichavant, M. (1985) Phase relations of tourmaline leucogranites and the significance of tourmaline in silicic magmas. J. Geol. 93, 271-291.

Beran, A. (1970) Ultrarotspektroskopischer nachweis von OH-gruppen en den mineralen der 'Al_2SiO_5', Modifikationen. Oesterr. Akad. Wiss., Math.-Naturw. KL. 107, 184-185.

______, and Zemann, J. (1969) Messun des ultrarot-pleochroismus von mineralen, VIII der pleochroismus der OH-streckfrequenz in andalusit. Tschermaks mineral. petrogr. Mitt. 13, 285-292.

______, Hafner, St., and Zemann, J. (1983) Untersuchungen über den einbau von hydroxilgruppen im edelstein-sillimanit. N. Jahrb. Mineral. Monat. 5, 219-226.

______, and Götzinger, M.A. (1987) The quantitative IR spectroscopic determination of structural OH groups in kyanites. Mineral. Petrol. 36, 41-49.

______, Grew, E.S, and Rossman, G.R. (1989) The hydrous component of sillimanite. Amer. Mineral. 74, 812-817.

Berg, J.H., and Docka, J.A. (1983) Geothermometry in the Kiglapait contact aureole, Labrador. Amer. J. Sci. 283, 414-434.

Bergström, L. (1960) An occurrence of kyanite in a pegmatite in western Sweden. Geol. Fören. Förh., Stockholm 82, 270-272.

Berkley, J.L., and Callender, J.F. (1979) Precambrian metamorphism in the Placitas-Juan Tabo area, northwestern Sandia Mountains, New Mexico. New Mex. Geol. Soc. Guidebook, 30th Field Conf., Santa Fe Country, 181-188.

Berman, R.G. (1988) Internally-consistent thermodynamic data for minerals in the system Na_2O-K_2O-CaO-MgO-FeO-Fe_2O_3-Al_2O_3-SiO_2-TiO_2-H_2O-CO_2. J. Petrol. 29, 445-522.

______, R.G., Engi, M., Greenwood, H.J., and Brown, T.H. (1986) Derivation of internally-consistent thermodynamic data by the technique of mathematical programming: a review with application to the system $MgO-SiO_2-H_2O$. J. Petrol. 27, 1331-1364.

Best, M.G. (1982) Igneous and Metamorphic Petrology. W.H. Freeman & Co., San Francisco. 630 pp.

______, and Weiss, L.E. (1964) Mineralogical relations in some pelitic hornfelses from the southern Sierra Nevada, California. Amer. Mineral. 49, 1240-1266.

Bhattacharya, A., and Sen, S.K. (1985) Energetics of hydration of cordierite and water barometry in cordierite-granulites. Contrib. Mineral. Petrol. 89, 370-378.

Bickle, M.J., and Archibald, N.J. (1984) Chloritoid and staurolite stability: implications for metamorphism in the Archaean Yilgarn Block, Western Australia. J. Metamorphic Geol. 2, 179-203.

Billings, M.P. (1937) Regional metamorphism of the Littleton-Moosilaude area, New Hampshire. Bull. Geol. Soc. Amer. 48, 463-566.

Bingen, B., Demaiffe, D., and Delhal, J. (1988) Aluminous granulites of the Archean craton of Kasai (Zaire): petrology and P-T conditions. J. Petrol. 29, 899-919.

Birch, F. (1966) Compressibility; elastic constants. In: Clark, S.P., Jr. (ed.) Handbook of Physical Constants. Geol. Soc. Amer. Memoir 97, 97-173.

Blasi, A. (1971) Genesi dei noduli a sillimanite nelle anatessiti del Mt. Pélago (Alpi Marittime) in rapporto al fenomeni di metamorfismo, piegamento e granitizzazione. Mem. Soc. Geol. Ital. 10, 167-190.

______, and Schiavinato, G. (1968) Significato petrologico dei noduli a sillimanite e dei noduli a cordierite diffusi nelle anatessi biotitiche del M. Pélago (Massiccio cristallino dell' Argentera). Boll. Soc. Geol. Italiana 87, 253-275.

Bloss, F.D. (1985) Labelling refractive index curves for mineral séries. Amer. Mineral. 70, 428-432.

Blümel, P., and Schreyer, W. (1976) Progressive regional low-pressure metamorphism in Moldanubian metapelites of the northern Bavarian Forest, Germany. Krystalinikum 12, 7-30.

______, and ______ (1977) Phase relations in pelitic and psammitic gneisses of the sillimanite-potash feldspar and cordierite-potash feldspar zones in the Moldanubicum of the Lam-Boenmais area, Bavaria. J. Petrol. 18, 431-459.

Bøe, P. (1974) Petrography of the Gula Group in Hessdalen, southeastern Trondheim region, with special reference to the paragonitization of andalusite pseudomorphs. Norges Geol. Unders. 304, 33-46.

Boettcher, A.L., Windom, K.E., Bohlen, S.R., and Luth, R.W. (1981) Low-friction, anhydrous, low- to high-temperature furnace sample assembly for piston-cylinder apparatus. Rev. Sci. Instrum. 52, 1903-1904.

Bohlen, S.R. (1984) Equilibria for precise pressure calibration and a frictionless furnace assembly for the piston-cylinder apparatus. N. Jahrb. Mineral. Monatsh. 9, 404-412.

______ (1987) Pressure-temperature-time paths and a tectonic model for the evolution of granulites. J. Geol. 95, 617-632.

______, Dollase, W.A., and Wall, V.J. (1986) Calibration and applications of spinel equilibria in the system $FeO-Al_2O_3-SiO_2$. J. Petrol. 27, 1143-1156.

______, Montana, A., and Kerrick, D.M. (ms.) A precise determination of the kyanite = sillimanite and andalusite = kyanite equilibria and a revised triple point for the aluminum silicate minerals.

Bol, L.C.G.M., and Maijer, C. (1989) Premetamorphic laterisation in Proterozoic metabasites of Rogaland, SW Norway. Contrib. Mineral. Petrol. 103, 306-316.

Boland, J.N., Hobbs, B.E., and McLaren, A.C. (1977) The defect structure in natural and experimentally deformed kyanite. Phys. stat. sol.(a) 39, 631-641.

Bosshart, E., Frankc E., Hanni, H.A., and Barot, N. (1982) Blue-colour-changing kyanite from East Africa. J. Gemmology 18, 205-212.

Bossière, G. (1988) Evolutions chimico-minéralogiques du grenat et de la muscovite au voisinage de l'isograde biotite-staurotide dans un métamorphisme prograde de type barrovien: un exemple en Vendée littorale (Massif Armoricain). Comptes Rendus l'Acad. Sci., Paris, Série II 306, 135-140.

Bottinga, Y.D., Weill, D.F., and Richet, P. (1981) Thermodynamic modeling of silicate melts. In: Newton, R.C., Navrotsky, A., and Wood, B.J. (eds.) Thermodynamics of Minerals and Melts. Adv. Physical Geochemistry, v. 1, Springer-Verlag, New York, 207-241.

Bowen, N.L. (1940) Progressive metamorphism of siliceous limestone and dolomite. J. Geol. 48, 225-274.

Bowers, J.A., Kerrick, D.M., and Furlong, K.P. (1990) Conduction model for the thermal evolution of the Cupsuptic aureole, Maine. Amer. J. Sci. 290, 644-665.

Bowman, A.F. (1975) An investigation of Al_2SiO_5 phase equilibrium utilizing the scanning electron microscope. M.S. Thesis, Univ. of Oregon, Eugene, OR, 80 pp.

Boyle, A.P. (1986) Metamorphism of basic and pelitic rocks at Sulitjelma, Norway. Lithos 19, 113-128.

Brace, W.F., Scholz, C.H., and La Mori, P.N. (1969) Isothermal compressibility of kyanite, andalusite and sillimanite from synthetic aggregates. J. Geophys. Res. 74, 2089-2098.

Bradbury, H.J. (1979) Migmatisation, deformation and porphyroblast growth in the Dalradian of Tayside, Scotland. In: Harris, A.L., Holland, C.H., and Leake, B.E. (eds.), The Caledonides of the British Isles, reviewed. Scottish Academic Press, Edinburgh. 351-356.

Bradshaw, J.Y. (1989) Origin and metamorphic history of an Early Cretaceous polybaric granulite terrain, Fiordland, southwest New Zeland. Contrib. Mineral. Petrol. 103, 346-360.

Brammall, A., and Harwood, H.F. (1932) The Dartmoor granites: their genetic relationships. J. Geol. Soc. London 88, 171-237.

Bramwell, M.G.. (1985) Metamorphic differentiation; a mechanism indicated by zoned kyanite crystals in some rocks from the Lukmanier region, Switzerland. Mineral. Mag. 49, 59-64.

Brandstätter, F., and Zemann, J. (1984) The chemical composition of andalusite in the peraluminous granite of Rásná near Telč (Czechoslovakia). Tschermaks mineral. petrogr. Mitt. 33, 131-134.

Breaks, F.W., and Shaw, D.M. (1973) The Silent Lake pluton, Ontario: a nodular, sedimentary, intrusive complex. Lithos 6, 103-122.

Brew, D.A., Ford, A.B., and Himmelberg, G.R. (1989) Evolution of the western part of the Coast plutonic-metamorphic complex, south-eastern Alaska, USA: a summary. In: Daly, J.S., Cliff, R.A., and Yardley, B.W.D. (eds.), Evolution of Metamorphic Belts, Geol. Soc. London Special Publication 43, 447-452.

Brindley, J.C. (1957) The aureole rocks of the Leinster granite in south Dublin, Ireland. Proc. Roy. Irish Acad. 59B, 1.

Brothers, S.C. (1986) A new geothermometer and oxygen barometer for metamorphic rocks. Abstr. Geol. Soc. Amer. 18, 551.

______ (1987) Theoretical phase relations in the assemblage: rutile-ilmenite(hematite)-aluminum silicate-quartz (RISQ) implications for geothermometry and oxygen barometry. M.S. Thesis, Univ. of New Mexico, 77p.

Brown, C.Q. (1976) Sillimanite in Pickens County, South Carolina. Geol. Notes, Divn. of Geol. South Carolina 10, no. 1.

Brown, G.C., and Fyfe, W.S. (1971) Kyanite-andalusite equilibrium. Contrib. Mineral. Petrol. 33, 227-231.

Brown, G.E., Jr., and Ewing, R.C. (1986) Introduction to the Jahns Memorial Issue. Amer. Mineral. 71, 233-238.

Brown, M. (1983) The petrogenesis of some migmatites ftom the Presqu'ile de Rhuys, southern Brittany, France. In: Migmatites, Melting and Metamorphism, M.P. Atherton and C.D. Gribble (eds.), Shiva, Nantwich, England, 174-200.

______, and Earle, M.M. (1983) Cordierite-bearing schists and gneisses from Timor, eastern Indonesia: P-T conditions of metamorphism and tectonic implications. J. Metamorphic Geol. 1, 183-203.

Brunet, W.M. (1981) Kyanite, Litchfield County, Connecticut. Rocks and Minerals 56, 159-161.

Bucher-Nurminen, K., and Droop, G. (1983) The metamorphic evolution of garnet-cordierite-sillimanite-gneisses of the Gruf-Complex, Eastern Pennine Alps. Contrib. Mineral. Petrol. 84, 215-227.

Bühl, H. (1981) Zur sillimanitbildung in den gneisen der Zone von Bellinzona. Schweiz. mineral. petrogr. Mitt. 61, 275-295.

Buick, I.S., and Holland, T.J.B. (1989) The *P-T-t* path associated with crystal extenstion, Naxos, Cyclades, Greece. In: Daly, J.S., Cliff, R.A., and Yardley, B.W.D. (eds.), Evolution of Metamorphic Belts, Geol. Soc. London Special Publication 43, 365-369.

Burg, J.P., Delor, C.P., Leyreloup, A.F., and Romney, F. (1989) Inverted metamorphic zonation and Variscan thrust tectonics in the Rouergue area (Massif Central, France): *P-T-t* record from mineral to regional scale. In: Daly, J.S., Cliff, R.A., and Yardley, B.W.D. (eds.), Evolution of Metamorphic Belts, Geol. Soc. London Special Publication 43, 423-439.

Burnham, C. Wayne (1981) The nature of multicomponent aluminosilicate melts. In: Rickard, D.T., and Wichman, F.E. (eds.) Chemistry and Geochemistry of Solutions at High Temperatures and Pressures, Physics and Chemistry of the Earth, v. 13 and 14. Pergamon Press, New York., 197-229.

Burnham, Charles W. (1963a) Refinement of the crystal structure of sillimanite. Zeits. Kristallogr. 118, 127-148.

______ (1963b) Refinement of the crystal structure of kyanite. Zeits. Kristallogr. 118, 337-360.

______ (1964a) Crystal structure of mullite. Carnegie Inst. Washington, Ann. Rept. Dir. Geophys. Lab. 1963-64, 223-227.

______ (1964b) Composition limits of mullite, and the sillimanite-mullite solid solution problem. Carnegie Inst. Washington, Ann. Rept. Dir. Geophys. Lab. 1963-64, 227-228.

______, and Buerger, M.J. (1961) Refinement of the crystal structure of andalusite. Zeits. Kristallogr. 115, 269-290.

Burns, R.G. (1970) Mineralogical Applications of Crystal Field Theory. Cambridge Univ. Press, Cambridge, England, 224 p.

Burt, D.M., and Stump, E. (1983) Mineralogical investigation of andalusite-rich pegmatites from Szabo Bluff, Scott Glacier area. Antarctic J. United States, Annual Rev. 18, 49-52.

Caillère, S., and Pobeguin, Th. (1972) Sur la presence de sillimanite à Bournac (Haute-Loire). Bull. Soc. franç. Min. Crist. 95, 411.

Cameron, W.E. (1976a) Coexisting sillimanite and mullite. Geol. Mag. 113, 497-513.

______ (1976b) A mineral phase intermediate in composition between sillimanite and mullite. Amer. Mineral. 61, 1025-1026.

______ (1976c) Exsolution in 'stoichiometric' mullite. Nature. 264, 736-738.

______ (1977a) Non-stoichiometry in sillimanite: mullite compositions with sillimanite-type superstructures. Phys. Chem. Minerals, 1, 265-272.

______ (1977b) Mullite: a substituted alumina. Amer. Mineral. 62, 747-755.

______, and Ashworth, J.R. (1972) Fibrolite, and its relationship to sillimanite. Nature 235, 134-136.

Campbell, I., and Wright, L.A. (1950) Kyanite paragenesis at Ogilby, California. [Abstr.] Bull. Geol. Soc. Amer. 61, 1520-1521.

Cardoso, G.M.(1927) Röntenographische feinbaustudien am cyanit und staurolith. Centr. Min., Abt. A, 384-387.

Carey, J.W., Rice, J.M., and Grover, T.W. (in press) Petrology of aluminous schist in the Boehls Butte region of northern Idaho. Part 1 Phase relations and petrogenesis. J. Petrol.

Carlino, P. (1972) L'andalusite di San Giorgio Morgeto (Reggio Calabria). Rend Soc. Italiana Min. Petr. 28, 413-421.

Carlson W.D., and Rossman, G.R. (1988) Vanadium- and chromium-bearing andalusite: occurrence and optical-absorption spectroscopy. Amer. Mineral. 73, 1366-1369.

Carmichael, D.M. (1969) On the mechanism of prograde metamorphic reactions in quartz-bearing pelitic rocks. Contrib. Mineral. Petrol. 20, 244-267.

______ (1970) Intersecting isograds in the Whetstone Lake area, Ontario. J. Petrol. 11, 147-181.

______ (1978) Metamorphic bathozones and bathograds: a measure of the depth of post-metamorphic uplift and erosion on the regional scale. Amer. J. Sci. 278, 769-797.

Carpenter, M.A. (1986) Some recent advances in the characterization of silicates. In: Freer, R., and Dennis, P.F. (eds.) Kinetics and Mass Transport in Silicate and Oxide Systems., Trans Tech., Rockport, MA. 19-34.

Carswell, D.A., Dawson, J.B., and Gibb, F.G.F. (1981) Equilibrium conditions of upper-mantle eclogites: implications for kyanite-bearing and diamondiferous varieties. Mineral. Mag. 44, 79-89.

Cartwright, I., and Barnicoat, A.C. (1986) The generation of quartz-normative melts and corundum-bearing restites by crustal anatexis: petrogenetic modelling based on an example from the Lewisian of North-West Scotland. J. Metamorphic Geol. 3, 79-99.

______, and ______ (1989) Evolution of the Scourian complex. In: Daly, J.S., Cliff, R.A., and Yardley, B.W.D. (eds.), Evolution of Metamorphic Belts, Geol. Soc. London Special Publication 43, 297-301.

Catlow, C.R.A., and Parker, S.C. (1985) Computer modelling of minerals. In: Schock, R.N. (ed.) Point Defects in Minerals. Amer. Geophys. Union, Geophysical Monograph 31, Mineral Physics 1, 26-35.

Černý, P. (1982) Petrogenesis of granitic pegmatites. In: Černý, P. (ed.), Short Course in Granitic Pegmatites in Science and Industry, Mineralogical Association of Canada 8, 405-461.

______, and Hawthorne, F.C. (1982) Selected peraluminous minerals. In: Černý, P. (ed.), Short Course in Granitic Pegmatites in Science and Industry, Mineralogical Association of Canada 8, 163-186.

Chakraborty, K.R., and Sen, S.K. (1967) Regional metamorphism of pelitic rocks around Kandra, Singhbhum, Bihar. Contrib. Mineral. Petrol. 16, 210-232.

Chamberlain, C.P. (1986) Evidence for the repeated folding of isotherms during regional metamorphism. J. Petrol. 27, 63-89.

______, and Lyons, J.B. (1983) Pressure, temperature and metamorphic zonation studies of pelitic schists in the Merrimack Synclinorium, south-central New Hampshire. Amer. Mineral. 68, 530-540.

______, and Rumble, D. (1988) Thermal anomalies in a regional metamorphic terrane: an isotopic study of the role of fluids. J. Petrol. 29, 1215-1232.

______, and ______ (1989) The influence of fluids on the thermal history of a metamorphic terrain: New Hampshire, USA. In: Daly, J.S., Cliff, R.A., and Yardley, B.W.D. (eds.), Evolution of Metamorphic Belts, Geol. Soc. London Special Publication 43, 203-213.

Chapman, C.A. (1950) Quartz veins formed by metamorphic differentiation of aluminous schists. Amer. Mineral. 35, 693-710.
Charlu, T.V., Newton, R.C., and Kleppa, O.J. (1975) Enthalpies of formation at 970K of compounds in the system MgO-Al_2O_3-SiO_2 from high temperature solution calorimetry. Geochim. Cosmochim. Acta 39, 1487-1497.
Chatterjee, N.D. (1972) The upper stability limit of the assemblage paragonite + quartz and its natural occurrences. Contrib. Mineral. Petrol. 34, 288-303.
______, and Johannes, W. (1974) Thermal stability and standard thermodynamic properties of synthetic $2M_1$-muscovite, $KAl_2[AlSi_3O_{10}(OH)_2]$. Contrib. Mineral. Petrol. 48, 89-114.
______, and Froese, E. (1975) A thermodynamic study of the pseudobinary join muscovite-paragonite in the system $KAlSi_3O_8$-$NaAlSi_3O_8$-Al_2O_3-SiO_2-H_2O. Amer. Mineral. 60, 985-993.
Chatterjee, S.K. (1931) Rocks bearing kyanite and sillimanite in the Bhandra District. Rec. Geol. Surv. India 65, 285.
Chinner, G.A. (1961) The origin of sillimanite in Glen Clova, Angus. J. Petrol. 2, 312-323.
______ (1962) Regional metamorphic sequences in the light of the Al_2SiO_5 polymorphs. Proc. Geol. Soc. London No. 1594.
______ (1965) The kyanite isograd in Glen Clova, Angus, Scotland. Min. Mag. 34, 132-143.
______ (1966a) The significance of the aluminum silicates in metamorphism. Earth-Sci. Rev. 2, 111-126.
______ (1966b) The distribution of pressure and temperature during Dalradian metamorphism. J. Geol. Soc. London 122, 159-186.
______ (1973) The selective replacement of aluminum silicates by white mica: a comment. Contrib. Mineral. Petrol. 41, 83-87.
______ (1974) Dalradian margarite: a preliminary note. Geol. Mag. 111, 75-78.
______ (1980) Kyanite isograds of Grampian metamorphism. J. Geol. Soc. London 137, 35-39.
______, and Sweatman, T.R. (1968) A former association of enstatite and kyanite. Min. Mag. 36, 1052-1060.
______, Smith, J.V., and Knowles, C.R. (1969) Transition-metal contents of Al_2SiO_5 polymorphs. Amer. J. Sci. 267A, 96-113.
______, and Haseltine, F.J. (1979) The Grampide andalusite/kyanite isograd. Scott. J. Geol. 15, 117-127.
Christy, A.G. (1989) The effect of composition, temperature and pressure on the stability of the 1Tc and 2M polytypes of sapphirine. Contrib. Mineral. Petrol. 103, 203-215.
Clark, S.P., Jr. (1961) A redetermination of the equilibrium relations between kyanite and sillimanite. Amer. J. Sci. 259, 641-650.
______, Robertson, E.C., and Birch, F. (1957) Experimental determination of kyanite-sillimanite equilibrium relations at high temperatures and pressures. Amer. J. Sci. 255, 628-640.
Clarke, D.B. (1981) The mineralogy of peraluminous granites: a review. Canadian Mineral. 19, 3-17.
______, McKenzie, C.B., Muecke, G.K., and Richardson, S.W. (1976) Magmatic andalusite from the South Mountain batholith, Nova Scotia. Contrib. Mineral. Petrol. 56, 279-287.
______, Guiraud, M., Powell, R., and Burg, J.P. (1987) Metamorphism in the Olary Block, South Australia: compression with cooling in a Proterozoic fold belt. J. Metamorphic Geol. 5, 291-306.
Clemens, J.D., and Wall, V.J. (1981) Origin and crystallization of some peraluminous (S-type) granitic magmas. Canadian Mineral. 19, 111-131.
______, and ______ (1984) Origin and evolution of a peraluminous silicic ignimbrite suite: the Violet Town Volcanics. Contrib. Mineral. Petrol. 88, 354-371.

Codoni, A. (1981) Geologia e petrografia della regione del Pizzo di Claro. Ph.D. Thesis, University of Zurich, Zurich, Switzerland. 197p.

Coetzee, C.B. (1940) Sillimanite-corundum rock: a metamorphosed bauxite in Namaqualand. Trans. Roy. Soc. South Africa 28, 199-205.

Compagnoni, R., and Prato, R. (1969) Paramorfosi de cianite su sillimanite in scisti pregranitici del Massiccio del Gran Paradiso. Boll. Soc. Geol. Ital. 88, 537-549.

Compagnoni, R., Lombardo, B., and Prato, R. (1974) Andalousite et sillimanite aux contacts du granite central de l'Argentera (Alps Maritimes). Rend. Soc. Italiana Min. Petr. 30, 31-54.

Compton, R.R. (1960) Contact metamorphism in the Santa Rosa Range, Nevada. Bull. Geol. Soc. Amer. 71, 1383-1416.

Connolly, J.A.D., and Kerrick, D.M. (1987) An algorithm and computer program for calculating composition phase diagrams. CALPHAD 11, 1-54.

Cook, R.B., Jr. (1985) The mineralogy of Graves Mountain, Lincoln County, Georgia. Mineralogical Record 16, 443-458.

Cooper, A.F. (1980) Retrograde alteration of chromian kyanite in metachert and amphibolite whiteschist from the Southern Alps, New Zeland, with implications for uplift on the Alpine Fault. Contrib. Mineral. Petrol. 75, 153-164.

Cooper, R.F., and Kohlstedt, D.L. (1982) Interfacial energies in the olivine-basalt system. In Akimoto, S., and Manghnani, M.H. (eds.) High-pressure Research in Geophysics, Advances in Earth and Planetary Sciences 12, 217-228.

Corbett, G.J., and Phillips, G.N. (1981) Regional retrograde metamorphism of a high grade terrain: the Willyama Complex, Broken Hill, Australia. Lithos 14, 59-73.

Corey, M.C. (1988) An occurrence of metasomatic aluminosilicates related to high alumina hydrothermal alteration within the South Mountain Batholith, Nova Scotia. Maritime Sed. and Atlantic Geol. 24, 83-95.

Crawford, M.L., and Mark, L.E. (1982) Evidence from metamorphic rocks for overthrusting. Pennsylvania Piedmont, U.S.A. Canadian Mineral. 20, 333-347.

Currie, K.L. (1971) the reaction 3 cordierite = 2 garnet + 4 sillimanite + 5 quartz as a geological thermometer in the Opicon Lake region, Ontario. Contrib. Mineral. Petrol. 33, 215-226.

______ (1974) A note on the calibration of the garnet-cordierite geothermometer and geobarometer. Contrib. Mineral. Petrol. 44, 35-44.

______, and Pajari, G.E. (1981) Anatectic peraluminous granites from the Carmanville area, Northeastern Newfoundland. Canadian Mineral. 19, 147-161.

______, and Gittins, J. (1988) Contrasting sapphirine parageneses from Wilson Lake, Labrador and their tectonic implications. J. Metamorphic Geol. 6, 603-622.

Dachs, E. (1986) High-pressure mineral assemblages and their breakdown-products in metasediments South of the Grossvenediger, Tauern Window, Austria. Schweiz. mineral. petrogr. Mitt. 66, 145-161.

Dahanayake, K., Liyanage, A.N., and Ranashinghe, A.P. (1980) Genesis of sedimentary gem deposits in Sri Lanka. Sed. Geol. 25, 105-115.

D'Amico, C., Rottura, A., Bargossi, G.M., and Nannetti, M.C. (1983) Magmatic genesis of andalusite in peraluminous granites. Examples from Eisgarn type granites in Moldanubikum. Rend. Soc. Italiana Mineral. Petrol. 38, 15-25.

Damman, A. (1988) Hydrothermal subsilicic sodium gedrite from the Gåsborn area, West Bergslagen, central Sweden. Mineral. Mag. 52, 193-200.

Dawson, J.B., and Smith, J.V. (1987) Reduced sapphirine granulite xenoliths from the Lace Kimberlite, South Africa; implications for the deep structure of the Kaapvaal Craton. Contrib. Mineral. Petrol. 95, 376-383.

Day, H.W. (1976) A working model of some equilibria in the system alumina-silica-water. Amer. J. Sci. 276, 1254-1284.

______, and Kumin, H.J. (1980) Thermodynamic analysis of the aluminum silicate triple point. Amer. J. Sci. 280, 265-287.

______, and Chamberlain (1989) Implications of thermal and baric structure for controls on metamorphism: northern New England, USA. In: Daly, J.S., Cliff, R.A., and Yardley, B.W.D. (eds.), Evolution of Metamorphic Belts, Geol. Soc. London Special Publication 43, 215-222.

Deer, W.A., Howie, R.A., and Zussman, J. (1982) Rock-forming Minerals, Vol. 1A, Orthosilicates. Longman, London, 919p.

De Keyser, W.L. (1951) Contribution to the study of sillimanite and mullite by X-rays. Trans. Brit. Ceram. Soc. 50, 349-364.

Delor, C.P., and Leyreloup, A.F. (1986) Chromium-rich kyanite in an eclogite from the Rouergue area, French Massif Central. Min. Mag. 50, 535-537.

Delor, C., Burg, J.-P., Guiraud, M., and Leyreloup, A. (1987) Les métapélites à phengite-chloritoide-grenat-staurotide-disthène de la klippe de Najac-Carmaux: nouveaux marqueurs d'un métamorphisme de haute pression varisque en Rouergue occidental. Comptes Rendus l'Acad. Sci., Paris, Série II 305, 589-595.

Demange, M. (1980/1981) Le métamorphisme mésozonal progressif des roches pélitiques sur le flanc nord du massif de l'Agout (Montage Noire). Fr., Bur. Rech. Geol. Minieres Bull. Section 1, 269-291.

______ (1985) The eclogite-facies rocks of the Montagne Noire, France. Chem. Geol. 50, 173-188.

______, and Gattoni, L. (1978) Le métamorphisme progressif des formations d'origine pélitique de flanc sud du massif de l'Agout (Montage Noire, France). Bull. Minéral. 101, 334-349.

______, Goutay, R., Issard, H., and Perrin, M. (1986) Présence de disthène épizonal dans la zone axiale de la Montagne Noire (Massif central, France). Bull. Soc. Géol. France 8, 525-526.

Dempster, T.J. (1985) Garnet zoning and metamorphism of the Barrovian type area, Scotland. Contrib. Mineral. Petrol. 89, 30-38.

Deng, J.-F. (1978) Thermodynamic calculations for metamorphic reactions in the formation of sillimanite-hornfels [in Chinese]. Chem. Abtracts 89, 200536d.

De Yoreo, J.J., Lux, D.R., Guidotti, C.V., Decker, E.R., and Osberg, P.H. (1989) The Acadian thermal history of western Maine. J. Metamorphic Geol. 7, 169-190.

Dickenson M.P. (1988) Local and regional differences in the chemical potential of water in amphibolite facies pelitic schists. J. Metamorphic Geol. 6, 365-381.

Dickerson, R.P., and Holdaway, M.J. (1989) Acadian metamorphism associated with the Lexington Batholith, Bingham, Maine. Amer. J. Sci. 289, 945-974.

Digel, S.G., Ghent, E.D., and Simony, P.S. (1989) Metamorphism and structure of the Mount Cheadle area, Monashee Mountains, British Columbia. In: Current Research, Part E, Geol. Survey Canada, Paper 89-1E, 95-100.

Dike, P.A. (1951) Kyanite pseudomorphs after andalusite from Delaware County, Pennsylvania. Amer. J. Sci. 249, 457-458.

Dipple, G.M., Wintsch, R.P., and Andrews, M.S. (in press) Identification of the scales of differential element mobility in a ductile fault zone through multi-sample mass balance. J. Metamorphic Geol.

Dodge, F.C.W. (1971) Al_2SiO_5 minerals in rocks of the Sierra Nevada and Inyo Mountains, California. Amer. Mineral. 56, 1443-1451.

Dostál, J. (1966) Mineralogische und petrographische verhältnisse von chrysoberyl-sillimanit-pegmatit von Maršíkov. Acta Univ. Carolinae-Geol. 4, 271-287.

______ (1975) The origin of garnet-cordierite-sillimanite bearing rocks from Chandos Township, Ontario. Contrib. Mineral. Petrol. 49, 163-175.

Doukhan, J.C., and Christie, J.M. (1982) Plastic deformation of sillimanite-Al_2SiO_5 single crystals under confining pressure and TEM investigation of the induced defect structure. Bull. Minéral. 105, 583-589.

______, and Paquet, J. (1982) Plastic deformation of andalusite single crystal Al_2SiO_5. Bull. Minéral. 105, 170-175.

______, Doukhan, N., Koch, P.S., and Christie, J.M. (1985) Transmission electron microscopy investigation of lattice defects in Al_2SiO_5 polymorphs and plasticity induced polymorphic transformations. Bull. Minéral. 108, 81-96.

Dowty, E. (1976a) Crystal structure and crystal growth: I. The influence of internal structure on morphology. Amer. Mineral. 61, 448-459.

______ (1976b) Crystal structure and crystal growth: II. Sector zoning in minerals. Amer. Mineral. 61, 460-469.

Droop, G.T.R., and Treloar, P.J. (1981) Pressures of metamorphism in the thermal aureole of the Etive Granite Complex. Scott. J. Geol. 17, 85-102.

______, and Charnley, N.R. (1985) Comparative geobarometry of pelitic hornfelses associated with the Newer Gabbros: a preliminary study. J. Geol. Soc. London 142, 53-62.

Dunn, J.A. (1929) The aluminous refractory materials: kyanite, sillimanite and corundum in northern India. Mem. Geol. Surv. India 52, pt.2.

Durovic, S. (1963) Die kristallstruktur der mullitmischkristallreihe und ihre beziehung zur struktur des sillimanits. Ber. Dtsch. Keram. Gesell. 40, 287-293.

______, and Fejdi, P. (1976) Synthesis and crystal structure of germanium mullite and crystallochemical parameters of D-mullite. Silikáty 20, 97-110.

Dyar, M.D. (1990) Mössbauer spectra of biotites from metapelites. Amer. Mineral. 75, 656-666.

______, Grover, T.W., and Guidotti, C.V. (ms.), Effects of Fe^{3+} and ordering on geothermometers for garnet-biotite pairs from pelitic schists in northwestern Maine.

Dziewański, J., Heflik, W. and Peitrzyk-Sokulska, E. (1983) New data on the petrogenesis of metamorphic rocks of the Bystrzyckie Mts. Bull. Polish Acad. Sci., Earth Sci. 31, 1-8.

Efimova, M.I., Blagodereva, N.S., Fedchina, G.N., and Vasilenko, G.P. (1984) Post-magmatic alterations of porphyry rocks of the ore fields Sikhote-Alin Mts, USSR [in Polish]. Archi. Mineral., Polska Akad. Nauk 39, 101-104.

Eggert, R.G., and Kerrick, D.M. (1981) Metamorphic equilibria in siliceous dolomites: 6 kb experimental data and geologic implications. Geochim. Cosmochim. Acta 45, 1039-1049.

Eigenfeld, I., and Machatski, F. (1957) The supposed alkali content of kyanite. Osterr. Akad. Wiss. Math.-naturw. Kl. Anz. 94, 151-152.

Elan, R. (1985) High grade contact metamorphism at the Lake Isabella north shore roof pendant, southern Sierra Nevada, California. M.S. Thesis, Univ. of Southern California, Los Angeles, CA, 202 p.

Enami, M., and Zang, Q. (1988) Magnesian staurolite in garnet-corundum rocks and eclogite from the Donghai district, Jiangsu province, east China. Amer. Mineral. 73, 48-56.

Espenshade, G.H., and Potter, D.B. (1960) Kyanite, sillimanite and andalusite deposits of southeastern States. U.S. Geol. Surv. Prof. Paper 336.

Essene, E.J. (1989) The current status of thermobarometry in metamorphic rocks. In: Daly, J.S., Cliff, R.A., and Yardley, B.W.D. (eds.), Evolution of Metamorphic Belts, Geol. Soc. London Special Publication 43, 1-44.

Eugster, H.P. (1970) Thermal and ionic equilibria among muscovite, K-feldspar and aluminosilicate assemblages. Fortschr. Mineral. 47, 106-123.

Evans, B.W. (1965) Application of a reaction-rate method to the breakdown equilibria of muscovite and muscovite plus quartz. Amer. J. Sci. 263, 647-667.

______, and Guidotti, C.V. (1966) The sillimanite-potash feldspar isograd in Western Maine. U.S.A. Contrib. Mineral. Petrol. 12, 25-62.

Evans, N.H., and Speer, J.A. (1984) Low-pressure metamorphism and anatexis of Carolina Slate Belt phyllites in the contact aureole of the Lilesville Pluton, North Carolina, USA. Contrib. Mineral. Petrol. 87, 297-309.

Evers, Th.J.J.M., and Wevers. J.M.A. (1984) The composition and related optical axial angle of sillimanites sampled from the high-grade metamorphic Precambrian of Rogland, SW Norway. N. Jahrb. Mineral. Monatsh. 2, 49-60.

Evirgen, M.M., and Ashworth, J.R. (1984) Andalusitic and kyanitic facies series in the central Menderes Massif, Turkey. N. Jahrb. Miner. Monatsh. 5, 219-227.

Exley, C.S., and Stone, M. (1966) The granitic rocks of south-west England. In: Hosking, K.F.G., and Shrimpton, G.J. (eds.), Present Views of Some Aspects of the Geology of Cornwall and Devon. Penzance, Engl.: Royal Geol. Soc. Cornwall, 131-184.

Faye, G.H., and Harris, D.C. (1969) On the origin of colour and pleochroism in andalusite from Brazil. Canadian Mineral. 10, 47-56.

______, and Nickel, E.H. (1969) On the origin of color and pleochroism of kyanite. Canadian Mineral. 10, 35-46.

Fediuková, E. (1971) Andalusite in granulites of the Bohemian Massif. Tschermaks Mineral. Petr. Mitt. 15, 249-257.

Feininger, T. (1980) Eclogite and related high-pressure regional metamorphic rocks from the Andes of Ecuador. J. Petrol. 21, 107-140.

Ferguson, C.C., and Al-Ameen, S.I. (1986) Geochemistry of Dalradian pelites from Connemara, Ireland: new constraints on kyanite genesis and conditions of metamorphism. J. Geol. Soc. London 143, 237-252.

Ferry, J.M. (1980) A comparative study of geothermometers and geobarometers in pelitic scists from south-central Maine. Amer. Mineral. 65, 720-732.

______ (1983) Regional metamorphism of the Vassalboro Formation, south-central Maine, USA: a case study of the role of fluid in metamorphic petrogenesis. J. Geol. Soc. London 140, 551-576.

______, and Spear, F.S. (1978) Experimental calibration of partitioning of Fe and Mg between biotite and garnet. Contrib. Mineral. Petrol. 66, 113-117.

Finger, L.W., and Prince, E. (1972) Neutron diffraction studies: Andalusite and sillimanite. Carnegie Inst. Wash. Year Book 71, 496-500.

Fisher, G.W. (1970) The application of ionic equilibria to metamorphic differentiation: an example. Contrib. Mineral. Petrol. 29, 91-103.

______ (1973) Nonequilibrium thermodynamics as a model for diffusion-controlled metamorphic processes. Amer. J. Sci. 273, 897-924.

______ (1977) Nonequilibrium thermodynamics in metamorphism. Chapter 19 in: Thermodynamics in Geology. D.G. Fraser, (ed.), Reidel Pub. Co., Boston, 381-403.

______, and Lasaga, A.C. (1981) Irreversible thermodynamics in petrology, Chapter 5 in: Kinetics of Geochemical Processes. Lasaga, A.C. and Kirkpatrick, R.J. (eds.), Reviews in Mineralogy 8, 171-209.

Fisher, J.R., and Zen, E-an (1971) Thermochemical calculations from hydrothermal phase equilibrium data and the free energy of H_2O. Amer. J. Sci. 270, 297-314.

Fleet, M.E., and Arima, M. (1985) Oriented hematite inclusions in sillimanite. Amer. Mineral. 70, 1232-1237.

Fleming, P.D. (1973) Mg-Fe distribution between coexisting garnet and biotite, and the status of fibrolite in the andalusite-staurolite zone of the Mt. Lofty Ranges, South Australia. Geol. Mag. 109, 477-482.

Fletcher, C.J.N., and Greenwood, H.J. (1979) Metamorphism and structure of Penfold Creek Area, near Quesnel Lake, British Columbia. J. Petrol. 20, 743-794.

Flood, R.H., and Vernon, R.H. (1978) The Cooma Granodiorite, Australia: an example of in situ crustal anatexis? Geology 6, 81-84.

Flöttmann, T. (in press) Fibrolitic sillimanite in retrograde shear zones of the Central Schwarzwald basement (Southwest Germany): deformation-dehydration interaction; microstructural implications. J. Metamorphic Geol.

Fonteilles, M. (1965) Sur la profondeur de formation des veines à disthène géodique de la région de Baud (Morbihan) et sur la signification des veines à disthène en général. Bull. Soc. franç. Minéral. Crist. 88, 281-289.

Foster, C.T. (1977) Mass transfer in sillimanite-bearing pelitic schists near Rangeley, Maine. Amer. Mineral. 62, 727-746.

______ (1978) Thermodynamic models of aluminum silicate reaction textures. [Abstr.] Geol. Soc. Amer. 10, 403.

______ (1981) A thermodynamic model of mineral segregations in the lower sillimanite zone near Rangeley, Maine. Amer. Mineral. 66, 260-277.

______ (1982) Textural variations of sillimanite segregations. Canadian Mineral. 20, 379-392.

______ (1983) Thermodynamic models of biotite pseudomorphs after staurolite. Amer. Mineral. 68, 389-397.

______ (1986) Thermodynamic models of reactions involving garnet in a sillimanite/staurolite schist. Mineral. Mag. 50, 427-439.

______ (1990) The role of biotite as a catalyst in reaction mechanisms that form fibrolite. [Abstr.] Geol. and Mineral. Assoc. Canada 15, A40.

Franceschelli, M., Pandeli, E., and Puxeddu, M. (1984) Kyanite-bearing early Alpine metapsammite in the Larderello Geothermal Region (Italy) and its implications to Alpine Metamorphism and Triassic Paleogeography. Schweiz. mineral. petrogr. Mitt. 64, 405-422.

______, Leoni, L., Memmi, I., and Puxeddu, M. (1986) Regional distribution of Al-silicates and metamorphic zonation in the low-grade Verrucano metasediments from the northern Appennines, Italy. J. Metamorphic Geol. 4, 309-321.

______, Memmi, I., Pannuti, F., and Ricci, C.A. (1989) Diachronous metamorphic equilibria in the Hercynian basement of northern Sardinia, Italy. In: Daly, J.S., Cliff, R.A., and Yardley, B.W.D. (eds.), Evolution of Metamorphic Belts, Geol. Soc. London Special Publication 43, 371-375.

French, B.M., and Meyer, H.O.A. (1970) Andalusite and 'β-quartz' in Macusani glass, Peru. Carnegie Inst. Wash. Ann. Rept. Dir. Geophys. Lab., 1968-1969, 339-342.

Frey, M. (1974) Alpine metamorphism of pelitic and marly rocks of the Central Alps. Schweiz. Mineralog. Petrogr. Mitt. 54, 489-506.

______, Hunziker, J.C., Frank, W., Bocquet, J., Dal Piaz, G.V., Jäger, E., and Niggli, E. (1974) Alpine metamorphism of the Alps: A review. Schweiz. mineral. petrogr. Mitt. 54, 247-290.

______, Bucher, K., Frank, E., and Mullis, J. (1980) Alpine metamorphism along the Geotraverse Basel-Chiasso: a review. Eclogae geologicae Helvetiae 73, 527-546.

Frick, C., and Coetzee, C.B. (1974) The mineralogy and the petrology of the sillimanite deposits west of Pofadder, Namaqualand. Trans. Geol. Soc. South Africa 77, 169-183.

Froese, E., and Gasparini, E. (1975) Metamorphic zones in the Snow Lake area, Manitoba. Canadian Mineral. 13, 162-167.

Fry, N., and Barnicoat, A.C. (1987) The tectonic implications of high-pressure metamorphism in the western Alps. J. Geol. Soc. London 144, 656-659.

Fund, B.G. (1940) The sillimanite minerals: A summary. The Mineralogist 8, 129-132, 200-101.

Fyfe, W.S. (1967) Stability of Al_2SiO_5 polymorphs. Chem. Geology 2, 67-76.

______ (1969) Some second thoughts on Al_2O_3-SiO_2. Amer. J. Sci. 267, 291-296.

______, Turner, F.J., and Verhoogen, J. (1958) Metamorphic reactions and metamorphic facies. Geol. Soc. Amer. Memoir 73, 259 pp.

______, and Mackenzie, W.S. (1969) Some aspects of experimental petrology. Earth Sci. Rev. 5, 185-215.

______, Price, N.J., and Thompson, A.B. (1978) Fluids in the Earth's Crust. Elsevier, New York, 383 p.

Ganguly, J., and Saxena, S.K. (1984) Mixing properties of aluminosilicate garnets: Constraints from natural and experimental data, and applications to geothermobarometry. Amer. Mineral. 69, 88-97.

______, and Saxena, S.K. (1985) Mixing properties of aluminosilicate garnets: Constraints from natural and experimental data, and applications to geothermobarometry: Clarifications. Amer. Mineral. 70, 1320.

Garifulin, L.L. (1967) Paramourphs of kyanite on andalusite (chiastolite) from the Voronja tundra region [in Russian]. J. Min. Sborn., Lvov. 21, 383-385.

Ghent, E.D. (1968) Petrology of metamorphosed pelitic rocks and quartzites, Pikikiruna Range, north-west Nelson, New Zealand. Trans. Royal Soc. New Zeland 5, 193-213.

______ (1976) Plagioclase-garnet-Al_2SiO_5-quartz: a potential geobarometer-geothermometer. Amer. Mineral. 61, 710-714.

______, Robbins, D.B., and Stout, M.Z. (1979) Geothermometry, geobarometry, and fluid compositions of metamorphosed calc-silicates and pelites, Mica Creek, British Columbia. Amer. Mineral. 64, 874-885.

______, Simony, P.S., and Knitter, C.C. (1980) Geometry and pressure-temperature significance of the kyanite-sillimanite isograd in the Mica Creek area, British Columbia. Contrib. Mineral. Petrol. 74, 67-73.

______, Knitter, C.C., Raeside, R.P., and Stout, M.Z. (1982) Geothermometry and geobarometry of pelitic rocks, upper kyanite and sillimanite zones, Mica Creek area, British Columbia. Canadian Mineral. 20, 295-305.

Ghera, A., Graziani, G., and Lucchesi, S. (1986) Uneven distribution of blue colour in kyanite. N.Jahrb. Mineral. Abh. 155, 109-127.

Ghose, S., and Tsang, T. (1973) Structural dependence of quadrupole coupling constant e^2qQ/h for ^{27}Al and crystal field parameter D for Fe^{3+} in aluminosilicates. Amer. Mineral. 58, 748-755.

Giannini, W.F. (1984) Characteristics and economic potential of an upland fluvial terrace, Buckingham County, Virginia. Virginia Minerals 30, 42-46.

______, Penick, D.A., Jr., and Mitchell, R.S. (1986) Large andalusite crystals from Virginia. Virginia Minerals 32, 43-44.

Gibson, G.M. (1979) Margarite in kyanite-and corundum-bearing anorthosite, amphibolite, and hornblendite from Central Fiordland, New Zeland. Contrib. Mineral. Petrol. 68, 171-179.

Gil Ibarguchi, J.I., and Martinez, F.J. (1982) Petrology of garnet-cordierite-sillimanite gneisses from the El Tormes thermal dome, Iberian Hercynian foldbelt (W Spain). Contrib. Mineral. Petrol. 80, 14-24.

Gimmel'farb, G.B., Nikolayev, Yu. T., and Belonozkho, L.B. (1968) Origin of quartz-sillimanite inclusions in biotite gneiss of the lower Proterozoic Udokan Series (Olekma-Vitim Highlands). Doklady Akad. Nauk USSR 182, 411-414.

Ginsburg, A.I., Timofeyev, I.N., and Feldman, L.G. (1979) Principles and geology of the granitic pegmatites. Nedra Moscow, 296 pp.

Givargizov, E.I. (1987) Highly Anisotropic Crystals. In: Senechal, M. (ed.) Materials Sciences of Minerals and Rocks. D. Reidel Publishing Co., Dordrecht, Holland, 394 p.

Glen, R.A. (1979) Evidence for cyclic reactions between andalusite, "sericite" and sillimanite, Mount Franks area, Willyama Complex, N.S.W. Tectonophysics 58, 97-112.

Godovikov, A.A., and Kennedy, G.C. (1968) Kyanite eclogites. Contrib. Mineral. Petrol. 19, 169-176.

Goel, O.P. (1983) Compositional variations in biotites from sillimanite-muscovite and sillimanite-orthoclase zones of Kuanthal region, Udaipur district. Indian Mineral. 22, 44-59.

______, and Chaudhari, M.W. (1979) Compositional restraints on the sillimanite paragenesis in metapelites from Kuanthal, district Udaipur, India. Lithos 12, 153-158.

Goffe, B., and Chopin, C. (1986) High-pressure metamorphism in the Western Alps: zoneography of metapelites, chronology and consequences. Schweiz. mineral. petrogr. Mitt. 66, 41-52.

Golani, P.R. (1989) Sillimanite-corundum deposits of Sonapaher, Meghalaya, India: a metamorphosed Precambrian paleosol. Precambrian Research 43, 175-190.

Gomez-Pugnaire, M.T., Visona, D., and Franz, G. (1985) Kyanite, margarite and paragonite in pseudomorphs in amphibolitized eclogites from the Betic Cordilleras, Spain. Chem. Geol. 50, 129-141.

Gordon, T.M. (1973) Determination of internally consistent thermodynamic data from phase equilibrium experiments. J. Geol. 81, 199-208.

______ (1989) Thermal evolution of the Kisseynew sedimentary gneiss belt, Manitoba: metamorphism at an early Proterozoic accretionary margin. In: Daly, J.S., Cliff, R.A., and Yardley, B.W.D. (eds.), Evolution of Metamorphic Belts, Geol. Soc. London Special Publication 43, 233-243.

Grambling, J.A. (1981) Kyanite, andalusite, sillimanite, and related mineral assemblages in the Truchas Peaks region, New Mexico. Amer. Mineral. 66, 702-722.

______ (1983) Reversals in Fe-Mg partitioning between chloritoid and staurolite. Amer. Mineral. 68, 373-388.

______ (1984) Coexisting paragonite and quartz in sillimanitic rocks from New Mexico. Amer. Mineral. 69, 79-87.

______ (1986) A regional gradient in the composition of metamorphic fluids in pelitic schist, Pecos Baldy, New Mexico. Contrib. Mineral. Petrol. 94, 149-164.

______, and Williams, M.L. (1985) The effects of Fe^{3+} and Mn^{3+} on aluminum silicate phase relations in North-Central New Mexico, U.S.A. J. Petrol. 26, 324-354.

______, Williams, M.L., Smith, R.F., and Mawer, C.K. (1989) The role of crustal extension in the metamorphism of Proterozoic rocks in northern New Mexico. In: Grambling, J.A., and Tewksbury, B.J. (eds.) Proterozoic Geology of the Southern Rocky Mountains. Geol. Soc. Amer. Special Paper 235, 87-109.

______, Williams, M.L., Mawer, C.K., and Smith, R.F. (1989) Metamorphic evolution of Proterozoic rocks in New Mexico. In: Daly, J.S., Cliff, R.A., and Yardley, B.W.D. (eds.), Evolution of Metamorphic Belts, Geol. Soc. London Special Publication 43,, 461-467.

Grant, J.A. (1973) Phase equilibria in high-grade metamorphism and partial melting of pelitic rocks. Amer. J. Sci. 273, 289-317.

______ (1985) Phase equilibria in low-pressure partial melting of pelitic rocks. Amer. J. Sci. 285, 409-435.

______, and Frost, B.R. (ms.) Contact metamorphism and partial melting of pelitic rocks in the aureole of the Laramie Anorthosite Complex, Morton Pass, Wyoming.

Grapes, R.H. (1987) Composition and melting relationships of andalusite in a schist xenolith, Wehr Volcano, East Eifel. N. Jahrb. Mineral. Monatsh. 12, 550-556.

Green, J.C. (1963) High-level metamorphism of pelitic rocks in northern New Hampshire. Amer. Mineral. 48, 991-1023.

Green, T.H. (1967) An experimental investigation of sub-solidus assemblages formed at high pressure in high-alumina basalt, kyanite eclogite and grospydite compositions. Contrib. Mineral. Petrol. 16, 84-114.

______ (1969) The diopside-kyanite join at high pressures and temperatures. Lithos 2, 333-341.

Greenwood, H.J. (1972) Al^{IV}-Si^{IV} disorder in sillimanite and its effect on phase relations of the aluminum silicate minerals. Geol. Soc. Amer. Memoir 132, 553-571.

______ (1976) Metamorphism at moderate temperatures and pressures. In: Bailey, D.K., and Macdonald, R. (eds.), The Evolution of Crystalline Rocks, Academic Press, London, 187-259.

Gresens, R.L. (1971) Application of hydrolysis equilibria to the genesis of pegmatite and kyanite deposits in northern New Mexico. Mountain Geologist 8, 3-16.

Grew, E.S. (1976) Blue sillimanite. [Abstr.] Trans. Amer. Geophys. Union 57, 1019.

______ (1977) First report of kyanite-sillimanite association in New Zeland (Note). N.Z. J. Geol. Geophys. 20, 797-802.

______ (1979) Al-Si disorder of K-feldspar in crustal xenoliths at Kilbourne Hole, New Mexico. Amer. Mineral. 64, 912-916.

______ (1980) Sillimanite and ilmenite from high-grade metamorphic rocks of Antarctica and other areas. J. Petrol. 21, 39-68.

______ (1983) A grandidierite-sapphirine association from India. Mineral. Mag. 47, 401-403.

______ (1986) Petrogenesis of kornerupine at Waldheim (Sachsen) German Democratic Republic. Zeits. Geolog. Wissenschaften 14, 525-558.

______ (1989) A second occurrence of kornerupine in Waldheim, Saxony, German Democratic Republic. Zeits. Geolog. Wissenschaften 17, 67-76.

______, and Day, H.W. (1972) Staurolite, kyanite, and sillimanite from the Narragansett Basin of Rhode Island. U.S. Geol. Survey Prof. Paper 800-D, D151-D157.

______, and Rossman, G.R. (1976a) Iron in some Antarctic sillimanites. Trans. Amer. Geophys. Union 57, 337.

______, and Rossman, G.R. (1976b) Color and iron in sillimanite. Abstracts, 25th Int'l. Geol. Congr. 2, 564-565.

______, and Hinthorne, J.R. (1983) Boron in sillimanite. Science 221, 547-549.

______, and Rossman, G.R. (1985) Co-ordination of boron in sillimanite. Mineral. Mag. 49, 132-135.

______, Abraham, K., and Medenbach, O. (1987) Ti-poor hoegbomite in kornerupine-cordierite-sillimanite rocks from Ellermmankovilpatti, Tamil Nadu, India. Contrib. Mineral. Petrol. 95, 21-31.

______, Belakovskiy, D.I., and Leskova, N.V. (1988) Phase equilibria in talc-kyanite-hornblende rocks (with andalusite and sillimanite) from Kugi-Lal, southwestern Pamirs. Doklady Akad. Nauk USSR 299, 1222-1226.

______, Chernosky, J.V., Werding, G., Abraham, K., Marquez, N., and Hinthorne, J.R. (in press) Chemistry of kornerupine and associated minerals, a wet chemical, ion microprobe, and X-ray study emphasizing Li, Be, B and F contents. J. Petrol.

Griffin, W.L. (1987) 'On the eclogites of Norway' - 65 years later. Mineral. Mag. 51, 333-343.

Grigor'yev, A.P. (1973) Kinetics of crystallization of sillimanite from corundum and quartz. Dokl. Acad. Sci. USSR, Earth Sci. Sect., 208, 148-151.

Griggs, D.T., and Kennedy, G.C. (1956) A simple apparatus for high pressures and temperatures. Amer. J. Sci. 254, 722-735.

Grosemans, P. (1948) Un filon de quartz à andolousite de la région de Mandwe (Kibara). Bull. Soc. Belge Géol. 57, 148-150.

Gross, E.B., and Parwel. A. (1968) Rutile mineralization at the White Mountain andalusite deposits, California. Arkiv. Mineral. Geol. Stockholm 4, 493-497.

Gross, K.A. (1965) X-ray line broadening and stored energy in deformed and annealed calcite. Philos. Mag, 8th ser., 12, 801-813.

Guerirard, S., and Pailler, A. (1971) Coexistence des trois polymorphs, disthène-andalusite, sillimanite, dans les micaschistes á staurotide et grenat de la chaine cótiere des Maures (Var, France). C.R. Acad., Paris 272, ser. D, 193-195.

Guidotti, C.V. (1963) Metamorphism of pelitic schists in the Bryant Pond quadrangle, maine. Amer. Mineral. 48, 772-791.

______ (1970) The mineralogy and petrology of the transition from the lower to upper sillimanite zone in the Oquossoc area, Maine. J. Petrol. 11, 277-336.

______ (1974) Transition from staurolite to sillimanite zone, Rangeley Quadrangle, Maine. Bull. Geol. Soc. Amer. 85, 475-490.

______ (1978) Muscovite and K-feldspar from two-mica adamellite in northwestern Maine: composition and petrogenetic implications. Amer. Mineral. 63, 750-753.

______, and Cheney, J.T. (1976) Margarite pseudomorphs after chiastolite in the Rangeley area, Maine. Amer. Mineral. 61, 431-434.

______, Post, J.L., and Cheney, J.T. (1979) Margarite pseudomorphs after chiastolite in the Georgetown area, California. Amer. Mineral. 64, 728-732.

______, and Dyar, M.D. (in press) Ferric iron in metamorphic biotite and its petrologic and crystallochemical implications. Amer. Mineral.

Guitard, C. (1987) Coexistence de la paragonite et de la muscovite dans les métamorphites hercyniens des Pyrénées orientales franco-espagnoles. Comptes Rend. l'Acad. Sci., Paris, Série II 261 ser. D, 5161-5164.

Guitard, G. (1965) Les types de métamorphisme régional à andalousite, cordiérite et almandin, et à andalousite, cordiérite, almandin et staurotide, dans la zone axiale des Pyrénees-Orientales: contributions à l'étude des types de metamorphisme de basse pression. C.R. Acad. Sci., Paris 261, ser. D, 5161-5164.

______ (1969) Mise en évidence de la reaction grenat + muscovite = andalousite + biotite + quartz dans les micaschistes mésozonaux des massifs du Canigou et du Roc de France (Pyrénées-Orientales); relations entre l'andalousite, la staurotide et le grenat. C.R. Acad. Sci., Paris, 269, ser. D, 1159-1162.

Gunter, M., and Bloss, F.D. (1982) Andalusite-kanonaite series: lattice and optical parameters. Amer. Mineral. 67, 1218-1228.

Guse, W., Saalfeld, H., and Tjandra, J. (1979) Thermal transformation of sillimanite single crystals. N. Jahrb. Mineral. Monatsh. 4, 175-181.

Gyepesová D., and Ďurovič, S. (1977) Single-crystal study of thermal decomposition of sillimanite. Silicáty 2, 147-150.

Haas, H., and Holdaway, M.J. (1973) Equilibria in the system Al_2O_3-SiO_2-H_2O involving the stability limits of pyrophyllite, and thermodynamic data of pyrophyllite. Amer. J. Sci. 273, 449-464.

Halbach, H., and Chatterjee, N.D. (1984) An internally consistent set of thermodynamic data for twenty-one CaO-Al_2O_3-SiO_2-H_2O phases by linear parametric programming. Contrib. Mineral. Petrol. 88, 14-23.

Hålenius, U. (1978) A spectroscopic investigation of manganian andalusite. Canadian Mineral. 16, 567-575.

______ (1979) State and location of iron in sillimanite. N. Jahrb. Mineral. Monatsh. 4, 165-174.

Hamid Rahman, S. (1987) HRTEM observation of disorder in andalusite, Al_2SiO_5. Zeits. Kristallogr. 181, 127-133.

Hänni, H.A. (1983) Weitere Untersuchungen an einigen farbwechselnden Edelsteinen. Zeits. Deutschen Gemmologischen Gesellschaft 32, 99-106.

Hariya, Y., Dollase, W.A., and Kennedy, G.C. (1969) An experimental investigation of the relationship of mullite to sillimanite. Amer. Mineral. 54, 1419-1441.

Harker, A. (1950) Metamorphism: A Study of the Transformations of Rock-Masses. 3rd Edition, Methuen and Co. Ltd., London, England, 362 p.

Harley, S.L. (1986) A sapphirine-cordierite-garnet-sillimanite granulite from Enderby Land, Antarctica: implications for FMAS petrogenetic grids in the granulite facies. Contrib. Mineral. Petrol. 94, 452-460.

Harris, N.B.W. (1974) The petrology and petrogenesis of some muscovite granite sills from the Barousse Massif, Central Pyrenees. Contrib. Mineral. Petrol. 45, 215-230.

Harte, B., and Johnson, M.R.W. (1969) Metamorphic history of Dalradian rocks in Glen Clova, Esk and Lethonot, Angus, Scotland. Scott. J. Geol. 5, 54-80.

______, and Hudson, N.F.C. (1979) Pelitic facies series and the temperatures and pressures of Dalradian metamorphism in E Scotland. In: Harris, A.L., Holland, K.E., and Leake, B.E. (eds.) The Caledonides of the British Isles, reviewed. Scott. Acad. Press, Edinburgh, Scotland. 323-337.

Hârtopanu, I. (1978) A particular fabric of the kyanite and staurolite-bearing rocks from the Sebeş Mountains. Genetical considerations. Rev. Roum. Géol. Géophys. Géogr., Géologie 22, 175-183.

Harwood, D.S., and Larson, R.R. (1969) Variations in the delta index of cordierite around the Cupsuptic Pluton, west-central Maine. Amer. Mineral. 54, 896-908.

Haslam, H.W. (1971) Andalusite in the Mullach nan Coirean granite, Inverness-shire. Geol. Mag. 108, 97-102.

Hattori, H. (1967) Occurrence of sillimanite-garnet-biotite gneisses and their significance in metamorphic zoning in the South Island, New Zealand. N. Z. J. Geol. Geophys. 10, 269-299.

Hawthorne, F.C. (ed.) (1988) Spectroscopic Methods in Mineralogy and Geology. Vol. 18, Reviwes in Mineralogy, 698 p.

Hay, R.S., and Evans, B. (1988) Intergranular distribution of pore fluid and the nature of high-angle grain boundaries in limestone and marble. J. Geophys. Res. 93B, 8959-8974.

Heim, R.C. (1952) Metamorphism in the Sierra de Guadarrama. Ph.D. Thesis, Univ. of Utrecht, Utrecht, Holland.

Heinrich, C.A. (1986) Eclogite facies regional metamorphism of hydrous mafic rocks in the Central Alpine Adula Nappe. J. Petrol. 27, 123-154.

Heinrich, E.W., and Corey, A.F. (1959) Manganian andalusite from Kiawa Mountains, Rio Arriba County, New Mexico. Amer. Mineral. 44, 1261-1271.

______, and Buchi, S.H. (1969) Beryl-chrysoberyl-sillimanite paragenesis in pegmatites. Indian Mineral. 10, 1-7.

Helgeson H.C., Delany, J.M., Nesbitt, H.W., and Bird, D.K. (1978) Summary and critique of the thermodynamic properties of the rock forming minerals. Amer. J. Sci. 278A, 229p.

Hemingway, B.S., Robie, R.A., Evans, H.T., Jr., and Kerrick, D.M. (ms.) Heat capacities and entropies of sillimanite, fibrolite, andalusite, kyanite, Brazilian quartz and Arkansas novaculite, and the Al_2SiO_5 phase equilibrium diagram.

Hemley, J.J. (1967) Stability relations of pyrophyllite, andalusite and quartz at elevated pressures and temperatures. [Abstr.] Am. Geophys. Union Trans. 48, 224.

______, Montoya, J.W., Marinenko, J.W., and Luce, R.W. (1980) Equilibria in the system Al_2O_3-SiO_2-H_2O and some general implications for alteration/mineralization processes. Econ. Geol. 75, 210-228.

Heninger, S.G. (1984) Hydrothermal experiments on the andalusite-sillimanite equilibrium. M.S. Paper, The Pennsylvania State Univ., 42 pp.

Henriques, Å. (1957) The alkali content of kyanite. Arkiv. Mineral. Geol. 2, 271.

Heritsch, H. (1984) Über das mögliche auftreten von sillimanit in den gneisen der Koralpe, Steiermark un Kärnten; ein kurzbericht. Österr. Akad. Wissen., Math. - Naturwissen. Klasse, Anzeiger 121, 75-77.

Herz, N., and Dutra, C.V. (1964) Geochemistry of some kyanites from Brazil. Amer. Mineral. 49, 1290-1305.

Hess, D.F. (1983) Further data on andalusite and kyanite pseudomorphs after andalusite from Delaware County, Pennsylvania. Friends Mineral., Penn. Chapt., Newsletter, 11, 2-4.

Hess, P.C. (1969) The metamorphic paragenesis of cordierite in pelitic rocks. Contrib. Mineral. Petrol. 24, 191-207.

Hey, J.S., and Taylor, H.W. (1931) The co-ordination number of aluminium in the aluminosilicates. Zeits. Krist. 80, 428-441.

Hietanen, A. (1956) Kyanite, andalusite and sillimanite in the schist in Boehls Butte quadrangle, Idaho. Amer. Mineral. 41, 1-27.

______ (1959) Kyanite-garnet gedrite from Orofino, Idaho. Amer. Mineral. 44, 539-564.

______ (1961) Metamorphic facies and style of folding of the Belt series northwest of the Idaho batholith. Comm. Géol. Finlande Bull. 196, 73-103.

______ (1962) Metasomatic metamorphism in Western Clearwater County, Idaho. U.S. Geol. Survey Prof. Paper 344-A, A1-A116.

______ (1967) On the facies series in various types of metamorphism. J. Geol. 75, 187-214.

______ (1968) Belt series in the region around Snow Peak and Mallard Peak, Idaho. U.S. Geol. Survey Prof. Paper 344-E (p. E1-E34).

Hildebrand, F.A. (1961) Andalusite-topaz greisen near Caguas, east-central Puerto Rico. U.S. Geol. Survey Prof. Paper 424-B, 222-224.

Hills, E.S., (1938) Andalusite and sillimanite in uncontaminated igneous rocks. Geol. Mag. 75, 296-304.

Hiroi, Y., Shiraishi, K., Nakai, Y., Kano, T., and Yoshikura, S. (1983) Geology and petrology of Prince Olav Coast, East Antarctica. In: Oliver, R.L., James, P.R., and Jago, J.B. (eds.) Antarctic Earth Science. 4th Int'l. Symposium. Cambridge Univ. Press, Cambridge, England. 32-35.

______ and Kishi, S. (1989) Staurolite and kyanite in the Takanuki pelitic gneisses from the Abukuma metamorphic terrain, northeast Japan [in Japanese]. J. Mineral. Petrol. Econ. Geol. 84, 141-151.

Hodges, K.V., and Spear, F.S. (1982) Geothermometry, geobarometry and the Al_2SiO_5 triple point at Mt. Moosilauke, New Hampshire. Amer. Mineral. 67, 1118-1134.

Holdaway, M.J. (1971) Stability of andalusite and the aluminum silicate phase diagram. Amer. J. Sci. 271, 97-131.

______ (1978) Significance of chloritoid-bearing and staurolite-bearing rocks in the Picurus Range, New Mexico. Bull. Geol. Soc. Amer. 89, 1404-1414.

______, and Lee, S.M. (1977) Fe-Mg cordierite stability in high-grade pelitic rocks based on experimental, theoretical, and natural observations. Contrib. Mineral. Petrol. 63, 175-198.

______, Guidotti, C.V., Novak., J.M., and Henry, W. (1982) Polymetamorphism in medium-to high-grade pelitic metamorphic rocks, West-Central Maine. Bull. Geol. Soc. Amer. 93, 572-584.

______, Dutrow, B.L., and Hinton, R.W. (1988) Devonian and Carboniferous metamorphism in west-central Maine: the muscovite-almandine geobarometer and the staurolite problem revisited. Amer. Mineral. 73, 20-47.

Holder, M.T. (1979) An emplacement mechanism for post-tectonic granites and its implications for their geochemical features. In: Atherton, M.P., and Tarney, J. (eds.) Origin of Granite Batholiths: Geochemical Evidence. Shiva Publishing Ltd., Orpington, England, p. 116-128.

Holland, T.J.B. (1979a) High water activities in the generation of high pressure kyanite eclogites of the Tauern Window, Austria. J. Geol. 87, 1-27.

______ (1979b) Experimental determination of the reaction paragonite = jadeite + kyanite + H_2O, and internally consistent thermodynamic data for part of the system Na_2O-Al_2O_3-SiO_2-H_2O, with applications to eclogites and blueschists. Contrib. Mineral. Petrol. 68, 293-301.

______, and Powell, R. (1985) An internally consistent thermodynamic dataset with uncertainties and correlations: 2. Data and results. J. Metamorphic Geol. 3, 343-370.

______, and Carpenter, M.A. (1986) Aluminium/silicon disordering and melting in sillimanite at high pressures. Nature 320, 151-153.

Hollister, L.S. (1969) Metastable paragenetic sequence of andalusite, kyanite, and sillimanite, Kwoiek area, British Columbia. Amer. J. Sci. 267, 352-370.

______, and Bence, A.E. (1967) Staurolite: sectoral compositional variations. Science 158, 1053-1056.

Hollocher, K. (1987) Systematic retrograde metamorphism of sillimanite-staurolite schists, New Salem area, Massachusetts. Geol. Soc. Amer. Bull. 98, 621-634.

Holm, J.L., and Kleppa, O.J. (1966) The thermodynamic properties of the aluminum silicates. Amer. Mineral. 51, 1608-1627.

Holuj, F., Thyer, J.R., and Hedgecock, N.E. (1966) ESR spectra of Fe^{+3} in single crystals of andalusite. Canadian J. Phys. 44, 509-523.

Huang, W.L., and Wyllie, P.J. (1981) Phase relationships of S-type granite with H_2O to 35 kbar: muscovite granite from Harney Peak, South Dakota. J. Geophys. Res. 86B, 10515-10529.

Huang, W.T. (1957) Origin of sillimanite rocks by alumina metasomatism, Witchita Mountains, Oklahoma. Abstr. Bull. Geol. Soc. Amer. 68, 1748.

Hubbard, M.S. (1989) Thermobarometric constraints on the thermal history of the Main Central Thrust Zone and Tibetan Slab, eastern Nepal Himalaya. J. Metamorphic Geol. 7, 19-30.

Hudson, N.F.C. (1980) Regional metamorphism of some Dalradian pelites in the Buchan Area, N.E. Scotland. Contrib. Mineral. Petrol. 73, 39-51.

______ (1985) Conditions of Dalradian metamorphism in the Buchan area, NE Scotland. J. Geol. Soc. London 142, 63-76.

Huebner, J.S., and Voigt, D.E. (1988) Electrical conductivity of diopside: Evidence for oxygen vacancies. Amer. Mineral. 73, 1235-1254.

Hull, D., and Bacon, D.J. (1984) Introduction to Dislocations. Pergamon Press, Oxford, England, 257 p.

Hunt, J.A., and Kerrick, D.M. (1977) The stability of sphene: experimental redetermination and geologic implications. Geochim. Cosmochim Acta 41, 279-288.

Hutcheon, I., Froese, E., and Gordon, T.M. (1974) The assemblage quartz-sillimanite-garnet-cordierite as an indicator of metamorphic conditions in the Daly Bay complex, N.W.T. Contrib. Mineral. Petrol. 44, 29-34.

Hutton, D.R., and Troup, G.J. (1964) Paramagnetic resonance of Cr^{3+} in kyanite. Brit. J. Appl. Physics 15, 275-280.

Iishi, K., Salje, E., and Werneke, Ch. (1979) Phonon spectra and rigid-ion model calculations on andalusite. Phys. Chem. Minerals 4, 173-188.

Irouschek-Zumthor, A., and Armbruster, T. (1985) Wagnerite from a metapelitic rock of the Simano Nappe (Lepontine Alps, Switzerland) Part I: mineralogy and geochemistry. Schweiz. mineral. petrogr. Mitt. 65, 137-151.

Jacob, J. (1937) Über den alkaligehalt der disthene. Schweiz. mineral. petrogr. Mitt. 17, 214-219.

______ (1940) Über den chemismus des andalusit. Schweiz. Mineral. Petr. Mitt. 20, 8-10.

______ (1941) Chemische und strukturelle untersuchungen am disthene. Schweiz. mineral. petrogr. Mitt. 21, 131-135.

Jacobs, G.K., and Kerrick, D.M. (1981) Devolatilization equilibria in H_2O-CO_2 and H_2O-CO_2-NaCl fluids: an experimental and thermodynamic evaluation at elevated pressures and temperatures. Amer. Mineral. 66, 1135-1153.

Jacobson, R., and Webb, J.S. (1948) The occurrence of nigerite, a new tin mineral in quartz-sillimanite rocks from Nigeria. Mineral. Mag. 28, 118-128.

Jamieson, R.A. (1984) Low pressure cordierite-bearing migmatites from Kelly's Mountain, Nova Scotia. Contrib. Mineral. Petrol. 86, 309-320.

Janák, M., Kahan, S., and Jancčula, D. (1988) Metamorphism of pelitic rocks and metamorphic zones in SW part of western Tatra Mts. crystalline complexes. Geologica Carpathica, 39, 455-488.

Janes, N., and Oldfield, E. (1985) Prediction of silicon-29 nuclear magnetic resonance chemical shifts using a group electronegativity approach: Applications to silicate and aluminosilicate structures. J. Amer. Chem. Soc. 107, 6769-6775.

Jansen, J.B.H., and Schuiling, R.D. (1976) Metamorphism on Naxos: petrology and geothermal gradients. Amer. J. Sci. 276, 1225-1253.

Jantzen, C.M., and Herman, H. (1979) Phase equilibria in the system SiO_2-Al_2O_3. J. Amer. Ceram. Soc. 62, 212-214.

Jaques, A.L., Hall, A.E., Sheraton, J.W., Smith, C.B., Sun, S-S., Drew, R.M., Foudoulis, C., and Ellinsen, K. (1989) Composition of crystalline inclusions and C-isotopic composition of Argyle and Ellendale diamonds. In: Ross, J. (ed.) Kimberlites and Related Rocks, Volume 2, Their Mantle/Crust Setting, Diamonds and Diamond Exploration. Geol. Soc. America Special Pub. 14, 967-989.

Johannes, W. (1978) Pressure comparing experiments with NaCl, AgCl, talc, and pyrophyllite assemblies in a piston cylinder apparatus. N. Jahrb. Mineral. Monatsh. 2, 84-92.

______ (1988) What controls partial melting in migmatites? J. Metamorphic Geol. 6, 451-465.

______, Bell, P.M., Mao, H.K., Boettcher, A.L., Chipman, D.W., Hays, J.F., Newton, R.C., and Seifert, F. (1971) An interlaboratory comparison of piston-cylinder pressure calibration using the albite-breakdown reaction. Contrib. Mineral. Petrol. 32, 24-38.

Johnson, M.R.W. (1963) Some time relationships of movement and metamorphism in the Scottish Highlands. Geol. en. Mijnb. 42, 121-142.

Jones, I.L., Heine, V., Leslie, M., and Price, G.D. (ms.) A new approach to simulating disorder in crystals.

Kano, H., and Kuroda, Y. (1968) On the occurrences of staurolite and kyanite from the Abukuma pleateau, northeastern Japan. Proc. Japan Acad. 44, 77-82.

Karpoff, R. (1946) Sur la présence de disthène dans une pegmatite à la frontière Angéro-Soudanaise. Compt. Rend. Acad. Sci. Paris 233, 1154-1155.

Karsakov, L.P. (1973) Pyrope-bronzite-sillimanite schist of eastern Stanovik and the pressure-temperature conditions during its metamorphism. Dokl. Acad. Sci. USSR, Earth Sci. Sect., 210, 171-173.

Kataria, P. and Chaudhari, M.W. (1988) Petrochemical study of pelitic rocks from the Banded Gneissic complex of Amet region, Rajasthan. In: Roy, A.B. (ed.) Precambrian of the Aravalli Mountains, Rajasthan, India. Geol. Soc. India Mem. 9, 285-295.

Keller, F. (1968) Mineralparagenesen und geologie der Campo Tencia-Pizzo Forno-Gebirsgruppe. Beitr. Geol. Karte d. Schweiz, N.F. 135, 72p.

Kempe, D.R.C. (1967) Some topaz-, sillimanite-, and kyanite-bearing rocks from Tanganyika. Mineral. Mag. 36, 515-521.

Kennan, P.S. (1972) Exsolved sillimanite in granite. Mineral. Mag. 38, 763-764.

Kennedy, G.C. (1955) Pyrophyllite-sillimanite-mullite equilibrium relations to 20,000 bars and 800°C. Bull. Geol. Soc. Amer. 66, 1584.

Kerr, P.F. (1932) The occurrence of andalusite and related minerals at White Mountain, California. Econ. Geol. 27, 614-643.

______ (1977) Optical Mineralogy. 4th Ed., McGraw Hill, New York. 492 p.

______, and Jenny, P. (1935) The dumorierite-andalusite mineralization at Oreana, Nevada. Econ. Geol. 30, 287-300.

Kerrick, D.M. (1968) Experiments on the upper stability limit of pyrophyllite at 1.8 kilobars and 3.9 kilobars water pressure. Amer. J. Sci. 266, 204-214.

______ (1972) Experimental determination of muscovite + quartz stability with $P_{H_2O} < P_{total}$. Amer. J. Sci. 272, 946-958.

______(1977) The genesis of zoned skarns in the Sierra Nevada, California. J. Petrol. 18, 144-181.

______(1986) Dislocation strain energy in the Al_2SiO_5 polymorphs. Phys. Chem. Minerals 13, 221-226.

______(1987a) Cold-seal systems. In: Ulmer, G.C., and Barnes, H.L. (eds.) Hydrothermal Experimental Techniques. John Wiley & Sons, New York. p. 293-323.

______(1987b) Fibrolite in contact aureoles of Donegal, Ireland. Amer. Mineral. 72, 240-254.

______(1988) Al_2SiO_5-bearing segregations in the Lepontine Alps, Switzerland: aluminum mobility in metapelites. Geology 16, 636-640.

______, and Darken, L.S. (1975) Statistical thermodynamic models for ideal oxide and silicate solid solutions, with application to plagioclase. Geochim. Cosmochim. Acta 39, 1421-1442.

______, and Ghent, E.D. (1979) P-T-X_{CO_2} relations of equilibria in the system CaO-Al_2O_3-SiO_2-CO_2-H_2O. In: Zharikov, V.A., Fonarev, V.I., and Korikovskii, S.P. (eds.) Problems of Physico-chemical Petrology, D.S. Korzhinskii Commemorative Volume II, p. 32-52.

______, and Jacobs, G.K.(1981) A modified Redlich-Kwong equation of state for H_2O, CO_2, and H_2O-CO_2 mixtures at elevated pressures and temperatures. Amer. J. Sci. 281, 735-767.

______, and Heninger, S.G. (1984) The andalusite-sillimanite equilibrium revisited. Geol. Soc. Amer. Abstr. 16, 558.

______, and Speer, J.A. (1988) The role of minor element solid solution on the andalusite-sillimanite equilibrium in metapelites and peraluminous granitoids. Amer. J. Sci. 288, 152-192.

______, and Woodsworth, G.J. (1989) Aluminum silicates in the Mount Raleigh pendant, British Columbia. J. Metamorphic Geol. 7, 547-563.

______, Benesi, A.J., and Bluth, V.S. (1990) Al^{IV}-Si^{IV} disorder in sillimanite and fibrolite: Greenwood's demon revisited. Abstr. Geol. and Mineral. Assoc. Canada 15, A69.

______, Peterson, R.C., Xie, Q., and Ross, F. (ms.) Neutron diffraction study of Al^{IV}-Si^{IV} disorder in Sri Lanka sillimanite.

Khitarov, N.I., Pugin, V.A., Chzao-Bin, P., and Slutsky, A.B. (1963) Relations between andalusite, kyanite and sillimanite at moderate temperatures and pressures: Geochemistry 235-244.

Khostov, I., and Petrusenko, S. (1965) Andalusite paramourphs in pegmatoid lenses [in Bulgarian]. Trav. Géol. Bulgarie, Sér. Geoch., Miner., Petr. 5, 85-97.

Khvostova, V.A., and Feodot'ev, K.M. (1961) Andalusite in pegmatites of East Sayan [in Russian]. Trans. Mineral. Mus. Acad. Sci. USSR, no. 12, 239-243.

Kirkpatrick, R.J., Smith, K.A., Schramm, S., Turner, G., and Yang, W.-H. (1985) Solid-state nuclear magnetic resonance spectroscopy of minerals. Ann. Rev. Earth Planet. Sci. 13, 29-47.

Kiseleva, I.A., Ostapenko, G.T., Ogorodova, L.P., Topor, N.D., and Timoshkova, L.P. (1983) High-temperature calorimetry data on the equilibrium between andalusite, kyanite, sillimanite, and mullite. Geochem. Int'l. 5, 17-26.

Klaper, E.M. (1986a) The metamorphic evolution of garnet-cordierite-sillimanite gneisses of NW Spitsbergen (Svalbard). Schweiz. mineral. petrogr. Mitt. 66, 295-313.

______(1986b) Deformation und metamorphose im gebiet des Nufenenpasses, Lepontinische Alpen. Schweiz. mineral. petrogr. Mitt. 66, 115-128.

______, and Bucher-Nurminen, K. (1987) Alpine metamorphism of pelitic schists in the Nufenen Pass area, Lepontine Alps. J. Metamorphic Geol. 5, 175-194.

Klein, H. (1984) Eclogites and their retrograde transformation in the Schwarzwald (Fed. Rep. Germany). N. Jahrb. Mineral. Monatsh. 1, 25-38.

Klein, H.-H. (1976a) Alumosilikatführende knauern im Lepontin. Schweiz. mineral. petrogr. Mitt. 56, 435-456.

______(1976b) Metamorphose von peliten zwischen Reinwaldhorn und Pizzo Paglia (Adula-un Simano-Decke). Schweiz. mineral. petrogr. Mitt. 56, 457-479.

Klemd, R. (1989) P-T evolution and fluid inclusion characteristics of retrograded eclogites, Münchberg Gneiss Complex, Germany. Contrib. Mineral. Petrol. 102, 221-229.

Klemm, G. (1911) Über viridin, eine abart des andalusites. Notizbl. Ver. Erdk. Geol. Darmstadt. 32, 4-13.

Koch, E. (1982) Mineralogie und plurifazielle metamorphose der pelite in der Adula-Decke (Zentralalpen). Ph.D. Thesis, Univ. Basel, Basel, Switzerland. 201 pp.

Kolobov, V. Yu. (1983) The formation of andalusite and sillimanite in the contact aureole of the Ular granitoid massif (Sangilen, Tuva USSR). [Russian with English summary] Mineralogicheskii Zhurnal 5, 54-64.

Koons, P.O., and Thompson, A.B. (1985) Non-mafic rocks in the greenschist, blueschist and eclogite facies. Chem. Geol. 50, 3-30.

Korikovskiy, S.P. (1965) Quartz-sillimanite facies of acid leaching in granite-gneiss complexes. Dokl. Acad. Sci. USSR, Earth Sci. Sect., 152, 141-143.

Korzhinskiy, M.A. (1987) The solubility of corundum in an HCl fluid and forms taken by Al. Geochem. Int'l. 24, 105-110.

Koziol, A.M., and Newton, R.C. (1988) Redetermination of the anorthite breakdown reaction adn improvement of the plagioclase-garnet-Al_2SiO_5-quartz geobarometer. Amer. Mineral. 73, 216-223.

Kramm, U. (1979a) Formation and stability of viridine in microcline leptite from Ultevis District, northern Sweden. Contrib. Mineral. Petrol. 69, 143-150.

______(1979b) Kanonaite-rich viridines from the Venn-Stavelot Massif, Belgian Ardennes. Contrib. Mineral. Petrol. 69, 387-395.

Kretz, R., Loop, J., and Hartree, R. (1989) Petrology and Li-Be-B geochemistry of muscovite-biotite granite and associated pegmatite near Yellowknife, Canada. Contrib. Mineral. Petrol. 102, 174-190.

Kröner, A. (1971) A preliminary account of the growth of kyanite under conditions of very low temperature and pressure. N. Jahrb. Mineral. Monatsh. 8, 370-378.

Krough, E.J. (1982) Metamorphic evolution of Norwegian country-rock eclogites, as deduced from mineral inclusions and compositional zoning in garnets. Lithos 15, 305-321.

Kumagai, N., and Ito, H. (1959) On the P-T diagrams of the trimorphic minerals with a special reference to andalusite and cyanite as a geological thermometer or a geological piezometer. Mem. Coll. Sci. Kyoto B, 26, 215.

Kundig, R. (1989) Domal structures and high-grade metamorphism in the Higher Himalayan Crystalline, Zanskar Region, north-west Himalaya, India. J. Metamorphic Geol. 7, 43-55.

Kwak, T.A.P. (1971a) Compositions of natural sillimanites from volcanic inclusions and metamorphic rocks. Amer. Mineral. 56, 1750-1759.

______(1971b) The selective replacement of the aluminum silicates by white mica. Contrib. Mineral. Petrol. 32, 193-210.

______(1973) Compositions of natural sillimanites from volcanic inclusions and metamorphic rocks: a reply. Amer. Mineral. 58, 558-559.

Labotka, T.C. (1981) Petrology of an andalusite-type regional metamorphic terrane, Panamint Mountains, California. J. Petrol. 22, 261-296.

Lal, R.K. (1969a) Retrogression of cordierite to kyanite and andalusite at Fishtail Lake, Ontario, Canada. Mineral. Mag. 37, 446-471.

______(1969b) Paragenetic relations of aluminosilicates and gedrite from Fishtail Lake, Ontario, Canada. Lithos 2, 187-196.

______, and Ackermand, D. (1979) Coexisting chloritoid-staurolite from the sillimanite (fibrolite) zone, Sini, district Singhbuhm, India. Lithos 12, 133-142.

Lambert, R. St. J., Winchester, J.A., and Holland, J.G. (1979) Time, space and intensity relationships of the Precambrian and lower Paleozoic metamorphisms of the Scottish Highlands. In: Harris, A.L., Holland, K.E., and Leake, B.E. (eds.) The Caledonides of the British Isles, reviewed. Scott. Acad. Press, Edinburgh, Scotland, 363-367.

Lambregts, P.J., and van Roermund, H.L.M. (1990) Deformation and recrystallization mechanisms in naturally deformed sillimanites. Tectonophysics 179, 371-378.

Lang, H.M., and Rice, J.M. (1985a) Regression modelling of metamorphic reactions in metapelites, Snow Peak, Northern Idaho. J. Petrol. 26, 857-887.

______ (1985b) Geothermometry and geobarometry and T-X(Fe-Mg) relations in metapelites, Snow Peak, northern Idaho. J. Petrol. 26, 889-924.

Langer, K. (1976) Synthetic $3d^{3+}$-transition metal bearing kyanites, $(Al_{2-x}M_x^{3+})SiO_5$ In: The Physics and Chemistry of Rocks and Minerals, R.G. Strens (ed.), John Wiley, New York. 389-402.

______ (1983) Blue andalusite from Ottré, Venn-Stavelot Massif, Belgium: a new example of intervalence charge-transfer in the aluminum silicate polymorphs. Fortsch. Mineral. 61, 126.

______, and Seifert, F. (1971) High pressure-high temperature-synthesis and properties of chromium kyanite, $(Al,Cr)_2SiO_5$. Zeits. Anorg. Allgem. Chem. 383, 29-39.

______, and Frentrup, K.R. (1973) Synthesis and some properties of iron-and vanadium bearing kyanites, $(Al,Fe^{3+})_2SiO_5$ and $(Al,V^{3+})_2SiO_5$. Contrib. Mineral. Petrol. 41, 31-46.

______, and Abu-Eid, R.M. (1977) Measurement of the polarized absorption spectra of synthetic transition metal-bearing silicate microcrystals in the spectral range 44000 - 4000 cm^{-1}. Phys. Chem. Minerals 1, 273-299.

______, Hålenius, E., and Fransolet, A-M. (1984) Blue andalusite from Ottré, Venn-Stavelot Massif, Belgium: a new example of intervalence charge-transfer in the aluminum silicate polymorphs. Bull. Minéral. 107, 587-596.

Lappin, M.A. (1960) On the occurrence of kyanite in the eclogites of the Selje and Åheim district, Nordfjord. Norsk Geol. Tidsskr. 40, 289-296.

Lasaga, A.C. (1981) The atomistic basis of kinetics: defects in minerals. In: Lasaga, A.C. and Kirkpatrick, R.J. (eds.), Kinetics of Geochemical Processes. Reviews in Mineralogy 8, 261-319.

______ (1986) Metamorphic reaction rate laws and development of isograds. Mineral. Mag. 50, 359-373.

Laz'ko, Ye[E)], Koptil', V.I., Serenko, V.P., and Tsepin, A.I. (1983) Fassaite clinopyroxenes from diamond-bearing kyanite eclogite xenoliths. Dolk. Acad. Sci. USSR, Earth Sci. Sect. 258, 138-142.

Le Breton, N., and Thompson A.B. (1988) Fluid-absent (dehydration) melting of biotite in metapelites in the early stages of crustal anatexis. Contrib. Mineral. Petrol. 99, 226-237.

Lefebvre, A. (1982a) Transmission electron microscopy of andalusite single crystals indented at room temperature. Bull. Minéral. 105, 347-350.

______ (1982b) Lattice defects in three structurally related minerals: kyanite, yoderite and staurolite. Phys. Chem. Minerals 8, 251-256.

______, and Ménard, D. (1981) Stacking faults and twins in kyanite, Al_2SiO_5. Acta Crystallogr. A37, 80-84.

______, and Paquet, J. (1983) Dissociation of c dislocations in sillimanite Al_2SiO_5. Bull. Minéral. 106, 287-292.

Lefebvre, J.-J., and Naert, K.A. (1978) Coexistence of sillimanite, andalusite and kyanite in felsic to intermediate metavolcanics in the Savant Lake area, district of Thunder Bay, Ontario. Abstr. Geol. Assoc. Canada, 442-443.

______, and Patterson, L.E. (1982) Hydrothermal assemblages of aluminian serpentine, florencite and kyanite in the Zairian copperbelt. Annals Soc. Géologique Belgique 105, 51-71.

LeFort, P. (1981) Manaslu leucogranite: a collision signature of the Himalaya, a model for its genesis and emplacement. J. Geophys. Res. 86B, 10545-10568.

Leisure, F.G., Grosy, A.E., Williams, B.B., and Gazdik, G.C. (1982) Mineral resources of the Craggy Mountain wilderness study area and extension, Buncombe County, North Carolina. U.S. Geol. Survey Bull. 1515, 27 p.

LeMarshall, J., Hutton, D.R., Troup, G.J., and Thyer, J.R.W. (1971) A paramagnetic resonance study of Cr^{3+} and Fe^{3+} in sillimanite. Phys. stat. sol.(a) 5, 769-773.

Lepezin, G.G., and Goryunov, V.A. (1988) Areas of application of minerals of the sillimanite group. Sov. Geol. Geophys. [New York] 29, 68-74.

Lobjoit, W.M. (1964) Kyanite produced in a granitic aureole. Mineral. Mag. 33, 804-808.

London, D. (1987) Internal differentiation of rare-element pegmatites: effects of boron, phosphorous, and fluorine. Geochim. Cosmochim. Acta 51, 403-420.

______, Morgan, G.B., and Hervig, R.L. (1989) Vapor-undersaturated experiments with Macusani glass + H_2O at 200 MPa and the internal differentiation of granitic pegmatites. Contrib. Mineral. Petrol. 102, 1-17.

______, Hervig, R.L., and Morgan, G.B. (1988) Melt-vapor solubilities and elemental partitioning in peraluminous granite-pegmatite systems: experimental results with Macusani glass at 200 MPa. Contrib. Mineral. Petrol. 99, 360-373.

Lonker, S.W. (1988) An occurrence of grandidierite, kornerupine, and tourmaline in southeastern Ontario, Canada. Contrib. Mineral. Petrol. 98, 502-516.

Loomis, T.P. (1972a) Contact metamorphism of pelitic rock by the Ronda ultramafic intrusion, southern Spain. Bull. Geol. Soc. Amer. 83, 2449-2474.

______ (1972b) Coexisting aluminum silicate phases in contact metamorphic aureoles. Amer. J. Sci. 272, 933-945.

______ (1979) A natural example of metastable reactions involving garnet and sillimanite. J. Petrol. 20, 271-292.

______ (1982) Numerical simulation of the disequilibrium growth of garnet in chlorite-bearing aluminous pelitic rocks. Canadian Mineral. 20, 411-423.

Loosveld, R.J.H. (1989) The synchronism of crustal thickening and high T/low P metamorphism in the Mount Isa Inlier, Australia 1. An example, the central Soldiers Cap belt. Tectonophys. 158, 173-190.

Lorenzoni, S., Messina, A., Russo, S., Stagno, F. and Zanettin Lorenzoni, E. (1979) The two-mica Al_2SiO_5 granites of the Sila (Calabria). N. Jahrb. Mineral. Monatsh. 9, 421-436.

Losert, J. (1968) On the genesis of nodular sillimanitic rocks. 23rd Int'l. Geol. Congress, 4, Prague, 109-122.

Lowenstein, W. (1954) The distribution of aluminum in the tetrahedra of silicates and aluminates. Amer. Mineral. 39, 92-96.

Lundgren, L.W., Jr. (1966) Muscovite reactions and partial melting in south-eastern Connecticut. J. Petrol. 7, 421-453.

Macaudière, J., and Touret, J. (1969) La fibrolitisation tectonique: un mécanisme possible de formation des gneiss nodulaires du Bamble. Sciences de la Terre 14, 199-214.

Macdonald, G.A., and Merriam , R. (1938) Andalusite in pegmatite from Fresno County, California. Amer. Mineral. 23, 588-594.

MacKenzie, W.S. (1949) Kyanite-gneiss within a thermal aureole. Geol. Mag. 86, 251-254.

Manning, D.A.C., and Pichavant, M. (1983) The role of fluorine and boron in the generation of granitic melts. In: Atherton, M.P., and Gribble, C.D. (eds.), Migmatites, Melting and Metamorphism, Shiva Publishing Ltd., Nantwich, England. 94-109.

Manning, P.G. (1970) Racah parameters and their relationship to lengths and covalencies of Mn^{2+}-and Fe^{3+}-oxygen bonds in silicates. Canadian Mineral. 10, 677-688.

Marakushev, A.A., and Kudryavtsev, V.A. (1965) Hypersthene-sillimanite paragenesis and its petrologic implication. Dokl. Acad. Sci. USSR, Earth Sci. Sect., 164, 145-148.

Mark, H., and Rosbaud, P. (1926) Ueber die structur der aluminium-silikate vom typus Al_2SiO_5, und des pseudobrookits. N. Jahrb. Mineral., Abt. A, 54, 127-164.

Massone, H.-J. (1989) The upper thermal stability of chlorite + quartz: an experimental study in the system MgO-Al_2O_3-SiO_2-H_2O. J. Metamorphic Geol. 7, 567-581.

______, Mirwald, P.W., and Schreyer, W. (1981) Experimentalle uberprufung der reaktionskurve chlorit + quarz = talk + disthen im system MgO-Al_2O_3-SiO_2-H_2O. Fortsch. Mineral. 59, Beiheft 1, 122-123.

______, and Schreyer, W. (1989) Stability field of the high-pressure assemblage talc + phengite and two new phengite barometers. Euro. J. Mineral. 1, 391-410.

Matsushima, S., Kennedy, G.C., Akella, J., and Haygarth, J. (1967) A study of equilibrium relations in the systems Al_2O_3-SiO_2-H_2O and Al_2O_3-H_2O. Amer. J. Sci. 265, 28-44.

Matthes, S., Richter, P., and Schmidt, K. (1970) Die eklogitvorkommen des kristallinen Grundgebirges in NE-Bayern. III. Der disthen (kyanit) der eklogite und eklogitamphibolite des Münchberger gneisgebietes. N. Jahrb. Mineral. Abh. 113, 111-137.

Mchedlof-Perosyan, O.P. (1954) Erroneous interpretation of differential thermal curves of kyanite and andalusite [in Russian]. Mém. Soc. Russe Mineral. 83, 159.

McKenna, L.W., and Hodges, K.V. (1988) Accuracy versus precision in locating reaction boundaries: implications for the garnet-plagioclase-aluminum silicate-quartz geobarometer. Amer. Mineral. 73, 1205-1208.

McKeown, D.A. (1989) Aluminum X-ray absorption near-edge spectra in some oxide minerals: calculation versus experimental data. Phys. Chem. Minerals 16, 678-683.

McLellan, E.L. (1983) Contrasting textures in metamorphic and anatectic migmatites: an example from the Scottish Caledonides. J. Metamorphic Geol. 1, 241-262.

______ (1985) Metamorphic reactions in the kyanite and sillimanite zones of the Barrovian type area. J. Petrol. 26, 789-818.

______ (1989) Sequential formation of subsolidus and anatectic migmatites in response to thermal evolution, eastern Scotland. J. Geol. 97, 165-182.

______, Linder, D., and Thomas, J. (1989) Multiple granulite-facies events in the southern Appalachians, USA. In: Daly, J.S., Cliff, R.A., and Yardley, B.W.D. (eds.), Evolution of Metamorphic Belts, Geol. Soc. London Special Publication 43, 309-314.

McLelland, J.M. (1989) Pre-granulite-facies metamorphism in the Adirondack Mountains, New York. In: Daly, J.S., Cliff, R.A., and Yardley, B.W.D. (eds.), Evolution of Metamorphic Belts, Geol. Soc. London Special Publication 43, 315-317.

McMillan, P.F., and Hofmeister, A.M. (1988) Infrared and raman spectroscopy. In: Hawthorne, F.C. (ed.), Spectroscopic Methods in Mineralogy and Geology. Reviews in Mineralogy 18, 99-159.

Meinhold, K.D., and Frisch, T. (1970) Manganese silicate-bearing metamorphic rocks from central Tanzania. Schweiz. Mineral. Petr. Mitt. 50, 493-507.

Menard, D., Doukan, J.C., and Paquet, J. (1977) Dissociation of dislocations in kyanite: (Al_2O_3-SiO_2). Materials Res. Bull. 12, 637-642.

______, and Doukhan, J.C. (1978) Défauts de réseau dans la sillimanite: Al_2O_3-SiO_2. J. Physique - Lettres. 39, L19-L22.

______, Doukhan, J.C., and Paquet, J. (1979) Uniaxial compression of kyanite Al_2O_3-SiO_2. Bull. Minéral. 102, 159-162.

Mengel, F., and Rivers, T. (1989) Thermotectonic evolution of Proterozoic and reworked Archaean terranes along the Nain-Churchill boundary in the Saglek area, northern Labrador. In: Daly, J.S., Cliff, R.A., and Yardley, B.W.D. (eds.), Evolution of Metamorphic Belts, Geol. Soc. London Special Publication 43, 319-324.

Meyer, H.O.A. (1987) Inclusions in diamond. In: Nixon, P.H., (ed.), Mantle Xenoliths, John Wiley & Sons Ltd., New York, 501-523.

______, and Brookins, D.G. (1976) Sapphirine, sillimanite, and garnet in granite xenoliths from Stockdale kimberlite, Kansas. Amer. Mineral. 61, 1194-1202.

Michel-Lévy, M.C. (1950) Reproduction artificielle de la sillimanite. Compt. Rend. Acad. Sci., Paris, 230, 2213-2214

Miller, C. (1986) Alpine high-pressure metamorphism in the Eastern Alps. Schweiz. mineral. petrogr. Mitt. 66, 139-144.

Milton, D.J. (1986) Chloritoid-sillimanite assemblage from North Carolina. Amer. Mineral. 71, 891-894.

Mitchell, R.S., Giannini, W.F., and Penick, D.A., Jr. (1988) Large andalusite crystals from Campbell County, Virginia. Rocks & Minerals 63, 446-453.

Mitra, S. (1972) Cavity-filling kyanites from Cuttack district, Orissa. J. Geol. Soc. India 13, 173-174.

Miyano, T., Kaneko, Y., Van Reenen, D.D., and Beukes, N.J. (1988) Geothemo-barometry of andalusite-bearing pelitic rocks in the Timeball Hill Formation at the Penge area, northeastern Transvaal, South Africa. Ann. Rept. Inst. Geosci. Univ. Tsukuba, 14, 69-73.

Miyashiro, A. (1949) The stability relation of kyanite, sillimanite and andalusite, and the physical conditions of the metamorphic processes. J. Geol. Soc. Jap. 55, 218-223.

______ (1951) Kyanites in druses in kyanite-quartz-veins from Saiho-ri in the Fukushinzan District, Korea. J. Geol. Soc. Japan 57, 218-223.

______ (1961) Evolution of metamorphic belts. J. Petrol. 2, 277-311.

Mlinac, R.J., and White, W.B. (1978) Raman spectra of the aluminum orthosilicate polymorphs. Abstr. Geol. Assoc. Canada, 458.

Moecher, D.P., Essene, E.J., and Anovitz, L.M. (1988) Calculation and application of clinopyroxene-garnet-plagioclase-quartz geobarometers. Contrib. Mineral. Petrol. 100, 92-106.

Mohan, A., Windley, B.F., and Searle, M.P. (1989) Geothermobarometry and development of inverted metamorphism in the Darjeeling-Sikkim region of the eastern Himalaya. J. Metamorphic Geol. 7, 95-110.

Montel, J.-M., Weber, C., and Pichavant, M. (1986) Biotite-sillimanite-spinel assemblages in high-grade metamorphic rocks: occurrences, chemographic analysis and thermobarometric interest. Bull. Mineral. 109, 555-573.

Moore, J.M., and Best, M.G. (1969) Sillimanite from two contact aureoles. Amer. Mineral. 54, 975-979.

Moore, J.M., adn Reid, A.M. (1989) A Pan-African zincian staurolite imprint on Namaqua quartz-gahnite-silllimanite assemblages. Mineral. Mag. 53, 63-70.

Morand, V.J. (1988) Vanadium-bearing margarite from the Lachlan Fold Belt, New South Wales, Australia. Mineral. Mag. 52, 341-345.

______ (ms.) Low-pressure regional metamorphism in the Omeo metamorphic complex, Victoria, Australia.

Morgan, G.B., and London, D. (1989) Experimental reactions of amphibolite with boron-bearing aqueous fluids at 200 MPa: implications for tourmaline stability and partial melting in mafic rocks. Contrib. Mineral. Petrol. 102, 281-297.

Morin, J.A., and Turnock, A.C. (1975) The clotty granite at Perrault Falls, Ontario, Canada. Canadian Mineral. 13, 352-357.

Morton, A.C. (1984) Stability of detrital heavy minerals in Tertiary sandstones from the North Sea basin. Clay. Mineral. 19, 287-308.

Motoyoshi, Y., Matsubara, S., and Matsueda, H. (1989) *P-T* evolution of the granulite-facies rocks of the Lützow-Holm Bay region, East Antarctica. In: Daly, J.S., Cliff, R.A., and Yardley, B.W.D. (eds.), Evolution of Metamorphic Belts, Geol. Soc. London Special Publication 43, 325-329.

______, Hensen, B.J., and Matsueda, H. (1990) Metastable growth of corundum adjacent to quartz in a spinel-bearing quartzite from the Archaean Napier Complex, Antarctica. J. Metamorphic Geol. 8, 125-130.

Mrumba, A.H., and Basu, N.K. (1987) Petrology of the talc-kyanite-yoderite schist and associated rocks of Mautia Hill, Mpwapwa District, Tanzania. J. African Earth Sci. 6, 301-311.

Mügge, H. (1883) Beiträge für kentniss der strukturflächen des kalkspathes und über die beziechungen derselben untereinander und der zwillingsbilden an kalkspath und einegen anderen mineralen. N. Jahrb. Geol. Paläntol. Abhandl. 1, 32.

Munz, I.A. (1990) Whiteschists and orthoamphibolite-cordierite rocks and the *P-T-t* path of the Modum Complex, South Norway. Lithos 24, 181-200.

Murata, M. (1982) S-type and I-type granitic rocks of the Ohmine district, central Kii Peninsula [in Japanese]. J. Jap. Assoc. Mineral., Petrol. Econ. Geol. 77, 267-277.

Murdoch, J. (1936) Andalusite in pegmatite. J. Mineral. Soc. Amer. 21, 68-69.

Naggar, A., and Atherton, M.P. (1970) The composition and metamorphic history of some aluminum silicate-bearing rocks from the aureoles of the Donegal granites. J. Petrol. 11, 549-589.

Náray-Szabo, St., Taylor, W.H., and Jackson, W.W. (1929) The structure of cyanite. Zeits. Kristallogr. 71, 117-130.

Nassau, K., and Jackson, K.A. (1970) Trapiche emeralds from Chivor and Muzo, Colombia. Amer. Mineral. 55, 416-427.

Navrotsky, A., and Newton, R.C. (1967) The thermodynamics of cation distributions in simple spinels. J. Inorg. Nuclear Chem. 29, 2701-2714.

______, Newton, R.C., and Kleppa, O.J. (1973) Sillimanite-disordering enthalpy by calorimetry. Geochim. Cosmochim. Acta 37, 2497-2508.

Ndiaye, P.M., Robineau, B., and Moreau, C. (1989) Deformation et metamorphisme des formations birrimiennes en relation avec la mise en place du granite eburneen de Saraya (Senegal oriental). Bull. Soc. Geol. France 5, 619-625.

Neiva, A.M.R. (1984) Chromium-bearing kyanite from Mozambique. Mineral. Mag. 48, 563-564.

Neumann, F. (1925) Über der stabilitätsverhältnisse der modifikationen im polymorphen system Al_2SiO_5. Zeits. anorg. Chemie 145, 193-208.

Newton, R.C. (1966a) Kyanite-andalusite equilibrium from 700°C to 800°C. Science 153, 170-172.

______ (1966b) Kyanite-sillimanite equilibrium at 750°C. Science 151, 1222-1225.

______ (1969) Some high-pressure hydrothermal experiments on severely ground kyanite and sillimanite. Amer. J. Sci. 267, 278-284.

______ (1987) Thermodynamic analysis of phase equilibria in simple mineral systems. Chapter 1 in: Thermodynamic Modeling of Geological Materials: Minerals, Fluids and Melts. I.S.E. Carmichael and H.P. Eugster, eds., Reviews in Mineralogy 17, 1-33.

______, and Haselton, H.T. (1981) Thermodynamics of the garnet-plagioclase-Al_2SiO_5-quartz geobarometer. Chapter 7 in: Advances in Physical Geochemistry, R.C. Newton, A. Navrotsky, and B.J. Wood (eds.), Vol. 1, 131-147, Springer-Verlag, New York.

Nicolas, A., and Poirier, J.P. (1976) Crystalline Plasticity and Solid State Flow in Metamorphic Rocks. John Wiley & Sons, New York. 444 p.

Nixon, P.H., Grew, E.S., and Condliffe, E. (1984) Kornerupine in a sapphirine-spinel granulite from Labwor Hills, Uganda. Mineral. Mag. 48, 550-552.

Noble, D.C., Vogel, T.A., Peterson, P.S., Landis, G.P., Grant, N.K., Jezek, P.A., and McKee, E.H. (1984) Rare-element-enriched S-type ash-flow tuffs containing phenocrysts of muscovite, andalusite, and sillimanite, southeastern Peru. Geology 12, 35-39.

Nockolds, S.R., Knox, R.W.O'B., and Chinner, G.A. (1978) Petrology for Students. Cambridge Univ. Press, Cambridge, England. 435 p.

Nordstrom, D.K., and Munoz, J.L. (1985) Geochemical Thermodynamics. The Benjamin/Cummings Publishing Co., Inc., Menlo Park, CA, 477p.

Novak, J.M., and Holdaway, M.J. (1981) Metamorphic petrology, mineral equilibria, and polymetamorphism in the Augusta quadrangle, south-central Maine. Amer. Mineral. 66, 51-69.

Ohmoto, H., and Kerrick, D.M. (1977) Devolatilization equilibria in graphitic systems. Amer. J. Sci. 277, 1013-1044.

Okay, A.I., Arman, M.B., and Göncüoglu, M.C. (1985) Petrology and phase relations of kyanite-eclogites from eastern Turkey. Contrib. Mineral. Petrol. 91, 196-204.

Okrusch, M. (1969) Die gneishornfelse um Steinach in der Oberpfalz. Contrib. Mineral. Petrol. 22, 32-72.

______ (1971) Garnet-cordierite-biotite equilibria in the Steinach Aureole, Bavaria. Contrib. Mineral. Petrol. 32, 1-23.

______, and Evans, B.W. (1970) Minor element relationships in coexisting andalusite and sillimanite. Lithos 3, 261-268.

Omori, K. (1961) Infrared absorption spectra of sillimanite-andalusite-kyanite and diaspore [in Japenese]. Jap. Assoc. Mineral. Petr. Econ. Geol. 46, 88-91.

Ondarroa, C., Pesquera, A., and Gil, P.P. (1988) Mineralogical and geochemical features of pegmatites from the Ursuya Massif (western Pyrenees, France). Rend. Soc. Ital. Mineral. Petrol. 43, 587-592.

Osanai, Y. (1985) Geology and metamorphic zoning of the main zone of the Hidaka metamorphic belt in the Shizunai River region, Hokkaido [in Japanese]. J. Geol. Soc. Japan 91, 259-278.

Osberg, P.H. (1971) An equilibrium model for Buchan-type metamorphic rocks, south-central Maine. Amer. Mineral. 56, 570-586.

Ostapenko, G.T., Timoshkova, L.P., and Tsymbal, S.N. (1977) Gibbs energy of sillimanite determined from its solubility in water at 530°C and 1300 bars [in Russian]. Zap. Vses. Mineral. Obschch. 106, 243-244.

______, ______, and Gorogotskaya, L.I. (1978) Phase equilibria in the system SiO_2-Al_2O_3-H_2O at 400-700°C and water-vapor pressures of 1-1000 bars [in Russian]. Geokhimiya ______, 248.

______, Gorogotskaya, L.I., Sepchenko, S.B., Timoshkova, L.P., Sharkin, O.P., and Gevork'yan, S.V. (1982) The nature of x-andalusite. [Russian with English summary] Mineralogicheskii Zhurnal 4, 3-13.

______, Ryzhenko, B.N., and Khitarov, N.I. (1987) Simulating the dissolution of Al_2SiO_5 polymorphs in HCl at high T and P in kyanite. Geochem. Int'l. 24, 83-88.

Outhuis, J.H.M. (1989) Hydrothermal andalusite and corundum in a potassic alteration zone around a Proterozoic gabbro-tonalite-granite intrusion NE of Persberg, central Sweden. Mineralogy and Petrology 40, 1-16.

Overbeek, P.W. (1981) The evaluation of an andalusite deposit from Kleinfontein Groot Marico. Rept. Council Mineral Tech. [South Africa] 2109D, 23pp.

______ (1983) The evaluation of drill cores and a bulk sample from a deposit of andalusite near Thabazimbi. Report of the Nat'l Inst. for Metallurgy (South Africa) 2106D, 19p.

Ozerov, K.N., and Bykhover, N.A. (1936) Corundum and kyanite deposits of the Verkhne-Timpton district of the Yukutian Autonomous Soviet Socialist Republic [in Russian]. Trans. Centr. Geol. Prosp. Inst., no. 82, 106 pp.

Padovani, E.R., and Carter, J.L. (1974) Blue sillimanite in garnet granulite xenoliths from Kilbourne Hole, New Mexico. [Abstr.] Trans. Amer. Geophys. Union, 55, 482.

Pankratz, L.B., and Kelley, K.K. (1964) High temperature heat contents and entropies of andalusite, kyanite, and sillimanite. U.S. Bur. Mines Rept. Invest. 6370, 7 pp.

Pannhorst, W., and Schneider, H. (1978) The high-temperature transformation of andalusite (Al_2SiO_5) into 3/2-mullite ($3Al_2O_32SiO_2$) and vitreous silica (SiO_2). Mineral. Mag. 42, 195-198.

Papike, J.J. (1987) Chemistry of the rock-forming silicates: ortho, ring, and single-chain structures. Rev. Geophysics 25, 1483-1526.

______, and Cameron, M. (1976) Crystal chemistry of silicate minerals of geophysical interest. Rev. Geophys. Space Phys. 14, 37-80.

Parkin, K.M., Loeffler, B.M., and Burns, R.G. (1977) Mössbauer spectra of kyanite, aquamarine, and cordierite showing intervalence charge transfer. Phys. Chem. Minerals 1, 301-311.

Parks, G.A. (1984) Surface and interfacial free energies of quartz. J. Geophys. Res. 89, 3997-4008.

Partridge, F.C. (1931) The andalusite sands of the western Transvaal. Bull. Geol. Survey South Africa, no. 2, 16 pp.

Pattison, D.R.M. (1989) P-T conditions and the influence of graphite on pelitic phase relations in the Ballachulish Aureole, Scotland. J. Petrol. 30, 1219-1244.

______, and Harte, B. (1985) A petrogenetic grid for pelites in the Ballachulish and other Scottish thermal aureoles. J. Geol. Soc. London 142, 7-28.

______, and Harte, B. (1988) Evolution of structurally contrasting anatectic migmatites in the 3-kbar Ballachulish aureole, Scotland. J. Metamorphic Geology 6, 475-494.

______, and Harte, B. (in press) Petrography and mineral chemistry of metapelites in the Ballachulish aureole. In: Voll, G., Töpel, J., Pattison, D.R.M., and Seifert, F. (eds.) Equilibrium and kinetics in contact metamorphism: The Ballachulish Igneous Complex and its aureole. Springer Verlag, New York.

Pearson, G.R., and Shaw, D.M. (1960) Trace elements in kyanite, sillimanite and andalusite. Amer. Mineral. 45, 808-817.

Pecher, A. (1989) The metamorphism in the Central Himalaya. J. Metamorphic Geol. 7, 31-41.

Perchuk, L.L. (1989) *P-T*-fluid regimes of metamorphism and related magmatism with specific reference to the granulite-facies Sharyzhalgay complex of Lake Baikal. In: Daly, J.S., Cliff, R.A., and Yardley, B.W.D. (eds.), Evolution of Metamorphic Belts, Geol. Soc. London Special Publication 43, 275-291.

Percival, J.A. (1979) Kyanite-bearing rocks from the Hackett River area, N.W.T.: implications for Archean geothermal gradients. Contrib. Mineral. Petrol. 69, 177-184.

Perkins, D., Essene, E.J., Westrum, E.F., Jr., and Wall, V.J. (1979) New thermodynamic data for diaspore and their application to the system Al_2O_3-SiO_2-H_2O. Amer. Mineral. 64, 1080-1090.

Perreault, S., and Martignole, J. (1988) High-temperature cordierite migmatites in north-eastern Grenville province. J. Metamorphic Geol. 6, 673-696.

Pesquera, A., Fontan, F., and Velasco, F. (1986) Occurrence of alluaudite from a peraluminous minerals-bearing pegmatite in Cinco Villas (Basque Pyrenees, Navarra, Spain). N. Jb. Mineral. Mohat. 2, 82-88.

Peterson, D.E., and Newton, R.C. (1990) Free energy and enthalpy of formation of kyanite. Abstr. Geol. Soc. Amer. 22, 342.

Peterson, E.U., Essene, E.J., Peacor, D.R., and Marcotty, L.A. (1989) The occurrence of högbomite in high-grade metamorphic rocks. Contrib. Mineral. Petrol. 101, 350-360.

Peterson R.C. (1980) Bonding in minerals: I. charge density of the aluminosilicate polymorphs. Ph.D. Thesis, Virginia Polytechnic Inst. and State Univ., Blacksburg, Virginia.

______, and McMullan, R.K. (1986) Neutron diffraction studies of sillimanite. Amer. Mineral. 71, 742-745.

Petreus, I. (1974) The divided structure of crystals II. Secondary structures and habits. N. Jahrb. Mineral. Abh. 122, 314-338.

Petrusenko, S. (1981) Andalusite, corundum and tourmaline from the Markova Trapeza pegmatite deposit, district of Samokov. [Bulgarian with English abstract] Geohimija, Mineralogija i Petrologija 14, 73-83.

______, Maysyuk, S., and Platonov, A. (1988) Kyanite types in Bulgaria distinguished by their optical spectra and specific coloration [in Bulgarian]. Geohimija, Mineral. Petrol. 24, 18-27.

Philbrick, S.S. The contact metamorphism of the Onawa plution, Piscataquis County, Maine. Amer. J. Sci. 31, 1-40.

Pichavant, M. (1987) Effects of B and H_2O on liquidus phase relations in the haplogranite system at 1 kbar. Amer. Mineral. 72, 1056-1070.

______, and Manning, D. (1984) Petrogenesis of tourmaline granites and topaz granites; the contribution of experimental data. Phys. Earth Planet. Inter. 35, 31-50.

______, Kontak, D.J., Valencia Herrera, J., and Clark, A.H. (1988a) The Miocene-Pliocene Macusani Volcanics, SE Peru I. Mineralogy and magmatic evolution of a two-mica aluminosilicate-bearing ignimbrite suite. Contrib. Mineral. Petrol. 100, 300-324.

______, Kontak, D.J., Briqueu, L., Valencia Herrera, J., and Clark, A.H. (1988b) The Miocene-Pliocene Macusani Volcanics, SE Peru II. Geochemistry and origin of a felsic peraluminous magma. Contrib. Mineral. Petrol. 100, 325-338.

Pigage, L.C. (1982) Linear regression analysis of sillimanite-forming reactions at Azure Lake, British Columbia. Canadian Mineral. 20, 349-378.

______, and Greenwood, H.J. (1982) Internally consistent estimates of pressure and temperature: the staurolite problem. Amer. J. Sci. 282, 943-969.

Pitcher. W.S. (1965) The aluminum silicate polymorphs. In: Pitcher, W.S., and Flinn, G.W. (eds.), Controls of Metamorphism, 327-341, John Wiley & Sons, New York.

______, and Read, H.H. (1963) Contact metamorphism in relation to manner of emplacement of the granites of Donegal, Ireland. J. Geol. 71, 261-296.

______, and Berger, A.R. (1972) The Geology of Donegal: A Study of Granite Emplacement and Unroofing. John Wiley, New York, 435p.

Platt, J.P., and Visser, R.L.M. (1980) Extensional structures in anisotropic rocks. J. Structural Geol. 2, 397-410.

Plint, H.E., and Jamieson, R.A. (1989) Microstructure, metamorphism and tectonics of the western Cape Breton Highlands, Nova Scotia. J. Metamorphic Geol. 7, 407-424.

Plummer, C.C. (1980) Dynamothermal contact metamorphism superposed on regional metamorphism in the pelitic rocks of the Chiwaukum Mountains area, Washington Cascades. Bull. Geol. Soc. Amer., Part II, 91, 1627-1668.

Pognante, U., and Lombardo, B. (1989) Metamorphic evolution of the High Himalayan Crystallines in SE Zanskar, India. J. Metamorphic Geol. 7, 9-17.

Poirier, J.-P. (1985) Creep of Crystals. Cambridge University Press, Cambridge, England. 260 p.

Polkanov, Yu. A. (1966) Sillimanite from Tertiary placers of the middle Dnieper area [in Russian]. Mineral. Sborn. Lvov. 20, 126-128.

Poulson, S.R., and Ohmoto, H. (1989) Devolatilization equilibria in graphite-pyrite-pyrrhotite bearing pelites with application to magma-pelite interaction. Contrib. Mineral. Petrol. 101, 418-425.

Powell, R., and Holland, T.J.B. (1985) An internally consistent thermodynamic dataset with uncertainties and correlations: 1. Methods and a worked example. J. Metamorphic Geol. 3, 327-342.

Powers, R.E., and Bohlen, S.R. (1985) The role of synmetamorphic igneous rocks in the metamorphism and partial melting of metasediments, Northwest Adirondacks. Contrib. Mineral. Petrol. 90, 401-409.

Price, R.C. (1983) Geochemistry of a peraluminous granitoid suite from north-eastern Victoria, south-eastern Australia. Geochim. Cosmochim. Acta 47, 31-42.

Propach, G., and Gillessen, B. (1984) Petrology of garnet-, spinel-, and sillimanite-bearing granites from the Bavarian Forest, West Germany. Tschermaks mineral. petrogr. Mitt. 33, 67-75.

______ (1985) Die entwicklung des chemismus peralumischer magmen im AFM-dreieck. Fort. Mineral. 63 Beiheft 1, 289.

Puga, E., Fontboté, J.M., and Martín-Vivaldi, J.L. (1975) Kyanite psuedomorphs after andalusite in polymetamorphic rocks of the Sierra Nevada (Betic Cordillera, Southern Spain. Schweiz. mineral. petrogr. Mitt. 55, 227-241.

Pugin, V.A., and Khitarov, N.I. (1968) Sistema Al_2O_3-SiO_2 v uloviyakh povyshennykh temperatur i davleniy (The system Al_2O_3 under high temperature and pressure). Geokhimiya 157-165 (English abstr. p. 165).

Purtscheller, F. (1969) Petrographische untersuchungen an alumosilikatgneisen des Ötztaler-Staubaier altkristallins. Tschermaks Mineral. petrogr. Mitt. 13, 35-54.

Purvis, A.C. (1984) Metamorphosed altered komatiites at Mount Martin, Western Australia - Archean weathering products metamorphosed at the aluminosilicate triple point. Australian J. Earth Sci. 31, 91-106.

Putnis, A., and McConnell, J.D.C. (1980) Principles of Mineral Behavior. Blackwell Scientific Publications, Oxford, England. 257p.

Raeside, R.P., Hill, J.D., and Eddy, B.G. (1988) Metamorphism of Meguma Group metasedimentary rocks, Whitehead Harbour area, Guysborough County, Nova Scotia. Maritime Sed. and Atlantic Geol. 24, 1-9.

Råheim, A., and Green, D.H. (1974) Talc-garnet-quartz schist from an eclogite-bearing terrane, western Tasmania. Contrib. Mineral. Petrol. 43, 223-231.

Raleigh, C.B. (1965) Glide mechanisms of experimentally deformed minerals. Science 150, 739-741.

Ralph, R.L., Finger, L.W., Hazen, R.M., and Ghose, S. (1984) Compressibility and crystal structure of andalusite at high pressure. Amer. Mineral. 69, 513-519.

Ramberg, H. (1956) Pegmatites in West Greenland. Geol. Soc. Amer. Bull. 67, 185-214.

Rao, B.B., and Johannes, W. (1979) Further data on the stability of staurolite + quartz and related assemblages. N. Jahrb. Mineral. Mh. 10, 437-447.

Rao, S.V.L.N. (1955) Sillimanite from Relli area, Vizagapatam District, with a note on beneficiation. Quart. J. Geol. Mining Metall. Soc. India 27, 151-156.

Rast, N. (1965) Nucleation and growth of metamorphic minerals. In: Pitcher, W.S., and Flinn, G.W. (eds.), Controls of Metamorphism. John Wiley and Sons, New York, 73-102.

Ravindra, K.L., and Ackermand, D. (1979) Coexisting chloritoid-staurolite from the sillimanite (fibrolite) zone, Sini, district Singhbhum, India. Lithos 12, 133-142.

Raymond, M., and Virgo, D. (1972) X-ray photoelectron spectroscopy study of sillimanite (Al_2SiO_5) Carnegie Inst. Wash. Year Book 71, 504-506.

Read, H.H. (1932) On quartz-kyanite rocks in Unst, Shetland Islands, and their bearing on metamorphic differentiation. Mineral. Mag. 23, 317-328.

Reinhardt, J., and Rubenach, M.J. (1989) Temperature-time relationships across metamorphic zones: evidence from porphyroblast-matrix relationships in progressively deformed metapelites. Tectonophys. 158, 141-161.

Remy, P. (1984) Mise en evidence d'un métamorphisme dynamothermal dans les sédiments au contact des ophiolites d'Oréokastro (Macédoine grecque). Compt. Rend. Acad. Sci., Paris, Série D, 299, 27-30.

Rhee, S.K. (1975) Theoretical base and experimental techniques for determination of surface energies of ceramic materials. J. Amer. Ceram. Soc. 58, 441-446.

Ribbe, P.H. (1980) Aluminum silicate polymorphs and mullite. In: P.H. Ribbe, (ed.), Orthosilicates. Reviews in Mineralogy 5, 189-214.

Richardson, S.W., Bell, P.M., and Gilbert, M.C. (1967) Kyanite-sillimanite relations. Carnegie Inst. Wash. Year Book 65, 247-248.

______, ______, ______ (1968) Kyanite-sillimanite equilibrium between 700° and 1500°C. Amer. J. Sci. 266, 513-541.

______, Gilbert, M.C., and Bell, P.M. (1969) Experimental determination of kyanite-andalusite and andalusite-sillimanite equilibria: the aluminum silicate triple point. Amer. J. Sci. 267, 259-272.

Rivers, T., and Wood, H. (ms.) Corundum, kyanite, zoisite and garnet inclusions in plagioclase from a metamorphosed gabbro - evidence for solid-solid and hydration reactions during metamorphism at high pressure.

Robie, R.A. (1987) Caliorimetry. In: Ulmer, G.C., and Barnes, H.L. (eds.) Hydrothermal Experimental Techniques. John Wiley & Sons, New York. 389-422.

______, Hemingway, B.S., and Fisher, J.R. (1978) Thermodynamic properties of minerals and related substances at 298.15°K and 1 bar (10^5 pascals) pressure and at higher temperatures. U.S. Geol. Survey Bull. 1452, 456 pp.

______, and Hemingway, B.S. (1984) Entropies of kyanite, andalusite, and sillimanite: additional constraints on the pressure and temperature of the Al_2SiO_5 triple point. Amer. Mineral. 69, 298-306.

Robinson, G.R., Jr., Haas, J.L., Jr., Schafer, C.M., and Haselton, H.T., Jr. (1982) Thermodyanamic and thermophysical properties of selected phases in the MgO-SiO_2-H_2O-CO_2, CaO-Al_2O_3-SiO_2-SiO_2-H_2O-CO_2, and Fe-FeO-Fe_2O_3-SiO_2 chemical systems, with special emphasis on the properties of basalt. U.S. Geol. Surv. Open-file Rep. 83-79, 429 pp.

Robinson, P., and Jaffe, H.W. (1969) Aluminous enclaves in gedrite-cordierite gneiss from southwestern New Hampshire. Amer. J. Sci. 267, 389-421.

Rock, H.M.S., Macdonald, R., Drewery, S.E., Pankhurst, R.J., and Brook, M. (1986) Pelites of the Glen Urquhart serpentinite-metamorphic complex, west of Loch Ness (anomalous local limestone-pelite successions within the Moine outcrop: III). Scot. J. Geol. 22, 179-202.

Rose, R.L. (1957) Andalusite- and corundum-bearing pegmatites in Yosemite National Park, California. Amer. Mineral. 42, 635-647.

Rosenfeld, J.L. (1969) Stress effects around quartz inclusions in almandine and the piezothermometry of coexisting aluminum silicates. Amer. J. Sci. 267, 317-351.

Rossman, G.R., Grew, E.S., and Dollase, W.A. (1982) The colors of sillimanite. Amer. Mineral. 67, 749-761.

Rossovskyi, L.N. (1963) Pegmatites in magnesian marbles from Kugi-Lyal in the south-west Pamir [in Russian]. Trudy Mineral. Museum Acad. Sci. USSR 14, 166-181.

Rost, F., and Simon, E. (1972) Zur geochemie and Färbun des cyanits. N. Jahrb. Mineral. Monatsh. 9, 383-395.

Roy, A. (1984) On the occurrence and prospect of graphite in graphite schist around Khepchishi in the central crystallines of Bhutan Himalaya. Indian J. Earth Sci. 11, 134-147.

Roy, D.M. (1954) Hydrothermal synthesis of andalusite. Amer. Mineral. 39, 140-143.

Roy, R., and Francis, E.F. (1953) On the distinction of sillimanite from mullite by infra-red techniques. Amer. Mineral. 38, 725-728.

Rubenach, M.J., and Bell, T.H. (1988) Microstructural controls and the role of graphite in matrix/porphyroblast exchange during synkinematic andalusite growth in a granitoid aureole. J. Metamorphic Geol. 6, 651-666.

Rubie, D.C., and Brearley, A.J. (1987) Metastable melting during the breakdown of muscovite + quartz at 1 kbar. Bull. Minéral. 110, 533-549.

Rumble, D. (1973) Andalusite, kyanite, and sillimanite from the Mount Moosilaüke region, New Hampshire. Bull. Geol. Soc. Amer. 84, 2423-2430.

Ruplinger, P.K. (1983) Topaz and andalusite mining in Brazil. J. Gemmology 18, 581-591.

Rutland, R.W.R. (1965) Tectonic overpressures. In: Pitcher, W.S., and Flinn, G.W. (eds.) Controls of Metamorphism, 119-139, John Wiley & Sons, New York.

Saalfeld, H. (1979) The domain structure of 2:1-mullite ($2Al_2O_3{\cdot}1SiO_2$). N. Jahr. Mineral. Abh. 134, 305-316.

______, and Guse, W. (1981) Structure refinement of 3:2 mullite ($3Al_2O_3{\cdot}2SiO_2$). N. Jb. Mineral. Mh. 4, 145-150.

______, and Junge, W. (1983) Thermal decomposition of kyanite single crystals. Tschermaks mineral. petrogr. Mitt. 31, 17-26.

Sadanaga, R., Tokonami, M, and Takéuchi, Y. (1962) Structure of mullite $2Al_2O_3SiO_2$, and relationship with the structures of sillimanite and andalusite. Acta. Crystallogr. 15, 65-68.

Sahl, K., and Seifert, F. (1973) Relationship of fibrolite to sillimanite. Nature 241, 46-47.

Saini, H.S., Srivastava, A.P., and Rajagopalan, G. (1983) Fission track dating and uranium concentration in kyanite from the Himalaya. Geol. Mag. 120, 341-348.

Salje, E. (1986) Heat capacities and entropies of andalusite and sillimanite: The influence of fibrolitization on the phase diagram of the Al_2SiO_5 polymorphs. Amer. Mineral. 71, 1366-1371.

______, and Werneke, Ch. (1982a) The phase equilibrium between sillimanite and andalusite as determined from lattice vibrations. Contrib. Mineral. Petrol. 79, 56-67.

______, and Werneke, Ch. (1982b) How to determine phase stabilities from lattice vibrations. In: Schreyer, W. (ed.), High-Pressure Researches in Geoscience, E. Schweizerbart'sche Verlagsbuchhandlung, Stuttgart, 321-348.

Sanders, I.S. (1988) Plagioclase breakdown and regeneration reactions in Grenville kyanite eclogite at Glenelg, NW Scotland. Contrib. Mineral. Petrol. 98, 33-39.

Sanders, L.D. (1954) The status of sillimanite as an index of metamorphism in the Kenya basement system. Geol. Mag. 91, 144-152.

Santallier, D.S. (1983) Les éclogites du Bas-Limousin, Massif Central français. Comportement des clinopyroxènes et des plagioclases antérieurement à l'amphibolitisation. Bull. Minéral. 106, 691-707.

Santosh, M. (1987) Cordierite gneisses of southern Kerala, India: petrology, fluid inclusions and implications for crustal uplift history. Contrib. Mineral. Petrol. 96, 343-356.

Sauniac, S., and Touret, J. (1983) Petrology and fluid inclusions of a quartz-kyanite segregation in the main thrust zone of the Himilayas. Lithos 16, 35-45.

Saxena, S.K. (1974) Order-disorder in sillimanite. Contrib. Mineral. Petrol. 45, 161-167.

Scharbert, H.G. (1971) Cyanit und sillimanit in moldanubischen granuliten. Tschermaks Mineral. Petr. Mitt. 16, 252-267.

Schenk, V. (1989) *P-T-t* path of the lower crust in the Hercynian fold belt of southern Calabria. In: Daly, J.S., Cliff, R.A., and Yardley, B.W.D. (eds.), Evolution of Metamorphic Belts, Geol. Soc. London Special Publication 43, 337-342.

Schmetzer, K. (1982) Absorptionsspektroskopie and farbe von V^{3+}-haltigen natürlichen oxiden und silikaten - ein beitrag zur kristallchemie des vanadiums. N. Jahrb. Mineral. Abh. 144, 73-106.

Schneider, H. (1979a) Thermal expansion of andalusite. J. Amer. Ceram. Soc. 62, 307.

______ (1979b) Deformation of shock loaded andalusite studied with x-ray diffraction techniques. Phys. Chem. Minerals 4, 245-252.

______, and Hornemann, U. (1977) The disproportion of andalusite (Al_2SiO_5) to Al_2O_3 and SiO_2 under shock compression. Phys. Chem. Mineral. 1, 257-264.

Schneider, M.E., and Eggler, D.H. (1986) Fluids in equilibrium with peridotite minerals: Implications for mantle metasomatism. Geochim. Cosmochim. Acta 50, 711-724. 32, 381-385.

Schmetzer, K. (1982) Absorption sspektroskopie und farbe von V^{3+}-haltigen natürlichen oxiden un silikaten - ein beitrag zur kristallchemie des vanadiums. N. Jahrb. Mineral. Abh. 144, 73-106.

Scholze, H. (1955) zum sillimanit-mullit-problem. Ber. Deutsch. Keram. Gesell.

Schramke, J.A., Kerrick, D.M., and Blencoe, J.G. (1982) Experimental determination of the brucite = periclase + water equilibrium with a new volumetric technique. Amer. Mineral. 67, 269-276.

______, Kerrick, D.M., and Lasaga, A.C. (1987) The reaction muscovite + quartz . andalusite + K-feldspar + water. Part 1. Growth kinetics and mechanism. Amer. J. Sci. 287, 517-559.

Schreurs, J. (1985) Prograde metamorphism of metapelites, garnet-biotite thermometry and prograde changes of biotite chemistry in high-grade rocks of West Uusimaa, southwest Finland. Lithos 18, 69-80.

Schreyer, W. (1977) Whiteschists: their pressure temperature regime based on experimental, field, and petrographic evidence. Tectonophysics 43, 127-144.

______ (1985) Metamorphism of crustal rocks at mantle depths: high-pressure minerals and mineral assemblages in metapelites. Forts. Mineral. 63, 227-261.

______ (1988a) Experimental studies on metamorphism of crustal rocks under mantle pressures. Mineral. Mag. 52, 1-26.

______ (1988b) A discussion of: "Corundum, Cr-muscovite rocks at O'Briens, Zimbabwe: the conjunction of hydrothermal desilicification and LIL-element enrichment-geochemical and isotopic evidence" by Kerrich et al. Contrib. Mineral. Petrol. 100, 552-554.

______, and Chinner, G.A. (1966) Staurolite-quartzite bands in kyanite quartzite at Big Rock, Rio Arriba County, New Mexico. Contrib. Mineral. Petrol. 12, 223-244.

______, and Abraham, K. (1976) Three-stage metamorphic history of a whiteschist from Sar e Sang, Afghansitan, as part of a former evaporite deposit. Contrib. Mineral. Petrol. 59, 111-130.

Schuiling, R.D. (1957) A geo-experimental phase-diagram of Al_2SiO_5 (sillimanite, kyanite, andalusite). Koninkl. Nederlandse Akad. Wetensch. Proc. Series B, 60, 220-226

______ (1958) Kyanite-sillimanite equilibrium at high temperatures and pressures- discussion. Amer. J. Sci. 256, 680-682.

______ (1962) Die petrogenetische bedeutung der drei modifikationen von Al_2SiO_5. N. Jahrb. Mineral. 9, 200-214.

Schumacher, J.C., and Robinson, P. (1987) Mineral chemistry and metasomatic growth of aluminous enclaves in gedrite-cordierite-gneiss from southwestern New Hampshire, USA. J. Petrol. 28, 1033-1073.

______, Schumacher, R., and Robinson, P. (1989) Acadian metamorphism in central Massachusetts and south-western New Hampshire: evidence for contrasting *P-T* trajectories. In: Daly, J.S., Cliff, R.A., and Yardley, B.W.D. (eds.), Evolution of Metamorphic Belts, Geol. Soc. London Special Publication 43, 453-460.

Seifert, F., and Langer, K. (1970) Stability relations of chromium kyanite at high pressures and temperatures. Contrib. Minera. Petrol. 28, 9-18.

Sen, R. (1971) Crystal chemistry of andalusite in different stages of their growth. N. Jahrb. Mineral. Mon., 542-546.

Serdyuchenko, D.P. (1949) On the chemical constitution of manganian andalusite [in Russian]. Mém. Soc. Russe Mineral. 78, 133-135.

Sevigny, J.H., Parrish, R.R., and Ghent, E.D. (1989) Petrogenesis of peraluminous granites, Monashee Mountains, southeastern Canadian Cordillera. J. Petrol. 30, 557-581.

Sgarlata, F. (1962) Sulla struttura cristallina dell'andalusite [in Italian]. Periodico. Mineral., Rome, 31, 43-61.

Shannon, E.V. (1920) Notes on anglesite, anthophyllite, calcite, datolite, sillimanite, stilpnomelane, tetrahedrite and triplite. Proc. U.S. Nat. Mus. 58, 437-453.

Sherriff, B.L., and Grundy, H.D. (in press) The relationship between ^{29}Si MAS NMR chemical shift and silicate structure. Amer. Mineral.

Sharma, R.S., and Windley, B.F. (1984) Mineral parageneses and metamorphic reactions in metasedimentary enclaves from the Archaean gneiss complex of north-west India. Mineral. Mag. 48, 195-209.

Shelley, D. (1968) A note on the relationship of sillimanite to biotite. Geol. Mag. 105, 543-545.

Shiba, M. (1988) Metamorphic evolution of the southern part of the Hidaka belt, Hokkaido, Japan. J. Metamorphic Geol. 6, 273-296.

Siivola, J. (1971) The aluminum K^{β}-band structure of andalusite, sillimanite and kyanite. Bull. Geol. Soc. Finland 43, 1-6.

Singh, S. (1968) The occurrence of sillimanite and its status as an index of metamorphic grade in Guyana Shield rocks of southern British Guiana. Trans. 4th Carribean Geol. Conf., Trinidad (1965), 395-399.

Skinner, B.J., Clark, S.P., Jr., and Appleman, D.E. (1961) Molar volumes and thermal expansions of andalusite, kyanite, and sillimanite. Amer. J. Sci. 259, 651-668.

Slaughter, J.A., Kerrick, D.M., and Wall, V.J. (1975) Experimental and thermodynamic study of equilibria in the system: CaO-MgO-SiO_2-H_2O-CO_2. Amer. J. Sci. 275, 143-162.

Slutskiy, A.B. (1968) Conductance variation during kyanite-sillimanite (Al_2SiO_5) polymorphism under high temperatures and pressures. Dokl. Acad. Sci. USSR, Earth Sci. Sect. 179, 199-200.

Smith, D.G.W., and McConnell, J.D.C. (1966) A comparative electron diffraction study of sillimanite and some natural and artificial mullites. Mineral. Mag. 35, 810-814.

Smith, G., and Strens, R.G.J. (1976) Intervalence transfer absorption in some silicate, oxide and phosphate minerals. In: The Physics and Chemistry of Minerals and Rocks. R.G.J. Strens (ed.), John Wiley & Sons, New York, 583-612.

______, Hålenius, U., and Langer, K. (1982) Low temperature spectral studies of Mn^{3+}-bearing andalusite and epidote type minerals in the range 30000-5000 cm^{-1}. Phys. Chem. Mineral. 8, 136-142.

Smith, K.L., Kirkpatrick, R.J., Oldfield, E., and Henderson, D.M. (1983) High-resolution silicon-29 nuclear magnetic resonance spectroscopic study of rock-forming silicates. Amer. Mineral. 68, 1206-1215.

Smith, L.L., and Newcome, R., Jr., (1951) Geology of kyanite deposits at Henry Knob, South Carolina. Econ. Geol. 46, 757-764.

Smyth, J.R., McCormick, T.C., and Caporuscio, F.A. (1984) Petrology of a suite of eclogitic inclusions from the Bobbejaan kimberlite I. Two unusual corundum-bearing kyanite eclogites. In: Kornprobst, J., (ed.), Kimberlites II: The Mantle and Crust-Mantle Relationships. Elsevier, Amsterdam, 109-119.

______, and Bish, D.L. (1988) Crystal Structures and Cation Sites of the Rock-Forming Minerals. Allen & Unwin, Boston, MA.

Sobolev, V.S., and Bazarova, T.Y. (1963) On the temperature of kyanite crystallization in pegmatites [in Russian]. Doklady Acad. Sci. USSR 153, 174-175.

Sobolev, N.V., Jr., Kuznetsova, I.K., and Zyuzin, N.I. (1968) The petrology of grospydite xenoliths from the Zagodochnaya kimberlite pipe in Yakutia. J. Petrol. 9, 253-280.

Soman, K., and Nair, N.G.K. (1985) Chatoyancy in chrysoberyl cat's-eye from pegmatites of Trivandrum district, southern India. J. Gemmology 19, 412-415.

Southwick, D.L. (1969) A note on the occurrence of kyanite and chloritioid in some metaconglomerates in Maryland. Geol. Soc. Amer. Bull. 80, 2645-2652.

Spaenhauser, F. (1933) Die andalusit-und disthenvorkommen der Silvretta. Schweiz. mineral. petrogr. Mitt. 13, 323-346.

Spear, F.S. (1988) Metamorphic fractional crystallization and internal metasomatism by diffusional homogenization of zoned garnets. Contrib. Mineral. Petrol.

______ (1989) Relative thermobarometry and metamorphic *P-T* paths. In: Daly, J.S., Cliff, R.A., and Yardley, B.W.D. (eds.), Evolution of Metamorphic Belts, Geol. Soc. London Special Publication 43, 63-81.

______, and Selverstone, J. (1983) Quantitative P-T paths from zoned minerals: theory and tectonic applications. Contrib. Mineral. Petrol. 83, 348-357.99, 507-517.

______, and Franz, G. (1986) P-T evolution of metasediments from the eclogite zone, south-central Tauern window, Austria. Lithos 19, 219-234.99, 507-517.

______, and Cheney, J.T. (1989) A petrogenetic grid for pelitic schists in the system SiO_2-Al_2O_3-FeO-MgO-K_2O-H_2O. Contrib. Mineral. Petrol. 101, 149-164.

Speer, J.A. (1982) Metamorphism of the pelitic rocks of the Snyder Group in the contact aureole of the Kiglapait layered intrusion, Labrador: Effects of buffering partial pressures of water. Canadian J. Earth Sci. 19, 1888-1909.

______ (1983) Crystal chemistry and phase relations of orthorhombic carbonates. In: Reeder, R.J. (ed.), Carbonates: Mineralogy and Chemistry. Reviews in Mineralogy 11, 145-190.

Spencer, K.J., and Lindsley, D.H. (1981) A solution model for coexisting iron-titanium oxides. Amer. Mineral. 66, 1189-1201.

Spry, A. (1969) Metamorphic Textures. Pergamon Press, Oxford, England, 350 p.

Staněk, J., and Miškovský, J. (1964) Iron-rich cordierite from the Dolní Bory pegmatite [in Czechoslovakian]. Čas. mineral. geol. 9, 191-192.

Stanton, R.L. (1983) The direct derivation of sillimanite from a kaolinitic precursor: evidence from the Geco mine, Manitouwadge, Ontario. Econ. Geol. 78, 422-437.

Stäubli, A. (1989) Polyphase metamorphism and the development of the Main Central Thrust. J. Metamorphic Geol. 7, 73-93.

Steefel, C.I., and Atkinson, W.W., Jr. (1984) Hydrothermal andalusite and corundum in the Elkhorn District, Montana. Econ. Geol. 79, 573-579.

Steltenpohl, M.G., and Bartley, J.M. (1987) Thermobarometric profile through the Caledonian nappe stack of Western Ofoten, North Norway. Contrib. Mineral. Petrol. 96, 93-103.

St-Onge, M.R. (1984) The muscovite-melt bathograd and low-P isograd suites in north-central Wopmay orogen, Northwest Territories, Canada. J. Metamorphic Geol 2, 315-326.

______ (1987) Zoned poikiloblasitc garnets: P-T paths and syn-metamorphic uplift through 30 km of structural depth, Wopmay Orogen, Canada. J. Petrol. 28, 1-21.

Stout, M.Z., Crawford, M.L., and Ghent, E.D. (1986) Pressure-temperature and evolution of fluid compositions of Al_2SiO_5-bearing rocks, Mica Creek, B.C., in light of fluid inclusion data and mineral equilibria. Contrib. Mineral. Petrol. 92, 236-247.

Strens, R.G.J. (1968) Stability of the Al_2SiO_5 solid solutions. Mineral. Mag. 36, 839-849.

Stuckey, J.L. (1935) Origin of cyanite. Econ. Geol., 30, 444-450.

Sturt, B.A. (1970) Exsolution during metamorphism with particular reference to feldspar solid solutions. Mineral. Mag. 37, 815-832.

Stüwe, K., and Powell, R. (1989) Metamorphic evolution of the Bunger Hills, East Antarctica: evidence for substantial post-metamorphic peak compression with minimal cooling in a Proterozoic orogenic event. J. Metamorphic Geol. 7, 449-464.

Swapp, S.M., and Onstott, T.C. (1989) P-T-time characterization of the Trans-Amazonian orogeny in the Imataca Complex, Venezuela. Precam. Res. 42, 293-314.

Switzer, G., and Melson, W.G. (1969) Partially melted kyanite eclogite from the Roberts Victor mine, South Africa. Smithsonian Contrib. Earth Sci., no. 1, 1-9.

Sykes, M.L., and Moody, J.B. (1978) Pyrophyllite and metamorphism in the Carolina slate belt. Amer. Mineral. 63, 96-108.

Taber, S. (1935) The origin of cyanite. Econ. Geol. 30, 923-924.

Tanner. S.B., Kerrick, D.M., and Lasaga, A.C. (1985) Experimental kinetic study of the reaction: calcite + quartz = wollastonite + CO_2, from 1 to 3 kilobars and 500° to 850°C. Amer. J. Sci. 285, 577-620.

Taylor, W.H. (1928) The structure of sillimanite and mullite. Zeits. Kristallogr. 68, 503-521.

______ (1929) The crystal structure of andalusite Al_2SiO_5. Zeits. Kristallogr. 71, 205-218.

Teale, G.S. (1979) Margarite from the Olary Province of South Australia. Mineral. Mag. 43, 433-435.

Temkin, M. (1945) Mixtures of fused salts as ionic solutions. Acta Physicochim. USSR 20, 411-420.

Temperley, B.N. (1953) Kyanite in Kenya, with an account of its occurrence in some other countries and a discussion on its origin. Geol. Survey Kenya Memoir No. 1, 87p.

Thomas, K.K. (1984) The origin of sillimanite in Essex, Connecticut. M.S. Thesis, Indiana Univ., 101p.

Thompson, A.B. (1976) Mineral reactions in pelitic rocks: Parts I and II. Amer. J. Sci. 276, 401-454.

______ (1982) Dehydration melting of pelitic rocks and the generation of H_2O-undersaturated granitic liquids. Amer. J. Sci. 282, 1567-1595.

Thompson, J.B. (1957) The graphical analysis of mineral assemblages in pelitic schists. Amer. Mineral. 42, 842-858.

______, and Norton, S.A. (1968) Paleozoic regional metamorphism in New England and adjacent areas. In: Zen, E-an, White, W.S., Hadley, J.B. and Thompson, J.B., eds., Studies of Appalachian Geology - Northern and Maritime. John Wiley and Sons, New York, 319-327.

Thompson, P.H. (1976) Isograd patterns and pressure-temperature distributions during regional metamorphism. Contrib. Mineral. Petrol. 57, 277-295.

______, (1989) An empirical model for metamorphic evolution of the Archaean Slave Province and adjacent Thelon Tectonic Zone, north-western Canadian Shield. In: Daly, J.S., Cliff, R.A., and Yardley, B.W.D. (eds.) Evolution of Metamorphic Belts. Geol. Soc. London Spec. Pub. 43, 245-263.

______, and Bard, J.-P. (1982) Isograds and mineral assemblages in the eastern axial zone, Montagne Noire (France): implications for temperature gradients and P-T history. Canadian J. Earth Sci. 19, 129-143.

Thorneley, P.C., and Taylor, W.H. (1939) The co-cordination of aluminum in andalusite. Mem. Manchester Lit. Phil. Soc. 83, 17-30.

Tilley, C.E. (1924) Contact metamorphism in the Comrie area of the Perthshire Highlands. Geol. Soc. London Quart. J. 80, 22-70.

______ (1935) The role of kyanite in the 'hornfels zone' of the Carn Chuinneag granite (Ross-shire). Mineral. Mag. 24, 92-97.

Todd, S.S. (1950) Heat capacities at low temperatures and entropies at 298.16K of andalusite, kyanite, and sillimanite. Amer. Chem. Soc. J. 72, 4742-4743.

Topor, L., Kleppa, O.J., Newton, R.C., and Kerrick, D.M. (1989) Molten salt calorimetry of Al_2SiO_5 polymorphs at 1000 K. [Abstr.] EOS, Trans. Amer. Geophys. Union 70, 493.

Tozer, C.F. (1955) The mode of occurrence of sillimanite in the Glen District, Co. Donegal. Geol. Mag. 92, 310-320.

Tracy, R.J. (1978) High grade metamorphic reactions and partial melting in pelitic schist, west-central Massachusetts. Amer. J. Sci. 278, 150-178.

______ (1985) Migmatite occurrences in New England. In: Ashworth, J.R. (ed.), Migmatites. Blackie, London, 204-224.

______, and Robinson, P. (1980) Evolution of metamorphic belts: information from detailed petrologic studies. In: D.R. Wones (ed.) The Caledonides in the USA, I.G.C.P. project 27: Caledonide Orogen. Virginia Polytechnic Inst. State Univ. Memoir No. 2, 189-195.

______ (1983) Acadian migmatite types in pelitic rocks of Central Massachusetts. In: Atherton, M.P., and Gribble, C.D. (eds.) Migmatites, Melting and Metamorphism. Shiva, Nantwich, 163-173.

______, and McLellan, E.L. (1985) A natural example of the kinetic controls of compositional and textural equilibration. In: Thompson, A.B., and Rubie, D.C. (eds.) Metamorphic Reactions Kinetics, Textures, and Deformation. Advances in Physical Geochemistry 4, Springer Verlag, New York, 118-137.

Treloar, P.J. (1985) Metamorphic conditions in central Connemara, Ireland. J. Geol. Soc. London 142, 77-86.640.

Trommsdorff, V. (1980) Alpine metamorphism and Alpine intrusions. In: Trümpy, R. (ed.) Geology of Switzerland - A Guidebook. Part A: An Outline of the Geology of Switzerland. Wepf & Co., Basel, Switzerland. p.82-87.

Troup, G.J., and Hutton, D.R. (1964a) Paramagnetic resonance of Cr^{3+} in kyanite. Brit. J. Appl. Physics 15, 275-280.

______, and Hutton, D.R. (1964b) Paramagnetic resonance of Fe^{3+} in kyanite. Brit. J. Appl. Physics 15, 1493-1499.

Trzcienski, W.E., Jr., and Marchildon, N. (1989) Kyanite-garnet-bearing Cambrian rocks and Grenville granulites from the Ayer's Cliff, Quebec, Canada, lamprophyre dike suite: Deep crustal fragments from the northern Appalachians. Geology 17, 637-

Turco, G. (1964) Formation d'un nouveau silicate d'aluminum de formule Al_2SiO_5 au cours de le synthèse hydrothermale de la zunyite. Comptes Rendus Acad. Sci. Paris 258, 3331-3334.

Turpin, L., Cuney, M, Friedrich, M., Bouchez, J.-L., and Aubertin, M. (1990) Meta-igneous origin of Hercynian peraluminous granites in N.W. French Massif Central: implications for crustal history reconstructions. Contrib. Mineral. Petrol. 104, 163-172.

Tyler, I.M., and Ashworth, J.R. (1982) Sillimanite-potash feldspar assemblages in graphitic pelites, Strontian area, Scotland. Contrib. Mineral. Petrol. 81, 18-29.

______, and Ashworth, J.R. (1983) The metamorphic environment of the Foyers Granitic Complex. Scott. J. Geol. 19, 271-285.

van Reenen, D.D. (1986) Hydration of cordierite and hypersthene and a description of the retrograde orthoamphibole isograd in the Limpopo belt, South Africa. Amer. Mineral. 71, 900-915.

Varley, E.R. (1965) Sillimanite. Overseas Geological Surveys Mineral Resources Division, Her Majesty's Stationary Office, London, 165p.

Vaughan, M.T., and Weidner, D.J. (1978) The relationship of elasticity and crystal structure in andalusite and sillimanite. Phys. Chem. Minerals 3, 133-144.

Velde, B. (1970) La margarite, minérai d'alteration de l'andalousite des Forges de Salles (Bretagne). Bull. Soc. Franç. Mineral. Crist. 93, 402-403.

Verma, P.K. (1989) The Himalayan metamorphism. In: Daly, J.S., Cliff, R.A., and Yardley, B.W.D. (eds.), Evolution of Metamorphic Belts, Geol. Soc. London Special Publication 43, 377-383.

Vernon, R.H. (1975) Microstructural interpretation of some fibrolitic sillimanite aggregates. Mineral. Mag. 40, 303-306.

______ (1978) Pseudomorphous replacement of cordierite by symplectitic intergrowths of andalusite, biotite and quartz. Lithos 11, 283-289.

______ (1979) Formation of late sillimanite by hydrogen metasomatism (base-leaching) in some high-grade gneisses. Lithos 12, 143-152.

______ (1987a) Growth and concentration of fibrous sillimanite related to heterogeneous deformation in K-feldspar-sillimanite metapelites. J. Metamorphic Geol. 5, 51-68.

______ (1987b) Oriented growth of sillimanite in andalusite, Placitas-Juan Tabo area, New Mexico, U.S.A. Canadian J. Earth Sci. 24, 580-590.

______ (1988) Sequential growth of cordierite and andalusite porphyroblasts, Cooma Complex, Australia: microstructural evidence of a prograde reaction. J. Metamorphic Geol. 6, 255-269.

______, and Flood, R.H. (1977) Interpretation of metamorphic assemblages containing fibrolitic sillimanite. Contrib. Mineral. Petrol. 59, 227-235.

______, Flood, R.H., and D'Arcy (1987) Sillimanite and andalusite produced by base-cation leaching and contact metamorphism of felsic igneous rocks. J. Metamorphic Geol., 5, 439-450.

______, and Collins, W.J. (1988) Igneous microstructures in migmatites. Geology 16, 1126-1129.

______, Clarke, G.L., and Collins, W.J. (ms.) Local, mid-crustal granulite facies metamorphism and melting: an example in the Mount Stafford area, central Australia.

Vidale, R.J. (1974) Vein assemblages and metamorphism in Dutchess County, New York. Geol. Soc. Amer. Bull. 85, 303-306.

Vielzeuf, D., and Holloway, J.R. (1988) Experimental determination of the fluid-absent melting relations in the pelitic system. Consequences for crustal differentiation. Contrib. Mineral. Petrol. 98, 257-276.

Voloshin, A.V., and Davidenko, I.V. (1972) Andalusite in granite pegmatites of the Kola Peninsula. Dokl. Acad. Sci. USSR, Earth Sci. Sect. 203, 116-117.

Vrána, S. (1973) A model of aluminum silicate accretion in metamorphic rocks. Contrib. Mineral. Petrol. 41, 73-82.

______, Prasad, R., and Fediuková, E. (1975) Metamorphic kyanite eclogites in the Lafilian Arc of Zambia. Contrib. Mineral. Petrol. 51, 139-160.

______, Rieder, M., and Podlaha, J. (1978) Kanonaite, $(Mn^{3+}_{0.76}Al_{0.23}Fe^{3+}_{0.02})^{[6]}Al^{[5]}[O/SiO_4]$, a new mineral isotypic with andalusite. Contrib. Mineral. Petrol. 66, 325-332.

Vuichard, J.P., and Ballevre, M. (1988) Garnet-chloritoid equilibria in eclogitic pelitic rocks from the Sesia zone (Western Alps): their bearing on phase relations in high pressure metapelites. J. Metamorphic Geol. 6, 135-157.

Waldbaum, D.R. (1965) Thermodynamic properties of mullite, andalusite, kyanite and sillimanite. Amer. Mineral. 50, 186-195.

Wallis, G.R., and Kennedy, D.R. (1965) Acacia Vale sillimanite deposit, Silverton, N.S.W. Rept. Geol. Survey New South Wales 39, 1-17.

Walther, J.V., and Helgeson H.C. (1977) Calculation of the thermodynamic properties of aqueous silica and the solubility of quartz and its polymorphs at high pressures and temperatures. Amer. J. Sci. 277, 1315-1351.

______ (1986) Mineral solubilities in supercritical H_2O solutions. Pure & Applied Chem. 58, 1585-1598.

______, and Wood, B.J. (1984) Rate and mechanism in prograde metamorphism. Contrib. Mineral. Petrol. 88, 246-259.

______, and Schott, J. (1988) The dielectric constant approach to speciation and ion pairing at high temperature and pressure. Nature 332, 635-638.

Warren, R.G., and Hensen, B.J. (1989) The *P-T* evolution of the Proterozoic Arunta Block, central Australia, and implications for tectonic evolution. In: Daly, J.S., Cliff, R.A., and Yardley, B.W.D. (eds.), Evolution of Metamorphic Belts, Geol. Soc. London Special Publication 43, 349-355.

Watson, E.B., and Brenan, J.M. (1987) Fluids in the lithosphere, l. Experimentally-determined wetting characteristics of CO_2-H_2O fluids and their implications for fluid transport, host-rock physical properties, and fluid inclusion formation. Earth & Planet. Sci. Let. 85, 497-515.

Watson, J. (1948a) Late sillimanite in the migmatites of Kildonan, Scotland. Geol. Mag. 85, 149-162.

Wayte, G.J., Worden, R.H., Rubie, D.C., and Droop, G.T.R. (1989) A TEM study of disequilibrium plagioclase breakdown at high pressure: the role of infiltrating fluid. Contrib. Mineral. Petrol. 101, 426-437.

Webb, R.W. (1943) Two andalusite pegmatites from Riverside County, California. Amer. Mineral. 28, 581-593.

Weber, C. Barbey, P., Cuney, M., and Martin, H. (1985) Trace element behaviour during migmatization. Evidence for a complex melt-residuuum-fluid interaction in the St. Malo migmatitic dome (France). Contrib. Mineral. Petrol. 90, 52-62.

______, and Barbey, P. (1986) The role of water, mixing processes and metamorphic fabric in the genesis of the Baume migmatites (Ardèche, France). Contrib. Mineral. Petrol. 92, 481-491.

Webster, J.D., Holloway, J.R., and Hervig, R.L. (1987) Phase equilibria of a Be, U and F-enriched vitrophyre from Spor Mountain, Utah. Geochim. Cosmochim. Acta 51, 389-402.

Weill, D.F. (1966) Stability relations in the Al_2O_3-SiO_2 system calculated from solubilities in the Al_2O_3-SiO_2-Na_3AlF_6 system. Geochim. Cosmochim. Acta 30, 223-237.

______, and Fyfe, W.S. (1961) A preliminary note on the relative stability of andalusite, kyanite, and sillimanite. Amer. Mineral. 46, 1191-1195.

______, ______ (1964) The 1010 and 800°C isothermal sections in the system Na_3AlF_6-Al_2O_3-SiO_2. J. Electrochem. Soc. 111, 582-585.

Weiss, Z., Bailey, S.W., and Rieder, M. (1981) Refinement of the crystal structure of kanonaite $(Mn^{3+},Al)^{[6]}(Al,Mn^{3+})^{[5]}O[SiO_4]$. Amer. Mineral. 66, 561-567.

Wells, P.R.A. (1979) P-T conditions in the Moines of the Central Highlands, Scotland. J. Geol. Soc. London 136, 663-671.

______, and Richardson, S.W. (1979) Thermal evolution of metamorphic rocks in the Central Highlands of Scotland. In: Harris, A.L., Holland, K.E., and Leake, B.E. (eds.) The Caledonides of the British Isles, reviewed. Scott. Acad. Press, Edinburgh, Scotland, 339-344.

Wenk, E. (1970) Zur regionalmetamorphose und ultrametamorphose im Lepontin. Fortschr. Mineral. 47, 34-51.

Wenk, H.-R. (1980) Defects along kyanite-staurolite interfaces. Amer. Mineral. 65, 766-769.

______ (1983) Mullite-sillimanite intergrowth from pelitic inclusions in Bergell tonalite. N. Jahrb. Mineral. Abh. 146, 1-14.

______, Wenk, E., and Wallace, J.H. (1974) Metamorphic mineral assemblages in pelitic rocks of the Bergell Alps. Schweiz. mineral. petrogr. Mitt. 54, 507-554.

White, E.W., and White, W.B. (1967) Electron microprobe and optical absorption study of colored kyanites. Science 158, 915-917.

White, J.C., and White, S.H. (1981) On the structure of grain boundaries in tectonites. Tectonophysics 78, 613-628.

Wickham, S.M. (1987) Crustal anatexis and granite petrogenesis during low-pressure regional metamorphism: the Trois Seigneurs Massif, Pyrenees, France. J. Petrol. 28, 127-169.

Wikström, A. (1970) Note on the alteration of kyanite in the eclogites from the Nordfjord area, Norway. Lithos 3, 184-186.

Wilkins, R.W.T., and Sabine, W. (1973) Water content of some nominally anhydrous silicates. Amer. Mineral. 58, 508-516.

Willner, A., Schreyer, W., and Moore, J.M. (1990) Peraluminous metamorphic rocks rom the Namaqualand Metamorphic Complex (South Africa): Geochemical evidence for an exhalation-related, sedimentary origin of a Mid-Proterozoic rift system. Chem. Geol. 81, 221-240.

Winchester, J.A. (1974) The control of the whole-rock content of CaO and Al_2O_3 on the occurrence of the aluminum silicate polymorphs in amphibolite facies pelites. Geol. Mag. 111, 205-211.

Winter, J.K., and Ghose, S. (1979) Thermal expansion and high-temperature crystal chemistry of the Al_2SiO_5 polymorphs. Amer. Mineral. 64, 573-586.

Wintsch, R.P. (1975) Solid-fluid equilibria in the system $KAlSi_3O_8$-$NaAlSi_3O_8$-Al_2SiO_5-SiO_2-H_2O-HCl. J. Petrol. 16, 57-59.

______, (1981) Syntectonic oxidation. Amer. J. Sci. 281, 1223-1239.

______, and Dunning, J.D. (1985) The effect of dislocation density on the aqueous solubility of quartz and some geologic implications: a theoretical approach. J. Geophys. Res. 90, 3649-3657.

______, and Andrews, M.S. (1988) Deformation induced growth of sillimanite: "stress" minerals revisited. J. Geol. 96, 143-161.

Wirth, R. (1985) Dehydration and thermal alteration of white mica (phengite) in the contact aureole of the Traversella intrusion. N. Jahrb. Mineral Abh. 152, 101-112.

______, (1986) Some observations concerning the growth kinetics of biotite during the thermally-induced transformation of white mica (phengite) in the contact aureole of the Traversella intrusion (N-Italy). In: Freer, R., and Dennis, P.F. (eds.) Kinetics and Mass Transport in Silicate and Oxide Systems. Trans Tech., Rockport, MA. 123-132.

Wodeyar, B.K., and Sengupta, D.K. (1982) Cordierite-sillimanite-bearing gneisses from Channapatna area, Karnataka State, India. J. Karnatak Univ., Science 27, 31-38.

Wood, B.J., and Walther, J.V. (1983) Rates of hydrothermal reactions. Science 222, 413-415.

Woodland, B.G. (1963) A petrographic study of thermally metamorphosed pelitic rocks in the Burke area, northeastern Vermont. Amer. J. Sci. 261, 354-375.

Woodland, A.B., and Walther, J.V. (1987) Experimental determination of the solubility of the assemblage paragonite, albite, and quartz in supercritical H_2O. Geochim. Cosmochim. Acta 51, 365-372.

Woodsworth, G.J. (1979) Metamorphism, deformation, and plutonism in the Mount Raleigh pendant, Coast Mountains, British Columbia. Geological Survey Canada Bull. 295.

Workman, D.R., and Cowperthwaite, I.A. (1963) An occurrence of kyanite pseudomorphing andalusite from southern Rhodesia. Geol. Mag. 100, 456-466.

Wülfing, E.A. (1917) Der viridin und seine beziehung zum andalusit. Sitzungsber. Heidelberg Akad. Wiss. Math.-nat. Kl., Abt. A, 12, 18 pp.

Wyckoff, D. (1952) Metamorphic facies in the Wissahickon schist near Philadelphia, Pennsylvania. Bull. Geol. Soc. Amer. 63, 24-58.

Yardley, B.W.D. (1977) The nature and significance of the mechanism of sillimanite growth in the Connemara schists, Ireland. Contrib. Mineral. Petrol. 65, 53-58.

______ (1978) Genesis of the Skagit Gneiss migmatites, Washington, and the distinction between posible mechanisms of migmatization. Bull. Geol. Soc. Amer. 89, 941-951.

______ (1980) Metamorphism and orogeny in the Irish Dalradian. J. Geol. Soc. London 137, 303-309.

______ (1989) An Introduction to Metamorphic Petrology. Longman Scientific & Technical, Harlow, England. 248 p.

______, Long, C.B., and Max, M.D. (1979) Patterns of metamorphism in the Ox Mountains and adjacent parts of Western Ireland. In: Richardson, S.W., Bell, P.M., and Gilbert, M.C. (eds.) Scott. Acad. Press Edinburgh, Scotland 369-374.

______, Leake, B.E., and Farrow, C.M. (1980) The metamorphism of Fe-rich pelites from Connemara, Ireland. J. Petrol. 21, 365-399.

______, and Long, C.B. (1981) Contact metamorphism and fluid movement around the Easky adamellite, Ox Mountains, Ireland. Mineral. Mag. 44, 125-131.

______, and Baltatzis, E. (1985) Retrogression of staurolite schists and the sources of infiltrating fluids during metamorphism. Contrib. Mineral. Petrol. 89, 59-68.

______, Barber, J.P., and Gray, J.R. (1987) The metamorphism of the Dalradian rocks of western Ireland and its relation to tectonic setting. Phil. Trans. Royal Soc. London A321, 243-370.

Yokoi, K. (1983) Fe_2O_3 content of co-existing andalusite and sillimanite in the Ryoke metamorphic rocks occurring in the Hiraoka-Kadoya area, central Japan. J. Japanese Assoc. Mineral. Petrol. Econ. Geol. 78, 246-254.

Young, E.D., Anderson J.L., Clarke, H.S., and Thomas, W.M. (1989) Petrology of biotite-cordierite-garnet gneiss of the McCullough Range, Nevada I. Evidence for Proterozoic low-pressure fluid-absent granulite-grade metamorphism in the Southern Cordillera. J. Petrol. 30, 39-60.

Zaleski, E. (1985) Regional and contact metamorphism within the Moy Intrusive Complex, Grampian Highlands, Scotland. Contrib. Mineral. Petrol. 89, 296-306.

Zeck, H.P. (1970) An erupted migmatite from Cerro del Hoyazo, SE Spain. Contrib. Mineral. Petrol. 26, 225-246.

______ (1973) Compositions of natural sillimanites from volcanic inclusions and metamorphic rocks: a discussion. Amer. Mineral. 58, 555-557.

Żelaźniewicz, A. (1984) Remarks on the origin of sillimanite from the Góry Sowie, Sudetes, SW Poland [in Polish]. Geol. Sudetica 19, 101-119.

Zen, E-an (1961) Mineralogy and petrology of the system Al_2O_3-SiO_2-H_2O in some pyrophyllite deposits of North Carolina. Amer. Mineral. 46, 52-66.

______ (1969) The stability relations of the polymorphs of aluminum silicates: a survey and some comments. Amer. J. Sci., 267, 297-309.

______ (1972) Gibbs free energy, enthalpy, and entropy of ten rock-forming minerals: calculations, discrepancies, implications. Amer. Mineral. 57, 524-553.

______ (1988) Phase relations of peraluminous granitic rocks and their petrogenetic implications. Ann. Rev. Earth Planet. Sci. 16, 21-51.

Zwart, H.J. (1958) Regional metamorphism and relation granitisation in the Valle de Arán (Central Pyrenees). Geol. en Mijnb. 20, 18.

______ (1962) On the determination of polymetamorphic mineral associations, and its application to the Bosost area (Central Pyrenees). Geol. Rundschau 52, 38-65.